Electric Machines

The ELECTRIC POWER ENGINEERING Series
Series Editor Leo L. Grigsby

Published Titles

Electric Drives
Ion Boldea and Syed Nasar

Linear Synchronous Motors: Transportation and Automation Systems
Jacek Gieras and Jerry Piech

Electromechanical Systems, Electric Machines, and Applied Mechatronics
Sergey E. Lyshevski

Electrical Energy Systems
Mohamed E. El-Hawary

Distribution System Modeling and Analysis
William H. Kersting

The Induction Machine Handbook
Ion Boldea and Syed Nasar

Power Quality
C. Sankaran

Power System Operations and Electricity Markets
Fred I. Denny and David E. Dismukes

Computational Methods for Electric Power Systems
Mariesa Crow

Electric Power Substations Engineering
John D. McDonald

Electric Power Transformer Engineering
James H. Harlow

Electric Power Distribution Handbook
Tom Short

Synchronous Generators
Ion Boldea

Variable Speed Generators
Ion Boldea

Harmonics and Power Systems
Francisco C. De La Rosa

Electric Machines
Charles A. Gross

Electric Machines

CHARLES A. GROSS
Auburn University
Auburn, Alabama, U.S.A.

CRC Press is an imprint of the
Taylor & Francis Group, an informa business

Published in 2007 by
CRC Press
Taylor & Francis Group
6000 Broken Sound Parkway NW, Suite 300
Boca Raton, FL 33487-2742

Printed in the United States of America on acid-free paper
10 9 8 7 6 5 4 3 2

International Standard Book Number-10: 0-8493-8581-4 (Hardcover)
International Standard Book Number-13: 978-0-8493-8581-0 (Hardcover)
Library of Congress Card Number 2006000146

Library of Congress Cataloging-in-Publication Data

Gross, Charles A.
Electric machines / author, Charles A. Gross.
p. cm. -- (Electric power engineering series)
Includes bibliographical references and index.
ISBN-10: 0-8493-8581-4
ISBN-13: 978-0-8493-8581-0
1. Electric machinery. I. Title. II. Series.

TK2000.G76 2006
621.31'042--dc22 2006000146

Visit the Taylor & Francis Web site at
http://www.taylorandfrancis.com

and the CRC Press Web site at
http://www.crcpress.com

Dedication

To Robert, who changed everything

Preface

There are two fundamental applications of electromagnetic phenomena:

- The processing and transmission of energy.
- The processing and transmission of information.

Although the discipline of electrical engineering has always claimed ownership of both subjects, prior to World War II, energy subjects were the primary emphasis of the profession. After World War II, advances in electronics and military needs shifted the emphasis to information processing. The confluence of several revolutionary discoveries[1] has driven the shift at an accelerating rate. Today, capabilities and applications of electronic information processing surpass even the imagination of most science fiction writers. And with no end in sight, society is in the midst of a true technological revolution.

Yet, in spite of the spectacular advances and accomplishments, information processing, in and of itself, is of secondary importance to the human condition. The primary needs of human society remain as they always have been: food, shelter, clothing, clean air, water, and transportation. To meet these primary needs, we must control energy. Indeed, for civilization to advance, we must control energy on a large scale, efficiently, economically, reliably, and safely, and in a manner that does not harm the environment. Energy in the electromagnetic form is of vital concern in that regard because of its amenability to control. Hence, the electromagnetic-mechanical (EM) energy conversion process and the devices or "machines," which implement that process, are of continuing fundamental importance to electrical engineering as well as other related disciplines within engineering.

This book is dedicated to the study of EM machines. The approach taken here views the EM machine as a part of an integrated system consisting of an electrical source, a controller, the motor, and a mechanical termination. The machine is the primary focus, but the source, controller, and mechanical termination are also studied.

One can approach the study of machines from a design, applications, or control perspective. Design issues include machine construction details, analysis of internal magnetic fields and forces, internal losses, machine circuit modeling, and circuit parameters from tests. Application issues include mode of operation (motor, generator, braking), proper selection of motor types for specific loads, sizing of motors, and performance assessment from equivalent circuit analysis. Control issues include options for speed control (voltage magnitude and frequency control, etc.), requisite controller hardware, modes of operation (motor, generator, braking), steady-state and dynamic (particularly starting) performance, and peripheral issues such as harmonic contamination, etc. Here, a balanced approach is attempted, moving from the basic machine physics and principles of operation, through realistic applications, considering relevant control issues.

---

[1] a. The invention of cheap reliable solid-state semiconductor devices,
b. The realization that all information could be represented with binary numbers,
c. The realization that (b) could be represented as an electrical signal,
d. The realization that (a) could be mass produced, with ever-decreasing size, cost, and power requirements,
e. The invention of monolithic "integrated" circuits, with ever-decreasing size and cost, and ever-increasing complexity.

This book is written at an advanced engineering undergraduate level. Readers are assumed to have a background in basic engineering physics and mathematics as well as basic electrical circuits. It is estimated that most[2] of the material may be covered in one three-credit semester course; however, if all of the material is to be covered in detail, or if fundamentals are to be reviewed, more time will be required. The book could prove useful in at least three ways: as the textbook for a single required or elective course in EM machines; as the textbook for a follow-up second elective course in EM machines, following a basic course in electric power engineering; and as a general reference for practicing engineers, with interest in selected topics covered in more analytical depth than that which is available from more traditional references. A proposed catalog description of a course for which this book would serve as an appropriate textbook might be:

> ***Electromagnetic Machines*** *Prerequisites: engineering physics, mathematics, electric circuits. An integrated treatment of all fundamental issues relevant to electromagnetic machines, with balanced emphasis on design, application, and control. Induction, DC, and synchronous machines are covered, as are single- and polyphase designs. Motor drives are considered, with attention given to basic solid-state converter circuits. Rotary and translational devices are considered with many practical applications. Transformers are also covered.*

The Table of Contents serves as a recommended first draft of a course syllabus, and is essentially self-explanatory. Instructors should be able to tailor this outline to their individual needs. Appendix A provides a convenient reference for SI units and selected conversion factors. Appendix B provides a convenient review of circuit concepts, including three-phase, and symmetrical components. Appendix C supplies a convenient review of harmonics. It is recommended that this material be included at the proper point, unless student background is unusually strong in these areas.

Experienced engineering educators know that students need to solve relevant problems to solidify their understanding of technical material. End-of-chapter problems are provided for that purpose. Each problem has its own unique set of educational objectives, and is intended to reinforce and augment the preceding chapter material as well as provide continuity to earlier material in previous chapters. Therefore, it is recommended that all problems be assigned. If it is desired to change the answers from term to term to prevent students from copying solutions from files, it is a simple matter to change certain key data elements to generate a different set of answers to every problem. To demonstrate:

> *4.3 Consider the motor of Example 4.3. Operating at balanced rated stator voltage at 60 Hz, and running at 1785 rpm with the rotor shorted, find all currents, powers, torques, the slip, the power factor, and the efficiency.*
>
> *Revised problem: Work Problem 4.3 for a speed of 1775 rpm.*

There is public domain software to support this book. Actually two engineering analysis programs are relevant:

> XFMR: Transformers and Magnet Core Analyses
> EMAP: Electric Machine Analysis Program

To access these programs go to the public ftp site:

> ftp://ftp.eng.auburn.edu/pub/grossca

---

[2] Instructors may elect to omit transformers (Chapter 2); unbalanced induction machine performance (Chapter 6); unbalanced synchronous machine performance (Chapter 8); translational machines (Chapter 10); and other selected material in a first course in machines.

Under "grossca" you will find a folder named "programs", under which there are folders "xfmr" and "emap". To run either of these on line, double click on the *.exe file.

The reader is free to make copies of all software.

No book is entirely the product of one individual, as certainly is the case with this work. Many individuals have contributed in various ways, including, but not limited to, W.M. Feaster, R. Mark Nelms, Dallas W. Russell, L.L. Grigsby, J.L. Lowry, George McPherson, Stephen L. Smith, and J.D. Irwin. The many contributions of Sandy Johnson in the editing process are particularly appreciated. The cooperation and support of many companies, including General Electric, Alabama Power Company, the Southern Company, Reliance Electric, Rockwell, ABB, and Siemens is gratefully acknowledged. Finally, the many sacrifices made by my family, particularly my wife Dodie, are gratefully noted.

Contents

Appendix C: Harmonic Concepts

1

Basic Electromagnetic Concepts

One might argue that "electricity" is the world's most useless form of energy, i.e., useless until it is converted into some other form. It is because electrical energy is so controllable, and that it is readily converted into heat, light, sound, and mechanical force and motion, which makes it important to modern technology and engineering. This book is concerned with engineering issues related to the conversion of energy from the mechanical into the electrical form, and vice versa. The "vice versa" comment in the previous statement implies that the conversion process is reversible, i.e., that energy can also be readily converted from the electrical into the mechanical form, which indeed can be, and in fact the same device can be typically used to convert energy either way. Such conversion processes inevitably involve electric charge in motion, or current, which in turn mandates the presence of magnetic fields.

These magnetic fields play a major, if intermediary, role in the electromechanical energy conversion process. The term "electro-magnetic-mechanical" is too cumbersome to be commonly used, but it is rich in meaning, for it accurately describes the magnetic field to be "in the middle" of the energy conversion process. We shorten the term to "electromechanical," and further abbreviate it to "EM." A device, which performs the electrical to mechanical or the mechanical to electrical, energy conversion process, shall be called an "EM machine." If an EM machine normally converts energy from the mechanical into the electrical form, it is usually called a "generator." Actually it would be more accurate to call it an EM machine operating in the generator mode, recognizing that there is nothing intrinsic to the device that makes it irreversibly a "generator," but rather the externally imposed operating environment. Likewise, an EM machine, which normally converts energy from the electrical into the mechanical form, is usually called "a motor."

We are surrounded by motors and generators. An average of 60 are used in the typical modern home. To name a few: mixers, washers, dryers, microwave ovens, vacuum cleaners, food processors, drills, furnace blowers, wall clocks, dishwashers, garbage disposals, and garage door openers, all contain one or more motor/generators. Most manufacturing processes would be impossible without electric motors. Although automobiles are still powered by internal combustion engines, they also contain several generators and motors. What the digital computer is to human brain power, the EM machine is to human brawn.

This book addresses the study of EM machines and associated systems at an advanced undergraduate engineering level. As such, it assumes that the reader has the normal background in physics, mathematics, electric circuits, and field theory. SI units, or their base-ten multiples, shall be used throughout, i.e., the default units implied in all equations are SI, unless otherwise stated.

1.1 Basic Magnetic Concepts

We begin our formal discussion of these issues with a study of the magnetic field because of its fundamental importance. We cannot (and will not) ignore the science and mathematics related to the EM energy conversion process. However, this is an engineering textbook, and does not pretend to provide a comprehensive rigorous discussion of all physical and mathematical aspects of EM phenomena, a task which is best left to physics books.

It was observed by many early scientists, but Andre Marie Ampere (1775–1836) in particular, that an electric current could produce a force on a magnetic compass needle located near the current. This phenomenon of "action at a distance" may be explained by creating a structure of "force (or flux) lines" around the current, which aid in visualizing the mechanism by which the current interacts with the needle. We say the current creates a "magnetic field" around itself. It was further observed that the strength of the field was proportional to the current, and it was directional, depending on the flow direction of the current.

These observations are elegantly summarized in an equation form known as Ampere's law, which provides a quantitative relationship between magnetic field intensity (H), in A/m, and the causal electric current (i), in A. Units for all variables are in SI unless specifically stated to be otherwise throughout this book:

$$\oint \hat{H} \cdot \mathrm{d}\hat{x} = i_{\text{enclosed}} \tag{1.1}$$

where the "enclosure" is defined by the path of integration. Since Equation 1.1 is quite general, we may apply it to the structure of Figure 1.1a, consisting of a ferromagnetic core with an N-turn coil. Suppose we select a path of integration down the centerline of the core structure, and that the path length is ℓ. Further suppose that H and $\mathrm{d}x$ are everywhere parallel, and that H is constant at all points along x. Thus Equation 1.1 becomes

$$H\ell = Ni \tag{1.2}$$

The flux density (B), in W/m (T), within the core is related to H through the expression

$$B = \mu H \tag{1.3}$$

where μ is the core permeability in H/m. The permeability is analogous to conductivity in an electric circuit, is a property of the material through which the magnetic field passes, and is a measure of how easy it is for the field to pass through the core material.

The magnetic core flux ϕ is related to flux density through

$$\phi = \int \hat{B} \cdot \mathrm{d}\hat{A} \tag{1.4}$$

If B is uniform and perpendicular to the surface everywhere parallel to $\mathrm{d}A$, Equation 1.4 simplifies to

$$\phi = BA \tag{1.5}$$

where ϕ is the magnetic flux in W. Substituting Equations 1.3, 1.4, and 1.5 into 1.2

$$Ni = \frac{\ell}{\mu A}\phi \tag{1.6a}$$

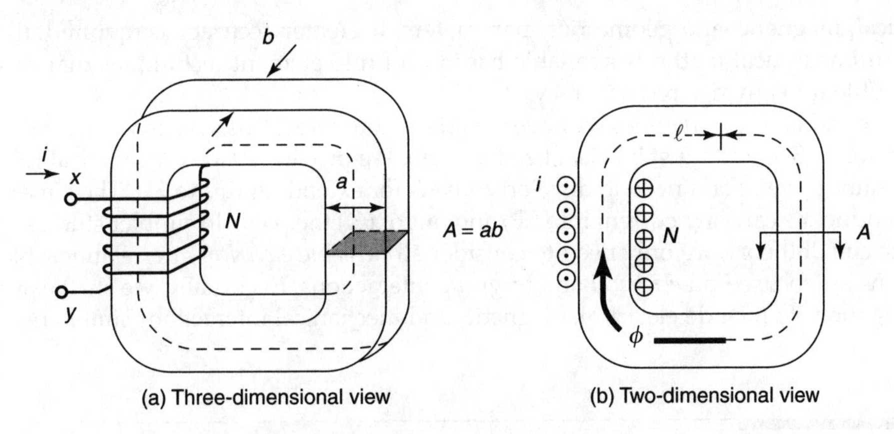

FIGURE 1.1. Simple magnetic system.

We define

$$\mathfrak{F} = Ni = \text{magnetomotive force (or mmf)} \tag{1.6b}$$

and

$$\mathfrak{R} = \frac{\ell}{\mu A} = \text{reluctance, in H}^{-1} \tag{1.6c}$$

and Equation 1.6 becomes

$$\mathfrak{F} = \mathfrak{R}\phi = \text{magnetomotive force (or mmf)} \tag{1.7a}$$

or

$$\phi = p\mathfrak{F} \tag{1.7b}$$

where

$$P = 1/\mathfrak{R} = \mu A/\ell = \text{permeance, in H} \tag{1.7c}$$

Equation 1.7 is fundamental to magnetic field production and is sometimes called "magnetic Ohm's law." It stated that mmf ($\mathfrak{F}$) causes flux (ϕ). How much flux is produced depends on $\mathfrak{R}$ (or P).

For strong fields, we need large μ and A, and small ℓ. Large μ requires that we select high μ core material. Large A and small ℓ values say something about the geometry of the core.

Let us consider the approximations we have made. We have assumed that the flux distributes itself uniformly throughout the core cross section, which in fact it does not.

Also the cross-sectional area presented to the flux path is neither constant, particularly as the flux rounds the 90° corners, nor does each flux line travel the length ℓ. In fact, some of the flux does not even stay in the core. And perhaps most serious of all, we shall learn that μ is not necessarily constant and determining a proper value is not a trivial problem. Yet, our results are still sufficiently accurate to provide insight into basic relations between electrical, magnetic, and geometrical parameters. If greater accuracy is required, there are powerful analytical methods available based on finite element techniques that can solve this problem to any desired accuracy.

Also, consider the difficulties we have already encountered. The system is inherently three-dimensional (3D) and must be visualized as such. We provide a two-dimensional (2D) view of the same system, and need to develop a knack for extending this to 3D. The circles represent conductors carrying current into (⊕) and out of (⊙) the page. In simpler studies, such as electric circuit theory, we only need to consider all possible *electrical* interrelations. Now our concerns are focused on *electrical* and *magnetic* interactions. Eventually, we shall extend our investigations to include electrical, magnetic, and mechanical interactions simultaneously.

1.2 Magnetically Linear Systems: Magnetic Circuits

If core permeability may be assumed constant, as we did in the previous section, then it is said to be "magnetically linear." When this is the case, it is possible to draw an analogy between DC electric and magnetic circuits. In a DC electric circuit, the electric forcing function ("source") is electromotive force, or emf (E), and the consequent flow variable is current (I), flowing through paths provided for that purpose ("conductors"). Analogously, in a magnetic circuit, the magnetic forcing function is magnetomotive force, or mmf ($\mathfrak{F}$), and the consequent flow variable is flux (ϕ), flowing through paths provided for that purpose

TABLE 1.1
Electric-Magnetic Circuit Analogs

Electric	Magnetic
I; E (+, −); R	ϕ; $\mathfrak{F}$ (N, S); $\mathfrak{R}$
Emf (E, voltage)	Mmf ($\mathfrak{F}$)
Current (I)	Flux (ϕ)
Conductivity (σ)	Permeability (μ)
$G = \sigma A/\ell$	$P = \mu A/\ell$
$R = 1/G$	$\mathfrak{R} = 1/P$
$V = E = RI$	$\mathfrak{F} = \phi\mathfrak{R}$
Around any path	
$\Sigma V = 0$	$\Sigma\mathfrak{F} = 0$
At any node	
$\Sigma I = 0$	$\Sigma\phi = 0$

("cores"). There is also a magnetic polarity north (N) and south (S), corresponding to electric polarity positive (+) and negative (−). Table 1.1 summarizes the electric-magnetic circuit analogs.

We have already noted that flux flowing through a reluctance causes an "mmf drop" across that reluctance. The analogs to the voltage and current laws are the facts that the mmf drops around any closed magnetic path must net out to 0, and that the sum of the fluxes out of any magnetic node must be 0, respectively. An example application will clarify matters.

Example 1.1

Consider the system shown in Figure 1.2a, with a coil current of 12 A.

(a) Draw the magnetic circuit.
(b) Solve for the core flux (ϕ) and flux density (B).
(c) Determine the mmf drops across the core and air gap.

SOLUTION

(a) The magnetic circuit is shown in Figure 1.2b.

The coil mmf $= \mathfrak{F} = NI = (200)(12) = 2400$ A turns

$$\mathfrak{R}_{\text{core}} = l/(\mu_{\text{core}} A) = (1.398)/((2000\mu_0)(0.01)) = 55.62\ (\text{mH})^{-1}$$

$$\mathfrak{R}_{\text{ag}} = g/(\mu_0 A) = 0.002/((\mu_0)(0.01)) = 159.2\ (\text{mH})^{-1}$$

(b) Flux and flux density are

$$\phi = \mathfrak{F}/(\mathfrak{R}_{\text{core}} + \mathfrak{R}_{\text{ag}}) = 2400/(55.62 + 159.2) = 11.174\,\text{mWb}$$

$$B = \phi/A = 1.117\,\text{T}$$

(c) The core and air gap mmf drops are

$$\mathfrak{F}_{\text{core}} = \phi\mathfrak{R}_{\text{core}} = 11.174\ (55.62) = 621.6\ \text{A turns}$$

$$\mathfrak{F}_{\text{ag}} = \phi\mathfrak{R}_{\text{ag}} = 11.174\ (159.2) = 1778.4\ \text{A turns}$$

Some general points relevant to magnetic circuits can be made with the help of Example 1.1. Flux flows out of the north ("N") terminal of the magnetic source, just as current flows out of the positive terminal of an electric source in an analogous electric circuit. But how do we determine which end of the coil is terminal "N"? One way to do this is to employ the so-called right-hand rule. Let the fingers of your right hand encircle the core in the same direction as does the coil current (go ahead — I'll wait). Notice that your thumb points up. This is the direction that the coil mmf attempts to force the flux out of the coil (i.e., out of the top); thus the "top" of the coil is effectively a north pole. Thus, two factors define the N, S mmf poles: the way the coil is wound, and the positive direction of coil current.

Parallel and series concepts relate analogously to magnetic circuits (the two reluctances are in series in Example 1.1, and may be added). Another issue is a "bulging" of the magnetic field flux as it leaves the core and enters the air gap. To account for this effect some investigators suggest increasing A by the gap width g in all directions. We shall not make this refinement in our calculations, on the premise that other approximations we have made are more serious anyway.

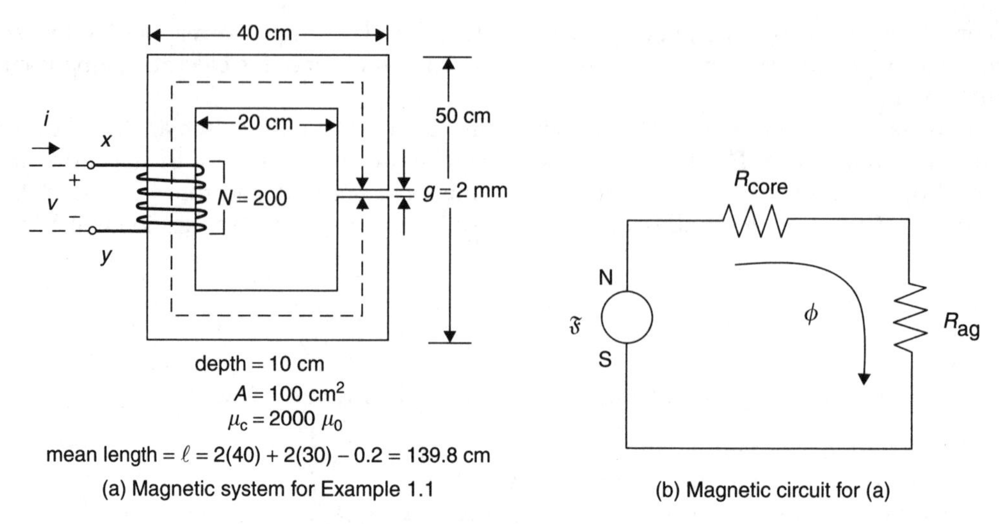

FIGURE 1.2. A magnetic system with air gap.

There is public domain software to support this book. Actually two engineering analysis programs are relevant:

XFMR: Transformers and Magnet Core Analyses
EMAP: Electric Machine Analysis Program

To access these programs go to the public ftp site:

ftp://ftp.eng.auburn.edu/pub/grossca

Under "grossca" you will find a folder named "programs", under which there are folders "xfmr" and "emap". To run either of these on line, double click on the *.exe file.

You are free to make copies of all software.

The XFMR solution to Example 1.1 follows.

Magnetic Core Parameters

Linear Analysis	**Core**	**Air Gap**
Length (cm)	139.8	0.200
Cross-sectional area (sq cm)	100.0	100.0
Relative permeability	2000.0	1.0
Reluctance (1/mH)	55.624	159.155
Mmf (A turns)	621.6	1778.4
H (kA/m)	0.4446	889.2
Flux density (B in mT)	1117.4	1117.4
$\mathfrak{F}_{core} + \mathfrak{F}_{ag}$ (A turns)	2400.0	
Magnetic flux (mWb)	11.174	
Reluctance (1/mH)	214.779	
Coil inductance (mH)	186.2	
Magnetic field energy (mJ)	13409.1	

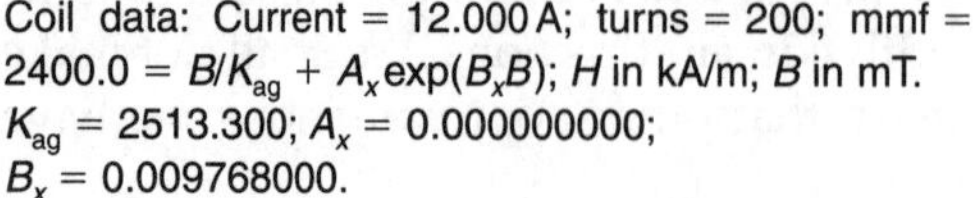
Coil data: Current = 12.000 A; turns = 200; mmf = 2400.0 = $B/K_{ag} + A_x \exp(B_x B)$; H in kA/m; B in mT. K_{ag} = 2513.300; A_x = 0.000000000; B_x = 0.009768000.

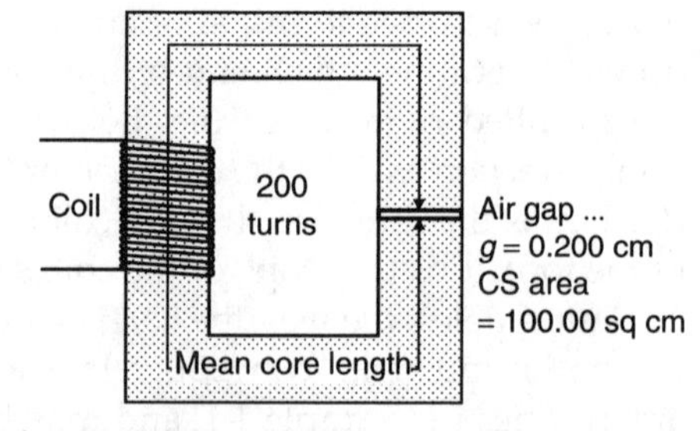

Core characteristics ...

Core length = 139.80 cm; CS area = 100.00 sq cm
$x = y/K_{ag} + A_x \exp(B_x y)$; $x = H$ in kA/m; $y = B$ in mT
K_{ag} = 2513.3; A_x = 0.000001661; B_x = 0.009768000
Unsaturated relative core permeability = 2000.0

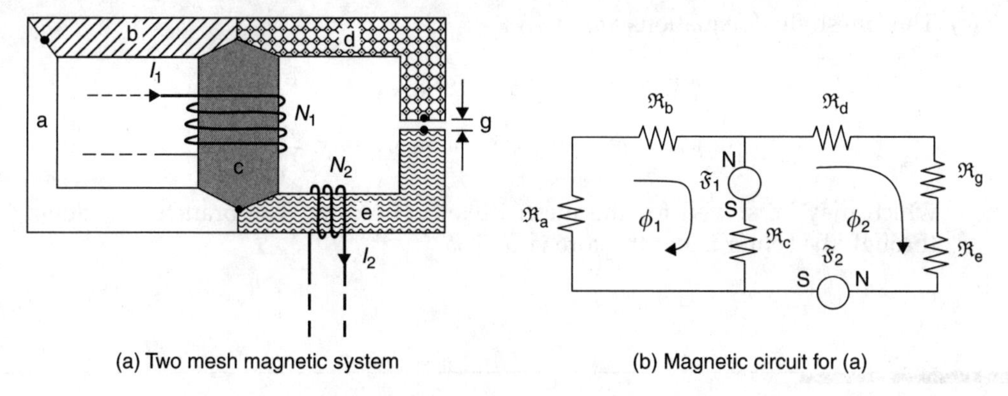

(a) Two mesh magnetic system

(b) Magnetic circuit for (a)

FIGURE 1.3. Magnetic system for Example 1.2.

We now consider the more complicated example of Figure 1.3.

Example 1.2
Consider the system shown in Figure 1.3a.

(a) Draw the magnetic circuit.
(b) Determine the reluctances required by the magnetic circuit.
(c) Write the "mesh flux" equations necessary to solve for all fluxes.

SOLUTION

(a) The core is first divided into sections of equal A and μ_0. The magnetic circuit is shown in Figure 1.3b.

$$\mathfrak{F}_1 = N_1 I_1, \quad \mathfrak{F}_2 = N_2 I_2$$

(b) Determine nodes located at the approximate geometric center of interface surfaces between sections. Along the geometric center of each section measure the mean section length. The requisite reluctances are computed for each section.

$$\Re_a = \frac{\ell_a}{\mu_a A_a}$$

where ℓ_a = mean length of core section a; μ_a = permeability of core section a; A = cross-sectional area of core section a.
Similarly

$$\Re_b = \ell_b/(\mu_b A_b), \quad \Re_c = \ell_c/(\mu_c A_c), \quad \Re_d = \ell_d/(\mu_d A_d),$$
$$\Re_e = \ell_e/(\mu_e A_e), \quad \Re_g = \ell_g/(\mu_0 A_g)$$

(c) The "mesh flux" equations are:

$$(\mathfrak{R}_a + \mathfrak{R}_b + \mathfrak{R}_c)\phi_1 - (\mathfrak{R}_c)\phi_2 = -\mathfrak{F}_1$$

$$-(\mathfrak{R}_c)\phi_1 + (\mathfrak{R}_c + \mathfrak{R}_d + \mathfrak{R}_e + \mathfrak{R}_g)\phi_2 = \mathfrak{F}_1 - \mathfrak{F}_2$$

which may be solved for the mesh fluxes ϕ_1 and ϕ_2. The branch flux down through the center leg of the core is $\phi_1 - \phi_2$.

1.3 Voltage, Current, and Magnetic Field Interactions

In the system of Figure 1.1, we have noted that coil current produces a magnetic field in the core. However, we have yet to consider what causes the current. For the DC case, the application of DC voltage (V) to the coil creates a current according to

$$i = I = \frac{V}{R}$$

where R is the electrical coil resistance in ohms. Now if the current is time varying, the magnetic field is also time varying, which raises a new issue. According to Faraday's law, any circuit (e.g., our coil) linked by a time-varying magnetic field will experience an internal induced voltage, equal to the time rate of change of that flux linkage. In equation form

$$v_i = \frac{d\lambda}{dt}$$

where $\lambda = N\phi$.

The polarity of this voltage is at first difficult to appreciate. In fact, the issue is addressed in a principle sometimes referred to as Lenz's law, which can be stated as follows:

The polarity of the induced voltage is such that it will attempt to create a current in the circuit that will oppose the change in magnetic field; i.e., the induced current will create a second field in opposition to the original field.

Continuing our analysis,

$$\lambda = N\phi = N\left(\frac{\mathfrak{F}}{\mathfrak{R}}\right)$$

$$= \frac{N^2 i}{\mathfrak{R}} = Li$$

and

$$v_i = L\frac{di}{dt}$$

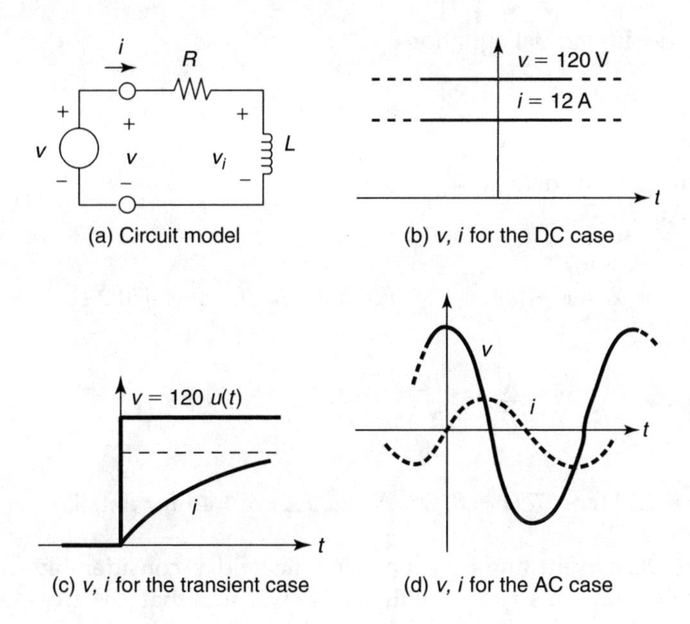

FIGURE 1.4. Electrical issues for the coil of the system of Figure 1.2a.

We define the constant $N^2/\mathfrak{R}$ to be the inductance L. The general relation for the coil v, i relationship is then

$$v = iR + L\frac{\mathrm{d}i}{\mathrm{d}t}$$

We now have the basis for a circuit model that includes both ohmic and Faraday effects, summarized in Figure 1.4a. An example will illustrate these important issues.

Example 1.3
Consider the magnetic system of Figure 1.2a. The coil resistance is 10 Ω.

(a) Determine the coil inductance and time constant.
Determine the coil current $i(t)$ if the applied coil voltage $v(t)$ is
(b) 120 V DC
(c) $120\,u(t)$ volts (suddenly applied at $t = 0$).
(d) 120 V rms AC, 60 Hz (i.e., $v(t) = 169.7\cos(377\,t)$ V)

SOLUTION

(a) $L = \dfrac{N^2}{\mathfrak{R}} = \dfrac{N^2}{\mathfrak{R}_c + \mathfrak{R}_{ag}} = \dfrac{(200)^2}{55.6 + 159.2} = 186.2$ mH
$\tau = L/R = 18.62$ ms
(b) Since $v(t)$ is constant in time, so is $i(t)$. Hence
$\mathrm{d}\lambda/\mathrm{d}t = 0$ and $i(t) = v(t)/R = 120/10 = 12$ A. See Figure 1.4b for details.
(c) For $t < 0$, $v = 0$. Hence $i(t) = 0$ A.
For $t > 0$, $v(t) = 120 = L(\mathrm{d}i/\mathrm{d}t) + Ri$

Solving the differential equation

$$i(t) = 12(1 - e^{-t/\tau})$$

See Figure 1.4c for details.

(d) For the AC case

$$\bar{Z} = R + j\omega L = 10 + j(377)(0.1862) = 10 + j70.2\ \Omega$$

$$\bar{I} = \frac{\bar{V}}{\bar{Z}} = \frac{120\angle 0^\circ}{10 + j70.2} = 1.692\angle -81.9^\circ\,\text{A}$$

and $i(t) = 2.393\cos(377t - 81.9^\circ)$ A. See Figure 1.4d for details.

Observe that the DC current, and thus the magnetic field, is considerably stronger than the AC current, for the same rms applied voltage. We will find that this gives DC machines an important advantage in energy conversion applications.

1.4 Magnetic Properties of Materials

To fully understand the electrical and magnetic properties of materials requires the physics of quantum mechanics. Fortunately, this detailed study is unnecessary for most engineering applications. A much simplified microscopic view of matter is that it is composed of atoms, each of which in turn is composed of a relatively dense charged spinning nucleus surrounded by orbiting electrons, each of which may also be rotating. The combination of these three motions of charged bodies ascribe to each atom a property called the "magnetic moment." Simply put, the "magnetic moment" property means that the atom behaves like an infinitesimally tiny permanent magnet (sometimes described as a "magnetic dipole"). The strength of the atomic magnetic moment for a given material depends on the particulars of the electron configuration, the electronic and nuclear spins, and spatial orientations.

On a macroscopic scale, a sample of a given material will contain many trillions of atoms whose magnetic moments are randomly aligned. A material may be classified magnetically according to how this random alignment is affected by an externally applied magnetic field intensity (H). An excellent relative scale is achieved by first considering empty space (a vacuum). Recall

$$B = \mu H \tag{1.8}$$

Here, B is the flux density produced by a given field intensity (H), μ, is the permeability, which tells us how much B we generate for a given H in a given material. When this "material" is a vacuum, $\mu = \mu_0 = 4\pi \times 10^{-7}$ H/m. Thus, μ becomes a measure of how "permeable" material is to a magnetic field. Relative permeability is defined as

$$\mu_r = \mu/\mu_0 \tag{1.9}$$

TABLE 1.2

Some Representative Magnetic Materials

Materials	Classification	μ_r (DC)
Silver	Diamagnetic	0.99998
Lead	Diamagnetic	0.99998
Copper	Diamagnetic	0.99999
Water	Diamagnetic	0.99999
Vacuum	Nonmagnetic	1.00000
Air	Paramagnetic	1.00000
Aluminum	Paramagnetic	1.00002
Palladium	Paramagnetic	1.00080
Cobalt	Ferromagnetic	250
Nickel	Ferromagnetic	600
Iron	Ferromagnetic	5,000
Silicon iron	Ferromagnetic	7,000
Mumetal	Ferromagnetic	100,000
Purified iron	Ferromagnetic	200,000
Supermalloy	Ferromagnetic	1,000,000

Hence the relative permeability of a vacuum is unity by definition. A material whose relative permeability is unity is defined as "nonmagnetic." Few, if any, materials are truly nonmagnetic. There are materials whose relative permeability is slightly <1. These are classified as "diamagnetic;" a common example is copper. Materials whose relative permeability is slightly >1 are classified as "paramagnetic;" a common example is aluminum.

There are a few materials whose relative permeability is much greater than 1. Such materials, classified as "ferromagnetic," not only have atomic structures with large magnetic moments, but also organize themselves on a molecular level into clusters of atoms, called "domains," with the individual moments aligned such that they reinforce. Large permeability is achieved because these domains readily align upon external field application. The most common example of ferromagnetic material is iron.

Materials with large domain moments, but do not readily align, such that their relative permeability is only slightly greater than 1 (like paramagnetic materials) are classified as "antiferromagnetic." For engineering purposes, paramagnetic, diamagnetic, and antiferromagnetic materials are usually treated as nonmagnetic, with a relative permeability of unity.

Ferromagnetic materials are normally reasonably good electric conductors. However, there are some high μ materials that are poor conductors, in fact considerably poorer than semiconductors. These materials are said to be "ferrimagnetic," and are called "ferrites." Ferrites are frequently used in high-frequency applications because of their low internal losses. Some typical magnetic materials are listed in Table 1.2.

Let us reconsider the ferromagnetic case. Recall that in response to an externally applied field, the domains respond by bringing their spatial orientation to bring into alignment with each other and the external field, resulting in an extremely high μ. This phenomenon occurs by degrees, and more or less linearly; that is, if the field is doubled, the number of aligned domains doubles approximately. However, if the field continues to increase, we will eventually reach a condition in which all of the domains are aligned. This condition is referred to as "saturation." The situation is illustrated in Figure 1.5a.

Hence, starting from zero external field, if H is monotonically increased, we experience a linear rise in B, then a nonlinear transition into saturation, which again is linear. If we still think in terms of the expression $B = \mu H$, clearly μ must be a variable! The μ values referred to earlier for ferromagnetic materials relate to the linear region; in saturation, μ will eventually converge to μ_0 for all materials.

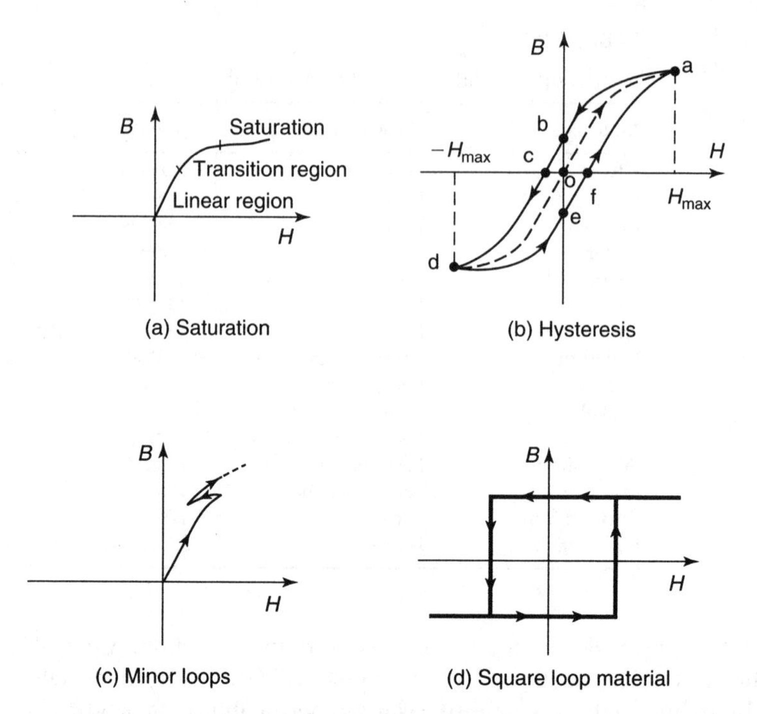

FIGURE 1.5. The magnetization characteristic for ferromagnetic materials.

The situation is even more complicated than first assumed. If we now reduce H, we follow a different path in the H–B plane, as shown in Figure 1.5b. If we continue to monotonically decrease H through zero on into hard negative saturation before we reverse the sense of change in H, we trace out the path oabcdef. This "memory" effect, that is the fact that what path is taken depends on the history of how we arrived the present location in the H–B plane, is called "hysteresis," and the closed path in the H–B plane is sometimes called "hysteresis loop." It is symmetrical about the origin, provided that $H = -H$, and that H is large enough to drive material into complete saturation. However, suppose we "reverse direction twice," as illustrated in Figure 1.5c, we create a so-called minor loop in the B–H characteristic. Much engineering effort has been invested in understanding and accurately modeling this complicated situation and its analytical treatment is beyond the scope of this book. Fortunately, an accurate analysis of this phenomenon is unnecessary to our work, which will be restricted to DC, AC, and dynamic performance of EM machines.

The size and shape of the hysteresis loop is dependent on the material. Consider the extreme case of so-called square-loop material shown in Figure 1.5d. The interesting thing to note is that B is substantial even for zero H. A core constructed of such material is called a "permanent magnet." Two applications are particularly noteworthy. If we need a "flux source," that is an element that supplies ϕ webers of flux at its poles independently of the external termination (provided that H_0 was not exceeded), the "permanent magnet" is appropriate. Also, if we need a memory device that stores the fact that a strong positive or negative H was last externally applied, a square loop core is ideal.

For situations where hysteresis can be neglected, but the basic nonlinearity cannot, we use the single-valued dashed nonlinear B–H characteristic (see Figure 1.5b). For best accuracy, graphical characteristics of the material should be used. However, graphical characteristics are sometimes awkward to use, particularly when analysis is done using a computer. Several attempts have been made to model the single-valued B–H characteristic.

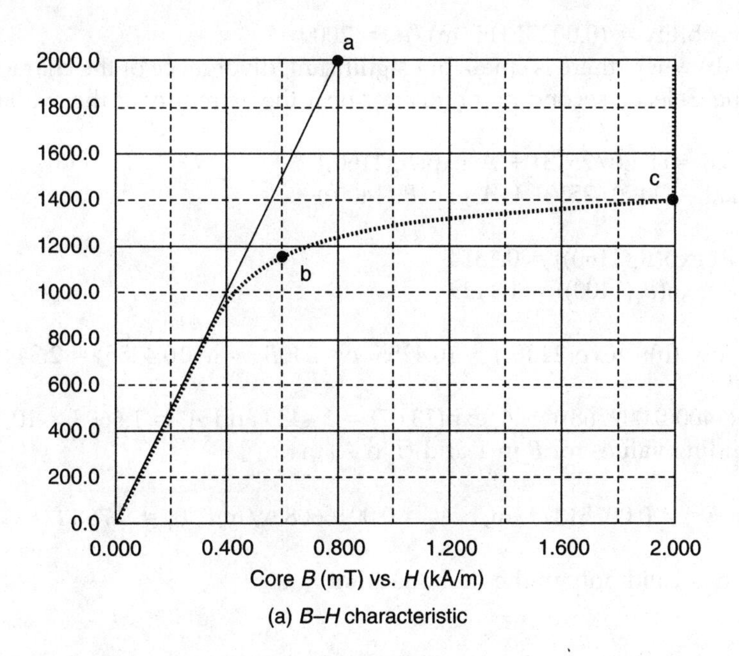

(a) *B–H* characteristic

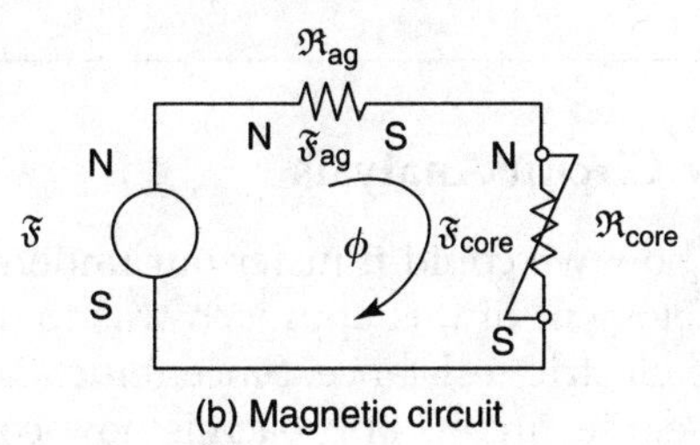

(b) Magnetic circuit

FIGURE 1.6. System of Figure 1.2a with nonlinear core.

One common expression is the empirical equation

$$H = \frac{B}{K_{ag}} + A_x\, e^{B_x B} \tag{1.10}$$

Note that the constant K_{ag} is in effect μ for the unsaturated portion of the curve. Hence define $\mu_{ag} = K_{ag}$. The expression gives a reasonably good fit over a fairly large range. Example 1.4 demonstrates how the constants K_{ag}, A_x, and B_x are determined.

Example 1.4
Consider the single-valued *B–H* characteristic in Figure 1.6a. Evaluate the constants K_{ag}, A_x, and B_x.

SOLUTION
Draw a linear extension of the linear portion of the *B–H* characteristic. This line (o–a, called the "air-gap line") is represented by

$$B = \mu_{ag} H = K_{ag} H$$

Hence $\mu_{ag} = K_{ag} = 2000/0.7958 = 2513\,\text{mH/km}$

Relative permeability = $(0.002513\,\text{H/m})/\mu_0 = 2000$
Select a point (b) where there is small, but significant, divergence of the characteristic from the air-gap line. Select a second point (c) at or near the extremity of the available data.

Point (b): $0.6 = (1160/2513) + A_x\exp(B_x(1160))$
Point (c): $2.0 = (1400/2513) + A_x\exp(B_x(1400))$
or
Point (b): $A_x\exp(B_x(1160)) = 0.1384$
Point (c): $A_x\exp(B_x(1400)) = 1.4429$

Dividing (c) by (b): $\exp(240B_x) = 10.4255$ or $240B_x = \ln(10.4255) = 2.3443$ and $B_x = 0.009768\,\text{mT}^{-1}$
Hence $A_x\exp(1400(0.009768)) = A_x\exp(13.67) = 1.4429$ and $A_x = 1.6608 \times 10^{-6}\,\text{kA/m}$.
The corresponding values for B in T and H in A/m...

$$K_{ag} = 0.002513\,\text{H/m};\quad A_x = 0.0016608\,\text{A/m};\quad B_x = 9.768\,\text{T}^{-1}$$

The expression is valid only in the first quadrant.

1.5 Nonlinear Magnetic Circuit Analysis

In Section 1.2 we have seen how we could transfer our understanding of electric circuits to magnetic systems. A key element of this approach was to define the concept of reluctance, which is analogous to electric resistance. Since reluctance depends on μ, which is variable in the nonlinear case, the circuits approach is now complicated, but still useful, particularly for visualizing the problem. The point may be made through an example analysis.

Example 1.5
Consider the magnetic system of Figure 1.2a. Assume that all system constants are the same, with one important exception. We will now consider a nonlinear core, whose B–H characteristic is provided in Figure 1.5a. The applied DC voltage is 120 V.

(a) Draw the magnetic circuit.
(b) Solve for the core flux (ϕ) and flux density (B).
(c) Determine the mmf drops across the core and air gap.

SOLUTION

(a) The system magnetic circuit is provided in Figure 1.5b.
(b) The coil mmf = $NI = (200)(12) = 2400$ A turns = 2.4 kA turns
$\mathfrak{R}_{ag} = g/(\mu_0 A) = 0.002/((\mu_0)(0.01)) = 159.2\,\text{H}^{-1}$
$\mathfrak{R}_{core} = \ell/(\mu_{core}A) = (1.398)/((\mu_{core})(0.01)) = 139.8/\mu_{core}$

and is unknown since μ_{core} is also unknown.

Now

$$\mathfrak{F} = 2400 = H(1.398) + H_{ag}\,(0.002) = (1.398)H + (B/\mu_0)(0.002)$$

or

$$H = 1716.7 - 1138.5\,B \text{ (}B\text{ in T; }H\text{ in A/m)}$$

But recall that also

$$H = B/K_{ag} + A_x \exp(B_x B)$$

$$K_{ag} = 0.002513\,\text{H/m}; \quad A_x = 0.0016608\,\text{A/m}; \quad B_x = 9.768\,\text{T}^{-1}$$

See Example 1.3 for details. Therefore

$$B/K_{ag} + A_x \exp(B_x B) = 1716.7 - 1138.5B$$

Solving for B (by calculator, MatLab, etc.), $B = 1077\,\text{mT}$

The flux is

$$\phi = BA = 1077(0.01) = 10.77\,\text{mWb}$$

(c) The air gap and core mmf drops are

$$\mathfrak{F}_{ag} = \phi\mathfrak{R}_{ag} = 10.77(159.2) = 1712.5\text{ A turns}$$

$$H_{core} = 1716.7 - 1138.5B = 0.4918\,\text{kA/m} = 491.8\,\text{A/m}$$

$$\mathfrak{F}_{core} = H_{core}\,l = 491.8(1.398) = 687.5\text{ A turns}$$

Check

$$\mathfrak{F}_{ag} + \mathfrak{F}_{core} = 1712.5 + 687.5 = 2400\text{ A turns}$$

Incidentally,

$$\mu_{core} = B/H = 1076/491.8 = 2.188\,\text{mH/m}$$

and

$$\mathfrak{R}_{core} = 139.8/\mu_{core} = 139.8/2.188 = 63.897\,\text{mH}^{-1}$$

Note that we were able to evaluate μ_{core} and R_{core}, but only after we already determined the core flux. Comparing the results of our analyses of comparable linear and nonlinear systems (Example 1.1 and Example 1.4, respectively) we draw the following conclusions. Saturation has

- degraded performance, giving us less flux for the same current,
- increased the core reluctance,
- reduced the core permeability, and
- made our analysis considerably more complicated.

All of these conclusions have general implications. For all EM machines, saturation is generally undesirable. The further we enter the saturated region, the more serious these issues become. And yet good design means that we should "get the most" out of our core material. It is normal practice to compromise, and design an EM system to operate well into the transition region, but short of "hard" saturation.

Again consider the general situation of Figure 1.2a, and the specific situation examined in Example 1.4d, where we determined the coil current in response to an AC voltage. It is clear that average power is dissipated in the linear case; note that the average power

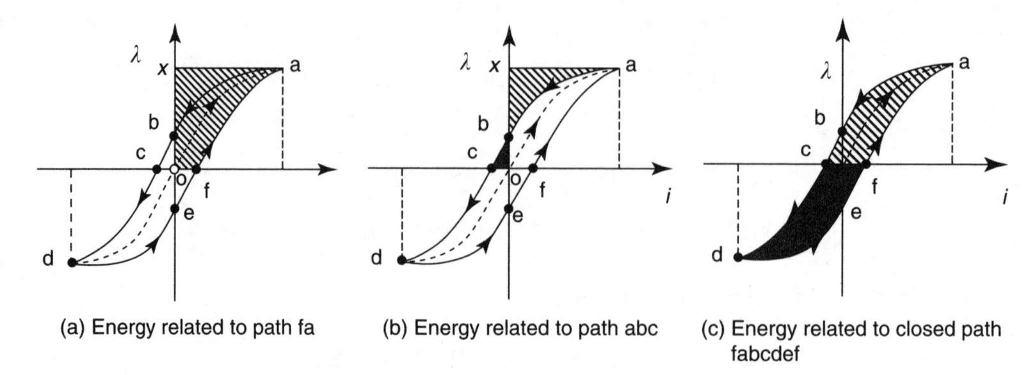

FIGURE 1.7. Graphical interpretation of core loss.

absorbed by the inductor was 0, due to the 90° phase shift between the current and inductor voltage. However, in the nonlinear case, the situation is more complicated. Focus on the energy involved in the inductive part of the equivalent circuit.

$$W = \int v_i(t) i \, dt = \int \frac{d\lambda}{dt} i \, dt = \int i \, d\lambda$$

Now consider the AC situation represented in Figure 1.7. For the AC operation, the system continually cycles around the path fabcdef. The energy required from the coil is equal to the positive cross-hatched area "faxo" in Figure 1.7a, as we move along the path f–a. As we continue on the path a–b, the corresponding area, and thus energy, is the cross-hatched area axb in Figure 1.7b, and is negative. Finally, movement from b to c, produces the solid area bco, which is positive. Summarizing, movement along the path fabc, requires an energy from the coil equal to the cross-hatched area fabc in Figure 1.7c. If we continue the analysis along path cdef, we compute energy as the solid area cdef, and arrive at the following conclusion.

> *AC operation of a nonlinear magnetic core requires a transfer of energy from the electric circuit into the core equal to the area enclosed by the λ–i characteristic ($W_{interior}$) in each AC cycle.*

Recalling that frequency is "cycles/sec," the energy required per second, or power, is

$$P_c = \text{core loss} = f\, W_{\text{interior}}, \text{ in W}$$

Core loss is a combination of two different physical phenomena. The first of these, called "hysteresis," is due to the fact that it requires energy to alter the orientation of the internal magnetic domains. If we determine the λ–*i* characteristic by setting *i* to a DC value, and then determining the corresponding DC λ, we form the DC λ–*i* characteristic, and the area enclosed is the energy per cycle required by hysteresis. Hysteresis is sometimes accounted for with the empirical expression

$$P_h = K_h B_m^s f, \quad \text{hysteresis loss, in W}$$

where

K_h = hysteresis constant, depending on the core material and construction, in J/T
B_m = maximum core flux density, in T
s = Steimetz exponent (ranging from 1.5 to 2.5), depending on the core material and construction.[1]
f = frequency, in Hz.

If we determine the λ–i characteristic at higher frequency (e.g., 60 Hz), we form the AC λ–i characteristic, and the area enclosed is the energy per cycle required by hysteresis, plus another phenomena. The additional effect is due to the fact that ferromagnetic cores are also good *electric* conductors. Thus, a time-varying core magnetic field will induce voltages, and therefore electric currents, within the core. These currents, called *eddy currents,* will have a corresponding i^2R power loss. One can derive the following equation:

$$P_e = K_e B_m^2 f^2, \quad \text{eddy current loss, in W}$$

where

K_e = eddy current constant, depending on the core material and construction, in W/T²
B_m = maximum core flux density, in T
f = frequency, in Hz.

The eddy current loss can be substantial in solid cores. If the core is constructed of thin sheets (laminations) of thickness d, it can be shown that the constant K_e, and hence P_e, is proportional to d^2. Therefore, for any core designed to carry an AC magnetic field, it is standard practice to use laminated cores to reduce eddy current loss. The sum of the hysteresis and eddy current losses is called *core loss* (P_c).

$$P_c = P_h + P_e, \quad \text{core loss, in W}$$

All ferromagnetic cores operating in an AC mode exhibit core loss.

Another important effect is waveform harmonic distortion. Consider a voltage $v(t) = V_m \sin(\omega t)$ applied to a zero resistance coil wrapped around a magnetic core. If the coil resistance is negligible,[2]

$$\lambda = \int v(t)\,dt = \int V_m \sin(\omega t)\,dt = \frac{-V_m}{\omega}\cos(\omega t)$$

Now if the magnetic core is linear,

$$i(t) = \frac{\lambda(t)}{L} = \frac{-V_m}{\omega L}\cos(\omega t)$$

---

[1] We would fail in our duties if we do not acknowledge many contributions in this area of Charles Proteus Steinmetz, a true pioneer of electrical engineering.

[2] Note that the waveform of $v(t)$ will now define the waveform of $\lambda(t)$. In particular, if $v(t)$ is sinusoidal, $\lambda(t)$ will be sinusoidal, *even if the core is magnetically nonlinear!*

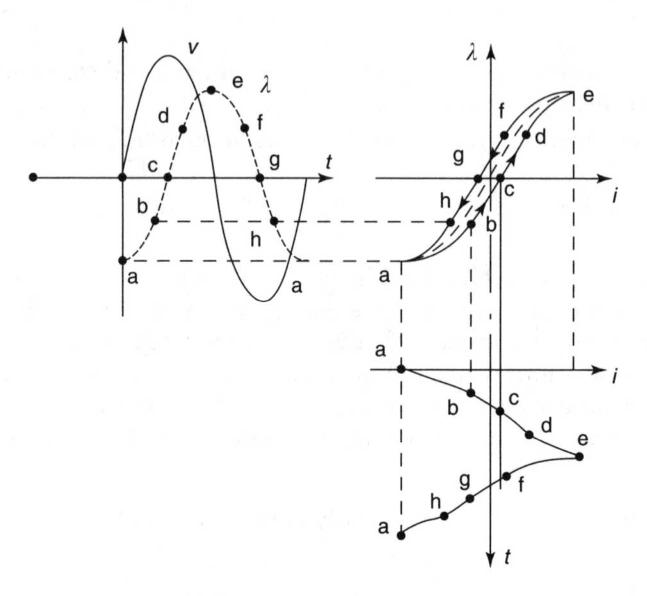

(a) Current harmonic distortion

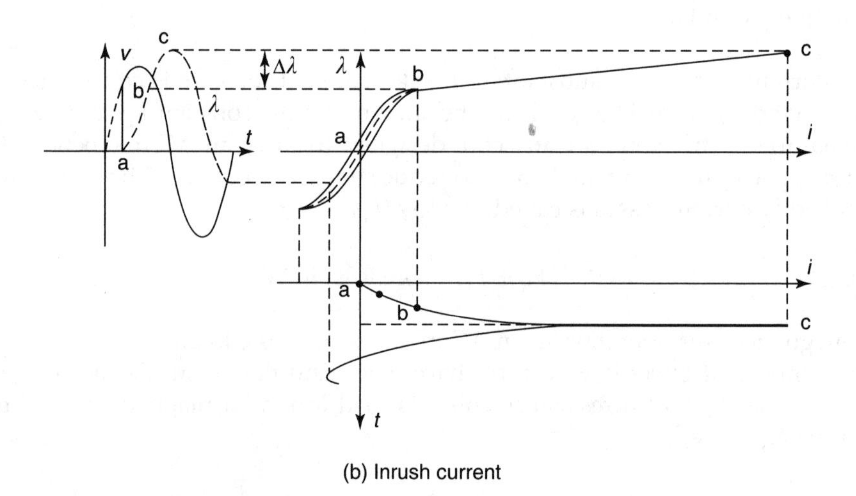

(b) Inrush current

FIGURE 1.8. Coil current in the nonlinear case.

and $v(t)$, $\lambda(t)$, and $i(t)$ are all sinusoidal. However, for the nonlinear case, $i(t)$ must be determined graphically from the λ–i characteristic, as shown in Figure 1.8a. Suppose we suddenly apply the sinusoidal voltage to the winding. We start at the first zero on the voltage waveform, corresponding to the negative maximum (point a) on the λ waveform at $t = 0$. Projecting λ to the left to the λ–i characteristic, we determine the corresponding point on $i(t)$ (point a). Projecting the current downward to point a (time t is plotted downwards), we determine the current at $t = 0$. Repeat this procedure for points bcdefgh, and plot a full cycle of current ($i(t)$ is plotted horizontally, and "t" downward). The key point to note is that *for sinusoidal* $\lambda(t)$, $i(t)$ *is nonsinusoidal.* The reverse is also true, i.e., for sinusoidal $i(t)$, $\lambda(t)$ is nonsinusoidal. Thus, one cause of harmonic distortion in EM machines is the nonlinearity of the magnetics.

Under AC steady conditions, the excursion of $\lambda(t)$ is symmetrical between $+V_m/\omega$ and $-V_m/\omega$. But now consider the transient situation, i.e., the case where we suddenly apply the voltage to the coil at some arbitrary point t_1 in the AC cycle:

$$\lambda = \int_{t_1}^{t} V_m \sin(\omega t)\,dt = \frac{-V_m}{\omega}[\cos(\omega t) - \cos(\omega t_1)] = \frac{-V_m}{\omega}\cos(\omega t) + \Delta\lambda$$

Note that the excursion of $\lambda(t)$ is asymmetrical, between $+((V_m/\omega) + \Delta\lambda)$ and $-((V_m/\omega) - \Delta\lambda)$. Because that excursion of $\lambda(t)$ beyond $+(V_m/\omega)$ drives the core into saturation, the corresponding current can be extremely large. The situation is illustrated in Figure 1.8b. This large transient current, called *inrush current*, persists for several cycles until it dies the natural death of all transients. However it can cause problems, particularly in protection of EM machines.

1.6 Permanent Magnets

It has been implied that magnetic core flux (ϕ) is exclusively caused by externally applied H fields, usually due to current-carrying coils wrapped on the core. When the externally applied H field is removed, the core flux (ϕ) drops to 0. However, it is observed that ϕ does not drop completely to zero, but rather to some small value, due to hysteresis. This phenomenon is described as the retentivity of the core. Some materials exhibit extremely strong retentivity, and are referred to as "permanent magnetics" (PMs). A linear magnetic circuit model of such a core is the same as that presented in Figure 1.2b, except that the internal mmf source is not "Ni", but rather an intrinsic property of the core.

Consider Figure 1.9, which serves to define several concepts. *Coercivity* (H_c) is the applied magnetic intensity in the reverse direction required to drive the core flux to zero. *Remanence* (or residual flux density B_r) is the flux density that exists on magnetic short circuit (when the PM is terminated in a zero reluctance path). *Maximum energy density* (BH_{MAX}) is the maximum value of the BH product as $-H_c < H < 0$. The *Curie point (temperature)* is the temperature above which a PM loses its self-magnetizing properties.

Some of the more common PM materials are:

- **AlNiCo**: An alloy consisting of aluminum, nickel, and cobalt. A common PM, used extensively from the 1940s into the 1970s. Has a very low temperature coefficient, high residual flux density, and low coercivity.
- **Ferrites**: Iron processed with powder metallurgy (usually combined with barium, stontitium, or lead), developed in the 1950s and in common use today. Has a fairly high–low temperature coefficient, fairly low residual flux density, and higher coercivity. Advantages include low cost and negligible eddy current losses.
- **SmCo**: A rare earth PM (samarium–cobalt) produced commercially since the 1970s, with low temperature coefficient, high residual flux density, and very high coercivity. The maximum operating temperature is around 300°C. Its chief disadvantage is its relatively high cost.
- **NdFeB**: A rare earth PM (neodimium–iron–boron) developed by the Japanese in the 1980s, with excellent PM properties. Unfortunately, its properties are strongly temperature-dependent, with a maximum operating temperature of around 170°C. It features high residual flux density and very high coercivity. Costs are less than SmCo.

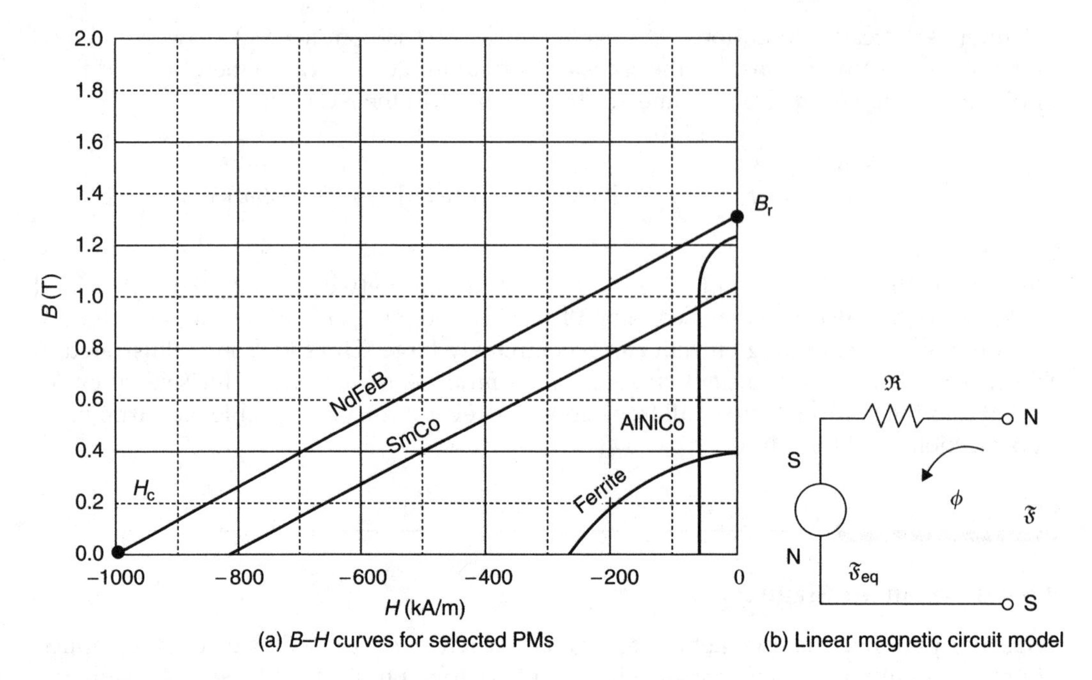

FIGURE 1.9. PM characteristics.

Some typical PM *B–H* characteristics are shown in Figure 1.9a. If the characteristic is linearized, it can be represented in magnetic circuit form, as shown in Figure 1.9b. On "open circuit" (zero flux), the mmf appearing at the magnetic "terminals" and reluctance are respectively,

$$\mathfrak{F}_{eq} = -H_c l_c, \quad \mathfrak{R} = \frac{-\mathfrak{F}_{eq}}{B_r A}$$

1.7 Superconducting Magnets

It is observed that the resistance of most electrical conductors decreases with decreasing temperature.[3] One early theory of electrical conduction argued that this trend should continue down to absolute zero (0 K), at which point the resistivity (ρ) would become 0. In 1910–1913, Kamerlingh-Onnes made measurements testing this theory, discovering that the resistivity of certain metals suddenly dropped to zero, significantly above 0 K (around the temperature of liquid helium, or about 4 K), creating a state called "superconducting." Later studies revealed that superconducting metals are not that rare; in fact nonsuperconducting metals are in the minority. More recent advances in "high-temperature" (above 70 K) superconductivity have advanced the technology further.

It is important to understand that ρ is not just small, but truly zero for superconductors. The conducting behavior of most metals is as depicted in Figure 1.10. Note the two points T_c

[3] Carbon and some semiconductors are notable exceptions.

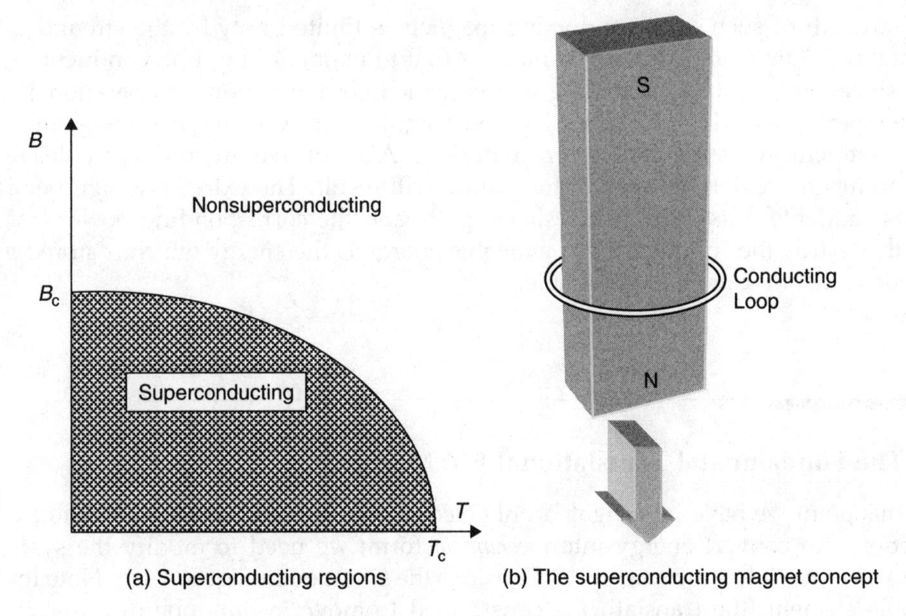

FIGURE 1.10. Superconductor magnets.

and B_c, called the critical temperature and the critical field, respectively. For most pure metals, these critical points are too low to render them useable for most practical applications;[4] however certain metallic compounds have values high enough to make them quite practical.[5]

To understand how the phenomenon can be used to make a magnet, suppose a magnet is inside the nonsuperconducting conducting loop as shown in Figure 1.10b, so that the flux linkage in the loop is λ. Now make the loop superconducting. Next pull the magnet down out of the loop.

How will the loop react? A Faraday voltage will be induced around the loop, so as to create a current that will circulate in a direction so as to oppose the change in flux linkage, i.e., to sustain the flux linkage at its original level λ. Since there is no resistance, this current will persist, and λ will be sustained at its original level after the magnet is removed. In fact, there is no way λ can be changed, as long as the loop is superconducting! If the magnet is re-inserted, the current will be driven to zero; if the magnet is reversed in polarity, and inserted, the current will be doubled, always to maintain constant flux linkage. The rate at which the magnet is inserted, or removed, is inconsequential: λ will always be sustained. We have created a constant flux magnet by a process that is essentially lossless.[6]

This system may be used in a fascinating demonstration. Suppose the magnet is remote from, and below, the ring, which is superconducting, and carrying zero initial current. Now move the magnet up toward the ring. As it gets close, a current will be induced in the ring to maintain zero flux linkage, creating a field that opposes the magnet's field. The two fields, being in opposition, will repel each other. As the magnet gets close to the loop, the force of repulsion is normally sufficient to levitate the loop, i.e., the loop will "float" above the magnet! In fact, this levitation process is used to support high-speed rail cars, as we shall see in Chapter 10.

---

[4] For example, for tin, T_c and B_c are about 3.8 K and 0.03 T, respectively.

[5] For example, for $Nb_{79}(Al_{73}Ge_{27})_{21}$, T_c and B_c are about 21 K and 45 T, respectively.

[6] Of course, conservation of energy is not violated. The energy stored in the magnetic field is precisely equal to the work done on the magnet as we pulled it out of the loop.

The strength of such superconducting magnets is limited only by the strength of the charging magnetic field. The disadvantage is that to maintain the superconducting state, we must stay in the superconducting region, which normally requires operation at cryogenic temperatures, which is expensive and complicated. Research continues into ever-higher temperature superconductor materials. Also, if we inadvertently leave the superconducting region, a catastrophic failure will result. The extremely high persistent currents suddenly must flow in a resistive path, and the corresponding power loss will normally destroy the conductor. The available energy is the energy which is stored in the magnetic field, and can be considerable.

1.8 The Fundamental Translational EM Machine

Up to this point we have investigated only electric to magnetic, and magnetic to electric, interactions. To convert energy into *mechanical* form, we need to modify the system of Figure 1.1 so that it has moving parts. Consider the situation in Figure 1.11a. Note that the moveable element (the translator) is constrained to move in only one direction, and is restrained by the spring, where $x_{min} \leq x \leq x_{max}$.

Define the *open* position as that where $x = x_{max}$, and *closed* where $x = x_{min}$. If we insert an incremental energy ΔW_e into the coil, by conservation of energy, it must appear in one of the four places:

$$\Delta W_e = \Delta W_m + \Delta W_f + \Delta W_{loss} + \Delta W_{rad}$$

where

ΔW_e = incremental electrical energy input to coil, in J
ΔW_m = incremental mechanical work done on translator, in J
ΔW_f = incremental increase in field energy, in J
ΔW_{loss} = incremental energy converted into heat, in J
ΔW_{rad} = incremental radiated energy out of the system, in J

For all of the EM machines treated in this book, the incremental radiated energy ΔW_{rad} is negligible. The losses of EM machines are important and will be studied in detail later. For now, however we will neglect them, and will define our system to be lossless. The field energy term ΔW_f includes both electric and magnetic field effects; however this book include only systems for which the electric field energy is much smaller than the magnetic field energy. Considering these points,

$$\Delta W_e = \Delta W_m + \Delta W_f$$

To determine the field energy, immobilize the translator, which forces ΔW_m to 0. Now apply a voltage $v(t)$. Since the coil has no resistance, the voltage is exclusively inductive, so that

$$W_f = W_e = \int i(t)v(t)\,dt = \int i(t)\frac{d\lambda}{dt}\,dt = \int i(\lambda)\,d\lambda$$

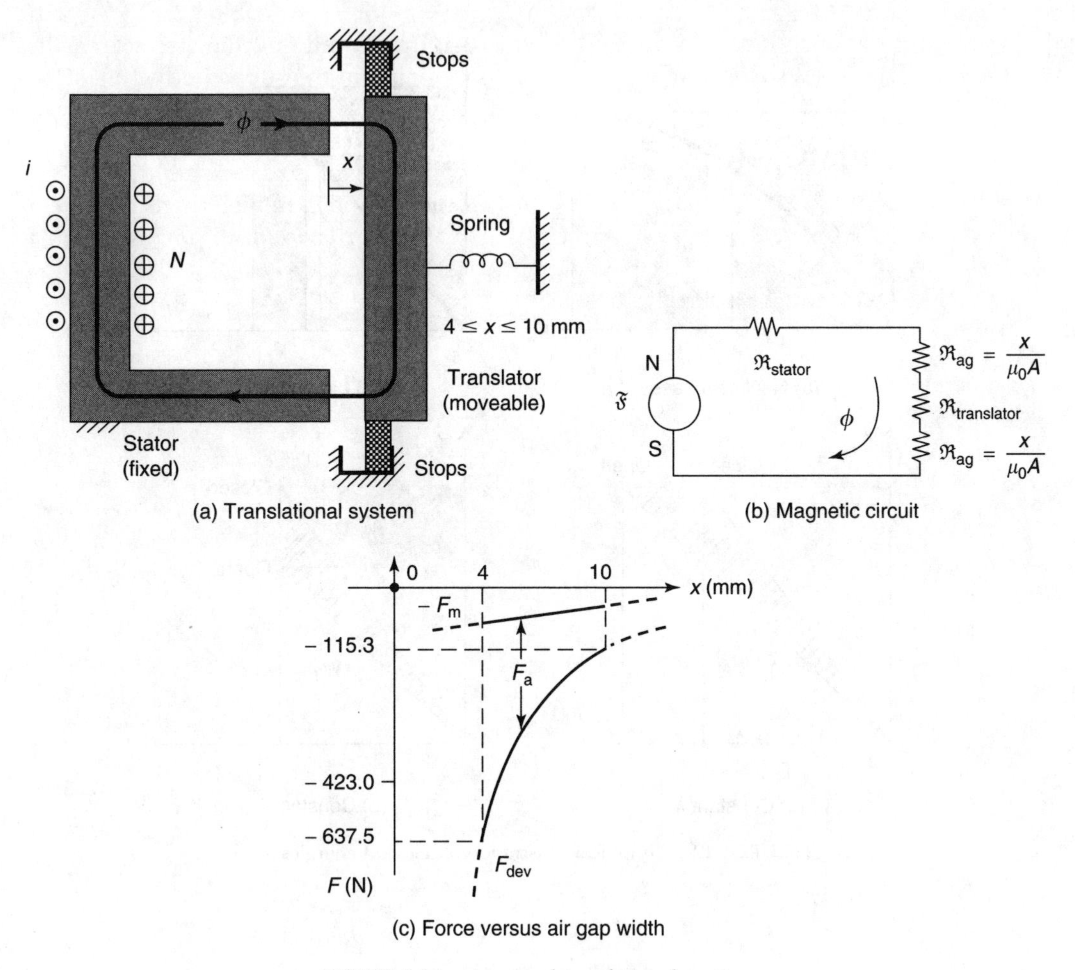

FIGURE 1.11. A general translational system.

It is expedient to define an "energy-like" term called coenergy:

$$W_f' = \int \lambda(i)\,di$$

Examine Figure 1.12a and Figure 1.12b, which illustrate the field energy W_f and coenergy W_f' graphically for the nonlinear and linear cases, respectively. Note that in the linear case,

$$W_f = W_f' = \frac{\lambda i}{2} = \frac{Li^2}{2} = \frac{\lambda^2}{2L}$$

Now consider the translator in motion, moving from the *open* to the *closed* position, separated apart by Δx. Since the magnetic path reluctance is *maximum* in the *open* position, likewise the field flux, and λ, is *minimum*, and vice versa. Consider the transition to occur under two idealized conditions.

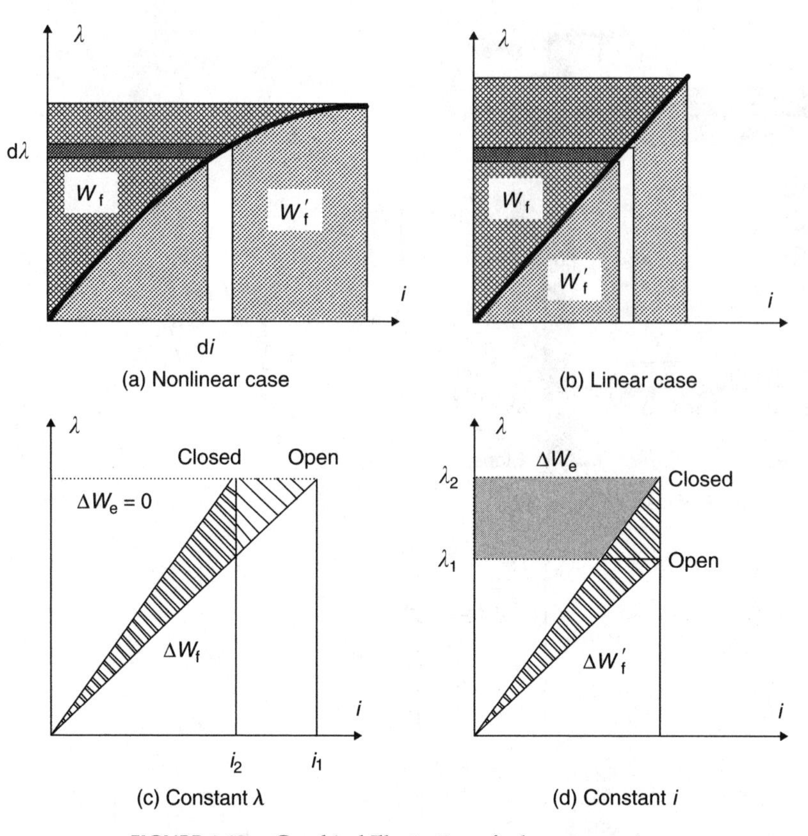

FIGURE 1.12. Graphical Illustration of relevant energies.

Constant flux linkage

The change in field energy can be computed as follows:

$$\Delta W_f = W_f(\text{CLOSED}) - W_f(\text{OPEN}) = \frac{\lambda(i_2 - i_1)}{2} = \frac{\lambda \Delta i}{2}$$

is equal to area between open and closed λ–i characteristics in Figure 1.12c. Since $\Delta\lambda = 0$, $d\lambda/dt = 0$, and $\Delta W_e = 0$; therefore $\Delta W_m = \Delta W_e - \Delta W_f = -\Delta W_f$.

Constant current

The change in field energy can be computed as follows:

$$\Delta W_f = W_f(\text{CLOSED}) - W_f(\text{OPEN}) = \frac{(\lambda_2 - \lambda_1)i}{2} = \frac{i\Delta\lambda}{2}$$

Also $\Delta W_e = \lambda_2 i - \lambda_1 i = i\Delta\lambda$ (gray area in Figure 1.12d). Therefore $\Delta W_m = \Delta W_e - \Delta W_f = \Delta W'_f$ (cross-hatched in Figure 1.12d).

In both cases the average EM force produced by the system on the translator is

$$F_{\text{average}} = \frac{\Delta W_{\text{m}}}{\Delta x} = -\frac{\Delta W_{\text{f}}}{\Delta x}\bigg|_{\lambda=\text{CONSTANT}} = +\frac{\Delta W_{\text{f}}'}{\Delta x}\bigg|_{i=\text{CONSTANT}}$$

Taking the limit as $\Delta x \rightarrow 0$ (and $F_{\text{average}} \rightarrow F(x)$)

$$F(x) = -\frac{\partial W_{\text{f}}(x,\lambda)}{\partial x} = +\frac{\partial W_{\text{f}}'(x,i)}{\partial x} \tag{1.11}$$

Equation 1.11 provides an elegant general method for determining the EM force in any EM system.[7] The problem is to formulate either $W_{\text{f}}(\lambda, x)$ or $W_{\text{f}}'(i, x)$ for the system under investigation, a task which can be quite difficult, particularly for nonlinear systems. The functional notation must be rigorously followed, i.e., $W_{\text{f}}(\lambda, x)$ must be expressed as a function *exclusively* of λ and x, and $W_{\text{f}}'(i, x)$ *exclusively* of i and x. For the linear case,

$$W_{\text{f}} = W_{\text{f}}' = \frac{\lambda i}{2} = \frac{Li^2}{2} = \frac{\lambda^2}{2L}$$

Observe that

$$F(x) = -\frac{\partial W_{\text{f}}(x,\lambda)}{\partial x} = -\frac{\partial}{\partial x}\left[\frac{\lambda^2}{2L}\right] = +\left[\frac{\lambda^2}{2L^2}\right]\frac{\partial L(x)}{\partial x} = +\left[\frac{i^2}{2}\right]\frac{\partial L(x)}{\partial x}$$

Also

$$F(x) = +\frac{\partial W_{\text{f}}'(x,i)}{\partial x} = +\frac{\partial}{\partial x}\left[\frac{Li^2}{2}\right] = +\left[\frac{i^2}{2}\right]\frac{\partial L(x)}{\partial x}$$

which demonstrates that the two approaches produce the same result for the linear case. It is straightforward (see Problem 1.) to show that

$$F(x) = +\left[\frac{\phi^2}{2}\right]\frac{\partial \mathfrak{R}(x)}{\partial x} \tag{1.12}$$

These equations elegantly provide general instructions as to how to compute force in EM systems. For an arbitrary system, all that is required is the formulation of the magnetic field energy, or coenergy, expressions, as a function of the spatial variable that defines the system movement. Subsequent partial differentiation with respect to that variable will identify the force in that direction. The positive sense of force is automatically defined by the positive sense of position. An example will illustrate the application of these concepts.

[7] Equation 1.11 can be further generalized to include multiple coils and mechanical degrees of freedom, in which case λ, i, and x all become vectors.

Example 1.6
Consider the magnetic system of Figure 1.11a, which may be assumed to be magnetically linear. System constants are

$$\mathfrak{R}_{STATOR} = 50\ (\text{mH})^{-1};\quad \mathfrak{R}_{TRANSLATOR} = 20\ (\text{mH})^{-1}$$

$$\text{Air gap: } A = 10\,\text{cm}^2 = 0.001\,\text{m}^2,\ 4 \times 10\,\text{mm}$$

$$N_{COIL} = 100\ \text{turns.}$$

(a) Draw the magnetic circuit.
(b) Determine and plot the force F_{dev} for 4×10 mm for a coil current of 10 A.

SOLUTION

(a) The system magnetic circuit is provided in Figure 1.11b.
(b) $\mathfrak{R}_{PATH} = \mathfrak{R}_{STATOR} + \mathfrak{R}_{TRANSLATOR} = 70\,(\text{mH})^{-1}$

The coil inductance L is

$$L = \frac{N^2}{\mathfrak{R}_{PATH} + \dfrac{2x}{\mu_0 A}} = \frac{\mu_0 A N^2}{\mu_0 A \mathfrak{R}_{PATH} + 2x}$$

$$\frac{\partial L}{\partial x} = \frac{N^2}{\mathfrak{R}_{PATH} + \dfrac{2x}{\mu_0 A}} = \frac{-2\mu_0 A N^2}{(\mu_0 A \mathfrak{R}_{PATH} + 2x)^2}$$

$$F_{dev} = \frac{-\mu_0 A N^2 i^2}{(\mu_0 A \mathfrak{R}_{PATH} + 2x)^2}$$

$$F_{dev} = \begin{cases} -637.5\,\text{N}, & x = 4\,\text{mm} \\ -115.3\,\text{N}, & x = 10\,\text{mm} \end{cases}$$

The plot appears in Figure 1.11c.

We next consider the motion of the translator in the system of Figure 1.11a. Like any body capable of motion, its movement is determined by the sum of forces acting on the translator. Since motion is possible only in the x direction, we need only consider forces in this direction. Note that movement to the right *increases* x, and that the negative EM forces computed in Example 1.6 tend to *close* the air gap, or move the translator to the *left*. However, the *positive sense* of F_{dev} is to *open* the air gap, since that is how positive x is defined.

The spring in the system of Figure 1.11a applies a force F_m on the translator to the right, is in tension over the range of interest, and works against F_{dev}. As x increases, the spring force decreases. By Newton's second law of motion,

$$F_{dev} + F_m = M\frac{du}{dt} = Mpu$$

where

F_{dev} = EM developed force, in N

F_m = externally applied mechanical force, in N

$p = \frac{\partial}{\partial t}(\cdots)$ = the "p operator".

$u = px$ = velocity, in m/s

M = mass of translator, in kg

Consider that the coil is driven by a current source. When $i = 0$, the spring applies a sustained force to the right, which moves the translator to the open position ($x = 10\,\text{mm}$) against the stop. If i steps up to 10 A, $F_{dev} = -115.3\,\text{N}$ (see Example 1.6), and provided $F_m < 115.3\,\text{N}$, the translator moves to the left. As long as $(-F_{dev}) > F_m$, the translator moves left. When the translator reaches the closed position ($x = 4\,\text{mm}$), its motion is arrested by the stop, and held there by a sustained force. The matter is examined further in Example 1.7

Example 1.7

The magnetic system of Example 1.6 has a specially designed spring such that for a coil current of 10 A, the accelerating force is constant at −100 N. If the translator mass is 10 kg, determine the translator position and velocity as it moves from the open to the closed position.

SOLUTION

$$M\,(d^2x/dt^2) = F_a = -100$$

$$d^2x/dt^2 = -10$$

Integrating:

$$\text{velocity} = dx/dt = -10t$$

$$\text{and position} = x(t) = -\,5t^2 + 0.010.$$

Time to reach closed position:

$$5t^2 = 0.010 - 0.004 = 0.006$$

and $t = 34.64\,\text{ms}$ with velocity $= -10t = -0.3464\,\text{m/s}$.

One final point: we have considered only a known coil *current* situation. Realistically, even for DC excitation, it is the coil *voltage* that is known.

$$v = Ri + \frac{d\lambda}{dt} = Ri + p\lambda = Ri + Lpi + i\frac{\partial L(x)}{\partial x}u$$

Thus, the general problem may be described as with a set of three nonlinear differential equations. In state variable form

$$pi = \left(\frac{-R}{L}\right)i + \left(\frac{-i}{L}\frac{\partial L(x)}{\partial x}\right)u + \left(\frac{1}{L}\right)v$$

$$pu = \frac{i}{2M}\frac{\partial L(x)}{\partial x}i + \left(\frac{1}{M}\right)F_m$$

$$px = u$$

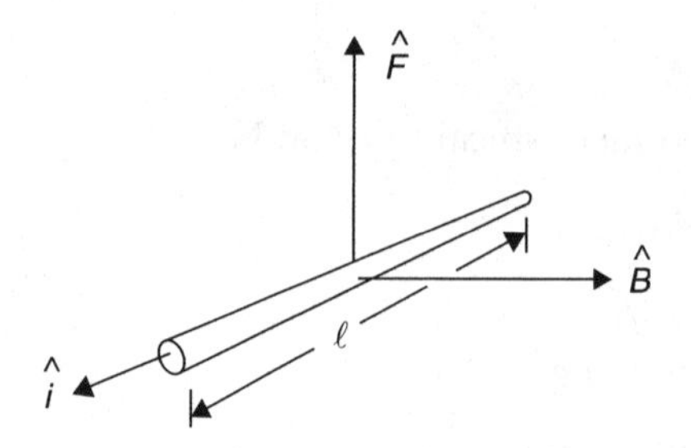

FIGURE 1.13. EM force on a length of conductor in a uniform magnetic field.

The state variables are the coil current (i), the translator position (x), and the translator velocity (u). The inputs are the coil voltage ($v(t)$) and the spring force ($F_m(x)$). It is not practical to solve the problem "by hand," but may be solved in MatLab©, MathCad©, or by other computer aided methods.

1.8.1 The Biot–Savart Law

In the previous section, we have noted the general relation between magnetic field energy and the electromagnetic force. Since the field energy is related to current, there should be a relation between current and force. Such a relation is the Biot–Savart law,

$$\hat{F} = (\hat{i} \times \hat{B})\ell$$

where space vector quantities are used for force, current, and magnetic field, with ℓ being the scalar length, as shown in Figure 1.13; $\times$ represents the vector cross-product. If force, current, and magnetic field are mutually perpendicular,

$$F = i \cdot B \cdot \ell$$

This approach to EM force determination is sometimes easier; however,

$$F_{\text{dev}} = -\frac{\partial W_{\text{f}}(x, \lambda)}{\partial x} = +\frac{\partial W_{\text{f}}'(x, i)}{\partial x}$$

is more general.

1.9 The Fundamental Rotational EM Machine

The electric and magnetic concepts discussed for the translational machine are directly applicable to the rotational machine. The mechanics are different but analogous. Table 1.3 should be helpful. The rotational equivalents of our basic energy conversion equations (Equation 1.11) are

$$T_{\text{dev}} = -\frac{\partial W_{\text{f}}(\theta, \lambda)}{\partial \theta} = +\frac{\partial W_{\text{f}}'(\theta, i)}{\partial \theta} = \text{EM developed torque applied to the rotor}$$

TABLE 1.3
Translational–Rotational System Analogs

Translational	Rotational
Position (x, m)	Angle (θ, rad)
Velocity ($u = dx/dt$, m/sec)	Angular velocity ($\omega = d\theta/dt$, rad/sec)
Acceleration ($du/dt = dx^2/dt^2$)	Angular acceleration ($d\omega/dt = d\theta^2/dt^2$)
Force (F, N)	Torque (T, N m)
Mass (M, kg)	Mass moment of inertia (J, kg m^2)

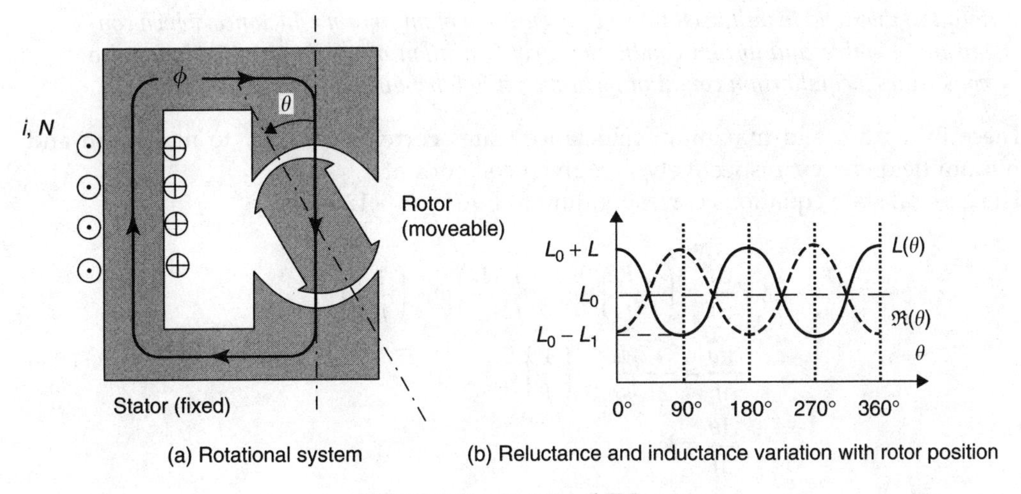

FIGURE 1.14. Rotational EM system.

The variable θ implies the movement of the moveable part of the system.

Now consider the rotational system of Figure 1.14a. Note that the reluctance of the magnetic path varies as the rotor rotates, minimizing at $\theta = 0°$ and 180°, and maximizing at $\theta = 90°$ and 270°. Since the inductance is inversely proportional to reluctance, the opposite effect is observed in $L(\theta)$. It is possible to design the shape of the rotor such that the variation is sinusoidal. Assuming this is the case,

$$L(\theta) = L_0 + L_1 \cos(2\theta)$$

The EM torque becomes

$$T_{\text{dev}} = +\frac{\partial W_f'(\theta, i)}{\partial \theta} = +\frac{i^2}{2} \cdot \frac{\partial L(\theta)}{\partial \theta} = -i^2 L_1 \sin(2\theta)$$

Note that θ (the rotor position) is defined positive in the counterclockwise (ccw) direction. Also observe that if θ is 0°, $T_{\text{dev}} = 0$, indicating an equilibrium condition. Now if θ is set to a small *positive* value ($\theta = 5°$, for example), T_{dev} is *negative*, indicating clockwise (cw) torque, which tends to move the rotor back to the equilibrium. Likewise, if θ is set to a small *negative* value (e.g., $\theta = -5°$), T_{dev} is *positive*, indicating ccw torque, again moving the rotor toward the vertical equilibrium. Thus the vertical position is referred to as a *stable* equilibrium, meaning that any small perturbation away from this point causes the system to react by producing a torque that restores the system to equilibrium.

Similarly, consider operation at $\theta = 90°$, where $T_{dev} = 0$, indicating an equilibrium. Now if the rotor is displaced by a small *positive* value from equilibrium ($\theta = 95°$, for example), T_{dev} is *positive* (ccw torque), which moves the rotor *further away* from equilibrium. Thus the horizontal position is referred to as an *unstable* equilibrium, meaning that any small perturbation away from this point causes the system to react by producing a torque that drives the system away from equilibrium.

A general principle may be derived from these observations:

> *In general, an EM system with moveable parts will inherently develop EM forces (or torques) that tend to pull itself into a configuration of minimum reluctance, which constitutes a stable equilibrium condition. A configuration of maximum reluctance also constitutes a equilibrium condition; however, it is inherently unstable.*

These minimum and maximum reluctance states correlate directly to maximum and minimum field energy, respectively, at a given coil current.

The general state equations corresponding to Equation 1.12 are

$$\begin{aligned}
\frac{di}{dt} &= \left(-\frac{R}{L}\right)i + \left(-\frac{i}{L}\frac{\partial L}{\partial \theta}\right)w + \left(\frac{1}{L}\right)v \\
\frac{dw}{dt} &= \frac{i}{2J}\frac{\partial L}{\partial \theta}i + \left(\frac{1}{J}\right)T_m \\
\frac{d\theta}{dt} &= W
\end{aligned} \tag{1.13}$$

The state variables are the coil current (i), the rotor position (θ), and the rotor velocity (ω). The inputs are the coil voltage ($v(t)$) and the externally applied rotor mechanical torque ($T_m(\theta)$). It is not practical to solve the problem "by hand," but it may be solved in MatLab©, MathCad©, or by other computer-aided methods.

1.10 Multiwinding EM Systems

Throughout our previous work, it was implied that there was only one coil, or winding, in the system. But what if there are several windings?

To investigate the situation consider the magnetic system of Figure 1.15. Assume

- magnetic linearity,
- flux densities are uniform everywhere, and
- no flux escapes the core paths.

Now consider the fluxes ϕ_a and ϕ_b *due to* i_1 *acting alone*, which we choose to denote as ϕ_{a1} and ϕ_{b1}, respectively. From the magnetic circuit

$$\phi_{a1} = \frac{\mathfrak{F}_1}{\mathfrak{R}_a + \dfrac{\mathfrak{R}_b\mathfrak{R}_c}{\mathfrak{R}_b + \mathfrak{R}_c}} = \frac{(\mathfrak{R}_b + \mathfrak{R}_c)\mathfrak{F}_1}{\mathfrak{R}_a\mathfrak{R}_b + \mathfrak{R}_a\mathfrak{R}_c + \mathfrak{R}_b\mathfrak{R}_c} = \frac{(\mathfrak{R}_b + \mathfrak{R}_c)\mathfrak{F}_1}{\Delta}$$

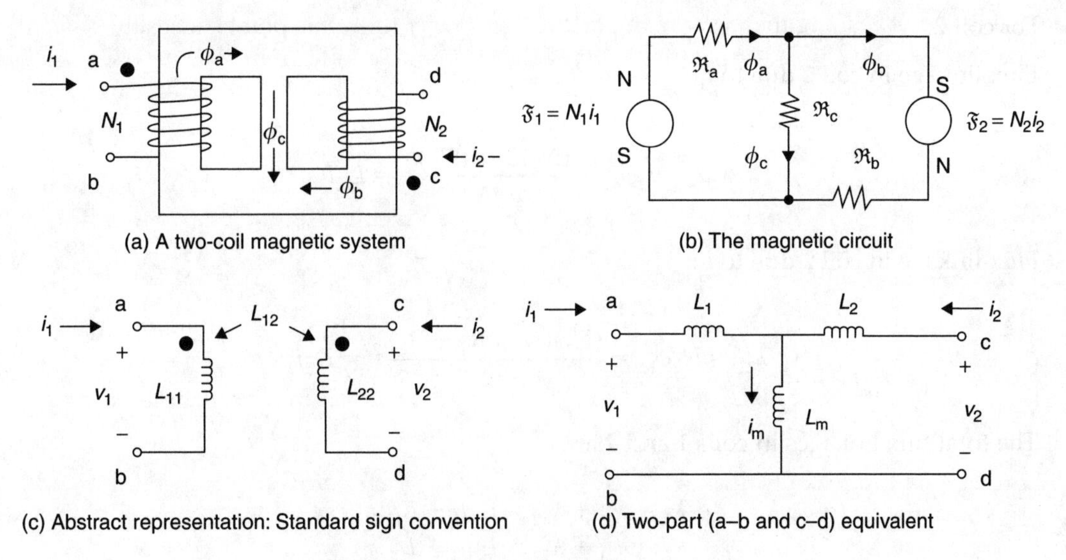

FIGURE 1.15. A two-winding magnetic system.

where

$$\Delta = \mathfrak{R}_a\mathfrak{R}_b + \mathfrak{R}_a\mathfrak{R}_c + \mathfrak{R}_b\mathfrak{R}_c$$

Likewise

$$\phi_{b1} = \frac{\mathfrak{R}_c\phi_{a1}}{\mathfrak{R}_b + \mathfrak{R}_c} = \frac{\mathfrak{R}_c\mathfrak{F}_1}{\Delta}$$

Similarly define the fluxes ϕ_a and ϕ_b *due to i_2 acting alone*, as ϕ_{a2} and ϕ_{b2}, respectively. From the magnetic circuit

$$\phi_{a2} = \frac{\mathfrak{R}_c\mathfrak{F}_2}{\Delta}$$
$$\phi_{b2} = \frac{(\mathfrak{R}_a + \mathfrak{R}_c)\mathfrak{F}_2}{\Delta}$$

Now define

Flux linkage in coil 1 due to

$$i_1 = \lambda_{11} = N_1\phi_{a1} = \frac{N_1(\mathfrak{R}_b + \mathfrak{R}_c)\mathfrak{F}_1}{\Delta} = \frac{(N_1)^2(\mathfrak{R}_b + \mathfrak{R}_c)i_1}{\Delta} = L_{11}i_1$$

Flux linkage in coil 1 due to

$$i_2 = \lambda_{12} = N_1\phi_{a2} = \frac{(N_1)(N_2)(\mathfrak{R}_c)i_2}{\Delta} = L_{12}i_2$$

For coil 2:

Flux linkage in coil 2 due to i_1:

$$\lambda_{21} = N_2\phi_{b1} = \frac{(N_1)(N_2)(\Re_c)i_2}{\Delta} = L_{12}i_2$$

Flux linkage in coil 2 due to i_2:

$$\lambda_{22} = N_2\phi_{b2} = \frac{(N_1)(N_2)(\Re_a + \Re_c)i_2}{\Delta} = L_{22}i_2$$

The total flux linkages in coils 1 and 2 are

$$\lambda_1 = \lambda_{11} + \lambda_{12} = L_{11}i_1 + L_{12}i_2$$
$$\lambda_2 = \lambda_{21} + \lambda_{22} = L_{21}i_1 + L_{22}i_2$$

To relate this approach to voltage, recall

$$v(t) = L\frac{\mathrm{d}i}{\mathrm{d}t} = Lpi \quad \text{where } p = \frac{\mathrm{d}}{\mathrm{d}t}(\ldots)$$

Therefore

$$v_1 = L_{11}pi_1 + L_{12}pi_2 = L_{11}\frac{\mathrm{d}i_1}{\mathrm{d}t} + L_{12}\frac{\mathrm{d}i_2}{\mathrm{d}t}$$
$$v_2 = L_{21}pi_1 + L_{22}pi_2 = L_{21}\frac{\mathrm{d}i_1}{\mathrm{d}t} + L_{22}\frac{\mathrm{d}i_2}{\mathrm{d}t}$$

For sinusoidal steady-state (AC) performance, the time-varying signals become phasors, and p becomes $j\omega$:

$$v_1 \to \bar{V}_1 \qquad v_2 \to \bar{V}_2$$
$$i_1 \to \bar{I}_1 \qquad i_2 \to \bar{I}_2$$
$$p \to j\omega$$

The corresponding frequency domain (phasor) equivalent expressions are

$$\bar{V}_1 = j\omega L_{11}\bar{I}_1 + j\omega L_{12}\bar{I}_2$$
$$\bar{V}_2 = j\omega L_{21}\bar{I}_1 + j\omega L_{22}\bar{I}_2$$

An alternate two-port equivalent circuit form may be derived.

$$v_1 = L_{11}pi_1 + L_{12}pi_2 = (L_{11} - L_{12})pi_1 + L_{12}p(i_1 + i_2)$$
$$v_2 = L_{21}pi_1 + L_{22}pi_2 = L_{21}p(i_1 + i_2) + (L_{22} - L_{12})pi_2$$

or

$$v_1 = L_1 p i_1 + L_m p i_m$$
$$v_2 = L_m p i_m + L_2 p i_2$$

where

$$L_m = L_{12} = L_{21}$$
$$L_1 = L_{11} - L_{12}$$
$$L_2 = L_{22} - L_{12}$$
$$i_m = i_1 + i_2$$

This is the so-called Tee equivalent circuit shown in Figure 1.15d. Care must be exercised in its application, for it is equivalent to Figure 1.15c only in a restricted two-port (a–b and c–d) sense. Note looking into port b–d in Figure 1.15d, one sees a short circuit, which is not the case in Figure 1.15c.

The above analysis may be extended to an n-winding situation:

$$\begin{aligned}
\lambda_1 &= \lambda_{11} + \lambda_{12} + \cdots + \lambda_{1n} = L_{11}i_1 + L_{12}i_2 + \cdots + L_{1n}i_n \\
\lambda_2 &= \lambda_{21} + \lambda_{22} + \cdots + \lambda_{2n} = L_{21}i_1 + L_{22}i_2 + \cdots + L_{2n}i_n \\
&\vdots \\
\lambda_n &= \lambda_{n1} + \lambda_{n2} + \cdots + \lambda_{nn} = L_{n1}i_1 + L_{n2}i_2 + \cdots + L_{nn}i_n
\end{aligned}$$

λ_i = total flux linking coil i
$\lambda_{ij} = L_{ij}i_j$ = flux linking coil i caused by current i_j
If $i = j$, L_{ii} is called the "self" inductance of coil "i"
If $i \neq j$, L_{ij} is called the "mutual" inductance between coils "i" and "j"

Observe that $L_{ij} = L_{ji}$. It is convenient to write such sets of equations in matrix form.
For example, for $n = 3$,

$$\begin{bmatrix} \lambda_1 \\ \lambda_2 \\ \lambda_3 \end{bmatrix} = \begin{bmatrix} L_{11} & L_{12} & L_{13} \\ L_{21} & L_{22} & L_{23} \\ L_{31} & L_{32} & L_{32} \end{bmatrix} \begin{bmatrix} i_1 \\ i_2 \\ i_3 \end{bmatrix}$$

or in compact notation

$$\hat{\lambda} = [L]\hat{\imath} \tag{1.14}$$

where in general

$\hat{\lambda} = n \times 1$ flux linkage vector, with general entry λ_i
$\hat{\imath} = n \times 1$ current vector, with general entry i_i
$[L] = n \times n$ inductance matrix, with general entry L_{ij}.

The reader may be, and in fact should be, concerned about an issue ignored to this point. By claiming that λ_i is the *sum* of the λ_{ij}'s, it clearly assumes that the field components must reinforce, or add up. But as we observed previously, flux is a directed quantity, and its direction is a function of the causing current direction. Therefore it is imperative that some system for coordinating current and field directions (as well as voltage polarities) must be established. A commonly used scheme is called the "dot" convention, and must be rigorously followed in a multicoil system (refer to Figure 1.15). The positive current directions must be defined into the dots for additive flux linkages in both windings (terminals a and c, or b and d, in this case). Also, voltages are defined positive at dots. Our two-winding system is represented in abstract form in Figure 1.15c, which also serves to define standard sign convention for multiwinding systems. We must understand that all equations presented in this section require conformity to this dot convention in order to be valid (i.e., all currents defined positive into dots and all voltages defined positive at the dots).

For the general case, the corresponding voltage equations are

$$\hat{v} = p\hat{\lambda} = p[L]\hat{\imath} = [L]\{p\hat{\imath}\}$$

or for AC analysis,

$$\hat{\bar{V}} = j\omega[L]\hat{\bar{I}}$$

The general system magnetic energy for n-winding linear systems is

$$W_f = \tfrac{1}{2}\hat{\imath}^t[L]\hat{\imath} = \frac{1}{2}\sum_{i=1}^{n}\sum_{j=1}^{n} i_i L_{ij} i_j \tag{1.15}$$

For the two winding case,

$$W_f = \tfrac{1}{2}L_{11}i_i^2 + L_{12}i_1 i_2 + \tfrac{1}{2}L_{22}i_2^2$$

For magnetically linear systems, the coenergy and the energy are the same:

$$W_f' = W_f$$

If the system has moving parts

$$L_{ij} = \begin{cases} L_{ij}(x), & \text{for the translational case} \\ L_{ij}(\theta), & \text{for the rotational case} \end{cases}$$

The general method for computing EM force (torque) in systems with one degree of motional freedom is

$$F(i_1, i_2, \cdots, i_n, x) = +\frac{\partial W_f'(i_1, i_2, \cdots, i_n, x)}{\partial x}, \quad \text{for the translational case}$$

$$T(i_1, i_2, \cdots, i_n, \theta) = +\frac{\partial W_f'(i_1, i_2, \cdots, i_n, \theta)}{\partial \theta}, \quad \text{for the rotational case}$$

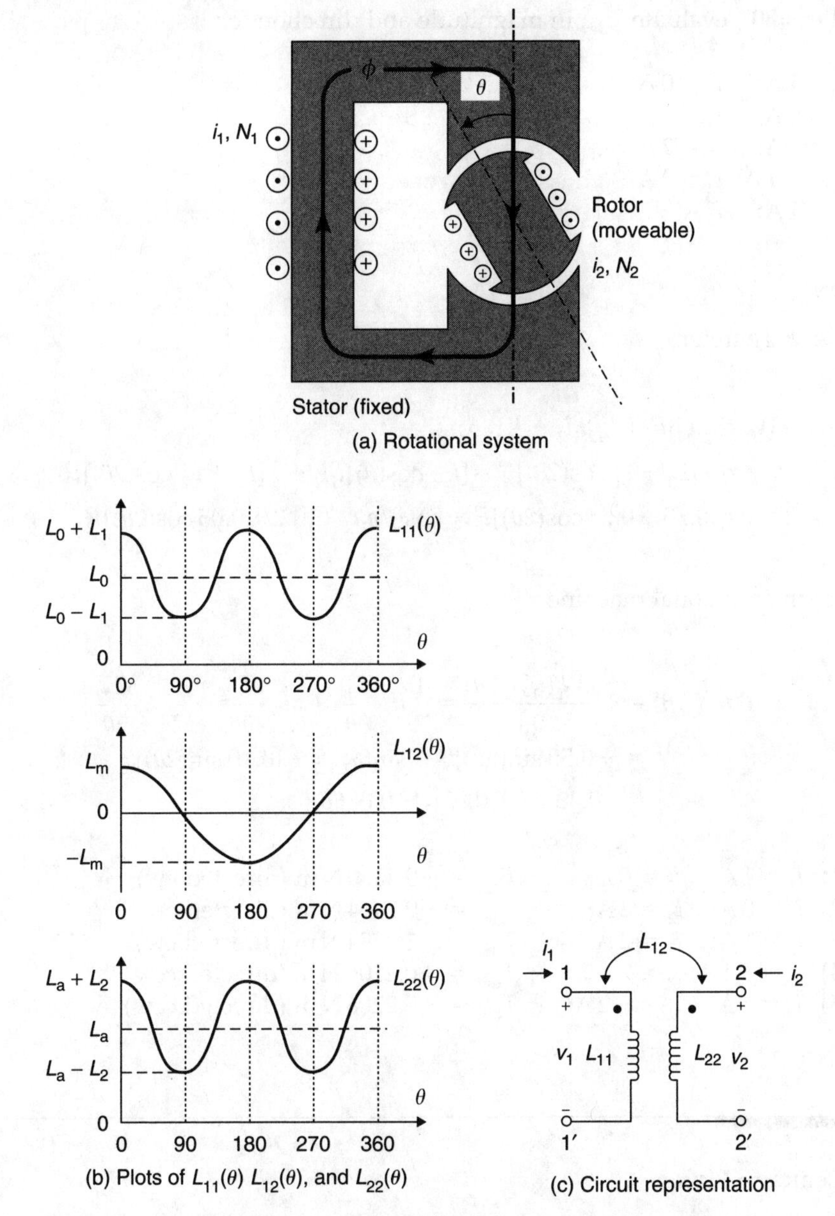

(a) Rotational system

(b) Plots of $L_{11}(\theta)$ $L_{12}(\theta)$, and $L_{22}(\theta)$

(c) Circuit representation

FIGURE 1.16. Two-winding rotation systems.

The extension to systems with any number of spatial variables is straightforward. An example will illustrate the concepts presented in this and the previous section.

Example 1.8
The rotational system of Figure 1.16a has two coils (the stator winding (1) and the rotor winding (2)). The inductances $L_{11}(\theta) = L_0 + L_1 \cos(2\theta)$, $L_{12}(\theta) = L_m \cos(\theta)$, and $L_{22} = L_a + L_2 \cos(2\theta)$ are shown in Figure 1.16b.

(a) Write the coenergy expression for the system.
(b) Compute the general EM torque T_{dev}.
(c) Suppose $L_0 = 1.5\,H$, $L_1 = 0.5\,H$, $L_m = 1\,H$, $L_a = 0.4\,H$, and $L_2 = 0.1\,H$.

For $\theta = +30°$, evaluate T_{dev} in magnitude and direction for

(1) $i_1 = 1\,\text{A}$; $i_2 = 0\,\text{A}$
(2) $i_1 = 0\,\text{A}$; $i_2 = 2\,\text{A}$
(3) $i_1 = 1\,\text{A}$; $i_2 = 2\,\text{A}$
(4) $i_1 = -1\,\text{A}$; $i_2 = 2\,\text{A}$
(5) $i_1 = 1\,\text{A}$; $i_2 = -2\,\text{A}$

SOLUTION

(a) $n = 2$. Therefore

$$
\begin{aligned}
W_f &= \tfrac{1}{2}L_{11}i_i^2 + L_{12}i_1 i_2 + \tfrac{1}{2}L_{22}i_2^2 \\
&= \tfrac{1}{2}[L_0 + L_1\cos(2\theta)]i_i^2 + [L_m\cos(\theta)]i_1 i_2 + \tfrac{1}{2}[L_a + L_2\cos(2\theta)]i_2^2 \\
&= [0.75 + 0.25\cos(2\theta)]i_i^2 + \cos(\theta)i_1 i_2 + [0.2 + 0.05\cos(2\theta)]i_2^2
\end{aligned}
$$

(b) For the rotational machine

$$
\begin{aligned}
T(i_1, i_2, \theta) &= +\frac{\partial W_f'(i_1, i_2, \theta)}{\partial \theta} = \frac{1}{2}i_1^2\frac{\partial L_{11}}{\partial \theta} + i_1 i_2\frac{\partial L_{12}}{\partial \theta} + \frac{1}{2}i_2^2\frac{\partial L_{22}}{\partial \theta} \\
&= [-0.50\sin(2\theta)]i_i^2 - \sin(\theta)i_1 i_2 - [0.10\sin(2\theta)]i_2^2 \\
&= -0.433\,i_i^2 - 0.5 i_1 i_2 - 0.0866 i_2^2
\end{aligned}
$$

(c) (1) $i_1 = 1\,\text{A}$; $i_2 = 0\,\text{A}$; $T_{dev} = -0.4330\,\text{N m}$ (directed cw)
(2) $i_1 = 0\,\text{A}$; $i_2 = 2\,\text{A}$; $T_{dev} = -0.3464\,\text{N m}$ (directed cw)
(3) $i_1 = 1\,\text{A}$; $i_2 = 2\,\text{A}$; $T_{dev} = -1.7794\,\text{N m}$ (directed cw)
(4) $i_1 = -1\,\text{A}$; $i_2 = 2\,\text{A}$; $T_{dev} = +0.2206\,\text{N m}$ (directed ccw)
(5) $i_1 = 1\,\text{A}$; $i_2 = -2\,\text{A}$; $T_{dev} = +0.2206\,\text{N m}$ (directed ccw)

1.11 Leakage Flux

In an electric circuit, there is essentially a one-to-one correspondence between the circuit diagram using ideal elements and the physical circuit, in terms of the paths available to electric current. For practical purposes, electric insulation may usually be considered "perfect," in that it passes negligible current. However, magnetic cores, while having a much higher permeability than the surrounding material, are not perfect magnetic conductors, and some small, but noticeable, fraction of the flux "leaks" out of the core. It is possible to compute the magnetic field pattern in a given magnetic system to any desired degree of accuracy to account for this leakage flux; however, this is a formidable problem, requiring much detailed core data and the use of sophisticated computer programs, and is usually unnecessary.

To examine the problem, consider the system of Figure 1.15. Assume that the system was designed to magnetically link, or "couple," the two coils; hence, the main core path between coils is around the outside path, and the leakage flux is considered to be the flux which bypasses the main path down the center leg (ϕ_c). An index of the degree of coupling is

$$k = \text{ coefficient of coupling} = \frac{L_{12}}{\sqrt{L_{11}L_{22}}} \tag{1.16}$$

$$= \sqrt{\frac{(\mathfrak{R}_c N_1 N_2/\Delta)\,(\mathfrak{R}_c N_1 N_2/\Delta)}{((\mathfrak{R}_b + \mathfrak{R}_c)N_1^2/\Delta)\,((\mathfrak{R}_a + \mathfrak{R}_c)N_2^2/\Delta)}} = \sqrt{\frac{\mathfrak{R}_c^2}{(\mathfrak{R}_b + \mathfrak{R}_c)(\mathfrak{R}_a + \mathfrak{R}_c)}}$$

Consider two extremes:

No leakage: $\mathfrak{R}_c \to \infty, \quad k = 1$
100% leakage: $\mathfrak{R}_c \to 0, \quad k = 0$

Thus, "k" provides some measure of how "tightly" coils are coupled, ranges from 0 to 1, and provides a handy check on data for physical systems ($L_{12} \leq \sqrt{L_{11}L_{22}}$).

Finally, we will return to the simple coil and core of Figure 1.1. Combining all effects discussed in this chapter for magnetic systems with no moveable parts, we produce the approximate equivalent AC circuit shown in Figure 1.17. The resistor R_1 accounts for the coil ohmic resistance. The inductor $X_1 = \omega L_1$ accounts for the leakage flux (i.e., that portion which does not flow around the core). The resistor R_c accounts for the core magnetic losses (hysteresis and eddy current). The inductor $X_m = \omega L_m$ provides a path for the magnetizing current, which provides the mmf necessary to create the core flux.

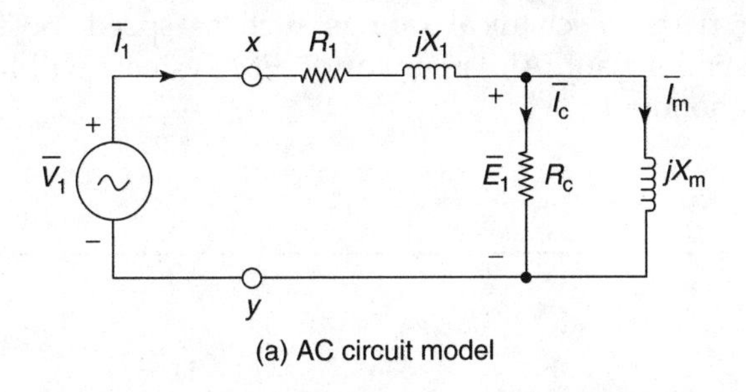

(a) AC circuit model

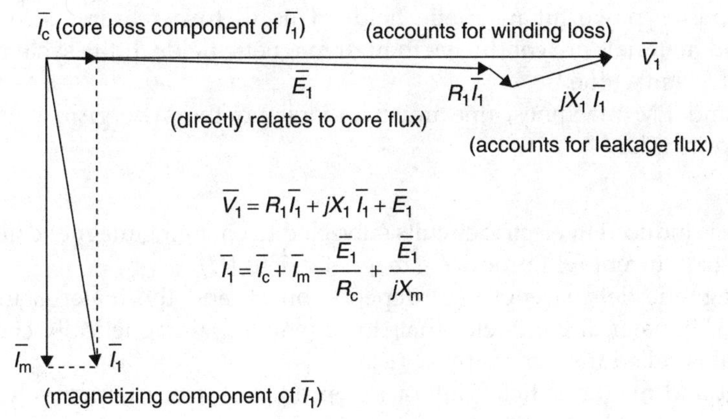

(b) Phasor diagram

FIGURE 1.17. Modeling the AC electrical performance of the system of Figure 1.1 driven from an AC voltage source.

1.12 The Concept of Ratings in EM Systems

We conclude by discussing a simple, but extremely important, topic. Short of superconductivity, any physical coil must have some upper limit on how much current it can safely carry. This is basically a thermal issue; heat is internally generated through I^2R losses, which raises the winding temperature. Higher temperatures can damage the winding insulation, and the core, and in extreme cases, melt the conductor. This upper limit is specified with a term called "rated current." For DC operation, it also correlates to a corresponding "rated" voltage.

Likewise, the core normally should not be driven very far into magnetic saturation, which corresponds to some upper limit B_{max}. This limit may also impact on the coil current rating. For AC operation, the voltage rating relates more to the B_{max} limit, since the major component of the coil voltage is normally inductive.

In general, ratings are understood to be maximum values which should not be exceeded under normal operating conditions. A second interpretation is that ratings are values at which a device was designed to operate continuously, near optimally, at no damage to itself. Ratings are normally conservative, and may be moderately exceeded under certain specific conditions, if one understands what they mean and how they were determined. For example, a common 120 V 60 W 865 lumen 1000 h incandescent lamp would be expected to operate satisfactorily, and near optimally, at 120 V, dissipating 60 W and with a light output of 865 lumen for a life of 1000 h. It does not mean that it cannot be operated at 115 or 125 V. In fact it can, but with some degradation in performance (less lumen output at low voltage and shortened life at high voltage).

For systems with moving parts, mechanical ratings such as speed, position, force, torque, and power may also be relevant. All "real-world" EM systems will have ratings that must be understood and honored.

1.13 Summary

All EM systems are designed to process either information or energy. Large-scale energy processing requires powerful magnetic fields. This in turn requires structures whose geometries and materials are conducive to high magnetic fields. If the system has moving parts, it is an EM "machine."

To understand EM machines, one must understand the synergism between several fundamental principles:

- Voltage is induced in electric circuits subjected to changing magnetic flux linkage.
- An electric current will produce a magnetic field (H).
- The magnetic field strength (B) depends on H, and the material in which H exists. The material parameter that documents how magnetically conductive a material is called the permeability (μ).
- Current and magnetic fields interact to produce force. Alternatively, EM force can be predicted by relating the change in EM energy to motion.
- The laws of circuit theory may be applied to all electric circuits within the system.
- The laws of mechanics may be applied to all moving parts of the system.

- EM systems have losses. These losses manifest themselves as heat. Ultimately heating effects will limit the capacity of any EM system.

Machines that convert energy from EM form into mechanical form are called "motors;" mechanical to EM are called "generators."

Problems

1.1 Consider the system shown in Figure 1.1, with dimensions:

$a = 10\,\text{cm}$;
$b = 5\,\text{cm}$;
mean length $= \ell = 160\,\text{cm}$;
core permeability $= 2000\,\mu_0$;
$N_{\text{COIL}} = 150$ turns;

(a) Draw the magnetic circuit.
(b) Compute the core reluctance ($\mathfrak{R}$).
(c) Compute the core permeance.

For a coil current of 8 A, solve for the

(d) Coil mmf ($\mathfrak{F}$),
(e) Core flux (ϕ),
(f) Core flux density (B).

1.2 Consider a system similar to that shown in Figure 1.2a, with revised dimensions:

width $= 30\,\text{cm}$;
height $= 40\,\text{cm}$;
cross-section $= 50\,\text{cm}^2$;
gap $= g = 1.5\,\text{mm}$;
depth $= 10\,\text{cm}$;
$N_{\text{COIL}} = 150$ turns;

(a) Draw the magnetic circuit.
(b) For a coil current of 8 A, solve for the core flux (ϕ) and flux density (B).
(c) Determine the mmf drops across the core and air gap.

1.3 Consider a system similar to that shown in Figure 1.2a, with revised dimensions

width $= 35\,\text{cm}$;
height $= 45\,\text{cm}$;
cross-section $= 60\,\text{cm}^2$;
gap $= g = 1.2\,\text{mm}$;
depth $= 10\,\text{cm}$;
$N_{\text{COIL}} = 150$ turns;

(a) Draw the magnetic circuit.
(b) For a coil current of 10 A, solve for the core flux (ϕ) and flux density (B).
(c) Determine the mmf drops across the core and air gap.

1.4 Consider a system similar to that shown in Figure 1.2a, with

$$\mathfrak{R}_{CORE} = 50\ (\text{mH})^{-1};\quad \mathfrak{R}_{AG} = 150\ (\text{mH})^{-1};$$
$$N_{COIL} = 150\ \text{turns};\quad R_{COIL} = 12\ \Omega$$

(a) Determine the coil inductance and time constant. Determine the coil current $i(t)$ if the applied coil voltage $v(t)$ is
(b) 120 V DC.
(c) $120\,u(t)$ V (suddenly applied at $t = 0$).
(d) 120 V rms AC, 60 Hz (i.e., $v(t) = 169.7\cos(377t)$ V).

1.5 Consider a system similar to that shown in Figure 1.2a, with

$$\mathfrak{R}_{CORE} = 60\ (\text{mH})^{-1};\quad \mathfrak{R}_{AG} = 200\ (\text{mH})^{-1};$$
$$N_{COIL} = 100\ \text{turns};\quad R_{COIL} = 15\ \Omega$$

(a) Determine the coil inductance and time constant. Determine the coil current $i(t)$ if the applied coil voltage $v(t)$ is
(b) 120 V DC.
(c) $120\,u(t)$ volts (suddenly applied at $t = 0$).
(d) 120 V rms AC, 60 Hz (i.e., $v(t) = 169.7\cos(377t)$ V).

1.6 Rework Example 1.4, moving point b to a new location ($H = 0.8$ kA/m; $B = 1240$ mT). Note that the results are slightly different.

1.7 Rework Example 1.5, after raising the applied DC voltage to 150 V, raising the coil current to 15 A.

1.8 Rework Example 1.5, after lowering the applied DC voltage to 100 V, lowering the coil current to 10 A.

1.9 A magnetic core has a coil, which is excited by an AC source. If the coil voltage is 120 V (rms) 60 Hz, the core eddy current and hysteresis losses are each 100 W. If the Steinmetz exponent is 2, and the coil has negligible resistance, determine the eddy current and hysteresis losses if the coil voltage is

(a) 132 V (rms) 60 Hz
(b) 120 V (rms) 66 Hz
(c) 132 V (rms) 66 Hz
(d) 109.1 V (rms) 60 Hz
(e) 109.1 V (rms) 66 Hz

1.10 Repeat Problem 1.9 if the Steinmetz exponent is 2.2.

1.11 Repeat Problem 1.9 if the Steinmetz exponent is 1.8.

1.12 We wish to design a small actuator of the geometry presented in Figure 1.11. Specifications are

- The coil wire is square in cross-section, 2 mm on a side, with an ampacity of 5 A.
- The translator moves so that (one) air gap varies from 3 (closed) to 8 mm (open).
- The device must develop at least 10 N in the open position.
- The center "hole" is square (closed position). The coil must fit in the hole, of course.
- The air gap dominates the magnetic circuit; neglect the core reluctance. However, the core flux density may never exceed 1.0 T.
- The device should be as small as possible.

Determine the number of coil turns and complete dimensions for your design.

1.13 Repeat Problem 1.12 if the core reluctance is not negligible, but may be assumed to be linear with $\mu_r = 2000$.

1.14 A conductor segment carrying a current of 10 A is oriented perpendicular to a magnetic field of 2 T. Determine the force on the segment and draw an appropriate 3D diagram.

1.15 Locate the missing dots. Add to the diagram canonical positive definitions for all currents and voltages.

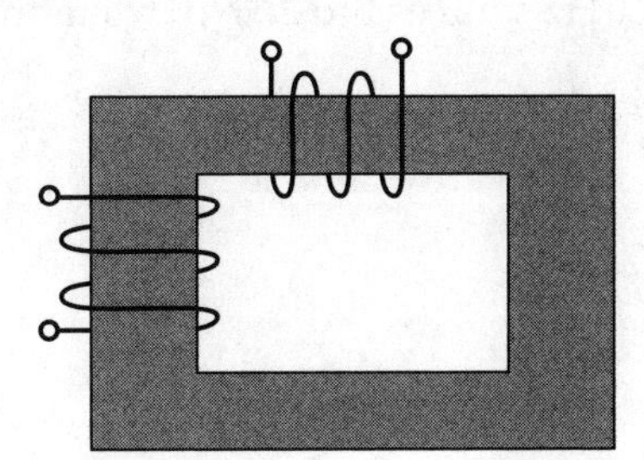

1.16 Consider the rotational machine of Example 1.8. Coil currents into dots located coil resistances are

$$R_1 = 10\,\Omega, \quad R_2 = 10\,\Omega$$

Determine all unknown voltages, currents, and the torque on the rotor if

(a) $v_1 = 10$ V DC;	coil 2 is open,	$\theta = 45°$
(b) $v_1 = 10$ V DC;	coil 2 is shorted,	$\theta = 0°$
(c) $v_1 = 10$ V DC;	coil 2 is shorted,	$\theta = 45°$
(d) Coil 1 is shorted;	$v_2 = 10$ V DC,	$\theta = 45°$
(e) $v_1 = 10\cos(2t)$ V AC;	coil 2 is open,	$\theta = 45°$
(f) $v_1 = 10\cos(2t)$ V AC;	coil 2 is shorted,	$\theta = 0°$
(g) $v_1 = 10\cos(2t)$ V AC;	coil 2 is shorted,	$\theta = 45°$
(h) $v_1 = 10\cos(2t)$ V AC;	coil 2 is shorted,	$\theta = 2t$
(i) $v_1 = 10\cos(2t)$ V AC;	coil 2 is shorted,	$\theta = 2t + \pi/4$
(j) $v_1 = 10\cos(2t)$ V AC;	coil 2 is shorted,	$\theta = 2t - \pi/4$

1.17 The system of Figure 1.15a is supplied with coils such that $N_1 = 100$ turns and $N_2 = 80$ turns. The system has negligible resistance. The following AC tests are made at 60 Hz.

With the secondary on open circuit: $V_1 = 120\,\text{V}$ (rms); $V_2 = 80\,\text{V}$; $I_1 = 10\,\text{A}$.
With the secondary on short circuit: $V_1 = 120\,\text{V}$ (rms); $I_1 = 15\,\text{A}$.

(a) Draw the Tee circuit as shown in Figure 1.15d, with all inductors evaluated.
(b) Draw the coupled circuit as shown in Figure 1.15c, with all inductors evaluated.
(c) Draw the magnetic circuit as shown in Figure 1.15b, with all reluctances evaluated.

1.18 The rotary system of Figure 1.16a is supplied with coils of negligible resistance. The following AC tests are made at 60 Hz.

With the secondary on open circuit, and $V_1 = 120\,\text{V}$ (rms):

(1) $I_1 = 4\,\text{A}$, when $\theta = 0$.
(2) $I_1 = 10\,\text{A}$, when $\theta = \pi/2$.

With the primary on open circuit, and $V_2 = 120\,\text{V}$ (rms):

(1) $I_2 = 4\,\text{A}$, when $\theta = 0$.
(2) $I_2 = 10\,\text{A}$, when $\theta = \pi/2$.

(a) Connect the primary and secondary windings in series (terminal $1'$ to 2), and apply a 120 V 60 Hz source (1 to $2'$). Now determine the current I_1 (rms A) and the torque.

(b) Connect the primary and secondary windings in series (terminals 1 to 2 and $1'$ to $2'$), and apply a 120 V 60 Hz source (1 to 2). Now determine the current I_1 (rms A) and the torque.

2

Transformers

It is necessary to change voltage levels efficiently, when processing bulk electrical energy. For example, it is convenient, economical, and safe to operate household appliances such as television sets and microwave ovens at a low voltage level (e.g. 120 V). However, it is not practical to transport any significant amount of power (over 10 kW, for example) more than a few dozen feet at that voltage because of the corresponding high current and associated I^2R winding loss. What is needed is a highly efficient and reliable device to convert electric power from one voltage level into another. The power transformer is just such a device. Strictly speaking, it fails to qualify as an electromagnetic (EM) machine, because it has no moving parts, and hence cannot perform the requisite electrical to mechanical energy conversion process. However, it performs electric-to-magnetic-to-electric energy conversion. This property, plus the fact that the power transformer is an important component found in almost all power systems, merits our study of this important device.

2.1 The Ideal *n*-Winding Transformer

Consider the three-winding situation in Figure 2.1a, assumed to be linear, with its corresponding magnetic circuit in Figure 2.1b.[1] It is common to distinguish windings using "1, 2, 3, etc.," and the terms primary, secondary, tertiary, etc. It is standard to associate the *primary* winding with the *source* side (power flows from primary to secondary, tertiary, etc.). From the magnetic circuit we write

$$N_1 i_1 + N_2 i_2 + N_3 i_3 = \mathfrak{R}_{\text{core}} \phi \tag{2.1}$$

where

$$\mathfrak{R}_{\text{core}} = \frac{\ell}{\mu_{\text{core}} A}$$

If we assume an "ideal" core, $\mu_{\text{core}} \to \infty$, and $\mathfrak{R} = 0$, so that

$$N_1 i_1 + N_2 i_2 + N_3 i_3 = 0 \tag{2.2}$$

[1] This magnetic circuit ignores leakage flux, assuming that the same flux links all windings. Since we intend to idealize the core, this is justifiable.

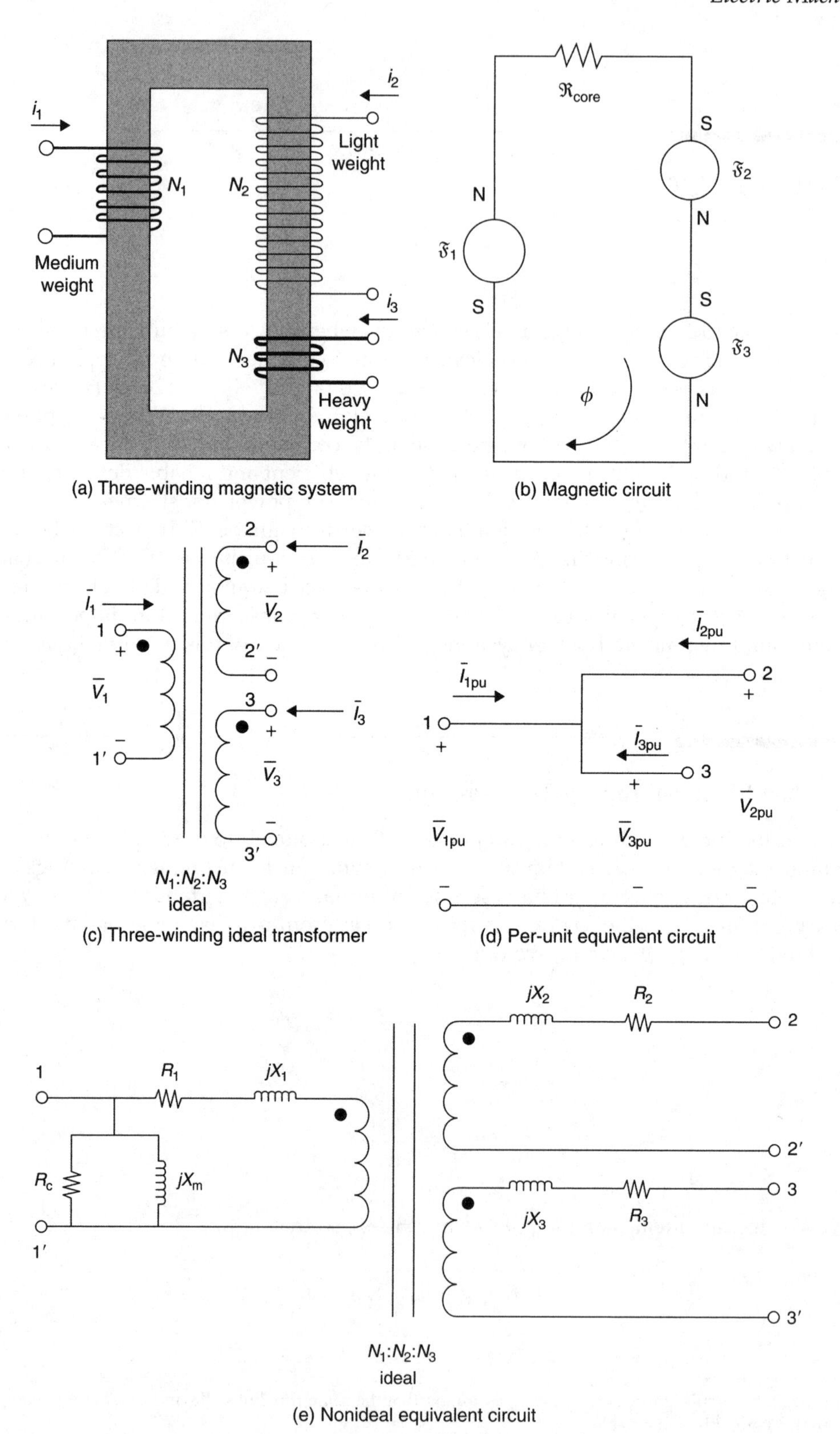

FIGURE 2.1. The three-winding power transformer.

Now, if we further assume "ideal" windings, $\sigma_{winding} \to \infty$, then winding resistances $R = 0$, so that winding voltage is exclusively inductive. Note that the same core flux ϕ links each winding, since leakage flux is impossible from our ideal core. Therefore

$$v_1 = p\lambda_1 = p(N_1\phi) \tag{2.3a}$$

$$v_2 = p\lambda_2 = p(N_2\phi) \tag{2.3b}$$

$$v_3 = p\lambda_3 = p(N_3\phi) \tag{2.3c}$$

Dividing Equations 2.3a by 2.3b, 2.3b by 2.3c, and 2.3c by 2.3a,

$$\frac{v_1}{v_2} = \frac{N_1}{N_2}; \qquad \frac{v_2}{v_3} = \frac{N_2}{N_3}; \qquad \frac{v_3}{v_1} = \frac{N_3}{N_1} \tag{2.4}$$

Equation 2.2 and Equation 2.4 completely define the v, i relations for an ideal component, the *ideal transformer*, the symbol for which appears in Figure 2.1c. Dividing Equation 2.2 by N_1

$$i_1 + \frac{N_2}{N_1} i_2 + \frac{N_3}{N_1} i_3 = 0 \tag{2.5}$$

Multiplying by v_1,

$$\begin{aligned} &v_1 i_1 + v_1 \frac{N_2}{N_1} i_2 + v_1 \frac{N_3}{N_1} i_3 = 0 \\ &v_1 i_1 + \frac{N_1}{N_2} v_2 \frac{N_2}{N_1} i_2 + \frac{N_1}{N_3} v_3 \frac{N_3}{N_1} i_3 = 0 \\ &v_1 i_1 + v_2 i_2 + v_3 i_3 = 0 \end{aligned} \tag{2.6}$$

For AC operation, these relations become

$$\begin{aligned} &\frac{\bar{V}_1}{\bar{V}_2} = \frac{N_1}{N_2}; \qquad \frac{\bar{V}_2}{\bar{V}_3} = \frac{N_2}{N_3}; \qquad \frac{\bar{V}_3}{\bar{V}_1} = \frac{N_3}{N_1} \\ &\bar{I}_1 + \frac{N_2}{N_1} \bar{I}_2 + \frac{N_3}{N_1} \bar{I}_3 = 0 \\ &\bar{V}_1 \bar{I}_1^* + \bar{V}_2 \bar{I}_2^* + \bar{V}_3 \bar{I}_3^* = 0 \end{aligned} \tag{2.7}$$

The implications of Equation (2.7) are profoundly important. They reveal that:

- Voltage is transformed according to the turns ratio.
- Current is transformed according to the inverse turns ratio.
- Power is transformed "not at all", i.e. the device is complex power invariant.

Whatever total complex power enters the device must flow out of the device, with zero losses.

Consider for a moment the positive definitions of currents and voltages in Figure 2.1 (currents into the dots; voltages, positive at the dots). The corresponding complex power terms are *inputs* to the windings. These must be strictly enforced if Equations 2.1 to 2.7 are to be used. If opposite current direction, or voltage polarity, is desired, one must draw a modified diagram, and replace any redefined variable by its negative in the impacted equation(s). The extension to n-windings is straightforward.

2.2 Transformer Ratings and Per-Unit Scaling

For an actual transformer, μ_{core} is large, but not infinite, of course. Likewise, $\sigma_{winding}$ is large, which produces small, but nonzero, winding resistances. Hence, the transformer has internal losses, which appear in the form of heat. This internal heat increases the temperature, which eventually can damage the device. Thus there are limiting operating levels, referred to as "ratings." Before we modify the ideal model to accurately account for losses, we shall consider transformer ratings and per-unit (pu) scaling.

In an actual n-winding transformer, each winding has two ratings: V_{rated} and S_{rated}.[2] From these data, the current rating for each winding can be derived:

$$I_{rated} = \frac{S_{rated}}{V_{rated}} \tag{2.8}$$

It was clear in Section 2.1 that the winding turns ratios are critically important data. In an actual power transformer, this information is provided indirectly through the voltage ratings. That is, we understand that

$$\frac{N_1}{N_2} = \frac{V_{1rated}}{V_{2rated}}; \qquad \frac{N_2}{N_3} = \frac{V_{2rated}}{V_{3rated}}; \qquad \frac{N_3}{N_1} = \frac{V_{3rated}}{V_{1rated}} \tag{2.9}$$

Thus the voltage ratings serve two purposes: (1) they identify "normal" levels at which the windings should operate, and (2) they provide the turns ratios. It is a common practice to scale transformer data, the scaled values said to be in "per-unit" (pu). The basic scaling equation is:

$$\text{scaled value in per-unit (pu)} = \frac{\text{actual value in SI units}}{\text{base value in SI units}} \tag{2.10}$$

Four quantities are normally scaled: voltage, power, impedance, and current. Therefore, four base values must be considered: V_{base}, S_{base}, Z_{base}, I_{base}. V_{base} and S_{base} are chosen arbitrarily, usually in compliance with an accepted convention, and the other two computed

$$\begin{aligned} I_{base} &= \frac{S_{base}}{V_{base}} \\ Z_{base} &= \frac{V_{base}}{I_{base}} = \frac{V_{base}^2}{S_{base}} \end{aligned} \tag{2.11}$$

In the case of transformers, a set of base values is selected for each winding. For example, for the three-winding case, a total of 12 base values are required, four for each winding. The normal practice is to select winding *base* voltages to be the winding *rated* voltages. The base value $S_{base} = S_{1base} = S_{2base} = S_{3base}$ is common to all windings, and may, or may not, be the rating of any particular winding. The consequences of these choices on Equation 2.7 are

$$\begin{aligned} \bar{V}_{1pu} &= \frac{\bar{V}_1}{V_{1base}} = \frac{\bar{V}_2(N_1/N_2)}{V_{2base}(N_1/N_2)} = \bar{V}_{2pu} \\ \bar{V}_{2pu} &= \frac{\bar{V}_2}{V_{2base}} = \frac{\bar{V}_3(N_2/N_3)}{V_{3base}(N_2/N_3)} = \bar{V}_{3pu} \end{aligned} \tag{2.12}$$

[2] For a more detailed discussion of ratings, refer to Section 1.10.

so that

$$\bar{V}_{1pu} = \bar{V}_{2pu} = \bar{V}_{3pu}$$

The consequences on Equation 2.7 are

$$\frac{\bar{I}_1}{I_{1base}} + \frac{(N_2/N_1)\bar{I}_2}{(N_2/N_1)I_{2base}} + \frac{(N_3/N_1)\bar{I}_3}{(N_3/N_1)I_{3base}} = \bar{I}_{1pu} + \bar{I}_{2pu} + \bar{I}_{3pu} = 0 \tag{2.13}$$

These results produce the primitive circuit diagram of Figure 2.1d. An example will demonstrate these relations.

Example 2.1
An ideal transformer has the following ratings:

- Primary: 15 kV, 100 MVA
- Secondary: 115 kV, 90 MVA
- Tertiary: 7.2 kV, 10 MVA

(a) Determine the turns ratios.
(b) Suppose S_{base} = 100 MVA. Using voltage ratings as bases, determine all base values in all windings.
(c) The transformer is terminated as follows:

- Primary: 15 kV AC 60 Hz voltage source
- Secondary: 50 MVA, pf = 0.8 lagging load
- Tertiary: 10 MVA, pf = 0.8 leading load

Solve for the primary *input* complex power ($\bar{S}_1 = \bar{V}_1(\bar{I}_1)^*$), the secondary *output* complex power ($\bar{S}_2 = \bar{V}_2(-\bar{I}_2)^*$), and the tertiary *output* complex power ($\bar{S}_3 = \bar{V}_3(-\bar{I}_3)^*$).

(d) Solve for the winding phasor currents.

SOLUTION
Figure 2.1a and Figure 2.1c define all winding voltages and currents.

(a) The rated voltage ratios reveal the turns ratios:

$$N_1/N_2 = V_{1rated}/V_{2rated} = 15/115 = 0.1304$$
$$N_2/N_3 = V_{2rated}/V_{3rated} = 115/7.2 = 15.97$$
$$N_3/N_1 = V_{3rated}/V_{1rated} = 7.2/15 = 0.4800$$

(b) The voltage ratings are the bases. Equation 2.2 applies, and the results are

Winding	V_{base} (kV)	S_{base} (MVA)	Z_{base} (Ω)	I_{base} (A)
1	15	100	2.25	6667
2	115	100	132.25	870
3	7.2	100	0.5184	13,889

(c) The voltages are:

$$\bar{V}_1 = 15\angle 0°\,\text{kV}$$
$$\bar{V}_2 = 115\angle 0°\,\text{kV}$$
$$\bar{V}_3 = 7.2\angle 0°\,\text{kV}$$

(c)
$$\bar{S}_2 = \bar{V}_2(-\bar{I}_2)^* = 50\angle{+36.9°}\ \text{MVA} = 40\ \text{MW} + j30\ \text{Mvar}$$
$$\bar{S}_3 = \bar{V}_3(-\bar{I}_3)^* = 10\angle{-36.9°}\ \text{MVA} = 8\ \text{MW} - j6\ \text{Mvar}$$
$$\bar{S}_1 = (40 + j30) + (8 - j6) = 48 + j24 = 53.66\angle{26.6°}\ \text{MVA}$$

(d) The phasor currents are:

$$\bar{I}_1 = (\bar{S}_1/\bar{V}_1)^* = 3578\angle{-26.6°}\ \text{A}$$
$$-\bar{I}_2 = (\bar{S}_2/\bar{V}_2)^* = 434.8\angle{-36.9°}\ \text{A}$$
$$-\bar{I}_3 = (\bar{S}_3/\bar{V}_3)^* = 1389\angle{+36.9°}\ \text{A}$$

2.3 The Nonideal Three-Winding Transformer

As noted, for an actual transformer, μ_{core} is large, but not infinite. Indeed, the actual core is magnetically nonlinear, complete with core losses. Therefore, if modeling the core effects has engineering relevance, the ideal transformer model presented in Figure 2.1c is inadequate. If we apply an AC source to the primary winding, with all other windings on open circuit (oc), we observe that the primary winding (1) draws a current and (2) absorbs average power. Note that in the ideal device, both the current and power are zero. We chose to add the shunt elements R_c and X_m, as shown in Figure 2.1e, sizing R_c to absorb the core loss, and sizing X_m to provide a path for the balance of the no load (NL) current. An example will demonstrate.

Example 2.2
Consider the transformer of Example 2.1 to be nonideal, with the same ratings. We apply a 15 kV AC 60 Hz voltage source to the primary, with the secondary and tertiary on oc. Determine the shunt elements R_c and X_m in Figure 2.1e, if the primary current and power are 333.3 A and 600 kW, respectively.

SOLUTION

$$R_c = V^2/P = (15)^2/0.6 = 375\ \Omega \quad (G_c = 2.667\ \text{mS})$$
$$Y = I/V = 333.3/15 = 22.22\ \text{mS}$$
$$B_m = \sqrt{Y^2 - (G_c)^2} = 22.06\ \text{mS} \quad (X_m = 45.33\ \Omega)$$

Once we acknowledge that the core is not a perfect magnetic conductor, we admit to the existence of leakage flux. As was discussed in Chapter 1, leakage flux can be accounted for with a series inductor in the appropriate winding.[3] Also, as was noted, $\sigma_{winding}$ is large, which produces small, but nonzero, winding resistances, which can be modeled as series resistors and reactors (see R_1, R_2, R_3, X_1, X_2, X_3, in Figure 2.1e). The next issue is to consider how we might obtain values for these additional components.

Consider terminating the primary winding in an AC source, the secondary winding in a short circuit, and the tertiary winding in an oc. If the transformer were ideal, the primary and secondary currents would be infinite for any finite primary voltage. For a physical transformer, the primary and secondary currents are typically large for a small primary voltage. Both of

[3] For multiple windings, this is not rigorously correct. Each winding has self-inductances, and a matrix of mutual inductances, accounting for couplings between all pairs of windings. However, unless extreme accuracy is required, the approximate Tee model is adequate, and is better than completely ignoring leakage flux.

these facts suggest that it would be reasonable to neglect the small currents that flow through the shunt elements R_c and X_m, that is, treat the shunt elements as opens, which we shall.

We now have a plan for evaluating R_1, R_2, R_3, X_1, X_2, X_3, in Figure 2.1e from measured *short circuit test* data. Define the following notation for a n-winding transformer ($n \geq 3$):

V_{ij} = voltage applied to winding i with winding j shorted; all other windings on OC
I_{ij} = current into winding i with winding j shorted; all other windings on OC
P_{ij} = power into winding i with winding j shorted; all other windings on OC

Arrange to measure V_{ij}, I_{ij}, and P_{ij} in winding i. Now supply an adjustable voltage source V_{ij}, starting from zero, and increasing until I_{ij} is maximized *without exceeding the current rating of windings i or j*. Record V_{ij}, I_{ij}, and P_{ij} and repeat the procedure for $i = 1$ to n. Convert all readings to pu to eliminate the turns ratios. Now

$$Z_{ij} = V_{ij}/I_{ij}, \qquad R_{ij} = \frac{P_{ij}}{I_{ij}^2}$$

$$X_{ij} = \sqrt{(Z_{ij})^2 - (R_{ij})^2}, \qquad \bar{Z}_{ij} = R_{ij} + jX_{ij}$$

Neglect the shunt elements R_c and X_m. Since we are in pu;

$$\bar{Z}_{ij} = \bar{Z}_i + \bar{Z}_j$$

For the n-winding case, we have n-independent equations. For example for the three-winding case, we have three independent equations:

$$\bar{Z}_{12} = \bar{Z}_1 + \bar{Z}_2$$
$$\bar{Z}_{23} = \bar{Z}_2 + \bar{Z}_3$$
$$\bar{Z}_{31} = \bar{Z}_3 + \bar{Z}_1$$

Solving for the unknowns $\bar{Z}_1, \bar{Z}_2$, and $\bar{Z}_3$,

$$\bar{Z}_1 = 0.5(\bar{Z}_{12} - \bar{Z}_{23} + \bar{Z}_{31})$$
$$\bar{Z}_2 = 0.5(\bar{Z}_{23} - \bar{Z}_{31} + \bar{Z}_{12})$$
$$\bar{Z}_3 = 0.5(\bar{Z}_{31} - \bar{Z}_{12} + \bar{Z}_{23})$$

A numerical example will demonstrate.

Example 2.3
Consider short-circuit test data for the transformer of Example 2.2:

	Winding					
Test	**1**	**2**	**3**	**Voltage**	**Current**	**Power**
OC	Meas	oc	oc	15,000 V	333 A	600,000 W
				1.000000 pu	0.050000 pu	0.006000 pu
SC	Meas	sc	oc	128 V	667 A	8500 W
				0.008500 pu	0.100000 pu	0.000085 pu
SC	oc	Meas	sc	886 V	87 A	11,550 W
				0.007700 pu	0.100000 pu	0.000116 pu
SC	sc	oc	Meas	79 V	1389 A	13,200 W
				0.011000 pu	0.100000 pu	0.000132 pu

Determine the series elements R_i and X_i ($i = 1, 2, 3$) in pu and Ω.

SOLUTION

The test data have been converted into pu (see Example 2.1 for base values). For example,

$$V_{12} = 128\text{ V}/15{,}000\text{ V} = 0.0085\text{ pu} \qquad I_{12} = 667\text{A}/6667\text{ A} = 0.10\text{ pu}$$
$$P_{12} = 8.5\text{ kW}/100{,}000\text{ kVA} = 0.000085\text{ pu}$$

Continuing,

$$Z_{12} = V_{12}/I_{12} = 0.0085/0.10 = 0.085\text{ pu}$$
$$R_{12} = P_{12}/(I_{12})^2 = 0.000085/(0.10)^2 = 0.0085\text{ pu}$$
$$X_{12} = \sqrt{(Z_{12})^2 - (R_{12})^2} = 0.08457\text{ pu}$$
$$\bar{Z}_{12} = R_{12} + jX_{12} = 0.0085 + j0.08457\text{ pu}$$

Complete pu results are presented in the following table:

	R	*X*	*Z*
Z_{12}	0.00850	0.08457	0.08500
Z_{23}	0.01155	0.07613	0.07700
Z_{31}	0.01320	0.10921	0.11000

Also

$$\bar{Z}_1 = 0.5(\bar{Z}_{12} - \bar{Z}_{23} + \bar{Z}_{31})$$
$$= 0.5(0.00850 + j0.08457 - (0.01155 + j0.07613 + 0.01320 + j0.10921)$$
$$= 0.00508 + j0.05883\text{ pu} = 0.508 + j5.883\%$$
$$\bar{Z}_2 = 0.5(\bar{Z}_{23} - \bar{Z}_{31} + \bar{Z}_{12}) = 0.00343 + j0.02575\text{ pu} = 0.343 + j2.575\%$$
$$\bar{Z}_3 = 0.5(\bar{Z}_{31} - \bar{Z}_{12} + \bar{Z}_{23}) = 0.00813 + j0.05038\text{ pu} = 0.813 + j5.038\%$$

Converting into Ω,

$$Z_1 = (0.00508 + j0.05883)Z_{\text{base}}$$
$$= (0.00508 + j0.05883)(2.25) = 0.01142 + j0.13236\ \Omega$$

Complete results in Ω are presented in the following table:

Ω	*R*	*X*	*Z*
Z_{12}	0.01913	0.19029	0.19125
Z_{23}	1.52749	10.06804	10.18325
Z_{31}	0.00684	0.05661	0.05702
Z_1	0.01142	0.13236	0.13285
Z_2	0.45296	3.40528	3.43527
Z_3	0.00421	0.02612	0.02645

The circuit of Figure 2.1e is an accurate and an useful model. However, like all models, it is not perfect and has the following intrinsic deficiencies:

- The shunt branches are not in the theoretical correct location, and should be moved inside the series element $\bar{Z}_1$ for slightly better accuracy. Likewise the procedure for evaluating the shunt elements R_c and X_m has a minor error. For example, a small part of the NL loss is primary winding I^2R loss.

- Whereas the series reactors X_i account for leakage flux, it is incorrect to assume that each X_i can be directly equated to the leakage flux for that specific winding. In fact, X_i can be, and frequently is, negative for actual transformers. Of course, neither R_{ij} nor X_{ij} can be negative for all i, j.
- Since it is a linear circuit, it can only predict the fundamental sinusoidal root mean square (rms) current and voltage. For example, the model cannot predict harmonic distortions in the magnetizing current.
- For transient and high-frequency analysis, the model should be refined to include inter- and intra-winding capacitance.

In spite of the model's shortcomings, the defects are usually negligible, and the model can be reliably used to predict transformer AC electrical performance for many engineering applications. Before we consider these, it is useful to consider the two-winding device.

2.4 The Nonideal Two-Winding Transformer

Although much of the discourse in the previous section was specific to the three-winding transformers, the extension to n-windings was clear. Likewise, it is straightforward to apply our development to the two-winding device. Consider a three-winding transformer with the tertiary on permanent oc, and with inaccessible tertiary terminals. Such a device becomes in effect a two-winding transformer and the circuit models are given in Figure 2.2. The following points should be noted:

- Since there are only two windings the complex power that enters the primary winding must exit the secondary winding (at least for the ideal device). Therefore, normally only *one* S rating is provided, which suffices for both primary and secondary windings.
- The shunt elements R_c and X_m are determined precisely as in Section 2.3.
- For the two-winding case, there are two possible short-circuit tests; however, they yield the same information ($\underline{Z}_{12} = \underline{Z}_{21}$, in pu) and it is not possible to separate $\underline{Z}_1$ and $\underline{Z}_2$. We change the notation to $\underline{Z}_{12} = \underline{Z}_{eq} = R_{eq} + jX_{eq}$, and speak of "the *transformer* series impedance", as opposed to "the *winding* series impedance".
- As in the n-winding case, the term "*primary*" is reserved for the *source* side, and the *secondary* refers to the *load* side. As such, it is "more natural" to reverse the positive assigned direction of the secondary current, as shown in Figure 2.2. If the high-voltage (HV) winding is the primary, the transformer is said to be operating "*step-down*." A "*step-up*" transformer uses the low voltage (LV) winding as the primary.

2.5 Transformer Efficiency and Voltage Regulation

Perhaps the single most important characteristic of any energy conversion device is its efficiency. The universally accepted definition of efficiency is output divided by input,

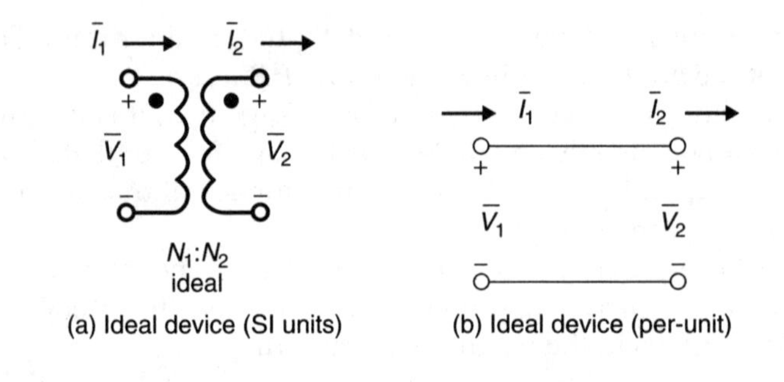

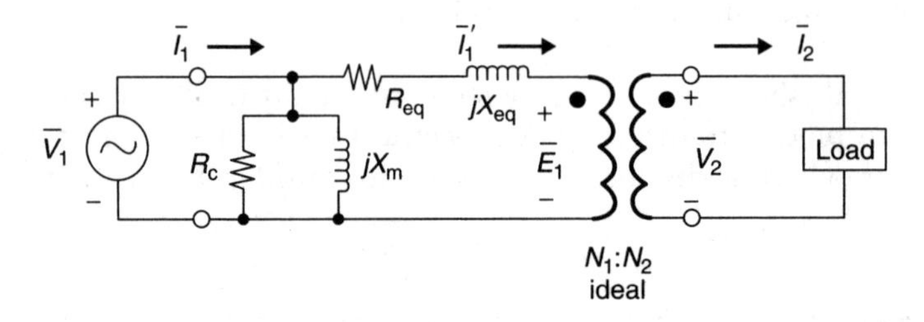

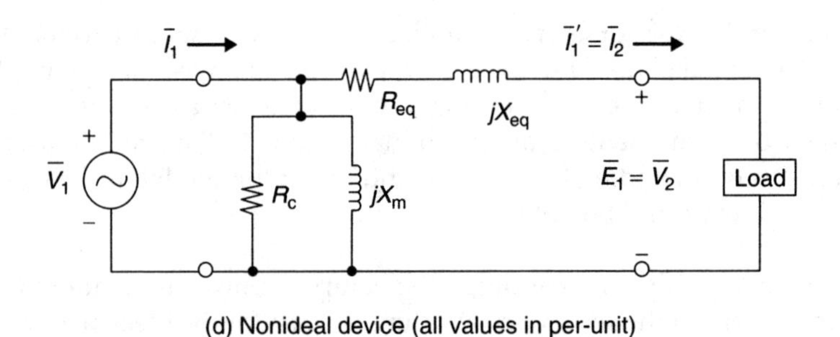

FIGURE 2.2. Two-winding power transformer AC circuit models.

expressed as a percentage. Unless otherwise specified, the physical quantity implied is *power*:

$$\text{efficiency} = \eta = \frac{P_{\text{out}}}{P_{\text{in}}} = \frac{P_{\text{out}}}{P_{\text{out}} + P_{\text{losses}}} \tag{2.14}$$

Thus, two points are clear. The output power of a given system must be clearly defined, and the total power losses must be determined for the corresponding output. Note that the input, by conservation of energy, is then the sum of the output and the losses. For the n-winding transformer, the output consists of the total of the powers (P) flowing *out of* the transformer. The losses consist of core loss plus the winding I^2R losses. For the two-winding transformer, the situation is simpler. The input and output are located at the primary and secondary, respectively.

A second important issue is the variation of output transformer voltage experienced under variable load conditions. To quantify this phenomenon, *voltage regulation* (VR) is defined. Considering the greatest possible fluctuation in load is from NL to full load (FL = rated load), we define

$$\text{voltage regulation (VR)} = \frac{V_{2NL} - V_{2FL}}{V_{2FL}} \tag{2.15}$$

where

V_{2FL} = rated secondary transformer voltage
V_{2NL} = secondary transformer voltage, after rated load at a given power factor is removed, assuming that the primary is driven from a voltage source

Voltage regulation is normally only computed for a two-winding transformer. Also, both efficiency and VR are usually expressed in percentage form. An example will illustrate transformer efficiency and VR calculations.

Example 2.4
A single-phase 60 Hz two-winding power transformer is rated at 2400 V:240 V 100 kVA and has the following equivalent circuit constants:

Values Referred To	HV Side (Ω)	LV Side (Ω)	pu
Series resistance	0.8064	0.008064	0.01400
Series reactance	4.9997	0.049997	0.08680
Shunt magnetizing reactance	1097.14	10.9714	19.05
Shunt core loss resistance	7314.29	73.1429	126.98

Suppose the transformer operates in the step-down mode, and serves rated load at rated voltage, pf = 0.8 lagging.

(a) Determine all circuit values in Figure 2.2c
(b) Determine the efficiency and VR for this operating condition

SOLUTION

(a) Step down means that the LV side is the secondary. Hence,

$$V_2 = 240\text{ V } S_2 = 100\text{ kVA}, \qquad V_2 = 240\angle 0°$$
$$\bar{S}_2 = 100\angle{+36.9°} = 80\text{ kW} + j60\text{ kvar}$$
$$\bar{I}_2 = (\bar{S}_2/\bar{V}_2)^* = 416.7\angle{-36.9°}\text{ A} \quad \bar{I}_1' = (N_2/N_1)\bar{I}_2 = 41.67\angle{-36.9°}\text{ A}$$
$$\bar{E}_1 = (N_1/N_2)\,\bar{V}_2 = 2400\angle 0°\text{ V}$$
$$\bar{V}_1 = (R_{eq} + jX_{eq})\,\bar{I}_1' + \bar{E}_1$$
$$= (0.8064 + j5)(41.67\angle{-36.9°}) + 2400 = 2556\angle 3.3°$$
$$\bar{I}_1 = (\bar{V}_1/R_c) + (\bar{V}_1/jX_m) + \bar{I}_1'$$
$$= (2556\angle 3.3°/7314) + (2556\angle 3.3°/j1097) + 41.67\angle{-36.9°}$$
$$= 43.46\angle{-38.9°}\text{ A}$$

(b) Transformer losses (P_{loss})

$$\text{Winding } I^2R \text{ loss} = (I_1')^2\,R_{eq} = (41.67)^2(0.8064) = 1400\text{ W}$$
$$\text{Core loss} = (V_1)^2/R_c = (2556)^2(7314) = 893.3\text{ W}$$

$$P_{loss} = 1400 + 893.3 = 2293.3\,W = 2.293\,kW$$
$$P_{in} = P_{out} + P_{loss} = 80 + 2.293 = 82.293\,kW$$
$$\eta = P_{out}/P_{in} = 80/82.293 = 0.9721 = 97.21\%$$

Voltage regulation: $V_{2FL} = 240\,V$. Now, the primary voltage for the given load was computed to be 2556 V. If the load is removed, $I_1' = 0$, and $E_1 = 2556\,V$. Therefore,

$$V_{2NL} = (N_2/N_1)E_1 = (240/2400)(2556) = 255.6\,V$$
$$VR = (255.6 - 240)/240 = 0.0650 = 6.50\%$$

The rated load efficiency and VR over a range of power factors are as follows:

Power Factor (---)		Efficiency (%)	VR (%)
0.0000	Lead	0.000	−8.6693
0.1000	Lead	82.920	−8.4686
0.2000	Lead	90.645	−8.1720
0.3000	Lead	93.546	−7.7760
0.4000	Lead	95.064	−7.2733
0.5000	Lead	95.995	−6.6518
0.6000	Lead	96.622	−5.8910
0.7000	Lead	97.072	−4.9550
0.8000	Lead	97.407	−3.7727
0.9000	Lead	97.663	−2.1603
1.0000	Unity	97.832	1.7708
0.9000	Lag	97.537	5.2901
0.8000	Lag	97.213	6.5031
0.7000	Lag	96.810	7.2989
0.6000	Lag	96.283	7.8615
0.5000	Lag	95.560	8.2623
0.4000	Lag	94.501	8.5374
0.3000	Lag	92.791	8.7076
0.2000	Lag	89.558	8.7852
0.1000	Lag	81.091	8.7778
0.0000	Lag	0.000	8.6890

2.6 Practical Considerations

Transformers come in a variety of shapes and sizes, depending on the application for which they were designed to serve. Although there is no universally accepted nomenclature for these devices, some of the applications and types include:

- *Battery chargers*. Wall-mounted small appliances (rechargeable flashlights, vacuum cleaners, power screwdrivers, etc.) typically use small transformers rated at 120 V and a few VA.
- *Electronic device power supplies*. Stereos, TV receivers and home computers have built-in power supplies, the main component of which is a transformer. Ratings range from 75 to 300 VA.
- *Distribution transformers*. Transformers are used to convert utility voltage in a suitable level for customer use, a common U.S. rating being 7.2 kV to 240/120 V. Power ratings range from 5 to 500 kVA.

- *Power transformers*. Transformers are used to convert bulk power from one voltage level into another within the utility. Ratings range from about 1 MVA to over 1000 MVA.
- *Unit transformers*. Utility generator outputs are generally connected directly to a transformer (the "unit transformer") to step the voltage up to transmission level. Ratings range from 10 MVA to over 1000 MVA.

It has been implied that windings are two-terminal structures, that is, the windings are electrically accessible at only two points (the ends). However, it is common to provide several accessible nodes (called "taps") at one end of a winding, as shown in Figure 2.3a. The connection used is usually defined in terms of percent. The 100% tap is that which correlates to rated winding voltage; the 105% tap correlates to 105% rated voltage; the 95% tap correlates to 95% rated voltage; and so forth. A common range of adjustment is ±10%, with 1.25% intervals between taps. "TCUL" designs permit "tap changing under load." It requires some physical effort to change the tap setting; hence motor-driven tap changers are common. If the device senses its own secondary voltage and automatically adjusts its tap to compensate for high or low voltage, it is called a *voltage regulator*.

The transformer core and insulated winding structure is encased in a protective steel container (sometimes called a "tank"), which is sometimes flooded with an insulating gas or liquid, giving rise to three types of transformers:

- Dry type
- Liquid-insulated type
- Gas-insulated type

There are two advantages that fluid (gas or liquid) types possess. The insulating system is "healing" in that if the winding insulation is somehow punctured, the fluid will flow into the breach and seal it. Also, with fluids, convection or forced cooling is possible; that is, hot fluid in direct contact with the windings and core may be circulated to cooler areas in the tank. In fact, radiating structures are sometimes designed as part of the tank to dissipate the heat due to winding and core loss.

Transformers are located in a variety of locations:

- Both indoor and outdoor locations are used.
- *Pad-mounted*. Transformers are frequently mounted free-standing on a concrete pad.
- *Pole-mounted*. Distribution transformers are frequently mounted on power utility poles for safety and security reasons.
- *Platform-mounted*. Large distribution transformers and small power transformers are sometimes mounted on raised platforms designed for that purpose.
- *Wall-mounted*. Small indoor transformers are sometimes wall-mounted.
- *Transformer vault*. For large indoor applications, a room especially designed for electrical equipment is sometimes provided.

Transformer winding polarity markings have extreme practical importance, but the dot notation is not generally used. The standard for designating HV and LV winding terminals, is by using the notation H1, H2 and X1, X2, respectively, where H1 and X1 correspond to the dots, as shown in Figure 2.3b.

Grounding of power equipment, and transformers in particular, is an important practical issue. "Grounding" or "earthing" means to provide an intentional low impedance electrical path from a selected point on the equipment to the earth. It is desirable for at least four reasons:

- *Safety to personnel.* If exposed parts of equipment are properly grounded, voltages from the case to local ground under normal and abnormal conditions will be negligible, so that persons in contact with equipment are at minimal shock and burn risk.
- *Fault detection and protection of equipment.* If equipment incurs a fault (short circuit), grounding simplifies both fault detection and subsequent rapid clearing of the fault.
- *Harmonic control.* The nonlinear magnetic core requires the flow of nonsinusoidal currents for sinusoidal voltage. If the harmonics of such currents are blocked, the voltage waveform will be distorted. Grounding provides a path for harmonic current flow.
- *Safety to equipment.* If equipment is subject to a sudden transient voltage, such as that produced by lightning, the magnitude of the transient will generally be minimized by grounding.

2.7 The Autotransformer

Transformers can be connected in all sorts of interesting ways, sometimes to great practical advantage. We will investigate this issue through an example.

Example 2.5

The transformer of Example 2.4 is to be used to supply a 2640 V load from a 2400 V source, if possible.

(a) If a solution exists, draw a circuit diagram.
(b) How much load (kVA) be supplied without overloading the transformer?
(c) For the maximum load of (b), compute the transformation efficiency. Neglect the ZI voltage drop (assume the source and load voltages are approximately 2400 V and 2640 V, respectively), and a pf of 1.0.

SOLUTION

(a) An appropriate connection is

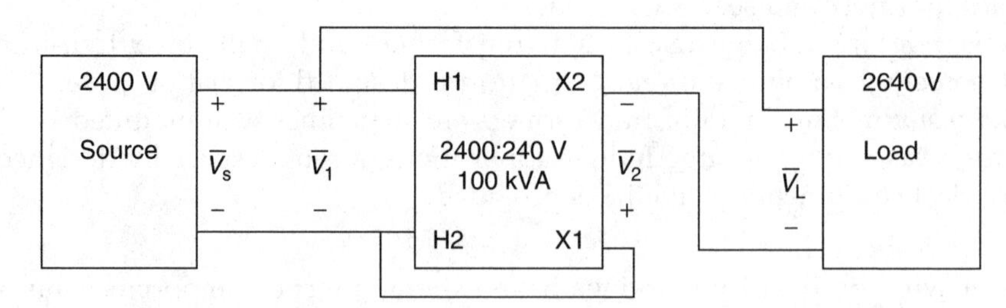

Consider the transformer to be ideal. By Kirchhoff's voltage law (KVL)

$$\bar{V}_S = \bar{V}_1 = 2400\angle 0° \text{ V}$$
$$\bar{V}_L = \bar{V}_1 + \bar{V}_2 = 2400\angle 0° + 240\angle 0° = 2640\angle 0° \text{ V}$$

(b) Consider the transformer to be ideal. If the transformer is loaded to capacity:

$$\bar{S}_2 = 100\angle\theta$$

where θ depends on the pf.

$$\bar{I}_2 = (\bar{S}_2/\bar{V}_2)^* = 416.7\angle{-\theta}\text{ A}, \quad \bar{I}_1 = (N_2/N_1)\bar{I}_2 = 41.67\angle{-\theta}\text{ A}$$
$$\bar{I}_S = \bar{I}_1 + \bar{I}_2 = 458.4\angle{-\theta}\text{ A}$$

The corresponding load and source power is

$$S_L = V_L I_L = 416.7(2.64\text{ kV}) = 1100\text{ kVA}$$

(c) Transformer losses (P_{loss})

$$\text{Winding } I^2R \text{ loss} = (I_1)^2 R_{\text{eq}} = (41.67)^2(0.8064) = 1400\text{ W}$$

$$\text{Core loss} = (V_1)^2/R_c = (2400)^2(7314) = 787.5\text{ W}$$

$$P_{\text{loss}} = 1400 + 787.5 = 2188\text{ W} = 2.188\text{ kW}$$

$$\text{At unity pf: } P_{\text{out}} = 1100\text{ kW}$$

$$\eta = \frac{P_{\text{out}}}{P_{\text{in}}} = \frac{1100}{1102.188} = 99.80\%$$

Consider the results of Example 2.4, which seem rather astounding. We managed to serve a 1100 kVA load with a 100 kVA transformer! Also there appears to be a dramatic improvement in efficiency! The secret is that not all of the load current or the load voltage was directly transformed. A portion of the current, and in one sense, the voltage, bypassed the transformer. The losses were essentially those incurred in transforming 100 kVA, and not the full load level of 1100 kVA.

Looking at the problem from another viewpoint, the situation is equivalent to having a single 2640 V winding, with a 2400 V tap, the 2400 V portion rated at 41.67 A, and the smaller 240 V portion rated at 416.7 A. Such a device, is called an *autotransformer*. The requisite core iron and winding material is equivalent to a 100 kVA device; consequently for a given kVA rating, autotransformers are smaller, lighter, and cheaper, and are extensively used, particularly where a small transformation ratio is required. The benefits dwindle as large transformation ratios are required. Likewise, since autotransformer windings are electrically interconnected, electrical isolation between windings is eliminated, which can be a serious disadvantage in some applications.

2.8 Operation of Transformers in Three-Phase Environments

If a transformer is to be used in a three-phase application, one can use three identical transformers, the primaries and secondaries of which can be interconnected in the standard wye or delta connections (see Figure 2.3a). However, one can also use an integrated three-phase design, which realizes economies in cost, size, and weight. Such a device is called a "three-phase transformer." See Figure 2.3b and Figure 2.3c for the distinctions.

It is assumed that the reader is familiar with three-phase AC circuit analysis. If this is not the case, the reader is referred to Appendix B, which provides a review of the subject,

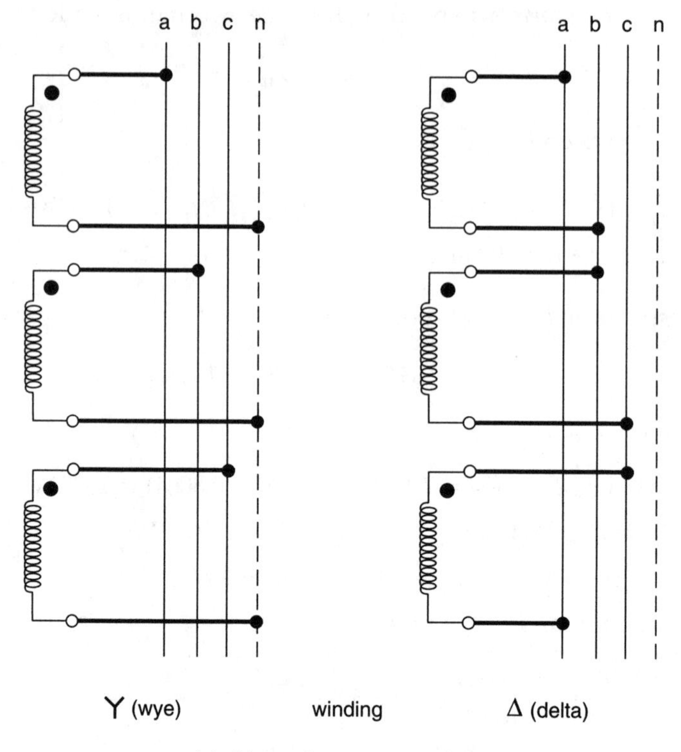

(a) 3ϕ transformer connections

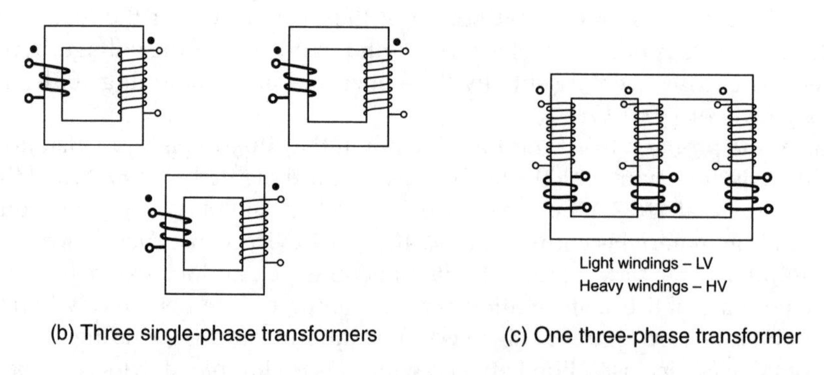

(b) Three single-phase transformers (c) One three-phase transformer

FIGURE 2.3. Transformers in three-phase environments.

including a treatment of the symmetrical component transformation (SCT). Even if one is familiar with the subject, a brief review of Appendix B is advisable, to familiarize the reader with the notation used in this book.

Let us consider four possible schemes for serving a three-phase load from a three-phase source, as shown in Figure 2.4. To simplify the situation, consider all transformers to be ideal, and that we are using three identical single-phase units. In each case, we are concerned with (1) the transformer kVA ratings; (2) the transformer voltage ratings; and (3) the current and voltage magnitudes everywhere.

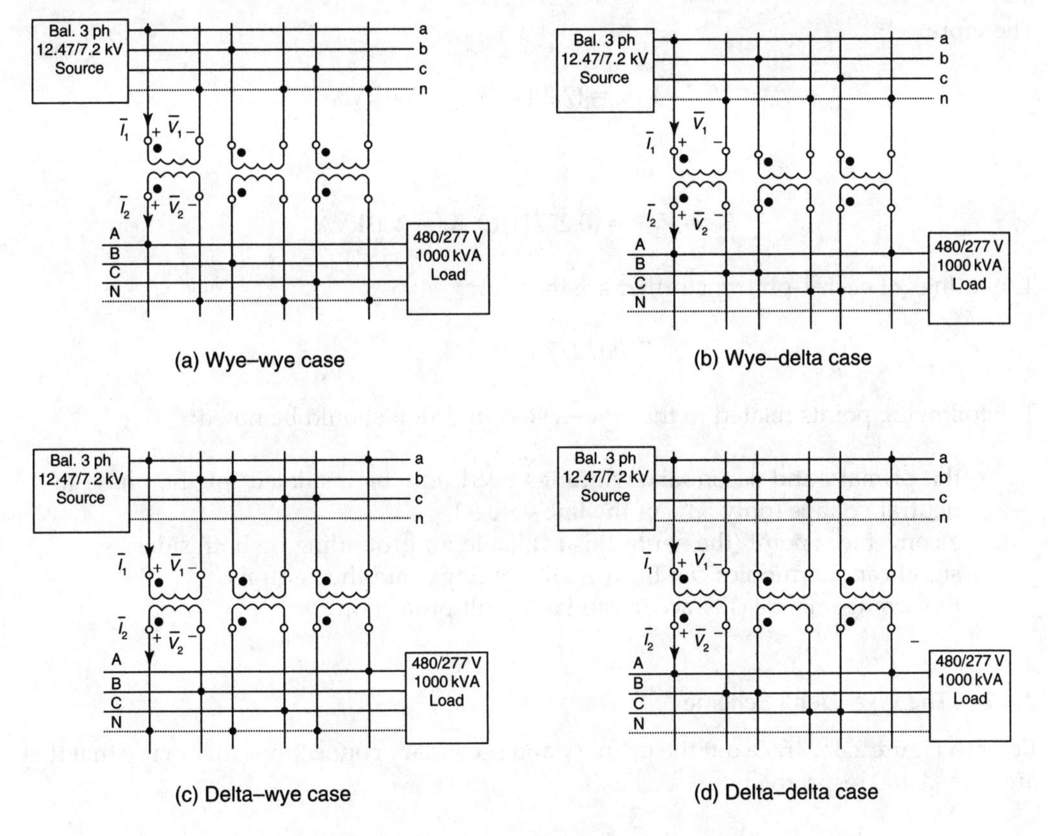

FIGURE 2.4. Transformers in polyphase connections.

For all schemes:

$$I_A = I_B = I_C = (S_{3ph}/3)/V_{AN} = 300/0.2771 = 1083\,\text{A}$$
$$V_{AB} = V_{BC} = V_{CA} = 480\,\text{V}$$
$$V_{AN} = V_{BN} = V_{CN} = (480/1.732) = 277.1\,\text{V}$$

and

$$I_a = I_b = I_c = (S_{3ph}/3)/V_{an} = 300/7.2 = 41.67\,\text{A}$$
$$V_{ab} = V_{bc} = V_{ca} = 12.47\,\text{kV}$$
$$V_{an} = V_{bc} = V_{ca} = (12.47/1.732) = 7.2\,\text{kV}$$

2.8.1 The Wye–Wye Scheme

Refer to Figure 2.4a. Trace out the primary and secondary connections to observe that they are wye, wye, respectively.

$$V_1 = V_{an} = 7.2\,\text{kV}, \qquad V_2 = V_{AN} = 277.1\,\text{V}$$
$$I_1 = I_a = 7.2\,\text{kV} \qquad I_2 = I_A = 1083\,\text{A}$$

Therefore

$$S_1 = V_1 I_1 = (7.2)(41.67) = 300\,\text{kVA}$$

also

$$S_2 = V_2 I_2 = (0.2771)(1083) = 300\,\text{kVA}$$

The ratings of each 1-ph transformer are therefore:

$$7200/277\,\text{V}, \quad 300\,\text{kVA}$$

The following points related to the wye–wye connection should be noted:

- the primary and secondary windings need only be insulated for the phase-to-neutral voltage (only 58% of the line values).
- a convenient point (the neutral) is available for grounding on both sides.
- significant harmonics can flow in all windings, and the neutrals.
- the arrangement easily accommodates fault protection.

2.8.2 The Wye–Delta Scheme

Refer to Figure 2.4b. Trace out the primary and secondary connections to observe that they are wye–delta, respectively:

$$V_1 = V_{an} = 7.2\,\text{kV}, \qquad V_2 = V_{AB} = 480\,\text{V}$$

$$I_1 = I_a = 41.47\,\text{A}, \qquad I_2 = (I_A/1.732) = 625\,\text{A}$$

Therefore

$$S_1 = V_1 I_1 = (7.2)(41.67) = 300\,\text{kVA}$$

also

$$S_2 = V_2 I_2 = (0.480)(625) = 300\,\text{kVA}$$

The ratings of each 1-ph transformer are therefore

$$7200/480\,\text{V}, \quad 300\,\text{kVA}$$

The following points related to the wye–delta connection should be noted:

- The primary winding need only be insulated for the phase-to-neutral voltage (only 58% of the line value); therefore the connection is frequently used for step-down applications.
- A convenient point (the neutral) is available for grounding on the primary, but not the secondary. Sometimes a center tap on one winding is grounded.
- Significant harmonics can flow in all windings and the neutrals.
- The arrangement easily accommodates fault protection on the primary, but is more complicated on the secondary since no zero sequence current can flow.

2.8.3 The Delta–Wye Scheme

Refer to Figure 2.4c. Trace out the primary and secondary connections to observe that they are delta–wye, respectively.

$$V_1 = V_{ab} = 12.47\,\text{kV}, \qquad V_2 = V_{NB} = V_{AN} = 277.1\,\text{V}$$

$$I_1 = I_{ab} = (41.67/1.732) = 24.06\,\text{A}, \qquad I_2 = |-I_B| = I_A = 1083\,\text{A}$$

Therefore

$$S_1 = V_1 I_1 = (12.47)(24.06) = 300\,\text{kVA}$$

also

$$S_2 = V_2 I_2 = (0.2771)(1083) = 300\,\text{kVA}$$

The ratings of each 1-ph transformer are therefore:

$$12470/277.1\,\text{V}, \quad 300\,\text{kVA}$$

The following points related to the delta–wye connection should be noted:

- The secondary winding need only be insulated for the phase-to-neutral voltage (only 58% of the line value).
- A convenient point (the neutral) is available for grounding on the secondary, but not the primary.
- Significant harmonics can flow in all windings, and the neutrals.
- The arrangement easily accommodates fault protection on the secondary, but is more complicated on the primary, since no zero sequence current can flow.

2.8.4 The Delta–Delta Scheme

Refer to Figure 2.4d. Trace out the primary and secondary connections to observe that they are delta–delta, respectively.

$$V_1 = V_{ab} = 12.47\,\text{kV}, \qquad V_2 = V_{AB} = 480\,\text{V}$$

$$I_1 = I_{ab} = (41.67/1.732) = 24.06\,\text{A}, \qquad I_2 = (I_A/1.732) = 625\,\text{A}$$

Therefore

$$S_1 = V_1 I_1 = (12.47)(24.06) = 300\,\text{kVA}$$

also

$$S_2 = V_2 I_2 = (0.480)(625) = 300\,\text{kVA}$$

The ratings of each 1-ph transformer are therefore

$$12.47/480\,\text{V}, \quad 300\,\text{kVA}$$

2.8.5 The Open Delta Scheme

The delta–delta bank can operate with any one transformer removed, in which case it is called an "open delta" connection. Therefore disconnect the right-most transformer (c-a; C-A).

Since each remaining transformer now carries the full line current:

$$I_1 = I_a = 41.67\,\text{A} \qquad I_2 = I_A = 1083\,\text{A}$$

Therefore

$$S_1 = V_1 I_1 = (12.47)(41.67) = 519.6\,\text{kVA}$$

also

$$S_2 = V_2 I_2 = (0.480)(1083) = 519.6\,\text{kVA}$$

Note that the two transformer ratings do *not* add up to 900 kVA! (why?). For the example situation, two transformers connected open delta should each be rated at 519.6 kVA to serve the 900 kVA load, or $1/\sqrt{3} = 57.7\%$ of the three-phase rating.

2.8.6 The Zig-Zag Scheme

A problem with the delta connection is that it results in a three-wire system, resulting in no access to the phase-to-neutral voltage, and provides no convenient point to ground and retain three-phase symmetry. There are clever transformer connections that can "derive" the neutral; i.e., convert a three-wire (abc) system to a four-wire (abcn) system. One such arrangement is the so-called "zig-zag" connection.

Refer to Figure 2.5a. Assume that all transformers are ideal, with 1:1 turns ratio, and that the system voltage is balanced three-phase. Since there are so many voltages to track, let us use a streamlined notation:

$$\text{Let "A}\alpha\text{"} = \bar{V}_{A\alpha}, \text{ etc.}$$

The source is balanced three-phase, and the system has a property defined as three-phase symmetry (i.e., the system appears to be the same electrically from the perspective of any one of the three phases). Hence, all triads of voltages will be balanced three-phase. Additionally, KVL requires that

$$\begin{aligned} A\alpha &= AN = Aa + a\alpha \\ B\beta &= BN = Bb + b\beta \\ C\gamma &= CN = Cc + c\gamma \end{aligned}$$

and

$$\begin{aligned} AB &= AN + NB \\ BC &= BN + NC \\ CA &= CN + NA \end{aligned}$$

Refer to Figure 2.5b for the relationship between all phasor voltages. Therefore node "N" is forced to locate at the system neutral, and we have a functional balanced three-phase four-wire system. An example will provide additional insight.

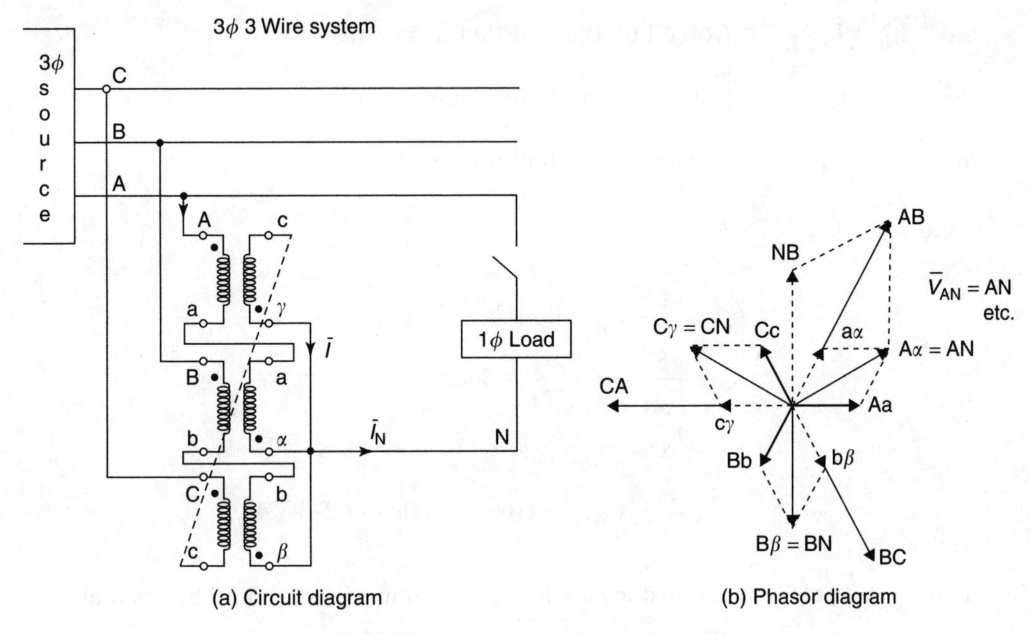

(a) Circuit diagram (b) Phasor diagram

FIGURE 2.5. Zig-zag transform connection.

Example 2.6
A three-phase three-wire (ABC) 480 V system is to be converted to four-wire (ABCN) using a zig-zag transformer connection.

(a) Show the proper connection.
(b) Draw the phasor diagram.
(c) Determine the voltage ratings of each transformer.
(d) If the maximum single-phase load to be served is 100 kVA, what is the required kVA rating of each transformer?

SOLUTION

(a) The circuit diagram appears in Figure 2.5a.
(b) The phasor diagram appears in Figure 2.5b.
(c) Since the situation is balanced three-phase:

$$\mathrm{AN} = \frac{\mathrm{AB}}{\sqrt{3}} = \frac{480}{\sqrt{3}} = 277.1\,\mathrm{V}$$

$$\mathrm{Aa} = \frac{\mathrm{AN}}{\sqrt{3}} = \frac{\mathrm{AB}}{3} = 160\,\mathrm{V}$$

Hence, each transformer should be rated 160 V : 160 V.

(d) Define:

$$\overline{I}_{\mathrm{Aa}} = \overline{I}$$

Then $\overline{I}_{c\gamma} = \overline{I}$ (forced by the top transformer)

But $\overline{I}_{a\alpha} = \overline{I}_{\mathrm{Aa}} = \overline{I}$ (forced by the series connection) and $\bar{I}_{Bb} = \bar{I}_{a\alpha} = \bar{I}$ (forced by the middle transformer)

But $\bar{I}_{Bb} = \bar{I}_{b\beta} = \bar{I}$ (forced by the series connection)

and $\bar{I}_{Cc} = \bar{I}_{b\beta} = \bar{I}$ (forced by the bottom transformer)

Finally,

$$\bar{I}_N = \bar{I}_{a\alpha} + \bar{I}_{b\beta} + \bar{I}_{c\gamma} = 3\bar{I}$$
$$I_N = \frac{S}{V_{AN}} = \frac{100}{0.2771} = 360.8$$
$$I = \frac{I_N}{3} = 120.3\text{ A} = I_{rating}$$
$$S_{rating} = V_{rating}I_{rating} = (160)(120.3) = 19.24\text{ kVA}$$

In general, the single-phase transformers in zig-zag connections should be rated at

$$\begin{aligned} V_{rating} &= \frac{V_{line}}{3} \\ S_{rating} &= \frac{\text{maximum single-phase load } (S)}{3\sqrt{3}} \end{aligned} \tag{2.16}$$

We have ignored phase relations between primary and secondary quantities. In the wye–wye and delta–delta cases, corresponding primary and secondary quantities phase (i.e., $\bar{V}_{an}$ is in phase with $\bar{V}_{AN}$, $\bar{I}_a$ is in phase with $\bar{I}_A$, etc.). However, in the wye–delta and wye–delta cases, this is not possible; the minimum phase shift between corresponding primary and secondary quantities is 30°. For these cases, it is desirable to

- Keep the phase shift between primary and secondary quantities predictable and uniform. That is, if $\bar{V}_{an}$ *leads* $\bar{V}_{AN}$ by 30°, then $\bar{V}_{bn}$ should lead $\bar{V}_{BN}$ by 30°, and so forth.
- HV quantities should lead (or lag) all corresponding LV by a fixed phase angle uniformly for all wye–delta or wye–delta connected transformers.

To achieve these objectives, a technical standard (IEEE/ANSI C 57.12) was adopted which states

For either wye–delta or wye–delta connections, phases shall be labeled in such a way that HV quantities shall lead their corresponding LV counterparts by 30° for normal balanced three-phase operation.

Consider Figure 2.6, which shows the desired phase relationship between all voltages. Observe from Figure 2.4b that $\bar{V}_{AB}$ is in phase with $\bar{V}_{an}$, as is $\bar{V}_{BC}$ with $\bar{V}_{bn}$, and $\bar{V}_{CA}$ with $\bar{V}_{cn}$. Likewise, observe from Figure 2.4c that $\bar{V}_{NC}$ is in phase with $\bar{V}_{ab}$, as is $\bar{V}_{NB}$ with $\bar{V}_{bc}$, and $\bar{V}_{NA}$ with $\bar{V}_{ca}$. Therefore $\bar{V}_{AN}$, $\bar{V}_{BN}$, and $\bar{V}_{CN}$ are located in their desired phase positions, and the standard is met.

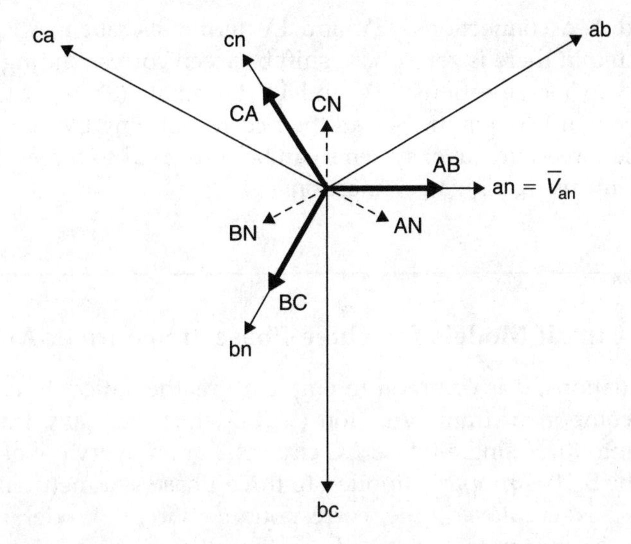

FIGURE 2.6. Standard phasor diagram for Y–Δ and Δ–Y connected transformer (phase seq abc; HV – lower case; LV – upper case).

There are literally dozens of transformer three-phase connections used for a variety of applications, particularly as we extend these ideas to multiwinding transformers (more than two winding devices).

To summarize the main issues relating to application of transformers in three-phase systems:

- There are basically two options with regard to transforming three-phase power
  - Use three (or two) single-phase transformers.
  - Use one three-phase transformer.
- There are two symmetrical connections
  - The wye connection.
  - The delta connection.
- In wye connections
  - The neutral is almost always grounded, since failure to do so will cause significant harmonic distortion.
  - Two voltage levels are available (phase-to-phase and phase-to-neutral).
  - Since the transformer winding is exposed only to the phase-to-neutral voltage, there is some economy to be realized at HV. Therefore, wyes are frequently used in the secondary of step-up applications, and the primary of step-down applications.
  - Protection is usually simpler because of access to the neutral.
- In delta connections
  - Since the neutral is not available, there is only one voltage level, which could be a disadvantage.
  - There is no obvious point to ground. For delta-connected secondaries, sometimes one phase is center-taped, and this center point grounded. Another scheme is to ground one phase ("corner" grounded).
  - If the transformer impedances are not perfectly matched, there will be unequal load sharing among transformers with delta connected secondaries, even for balanced external loads. The neutral is not available for grounding on either side.
  - The secondary voltage replicates the wave shape of the primary voltage. Significant triplen harmonic currents can and will flow internally in all windings.

- In Y–Y and Δ–Δ connections, HV and LV terminals (abcn; ABCN) should be labeled such that there is zero phase shift between corresponding quantities.
- In Y–Δ and Δ–Y connections, HV and LV terminals (abcn; ABCN) should be labeled such that HV quantities lead their corresponding LV quantities by 30°.
- Three-phase three-wire (abc) systems can be converted to three-phase four-wire (abcn) systems using zig-zag connections.

2.9 Sequence Circuit Models for Three-Phase Transformer Analysis

In three-phase situations, it is common to employ a mathematical transformation called the symmetrical component transformation (SCT), which basically transforms a three-phase AC circuit into three single-phase AC circuits. For an overview of the subject, refer to Appendix B. The SCT is properly applied to three-phase symmetric transformer situations.[4] If the SCT is employed, the corresponding circuit model consists of three "sequence" networks, distinguished one from the other as zero sequence (0), positive sequence (1), and negative sequence (2).

To comply with the three-phase symmetric constraint, we shall restrict our discussion to two cases:

- A three-phase three-winding transformer
- Three identical single-phase three-winding transformers

An important side note should be made. What is conventionally called a "three-phase three-winding" transformer actually has a three-phase primary, a secondary, and a tertiary, for a total of nine windings.

Also, it is required that each triad of windings (primary, secondary, and tertiary) be connected in either wye or delta, the only two possible three-phase symmetric connections.

The sequence circuits are presented in Figure 2.7. All values *must* be in per-unit. Study the positive and negative sequence circuits of Figure 2.7. Note that the circuits are identical (except for the variables, which are specific to the sequence).

Now focus on the zero sequence circuit. If the circuit models three identical single-phase three-winding transformers in three-phase connections, the eight elements (4 resistors and 4 reactors) are identical to the corresponding positive and negative sequence values. If the circuit models a three-phase three-winding transformer, the circuit elements in general are not equal to the corresponding positive and negative sequence values. However, they are close, and generally are assumed to be the same anyway. *But the most important feature involves the interior ports (1′-1″, 2′-2″, 3′-3″)!* They are connected according to the wye–delta phase connections. The general procedure is:

Wye, grounded with Z_g: connect "$3Z_g$" in the port.
Delta: connect the interior terminal to reference; leave the exterior terminal on oc.

Consult Figure 2.8 for illustration of this point.

We are now ready to attempt a demonstration analysis. So that we can demonstrate the approach, we shall restrict our loads to those which can be modeled as constant equal impedances, in four-wire grounded wye connections.[5] Consider Example 2.6.

[4] Actually, the SCT can be applied to any connection of transformers in three-phase situations, whether symmetrical or not. However, if the system is not three-phase-symmetric, the corresponding analysis is no simpler than that executed in phase quantities, defeating the purpose of using the SCT.

[5] To extend the work to more general loads, it would be necessary to digress and cover Appendix B in detail, a task which would distract us from our primary interests.

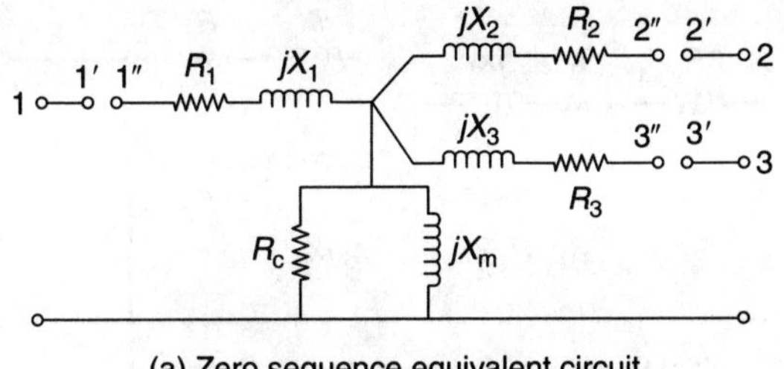

(a) Zero sequence equivalent circuit

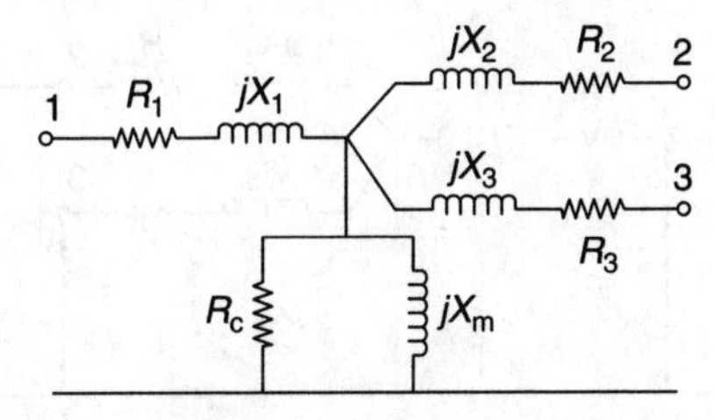

(b) Positive sequence equivalent circuit

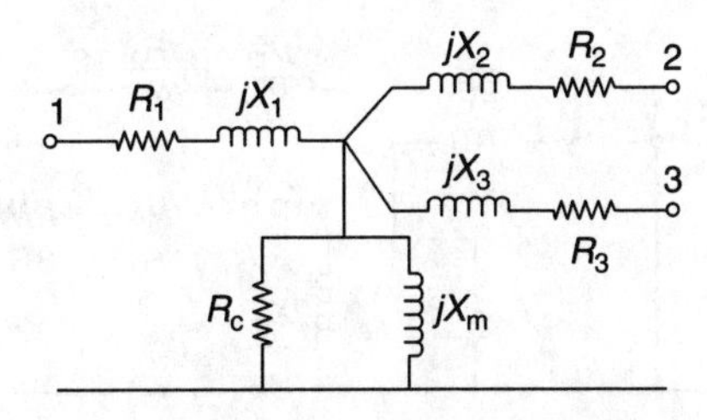

(c) Negative sequence equivalent circuit

FIGURE 2.7. Sequence circuits for three-phase three-winding transformers (all values must be in pu).

Example 2.7

A three-phase three-winding transformer has primary, secondary, and tertiary ratings of

$$13.8\,\text{kV Y}\ 100\,\text{MVA};\quad 138\,\text{kV Y}\ 90\,\text{MVA};\quad 4.157\,\text{kV}\ \Delta\ 20\,\text{MVA}.$$

The transformer is ideal except for the leakage reactances, which are all 5% on voltage ratings and a 100 MVA three-phase base. The transformer primary is grounded in 10% resistance. The secondary and tertiary supply balanced solid grounded wye-connected loads, which may be modeled as impedance, and are nominally:

Secondary:	138 kV,	75 MVA;	pf = 0.866 lagging
Tertiary:	4.157 kV,	15 MVA;	pf = 0.8 leading

(a) Determine the pu bases in each winding.
(b) Determine the load impedances.
(c) Draw the sequence circuit diagrams.
(d) Given that a balanced three-phase 13.8 kV source is applied to the primary, determine the primary currents:
(e) Repeat (d) for a unbalanced primary source of

$$\bar{V}_a = 8.007\angle 5.7^\circ\,\text{kV}$$
$$\bar{V}_b = 10.0457\angle -122.3^\circ\,\text{kV}$$
$$\bar{V}_c = 5.911\angle 116.1^\circ\,\text{kV}$$

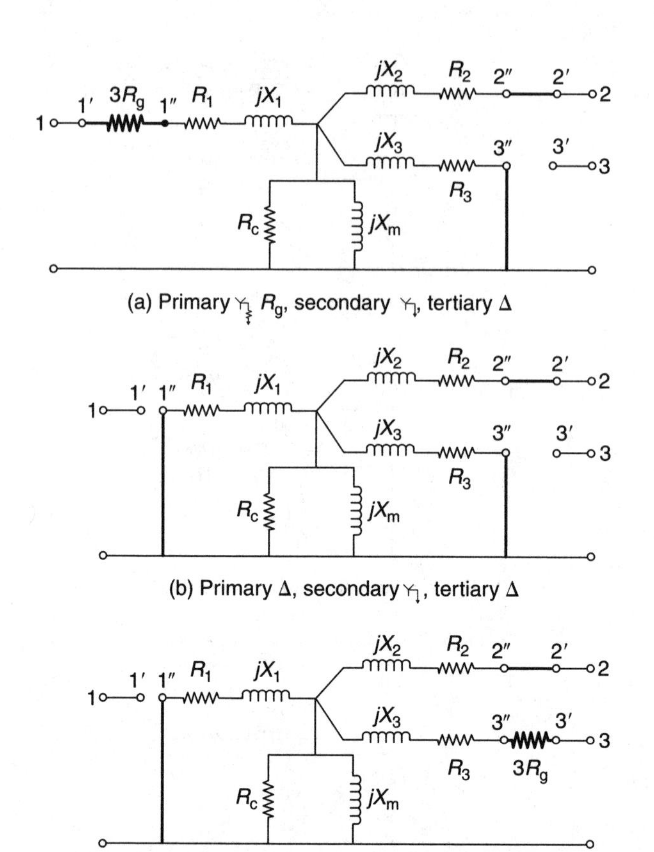

FIGURE 2.8. Zero sequence circuit for selected connection combinations.

SOLUTION

(a) The pu bases are:

$$S_{3\phi\,\text{base}} = 100\,\text{MVA}; \qquad S_{\text{base}} = 100/3 = 33.33\,\text{MVA} \quad (\textit{all windings})$$

Primary:
$$V_{\text{base}} = 13.8/\sqrt{3} = 7.967\,\text{kV}$$
$$I_{\text{base}} = S_{\text{base}}/V_{\text{base}} = 4.184\,\text{kA}$$
$$Z_{\text{base}} = V_{\text{base}}/I_{\text{base}} = 1.904\,\Omega$$

Secondary:
$$V_{\text{base}} = 138/\sqrt{3} = 79.67\,\text{kV}$$
$$I_{\text{base}} = S_{\text{base}}/V_{\text{base}} = 0.4184\,\text{kA}$$
$$Z_{\text{base}} = V_{\text{base}}/I_{\text{base}} = 190.4\,\Omega$$

Tertiary:
$$V_{\text{base}} = 4.157/\sqrt{3} = 2.4\,\text{kV}$$
$$I_{\text{base}} = S_{\text{base}}/V_{\text{base}} = 13.89\,\text{kA}$$
$$Z_{\text{base}} = V_{\text{base}}/I_{\text{base}} = 0.1728\,\Omega$$

(b) Loads are wye connected. Therefore:
Secondary

$$Z_{\text{Y}} = \frac{V_{\text{L}}^2}{S_{3\varphi}} = \frac{(138)^2}{75} = 253.9\,\Omega \qquad (Z_{\text{Y}} = Z_0 = Z_1 = Z_2)$$

$$\bar{Z}_{\text{Y}} = 253.9\angle 30^\circ\ \Omega = 219.9 + j127.0\,\Omega = 1.155 + j0.667\,\text{pu}$$

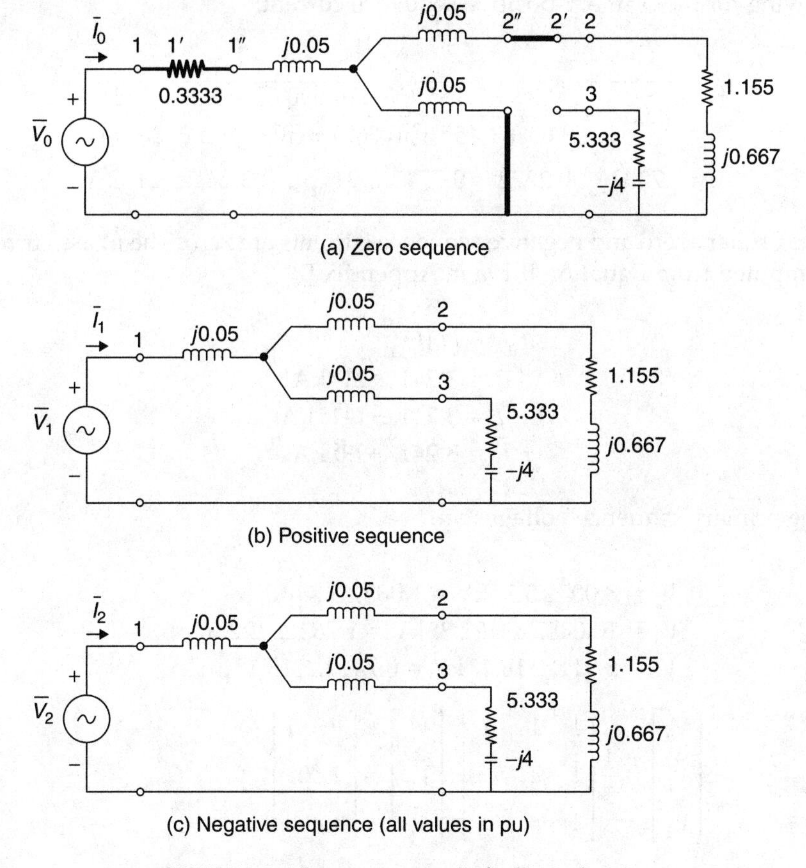

FIGURE 2.9. Sequence networks for Example 2.6.

Tertiary

$$Z_Y = \frac{V_L^2}{S_{3\varphi}} = \frac{(4.157)^2}{15} = 1.152\,\Omega$$

$$\bar{Z}_Y = 1.152\angle{-36.9^\circ}\,\Omega = 0.9216 - j0.6912\,\Omega = 5.333 - j4\text{ pu}$$

(c) The sequence circuits are provided in Figure 2.9.

(d) Observe that in the primary,

$$\bar{V}_{an} = \bar{V}_a = \frac{13.8}{\sqrt{3}}\angle 0^\circ = 7.967\angle 0^\circ\text{ kV} = 1.0\angle 0^\circ\text{ pu}$$

The primary sequence voltages are

$$\begin{bmatrix} \bar{V}_0 \\ \bar{V}_1 \\ \bar{V}_2 \end{bmatrix} = \frac{1}{3}\begin{bmatrix} 1 & 1 & 1 \\ 1 & a & a^2 \\ 1 & a^2 & a \end{bmatrix}\begin{bmatrix} \bar{V}_a \\ \bar{V}_b \\ \bar{V}_c \end{bmatrix} = \frac{1}{3}\begin{bmatrix} 1 & 1 & 1 \\ 1 & a & a^2 \\ 1 & a^2 & a \end{bmatrix}\begin{bmatrix} 1\angle 0^\circ \\ 1\angle{-120^\circ} \\ 1\angle{+120^\circ} \end{bmatrix} = \begin{bmatrix} 0 \\ 1\angle 0^\circ \\ 0 \end{bmatrix}$$

Solving for the primary positive sequence current:

$$\bar{I}_1 = \frac{1\angle 0^\circ}{j0.05 + \dfrac{(j0.05 + 1.155 + j0.667)(j0.05 + 5.333 - j4)}{(j0.05 + 1.155 + j0.667) + (j0.05 + 5.333 - j4)}}$$

$$= 0.7232 - j0.2772 = 0.7745\angle{-21^\circ}\text{ pu} = 3.241\angle{-21^\circ}\text{ kA}$$

The primary zero and negative sequence currents are zero. The phase currents are computed from Equation B.19a in Appendix B:

$$\hat{I}_{\text{abc}} = [T]\hat{I}_{012}$$
$$\bar{I}_{\text{a}} = 3.241\angle{-21^\circ}\text{ kA}$$
$$\bar{I}_{\text{b}} = 3.241\angle{-141^\circ}\text{ kA}$$
$$\bar{I}_{\text{c}} = 3.241\angle{+99^\circ}\text{ kA}$$

(e) The primary sequence voltages are

$$\bar{V}_{\text{a}} = 8.007\angle 5.7^\circ\text{ kV} = 1.005\angle 5.7^\circ\text{ pu}$$
$$\bar{V}_{\text{b}} = 10.0457\angle{-122.3^\circ}\text{ kV} = 1.262\angle{-122.3^\circ}\text{ pu}$$
$$\bar{V}_{\text{c}} = 5.911\angle 116.1^\circ\text{ kV} = 0.7419\angle 116.1^\circ\text{ pu}$$

$$\begin{bmatrix} \bar{V}_0 \\ \bar{V}_1 \\ \bar{V}_2 \end{bmatrix} = \frac{1}{3}\begin{bmatrix} 1 & 1 & 1 \\ 1 & a & a^2 \\ 1 & a^2 & a \end{bmatrix}\begin{bmatrix} \bar{V}_{\text{a}} \\ \bar{V}_{\text{b}} \\ \bar{V}_{\text{c}} \end{bmatrix} = \begin{bmatrix} -j0.1 \\ 1\angle 0^\circ \\ +j0.2 \end{bmatrix}$$

Solving for the primary sequence currents,

$$\bar{I}_0 = \frac{-j0.1}{0.333 + j0.05 + \dfrac{(j0.05 + 1.155 + j0.667)(j0.05)}{(j0.05 + 1.155 + j0.667) + (j0.05)}}$$

$$= -0.0812 - j0.2747 = 0.2864\angle{-106.5^\circ}\text{ pu} = 1.198\angle{-106.5^\circ}\text{ kA}$$

$$\bar{I}_1 = 0.7745\angle{-21^\circ}\text{ pu} = 3.241\angle{-21^\circ}\text{ kA}$$

$$\bar{I}_2 = \frac{0.2\angle 90^\circ}{j0.05 + \dfrac{(j0.05 + 1.155 + j0.667)(j0.05 + 5.333 - j4)}{(j0.05 + 1.155 + j0.667) + (j0.05 + 5.333 - j4)}}$$

$$= 0.1549\angle 69^\circ\text{ pu} = 0.6481\angle 69^\circ\text{ kA}$$

The phase currents are computed from Equation B.19a:

$$\hat{I}_{\text{abc}} = [T]\hat{I}_{012}$$
$$\bar{I}_{\text{a}} = 3.379\angle{-30.3^\circ}\text{ kA}$$
$$\bar{I}_{\text{b}} = 4.803\angle{-136.8^\circ}\text{ kA}$$
$$\bar{I}_{\text{c}} = 1.610\angle{+105.8^\circ}\text{ kA}$$

Example 2.7(d) illustrates an important general principle.

If balanced three-phase excitation is applied to a three-phase symmetrical system with balanced loads throughout, all zero and negative sequence voltages and currents are zero throughout the system, and the system may be modeled exclusively with the positive sequence equivalent circuit (a single-phase network). The quantities $\overline{V}_a$ and $\overline{I}_a$ are identical to $\overline{V}_1$ and $\overline{I}_1$, respectively.

That is, balanced three-phase analysis in three-phase symmetric systems reduce to single-phase circuits problem, the proper circuit being the positive sequence network!

Finally, sequence networks for transformers, particularly when they are scaled into pu, have a significant theoretical flaw that is not immediately apparent. They do not correctly predict the phase shift in corresponding quantities across wye–delta connections, a point that was discussed earlier. This is in part a labeling problem. Assume for the moment an ideal three-phase transformer with wye-connected primary and delta-connected secondary. The primary voltage on "a" phase (e.g., "an") is transformed to a secondary voltage (e.g., "AB"), which is forced to be in phase with "an". Hence the secondary voltage "AN" cannot be in phase with "an." For balanced operation recall that voltage "an" (or "AN") is also the positive sequence voltage. This phase shift, caused by the connection and the way we label the phases, is not predicted by the circuit model.

For some applications, this flaw is not important. Recall that, fundamentally, phase is always relative anyway. If one simply resets the phase reference as one moves from primary to secondary to tertiary, etc., the issue may be ignored. However, in other situations, we may be restricted to only one-phase reference per system.[6] To extend ANSI C57.12.70 to sequence networks:

For either wye–delta or delta–wye connections in ideal transformers, phases shall be labeled in such a way that HV positive sequence quantities shall lead corresponding LV positive sequence quantities by 30°. It follows that HV negative sequence quantities will lag corresponding LV negative sequence quantities by 30° (single-phase network). Corresponding zero sequence quantities remain in phase.

2.10 Harmonics in Transformers

Usually we are concerned with transformer operation at the normal power AC frequency (50 and 60 Hz). However, there are situations where it is important to consider that nonsinusiodal waveforms may be involved. The power system voltage is generally assumed to be sinusoidal. However, this is never actually the case. Close inspection of any actual system voltage will reveal that it is virtually always nonsinusoidal to some degree. Nonsinusoidal periodic functions may always be decomposed into a sum of sinusoidal signals, whose frequencies are integer multiples of the fundamental frequency (usually 50/60 Hz).[7] Since, our transformer circuit models are linear, we may employ superposition to analyze system performance one frequency at a time, and add, or superimpose the results, to obtain total system steady-state operation.[8]

If we critically examine the elements of our basic transformer circuit models, we realize that the shunt elements R_c and X_m are both linear approximations included to account for

[6] The critical issue is the presence of additional electrical paths between windings other than through the transformer.
[7] This process is sometimes called "Fourier analysis," and involves analytical tools such as the Fourier series and Fourier transform. See Appendix C for a basic treatment of the subject.
[8] Be advised that that the transformer equivalent circuit models presented in this chapter are not accurate at high frequencies. The model must be refined to include inter-turn and inter-winding capacitance.

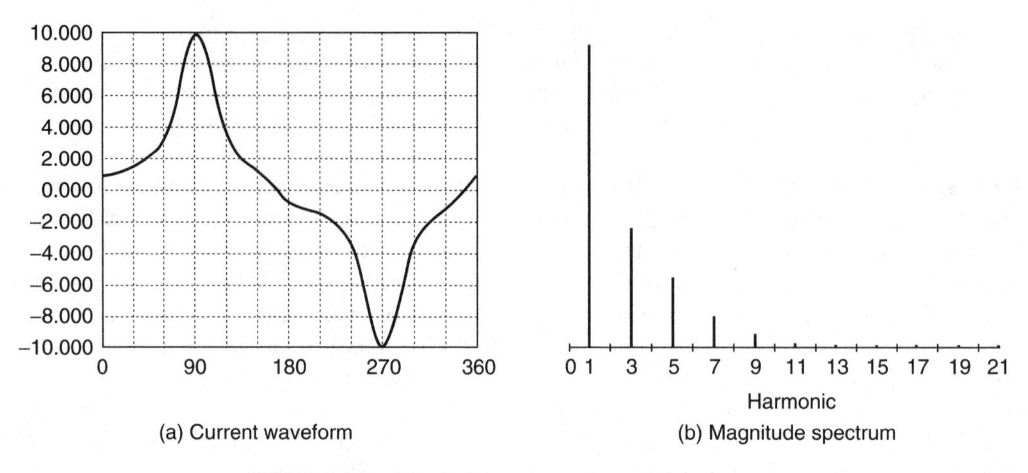

FIGURE 2.10. A typical transformer magnetizing current.

effects that are inherently nonlinear. Specifically, magnetic core excitation, which exhibits saturation and hysteretic effects, is directly involved. In fact, we investigated this issue in Section 1.4 and Section 1.5. Even if a perfectly sinusoidal voltage is applied to the transformer primary, the shunt magnetizing and core loss components of the primary current will be nonsinusoidal. As this nonsinusoidal current flows through linear circuit elements, it will produce nonsinusoidal voltage, which will contaminate other voltages (the sum of a sinusoidal voltage and a nonsinusoidal voltage will be nonsinusoidal). The waveform of a typical transformer magnetizing current was synthesized using the methods of Section 1.5, and is shown in Figure 2.10a.

Example 2.8

A typical magnetizing current waveform was derived from the presentation in Section 1.5, and is presented in Figure 2.10a. Analyze its spectral content.

The waveform was analyzed by the techniques presented in Appendix C, and implemented using a computer program. Results are presented in Table 2.1.

TABLE 2.1

Fourier Analysis of the Magnetizing Current Waveform of Figure 2.11a

Harm	An	Bn	Cn	Dn	Rms	phin
1	0.271	5.714	5.720	2.860	4.045	−87.3
3	0.358	−2.218	2.247	1.123	1.589	80.8
5	−0.044	1.269	1.270	0.635	0.898	−92.0
7	0.153	−0.524	0.546	0.273	0.386	73.7
9	0.006	0.186	0.186	0.093	0.132	−88.1
11	0.035	−0.037	0.051	0.026	0.036	46.8
13	0.023	0.017	0.028	0.014	0.020	−36.2
15	0.011	0.005	0.012	0.006	0.008	−22.9
17	0.016	0.006	0.017	0.009	0.012	−19.9
19	0.005	0.005	0.007	0.004	0.005	−42.7
21	0.011	0.005	0.012	0.006	0.009	−22.6

rms, 4.4564; sqr(rms), 19.860; 2 × Sum [sqr(Dn)], 19.859;
values: DC, 0.0; AC, 4.4564; fund, 4.0448; harm, 1.8705; THD = 46.24%.

The three forms of the *n*th term in the Fourier Series are:

Cosine, sine: An*cos(n*wo*x) + Bn*sin(n*wo*x)
Trigonometric: Cn*cos(n*wo*x + phin)
Exponential: Dn*exp(j*(n*wo*x + phin))

Example 2.8 illustrates that the current waveform has half-wave symmetry and zero DC content, which is typical of EM equipment. Also note that the current has substantial third harmonic (39% of the fundamental).

As an approximation, assume that the magnetizing characteristic is such that current is composed exclusively of two harmonics (the fundamental and the third harmonic)[9] for sinusoidal applied voltage. Perceive the following:

- If the winding is excited with a sinusoidal voltage, the current will be nonsinusoidal, composed of a first and third harmonic.
- If the winding is excited with a sinusoidal current, the voltage will be nonsinusoidal, with a strong first and third harmonic.

In other words, for the voltage to remain sinusoidal, the current third harmonic must be permitted to flow. Now consider a balanced three-phase situation. The fundamental currents are equal in magnitude, 120° out of phase. Therefore at the neutral in a wye connection, they sum to zero and in effect cancel each other. *However, the third-harmonic currents are* $3 \times 120° = 360°$ *out of phase, or in phase!* Therefore the neutral path must be available to allow the third-harmonic current to flow if the induced voltage is to be sinusoidal. But there is another option! In delta connection, each node joins the external line to two windings. Applying KCL to the node for the third harmonics, the current entering the node from one phase equals the current leaving the node from the other phase. Hence no current will flow to the "outside world," and the delta connection insures that third-harmonic current has an available path. These observations lead to the following rule:

> *In a three-phase transformer bank, there should be at least one wye connection with grounded neutral, and/or a delta connection, to provide a path for the triplen harmonic magnetizing currents to flow. Failure to consider this rule will render the core flux, and hence the induced winding voltages, nonsinusoidal.*

Consider an example.

Example 2.9
Sketch the waveform for the indicated quantities in an unloaded wye–delta transformer bank for three cases. The transformers are "near ideal" except that the magnetizing characteristic is nonlinear in such a way that only first and third harmonics are present at any time:

(a) Switch N open; switch D open.
(b) Switch N closed; switch D open.
(c) Switch N open; switch D closed.

Results are presented in Figure 2.11.

---

[9] Actually the following analysis can be a bit more general than this. The conclusions are valid for all so-called triplen harmonics. Triplen harmonic numbers are those which are integer divisible by 3 (3, 6, 9, 12, 15, etc.).

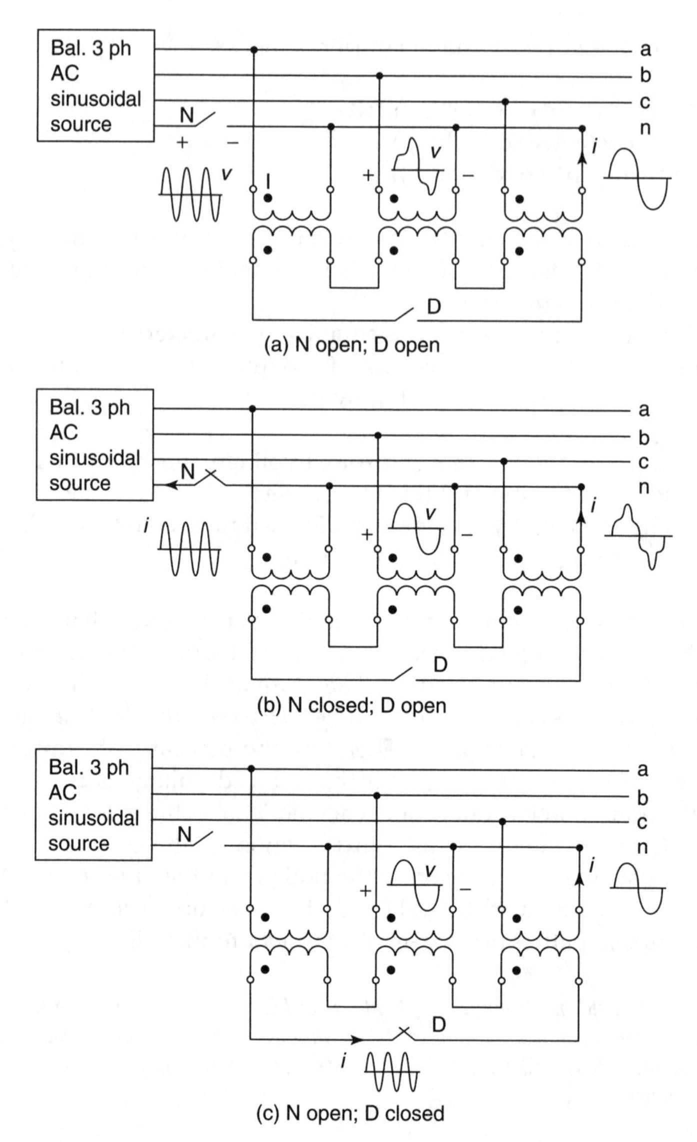

FIGURE 2.11. Harmonic waveforms in three-phase transformers.

2.11 Summary

Power transformers can convert AC voltage and current to different levels efficiently and reliably. They are extensively used throughout power systems, and constitute a vital and important system component. We considered the major engineering issues relating to the device, including ratings, pu scaling, circuit modeling, multiwinding devices, and various practical issues. We performed operational assessments, including circuit performance, efficiency, and VR. We examined loss mechanisms, including winding I^2R and core magnetic losses.

Extending our study to three-phase applications, we examined polyphase (wye, delta) connections. The zig-zag connection was an ingenious special connection, converting three- to four-wire systems. The autotransformer connection provided a very efficient design. For three-phase analysis, we applied symmetrical components to the transformer case, creating sequence circuit models. Finally, we examined harmonics in transformers, considering the inherent nonlinearity of the ferromagnetic core.

Formally, the power transformer does not qualify as an EM machine, because it cannot perform the requisite electrical to mechanical energy conversion process. However, it does perform electric-to-magnetic-to-electric energy conversion, and has important characteristics common to all EM machines. Also the power transformer is a major component found in almost all power systems, and is commonly used in conjunction with almost all types of motor and generators. Hence, its study is a major step in understanding the design and operation of EM machines, in general.

Problems

2.1 An ideal transformer has the following ratings:

primary: 12 kV 10 MVA
secondary: 120 kV 10 MVA

(a) Determine the turns ratio.
(b) Suppose $S_{base} = 10$ MVA. Using voltage ratings as bases, determine all base values in all windings.
(c) The transformer is terminated as follows:

primary: 12 kV AC 60 Hz voltage source
secondary: 5 MVA, pf = 0.8 lagging load

Solve for the primary *input* complex power, and the secondary *output* complex power.

(d) Solve for the winding phasor currents.

2.2 An ideal transformer has the following ratings:

primary: 12 kV 10 MVA
secondary: 120 kV 10 MVA
tertiary: 6 kV 1 MVA

(a) Determine the turns ratios.
(b) Suppose $S_{base} = 10$ MVA. Using voltage ratings as bases, determine all base values in all windings.
(c) The transformer is terminated as follows:

primary: 12 kV AC 60 Hz voltage source
secondary: 5 MVA, pf = 0.8 lagging load
tertiary: 1 MVA, pf = 0.8 leading load

Solve for the primary *input* complex power, the secondary *output* complex power, and the tertiary *output* complex power ($\bar{S}_3 = \bar{V}_3(-\bar{I}_3)^*$).

(d) Solve for the winding phasor currents.

2.3 Consider the transformer of Problem 2.2 to be nonideal, with the same ratings. We apply a 12 kV AC 60 Hz voltage source to the primary, with the secondary and tertiary on OC. Determine the shunt elements if the primary current and power are 5 A and 20 kW, respectively, in pu and in Ω, referred to the 12 kV winding.

2.4 Consider the transformer of Problem 2.3; the short-circuit test data for which are as follows:

	Winding					
Test	**1**	**2**	**3**	**Voltage**	**Current (A)**	**Power**
OC	Meas	oc	oc	12 kV	5	20 kW
SC	Meas	sc	oc	128 V	667	8,500 W
SC	oc	Meas	sc	886 V	87	11,550 W
SC	sc	oc	Meas	79 V	1389	13,200 W

Determine the series elements R_i and X_i ($i = 1, 2, 3$) in pu and Ω.

2.5 A single-phase 60 Hz two-winding power transformer is rated at 2400 V:480 V 150 kVA and has the following test data

	Test Data	
	Short Circuit (HV)	**Open Circuit (LV) Values**
Voltage (V)	196.31	480.0
Current (A)	62.50	14.644
Power (W)	1813.6	1020.2

(a) Determine the equivalent circuit with all values referred to the HV side, in SI units.
(b) Determine the equivalent circuit with all values referred to the LV side, in SI units.
(c) Determine the equivalent circuit with all values in pu, referred to ratings.

Note that the answers are available in Problem 2.6.

2.6 A single-phase 60 Hz two-winding 2400 V:480 V 150 kVA power transformer (the transformer of Problem 2.5) has the following equivalent circuit constants:

Values Referred To	**HV Side (Ω)**	**LV Side (Ω)**	**pu**
Series resistance	0.8064	0.008064	0.01400
Series reactance	4.9997	0.049997	0.08680
Shunt magnetizing reactance	1097.14	10.9714	19.05
Shunt core loss resistance	7314.29	73.1429	126.98

Suppose the transformer operates in the step-down mode and serves rated load at rated voltage at pf = 0.866 lagging.

(a) Draw the circuit with all values in SI units.
(b) Solve for the primary voltage, current, complex power, and power factor.
(c) Determine the VR.
(d) Determine the efficiency.

2.7 Repeat Problem 2.6 for step-up operation.

2.8 Rework Problem 2.6 in pu.

2.9 Rework Problem 2.7 in pu.

2.10 The transformer of Problem 2.6 is to used to supply a 2880 V load from a 480 V source. Model the transformer as ideal.

(a) Draw a circuit diagram.
(b) How much load (kVA) be supplied without overloading the transformer?

2.11 The transformer of Problem 2.6 is to be used to supply a 480 V load from a 2880 V source. Model the transformer as ideal. Repeat Problem 2.10.

(a) Draw a circuit diagram.
(b) How much load (kVA) can be supplied without overloading the transformer?

2.12 The transformer of Problem 2.6 is to be used to supply a 2880 V load from a 2400 V source. Model the transformer as ideal. Repeat Problem 2.10.

2.13 The transformer of Problem 2.6 is to be used to supply a 1920 V load from a 2400 V source. Model the transformer as ideal. Repeat Problem 2.10.

2.14 Repeat Problem 2.10, but this time model the transformer as nonideal. Assume that the output voltage is 480 V (the input voltage will be close to, but not exactly 2880 V) and the load pf is unity.

2.15 Continuing Problem 2.14, compute the transformation efficiency.

2.16 A standard American residential distribution transformer is rated at 7200 V: 120 V:120 V; 37.5 kVA; 18.75 kVA; 18.75 kVA. Model the transformer as ideal. It is used to serve two 120 V loads from a 7200 V source. The loads are:

(1) 120 V; 18 kVA, unity pf.
(2) 120 V; 15 kVA, pf = 0.8 lagging

(a) Draw a circuit diagram.
(b) Compute the primary current and pf.

2.17 Three identical single-phase transformers are to be used to supply a 900 kVA 480/277 V balanced three-phase load from a 4157/2400 V balanced three-phase source. Draw the correct circuit (as shown in Figure 2.4), determine the transformer voltage and kVA ratings, and determine the currents for:

(a) The delta–wye connection.
(b) The wye–wye connection.
(c) The wye–delta connection.
(d) The delta–delta connection.
(e) The open delta connection.

You may model the transformers as ideal.

2.18 Consider the single-phase transformer of Problem 2.6. Now consider three such transformers connected in a wye–wye three-phase connection. Both wyes are solid-grounded (zero impedance to ground). Draw the positive, negative, and zero sequence circuits (all values in pu).

2.19 Repeat Problem 2.18, except now the HV wye is grounded through a 10 Ω resistor, and the LV wye is ungrounded.

2.20 Repeat Problem 2.18, except now the connection is HV delta; LV wye is solid-grounded.

2.21 The system described in Problem 2.18 operates balanced three-phase step-down. The three-phase load is rated at 480/277 V, 600 kVA, pf = 0.866 lagging, and may be modeled as constant impedance, solid-grounded wye. The balanced three-phase source is ideal, and solid-grounded wye.

(a) Draw all three sequence networks (as in Problem 2.18), adding the source and load elements (all values in pu).
(b) Suppose the load voltage is 480 V. Find the primary and secondary *sequence* voltages and currents (in pu). Convert all results into SI units.
(c) Continuing (b), find the primary and secondary *phase* voltages and currents (in SI units).

2.22 Repeat Problem 2.21, except that the source is unbalanced, with primary sequence voltages as follows:

$$\bar{V}_0 = 100\angle 0^\circ \text{ V}$$
$$\bar{V}_1 = 2400\angle 0^\circ \text{ V}$$
$$\bar{V}_2 = 240\angle 0^\circ \text{ V}$$

2.23 Consider the nine waveforms in Figure 2.11. Provide a short (in one or two sentences) explanation of why each waveform is as shown.

2.24 A three-wire (abc) balanced 480 V three-phase system (phase sequence abc) is to be converted into a 480 V four-wire (abcn) system using zig-zag connected ideal transformers. The system serves a balanced 300 kVA, 0.8 pf lagging load, and a single-phase 100 kVA 277 V, pf = 0.8 leading load (a-n).

(a) Draw an appropriate circuit diagram.
(b) Determine the complex phasor currents everywhere, and place them on the circuit diagram of (a).

3

Basic Mechanical Considerations

The basic purpose of electromechanical (EM) machines is to convert energy from, or into, mechanical form. Therefore, it is essential to consider mechanical issues if we are to perform a comprehensive analysis of such devices. Since the overwhelming majority of applications are rotational, we will consider that configuration first. Likewise, whereas operation in the generator mode is important, and will be addressed in detail, motor mode operation is more ubiquitous, and probably of greater interest to readers of this book. Indeed, it is common to refer to EM machines that normally operate in the motor mode simply as "motors," and we will conform to that practice. Likewise, EM machines that normally operate in the generator mode are called "generators." Bear in mind, however, that the machine can operate in both modes if operating conditions require it. There are many types of EM machines and most of them will be discussed here. However, the subject of this chapter is not the EM machine, but the mechanical termination, called the "mechanical load" for motor applications, and the "prime mover" for generator applications.

3.1 Some General Perspectives

Consider the general situation in Figure 3.1a. Note that the flow of energy through the system is from left to right, or electrical to mechanical. Thus, it is natural to use the terminology, "electrical source" and "mechanical load" to describe the terminations. Also, note the positive definitions of currents, voltages, speed, and torques. These definitions are collectively called the "motor convention," and are logically used when motor applications are under study. Likewise, when generator applications are considered, the sign conventions of Figure 3.1b (called the generation convention) will be adopted. What this means is that variables will be positive under "normal" conditions (motors operating in the motor mode, generators in the generator mode), and negative under some "abnormal" conditions (motors running "backward," for example). We begin by focusing on the motor situation:

$$J\frac{d\omega_{rm}}{dt} = T_{dev} - (T_m + T_{RL}) = T_{dev} - T'_m \tag{3.1}$$

where

T_{dev} = EM torque, produced by the motor, in N m
T_m = torque absorbed by the mechanical load, in N m

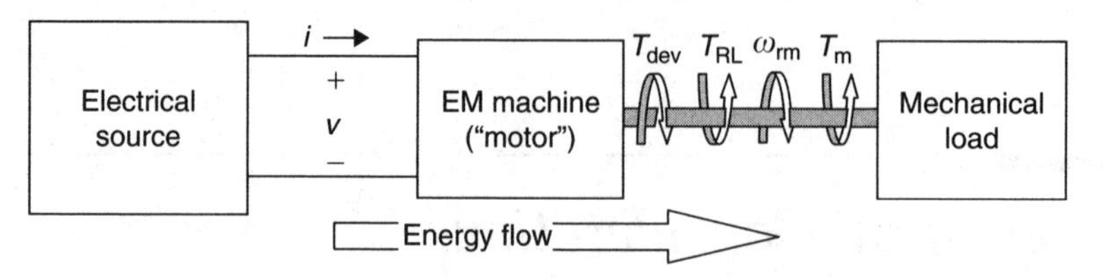

(a) The EM rotational machine; motor convention

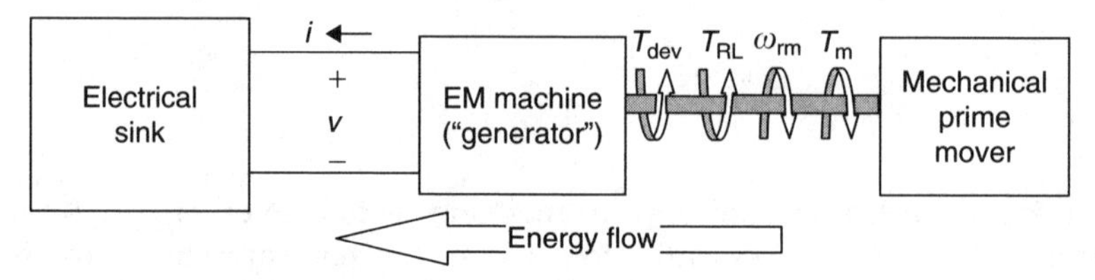

(b) The EM rotational machine; generator convention

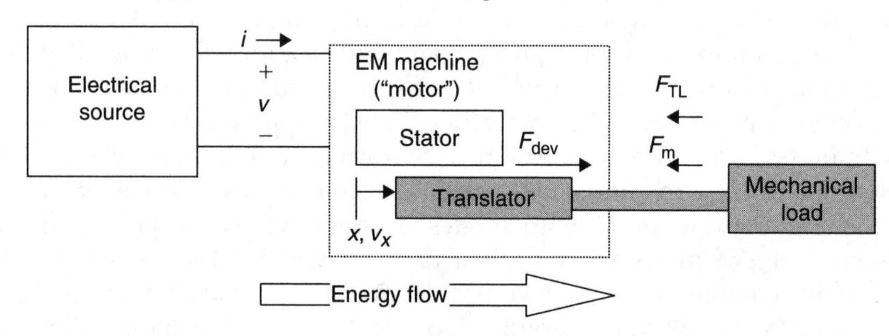

(c) The EM translational machine; motor convention

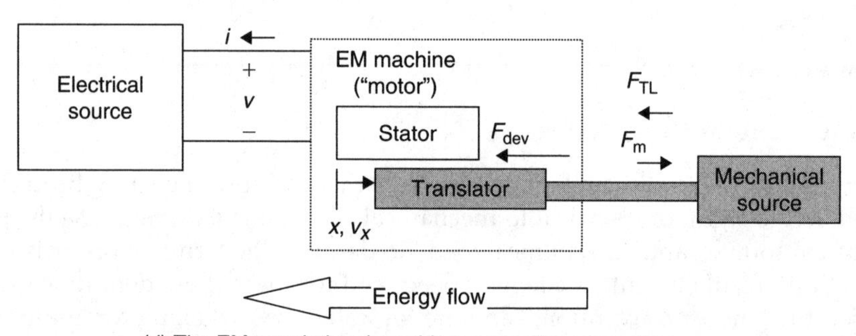

(d) The EM translational machine; generator convention

FIGURE 3.1. Motor and generator sign conventions: EM machines.

T_{RL} = rotational loss torque, internal to the motor, in N·m
$T'_m = T_m + T_{RL}$ = equivalent load torque, in N·m
J = mass polar moment of inertia of all rotating parts, in kg·m^2
ω_{rm} = angular velocity of rotating parts, in rad/s.

These variables deserve a comment. T_{dev} is a key concept. This is the electromagnetic torque produced internally in the motor by the interaction of the stator and rotor magnetic fields, and lies at the heart of the energy conversion process. The torque T_{RL} includes frictional-type motor losses, and opposes rotation regardless of the operating mode. T_m is torque absorbed by the mechanical load, including the load losses and that used for useful mechanical work. T'_{m} is a variable of convenience, lumping the motor rotational losses with the load. J is the mass polar moment of inertia of all parts of the system that rotate, including the shaft, the coupling, the motor rotor, and the load rotating elements. The overall J is the sum of J's for the individual parts.

It is important to consider some corresponding system powers:

$$P_{\mathrm{dev}} = T_{\mathrm{dev}}\omega_{\mathrm{rm}} \quad \text{(EM power, converted by the motor into mechanical form, in W)} \tag{3.2a}$$

$$P_{\mathrm{m}} = T_{\mathrm{m}}\omega_{\mathrm{rm}} \quad \text{(power absorbed by the mechanical load, including the load losses and that used for useful mechanical work, in W)} \tag{3.2b}$$

$$P_{\mathrm{RL}} = T_{\mathrm{RL}}\omega_{\mathrm{rm}} \quad \text{(rotational power loss, internal to the motor, in W)} \tag{3.2c}$$

We have raised a number of important issues, which can be expanded upon in the following example.

Example 3.1

Consider a motor–load system, whose torque–speed characteristics are shown in Figure 3.2a. $J = 2\,\mathrm{kg\,m^2}$. Determine

(a) the system speed transient, starting from zero speed.
(b) the system speed transient, starting from 200 rad/s.
(c) the steady-state speed in rad/s and rpm.
(d) the load power in kW and horsepower and torque in N m and ft lbs.

SOLUTION

(a) Applying Equation 3.1:

$$J\frac{\mathrm{d}\omega_{\mathrm{rm}}}{\mathrm{d}t} = T_{\mathrm{dev}} - T'_{\mathrm{m}} = (400 - 2\omega_{\mathrm{rm}}) - 0.6667\omega_{\mathrm{rm}} = 400 - 2.6667\omega_{\mathrm{rm}}$$

$$0.75\frac{\mathrm{d}\omega_{\mathrm{rm}}}{\mathrm{d}t} + \omega_{\mathrm{rm}} = 150$$

Solving for ω_{rm}: $\quad \omega_{\mathrm{rm}} = 150 + K\mathrm{e}^{-1.333t}$:

(a) If the initial speed is zero: $\omega_{\mathrm{rm}} = 150 - 150\mathrm{e}^{-1.333t}$.
(b) If the initial speed is 200 rad/s: $\omega_{\mathrm{rm}} = 150 + 50\mathrm{e}^{-1.333t}$. Results are plotted in Figure 3.2b.
(c) The steady-state speed is 150 rad/s for the situation in both (a) and (b). Converting to rpm: $(150\,\mathrm{rad/s})(60\,\mathrm{min/s})(1/(2\pi\,\mathrm{rad/rev}))\omega_{\mathrm{rm}} = (150\,\mathrm{rad/s})(60/2\pi) =$ 1432 rpm.
(d) At 1432 rpm,
$T_{\mathrm{dev}} = T'_{\mathrm{m}} = 100\,\mathrm{N\,m}$. Converting to ft lbs: $100\,\mathrm{N\,m}/(1.356\,\mathrm{N\,m/ft\,lb}) = 73.75\,\mathrm{ft\,lb}$.
The corresponding power is

$$P_{\mathrm{dev}} = T_{\mathrm{dev}}\omega_{\mathrm{rm}} = P'_{\mathrm{m}} = T'_{\mathrm{m}}\omega_{\mathrm{rm}} = (100)(150) = 15\,\mathrm{kW}$$

Converting to horsepower (hp) $= 15(1/0.746\,\mathrm{hp/kW}) = 20.11\,\mathrm{hp}$

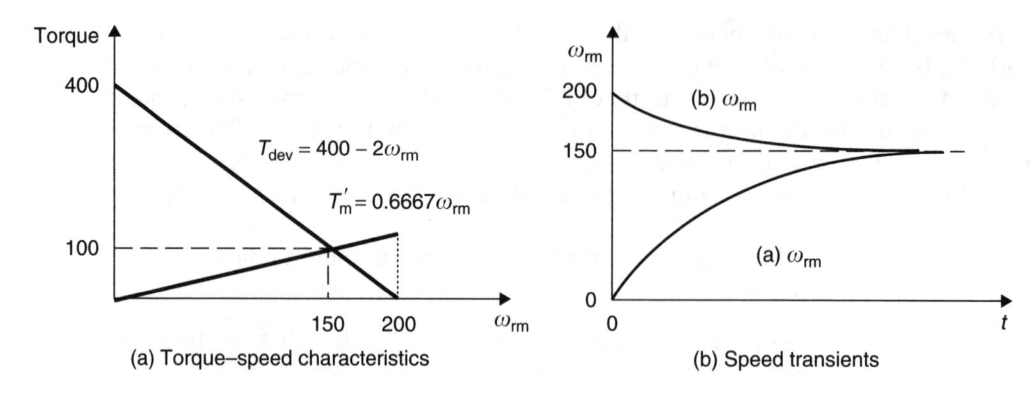

FIGURE 3.2. Motor–Load dynamic performance.

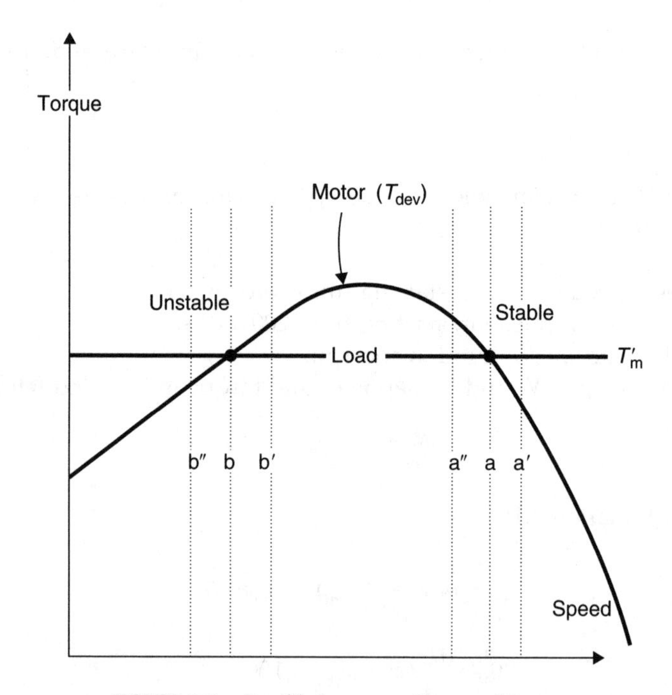

FIGURE 3.3. Equilibrium operating conditions.

The results of Example 3.1 lead to several important general conclusions. Observe that in part (a), starting from zero speed, $T_{dev} > T'_m$, and the system *accelerated* to a final speed of 150 rad/s. In part (b), starting from 200 rad/s, $T_{dev} < T'_m$, and the system *decelerated* to a final speed of 150 rad/s. In both cases, the system sought out the equilibrium condition of $T_{dev} = T'_m$, where the system was in torque balance. In general:

> *The steady–state running speed for any motor–load system occurs at the intersection of the motor and load torque–speed characteristics, i.e., where $T_{dev} = T'_m$. If $T_{dev} > T'_m$, the system is accelerating; for $T_{dev} < T'_m$, the system decelerates.*[1]

Consider the situation shown in Figure 3.3. Consider that the system operates at point a and some external disturbance forces the system to a higher speed (a′). Since the motor develops less torque than the load, the system slows down and returns to a. If some external

[1] Strictly true for stable equilibria only. Read on to consider this point.

disturbance forces the system to a lower speed (a″), the motor develops more torque than the load, and the system speeds up, and again returns to a. Point a is therefore called a *stable* equilibrium point. However, consider operation at point b and some external disturbance forces the system to a higher speed (b′). Now the motor develops more torque than the load, the system speeds up and moves further away from b. Likewise, if the system moves to a lower speed (b″), the motor develops less torque than the load, and the system slows down and cannot return to b. Point b is called an *unstable* equilibrium.

For technical work in the United States, it remains important to be able to convert speed, torque, and power from radian/second, newton-meter, and watts to revolutions per minute (rpm), foot-pounds (ft lb), and horsepower (hp), respectively. See Example 3.1, parts (c) and (d) for details. All equations in this book require SI units, unless specifically noted otherwise.

3.2 Efficiency

A simple, but fundamental and important, concept is that of efficiency. For any system or component designed to convert or process energy, efficiency provides a simple measure of how well the system or component achieves its intended purpose. Unless specifically stated otherwise, efficiency[2] is defined as

$$\text{Efficiency} = \eta = \frac{P_{OUT}}{P_{IN}}, \text{ normally expressed in \%}$$

where

P_{OUT} = designated total output power, in W
P_{IN} = designated total input power, in W.

Care must be taken to account for *all* of the input and output power. Claims are made from time to time by inventors that purport to have designed systems that create "free" energy, that is, systems that provide more output energy than is required at the input. Barring nuclear mass to energy conversion processes, this is in direct violation of the first law of thermodynamics, or conservation of energy.[3] Hence, efficiency can *never* exceed 100%. Results to the contrary normally mean that some elements of the input, or output, or both, have been overlooked, either accidently or otherwise.

An alternative definition of efficiency is in terms of energy:

$$\text{Energy efficiency} = \eta = \frac{W_{OUT}}{W_{IN}}, \text{ normally expressed in \%}$$

where

$$W_{OUT} = \int_0^T P_{OUT}(t)\,dt = \text{total output energy, in J}$$
$$W_{IN} = \int_0^T P_{IN}(t)\,dt = \text{total input energy, in J.}$$

[2] This is a technical definition, of course. Like most words, "efficiency" has other nontechnical definitions and usages, such as "Kevin does his job with high efficiency."

[3] Even nuclear processes are included, if the statement of the first law is properly generalized.

The two definitions are equivalent, if the input and output powers are constant over time interval T. However, in general they are not the same if the input and output powers vary in time. Although SI units are indicated, any correct units for power or energy may be used, as long as the same units are used for the input and output.

Example 3.2

A motor supplies a constant 10 hp to a mechanical load for one 30-day month, for which the electric utility bill is \$300. The energy rate is \$0.05 per kW h.

(a) Find the motor efficiency.
(b) If the motor is replaced by another with an efficiency of 93%, then determine the monthly savings.

SOLUTION

(a) Output power = $10 \times 0.746 = 7.46\,\text{kW}$
Input energy = $\$300/0.05 = 6000\,\text{kW h}$
Average electric power = $6000\,\text{kW h}/(24 \times 30) = 8.333\,\text{kW}$
Efficiency = $7.46/8.333 = 89.52\%$
(b) Output power = $10 \times 0.746 = 7.46\,\text{kW}$
Input power = $7.46/0.93 = 8.0215\,\text{kW}$
Average electric energy = $8.0125 \times (24 \times 30) = 5775.5\,\text{kW h}$
Monthly savings = $(6000 - 5775.5) \times 0.05 = \$11.23$

3.3 Load Torque–Speed Characteristics

A comprehensive engineering analysis of any motor–load system requires access to the load torque–speed characteristic. This may require some professional expertise and familiarity with the specific application under study. However, there are some general principles that may be employed. We write an empirical relation for load torque T_L as

$$T_L = A_0 + A_1\omega_L + A_2\omega_L^2 + \cdots + A_n\omega_L^n \tag{3.3}$$

n = order of load torque characteristic which is sufficiently general to handle most (but not all) mechanical loads. For applications where the motor shaft is directly connected to the load shaft ("direct drive" applications)

$$T_m = T_L \qquad \omega_{rm} = \omega_L$$

3.3.1 Constant Torque Loads ($n = 0$; $T_L = A_0$)

Examples of constant torque loads include conveyers (belts, screws, shakers), cranes (hoists, trolleys, bridges), crushers, elevators, extruders, kilns, looms, mills (rolling, rubber), positive displacement pumps, sanders, saws, shears, winches, and washers.

Constant torque loads impose particularly severe starting requirements, since the full load torque must be met, and exceeded, to start the system. We will investigate the elevator, a constant torque load, in detail later in this chapter.

3.3.2 Linear (First-Order) Torque Loads ($n = 1$; $T_L = A_1\omega_L$)

An example of a linear torque load is viscous friction. Linear torque loads impose less severe starting requirements, since the load torque is zero at the start. However, it is significant through the starting interval, and should be considered.

3.3.3 Parabolic (Second-Order) Torque Loads ($n = 2$; $T_L = A_2\omega_L^2$)

Common examples of second-order loads are fans and pumps. Fans are devices designed to move gases (usually air). They accomplish their function by creating a pressure drop ΔH between the inlet and the outlet. Assume that the fan output constitutes a column of air moving through sectional area A at velocity u. Then

$$\frac{dm}{dt} = Au\rho = \text{mass flow rate of air, in kg/s}$$

where

A = cross-section of wind intercepted by the turbine, in m^2

u = wind velocity, perpendicular to A, in m/s

ρ = density of air = 1.2 kg/m^3, as an estimated typical value.[4]

The corresponding wind power is

$$\text{Wind power} = P_W = \frac{dW}{dt} = \frac{1}{2}u^2\frac{dm}{dt} = \tfrac{1}{2}A\rho u^3, \text{ in W}$$

The fan must impart this power to the air flow, and does so at less than 100% efficiency. The input fan power is, therefore,

$$P_F = \eta P_W = \tfrac{1}{2}\eta A\rho u^3, \text{ in W}$$

where η is the fan efficiency.

The fan efficiency depends on fan blade design, and is a function of fan speed, with a theoretical maximum value of below 60%.[5] The velocity of the air stream (u) is directly proportional to the fan blade rotational velocity ω_{rm}. Therefore,

$$P_F = P_L = K\omega_L^3 = \text{input fan power, in W}$$

Converting into torque

$$T_L = \frac{P_L}{\omega_L} = A_2\omega_L^2 = \text{input fan torque, in N m}$$

Pumps are similar to fans, in that they create mass flow rate in a fluid. However, the fluid is in liquid form, which is normally incompressible, and therefore behaves quite differently

[4] Air density depends on temperature, pressure, and humidity. ρ = 1.225 kg/m^3 at 15°C, 760 mm Hg, 0.013 kg/m^3 water content.

[5] The German physicist Albert Betz proved that the theoretical upper limit was 16/27 = 0.5926 in his book on wind energy in 1926, a value which has become known as the "Betz limit."

from a gas. There are two basic types: positive displacement and centrifugal. Positive displacement pumps operate by forcing an incompressible fluid into a volume, and present a constant torque load.

Centrifugal pumps accomplish their function by creating a pressure drop ΔH between the inlet and the outlet. Then

$$\frac{\mathrm{d}m}{\mathrm{d}t} = Au\rho = \text{mass flow rate of fluid, in kg/s}$$

where
A = cross-section of fluid flow, in m^2
u = fluid velocity, perpendicular to A, in m/s
ρ = density of fluid, kg/m^3, water = $1000\,kg/m^3$.

The corresponding power is

$$P_W = \frac{\mathrm{d}W}{\mathrm{d}t} = \frac{1}{2}u^2\frac{\mathrm{d}m}{\mathrm{d}t} = \tfrac{1}{2}A\rho u^3, \text{ in W}$$

The pump must impart this power to the fluid flow, and does so at less than 100% efficiency. The input pump power is, therefore,

$$P_P = \eta P_W = \tfrac{1}{2}\eta A\rho u^3, \text{ in W}$$

where η is the pump efficiency.

The pump efficiency depends on impeller design, and is a function of speed. The velocity of the fluid (u) is directly proportional to the impeller rotational velocity ω_{rm}. Therefore,

$$P_P = P_L = K\omega_L^3 = \text{input power, in W}$$

Converting into torque

$$T_L = \frac{P_L}{\omega_L} = A_2\omega_L^2 = \text{input pump torque, in Nm}$$

If the pumps must operate against a back pressure, such as that caused by a gravitational gradient (i.e., the pump must pump "uphill"), a constant term may be added as well:

$$T_L = A_2\omega_L^2 + A_0$$

3.3.4 The General nth-Order Case ($T_L = A_0 + A_1\omega_L + A_2\omega_L^2 + \cdots + A_n\omega_L^n$)

If the load torque–speed characteristic is available, it is possible to use least mean-squared error techniques to fit the data with an nth-order curve. The following example will demonstrate.

Example 3.3
Suppose the following data are available for a particular load:

Speed (rpm)	Torque (N m)
0	20
200	35
400	53
600	62
800	78
1000	90
1200	112
1400	140
1600	175
1800	200

A least mean-squared error third-order fit to the data was computed. The coefficients are

$$A[0] = 2.1139860138\text{E}{+}01$$
$$A[1] = 7.1867360570\text{E}{-}01$$
$$A[2] = -2.0990490011\text{E}{-}03$$
$$A[3] = 1.8204919911\text{E}{-}05$$

For comparison, the original and calculated characteristics are plotted in Figure 3.4.

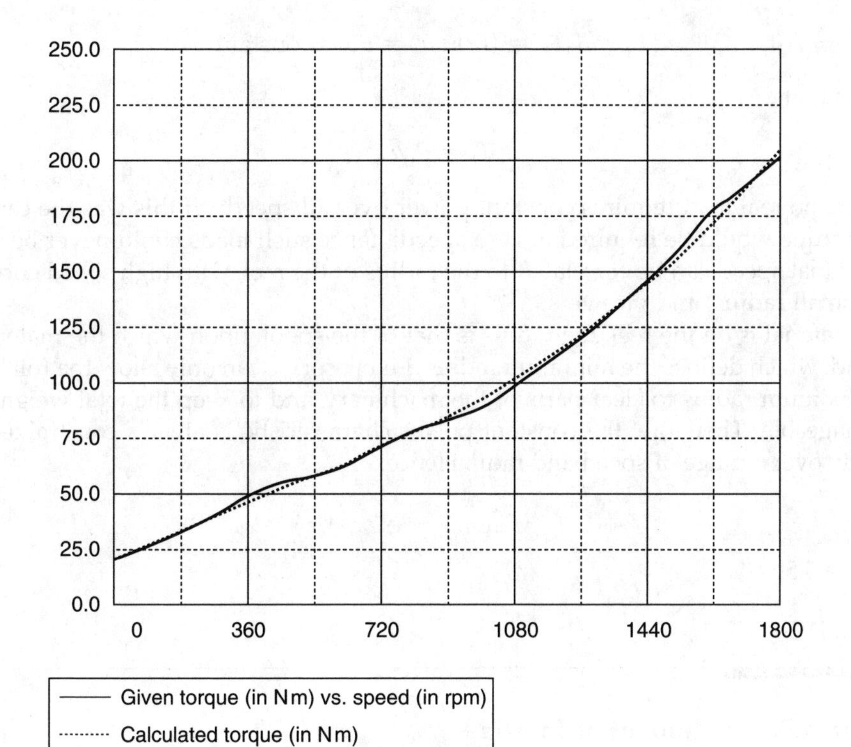

FIGURE 3.4. Third-order fit to nonlinear load torque–speed characteristic.

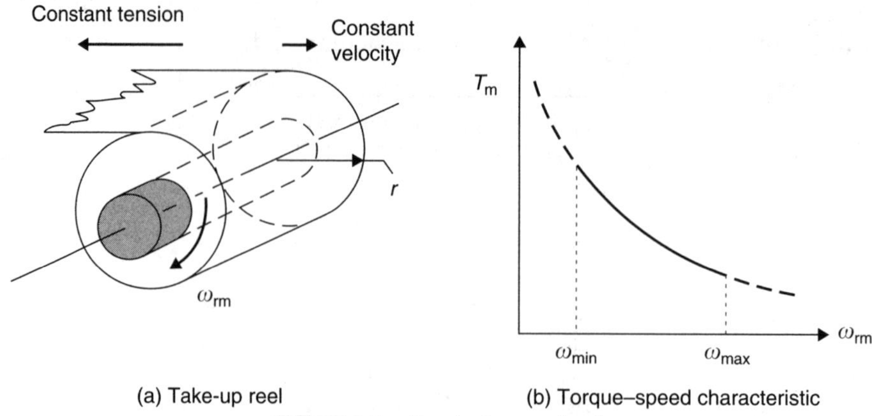

FIGURE 3.5. Constant power load.

3.3.5 The Constant Power Load

There are loads for which Equation 3.3 is not an appropriate model. An example is a re-winder, or take-up reel, similar to that used in the paper and textile industries. Such a load is shown in Figure 3.5a. To keep the process moving properly, it is desirable to hold both the tangential tension (F) and the velocity (v) constant, independent of the roll radius. Now

$$T_L = F(r) \quad \text{and} \quad \omega_L = v/r$$

Therefore,

$$P_L = T_L\omega_L = (Fr)(v/r) = Fv = \text{constant} \tag{3.4a}$$

It follows that

$$T_L = P_L/\omega_L \tag{3.4b}$$

Of course, no real load requires constant power over all speeds. If this was the case, then infinite torque would be required at zero speed: hence such loads could never be started. It is clear that speed can be correlated to the radius of the reel, with high speed corresponding to small radius, and vice versa.

If no material is on the reel, the radius is that of the spool, upon which the material will be wound, which defines the minimum radius. Likewise, we can only allow the roll to reach some maximum radius to clear parts of the machinery, and to keep the total weight of the roll manageable. Therefore, the constant power characteristic is always constrained to be valid only over a range of speed and radii. Hence,

$$r_{min} < r < r_{max}, \qquad \omega_{max} > \omega_L > \omega_{min}$$

See Figure 3.5b.

3.4 Mass Polar Moment of Inertia

It is true that the steady-state running speed for any motor–load system is exclusively determined by the motor and load torque–speed characteristics. However, if the mechanical

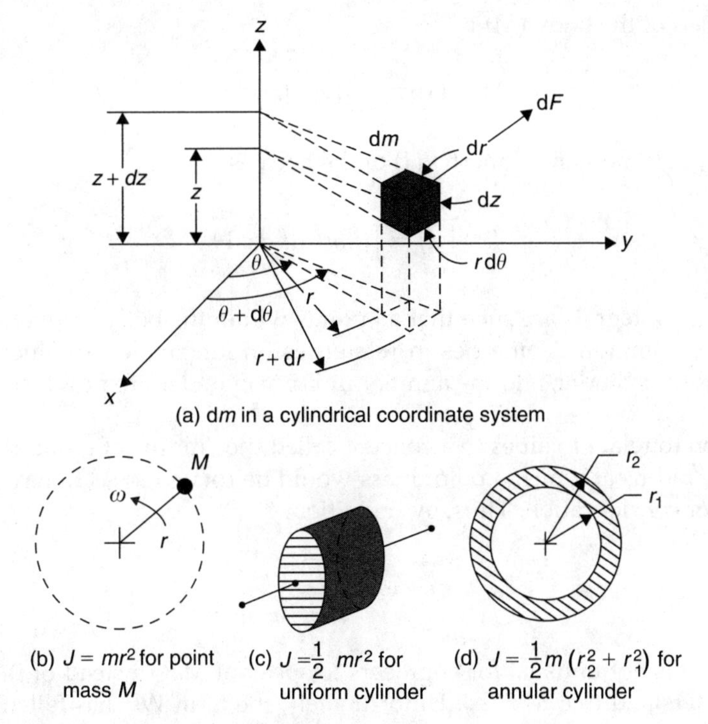

FIGURE 3.6. Inertial concepts.

dynamic performance of the system is to be analyzed, the system inertia must also be considered. To investigate inertia, consider an elemental mass rotating about the z-axis at radial distance r, as indicated in Figure 3.6a. Likewise, consider locating $\mathrm{d}m$ in cylindrical coordinates r, θ, and z.

To accelerate the mass in its circular orbit, apply a tangential force $\mathrm{d}F$ (perpendicular to r). By Newton's second law of motion,

$$\begin{aligned} \mathrm{d}F &= \mathrm{d}m\frac{\mathrm{d}v}{\mathrm{d}t} \\ &= \mathrm{d}m\frac{\mathrm{d}}{\mathrm{d}t}(r \cdot \omega) \\ &= r\mathrm{d}m\frac{\mathrm{d}\omega}{\mathrm{d}t} \end{aligned} \tag{3.5}$$

where v is the tangential velocity in the direction of $\mathrm{d}F$. The corresponding torque

$$\begin{aligned} \mathrm{d}T &= r\mathrm{d}F \\ &= (r^2\mathrm{d}m)\frac{\mathrm{d}\omega}{\mathrm{d}t} \\ &= \mathrm{d}J\,\frac{\mathrm{d}\omega}{\mathrm{d}t} \end{aligned} \tag{3.6}$$

where $\mathrm{d}J = r^2\,\mathrm{d}m$ is defined as the incremental mass polar moment of inertia, in kg m^2. Now consider $\mathrm{d}m$ to be an infinitesimal element within a rigid body of arbitrary geometry, free to rotate about the z-axis. Now,

$$\mathrm{d}m = \rho\,\mathrm{d}(\mathrm{vol}) = \rho(\mathrm{d}r)(\mathrm{d}z)(r\,\mathrm{d}\theta) = \rho r(\mathrm{d}r)(\mathrm{d}z)(\mathrm{d}\theta)$$

where ρ is the mass density in kg/m^3 and may be a function of r, z, and θ.

Then the mass of the body (M) is

$$M = \int dm = \iiint \rho r \, dr \, dz \, d\theta$$

and the mass polar moment of inertia (J) of the body is

$$J = \int dJ = \iiint \rho r^3 \, dr \, dz \, d\theta$$

The limits on the integrals are such that all points within the body are included. J is tabulated for many common geometries in references on kinematics; J values for the point mass, the uniform cylinder, and the annular uniform cylinder are provided in Figure 3.6b to Figure 3.6d.

It is common to relate J values to a concept called the "radius of gyration" (k), which is defined as the radius at which a point mass would be rotated so as to have the same J as the body under consideration. Thus, by definition,

$$k = \sqrt{\frac{J}{M}} \tag{3.7}$$

Since $J = Mk^2$, it is not unusual for engineers to speak of Mk^2 (instead of J), expecting the listener to understand that Mk^2 *is* J. Unfortunately, the term Wk^2 has fallen into common usage in the United States to represent J also, continuing the traditional confusion between weight and mass.

For a given application, the total J in Equation 3.1 must include all moving parts, which would include the motor rotor, the shaft, the coupling, and all load parts. For rigid bodies all rotating at the same speed, we may simply add the J values. Determination of J may not be straightforward in some cases. A general guiding principle is that if mass is accelerated as the system speed changes, that mass must contribute to J. Fortunately, for many loads, J values are available. An example should be helpful.

Example 3.4

A manufacturer's publication quotes a value of $Wk^2 = 232\,\text{lb}\,\text{ft}^2$ as the NEMA (National Electrical Manufacturers Association) maximum load inertia value that a 50 hp, 1800 rpm motor can safely start. (a) Determine the corresponding value of J in $\text{kg}\,\text{m}^2$. (b) If J for the motor is $0.5\,\text{kg}\,\text{m}^2$, what J should be used for dynamic analysis, if the load of (a) is to be studied.

SOLUTION

(a) Since $2.188\,\text{lb}_{\text{mass}} = 1\,\text{kg}$ and $3.281\,\text{ft} = 1\,\text{m}$, $J = 0.04246\,(Wk^2) = 9.851\,\text{kg}\,\text{m}^2$.

(b) $J = J_{\text{motor}} + J_{\text{load}} = 9.851 + 0.5 = 10.351\,\text{kg}\,\text{m}^2$. Inertia is additive.

3.5 Gearing

For direct drive systems, the motor shaft is rigidly connected to the load shaft, so that they turn at the same speeds. However, for many other applications, the desired load speed range may be quite different from that over which the motor can efficiently operate.

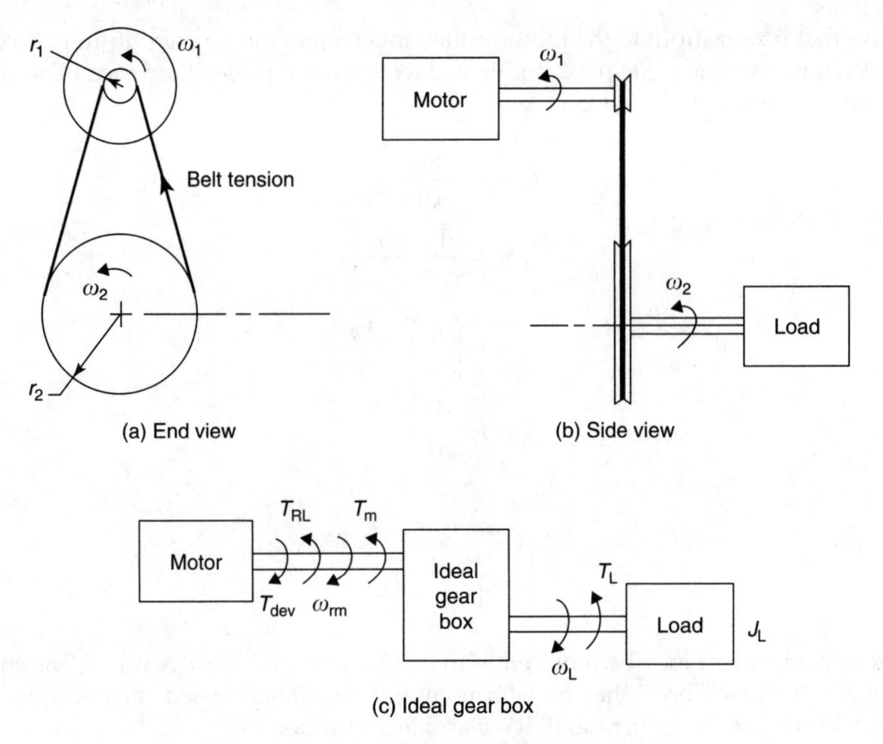

FIGURE 3.7. Gearing.

For example, consider an amusement park carousel (merry-go-round) that turns at only 3 or 4 rpm when driven frequently by an 1800 rpm motor. Thus, there is a need for a mechanism to interpose between the motor and load to perform the desired speed transformation.

Consider the situation in Figure 3.7, which shows a load driven by a motor using a belt–pulley arrangement. Ignore belt slippage or deformation and pulley frictional losses. If the motor, or primary, pulley (#1) turns through an angle θ_1, a length of belt $r_1\theta_1$ enters the pulley groove. But that same length must leave the load, or secondary, pulley groove (since the length outside both pulley grooves is constant). Therefore,

$$\theta_2 = (r_1/r_2)\theta_1 \tag{3.8a}$$

Differentiating Equation 3.8a,

$$\omega_2 = (r_1/r_2)\omega_1 \tag{3.8b}$$

The motor exerts a torque (T_1) on the motor pulley, which in turn exerts a tangential force on the belt, which appears as belt tension. Likewise, the belt tension exerts *the same* tangential force on the load pulley, which applies a torque T_2 to the load:

$$\text{Belt tension} = (T_1/r_1) = (T_2/r_2)$$

or

$$T_2 = (r_2/r_1)T_1 \tag{3.9a}$$

Note that

$$P_2 = \omega_2 T_2 = \omega_1 T_1 = P_1 \tag{3.9b}$$

which says that power input to the motor pulley must equal the power output to the load pulley. Also note that the *high*-speed pulley is the *low* torque pulley (and vice versa).

Now consider a pure inertial load

$$T_2 = J_2 \cdot \frac{d\omega_2}{dt}$$

$$\frac{r_2}{r_1} \cdot T_1 = J_2 \frac{d}{dt}\left(\frac{r_1 \omega_1}{r_2}\right)$$

$$T_1 = J_2 \left(\frac{r_1}{r_2}\right)^2 \cdot \frac{d\omega_1}{dt}$$

$$= J_1 \cdot \frac{d\omega_1}{dt}$$

so that

$$J_1 = \left(\frac{r_1}{r_2}\right)^2 \cdot J_2 \tag{3.10}$$

The pulley arrangement may be replaced with a more general "ideal gear box," as shown in Figure 3.7c, a "black box" that transforms motor–load shaft speeds and torques in a fixed ratio. We define the gear ratio (GR) of an ideal gear box as

$$\text{GR} = [\text{primary (motor) shaft speed}]/[\text{secondary (load) shaft speed}]$$
$$= \frac{\omega_{\text{rm}}}{\omega_{\text{L}}} \tag{3.11}$$

Therefore,

$$T_{\text{m}} = \frac{1}{\text{GR}} \cdot T_{\text{L}} \tag{3.12}$$

$$P_{\text{m}} = \omega_{\text{rm}} T_{\text{m}} = \omega_{\text{L}} T_{\text{L}} = P_{\text{L}} \tag{3.13}$$

$$J_{\text{m}} = \left(\frac{1}{\text{GR}}\right)^2 \cdot J_{\text{L}} \tag{3.14}$$

Example 3.5

Consider an amusement park carousel (merry-go-round) that turns at 3 rpm when driven by an 1800 rpm, 25 hp motor. The ride platform is attached to a fixed pulley #3 of radius 3 m, with a belt drive to the outer groove of pulley #2 (radius 10 cm). The inner groove of pulley #2 (radius 150 cm) has a second belt drive to pulley #1, which is on the motor shaft.

(a) What is the radius of the motor pulley?
(b) If the motor runs at 1800 rpm and 25 hp, what is the torque on the motor pulley?
(c) If the motor runs at 1800 rpm and 25 hp, what is the torque on the platform pulley?

SOLUTION

(a)

$$\omega_2 = \frac{300}{15}\omega_3 = 60\,\text{rpm}; \qquad r_1 = \frac{60}{1800}(150) = 5\,\text{cm}$$

(b) 1800 rpm = 188.5 rad/s.
(c) 3 rpm = 0.3412 rad/s; P = 18.65 kW

$$T_1 = \frac{P}{\omega_1} = \frac{18.65}{188.5} = 98.94\ \text{N m} \qquad \text{(on the motor shaft)}$$

$$T_3 = \frac{P}{\omega_3} = \frac{18.65}{0.3142} = 59.36\ \text{kN m} \qquad \text{(on the platform shaft)}$$

3.6 Operating Modes

So far, our focus has been on motor operation. We have implied that only *positive* values of speed are considered (in Equation 3.3, for example). What would *negative* speed values mean, from a physical perspective? The short answer is "the opposite, or reverse of positive." We define the term "forward" to mean rotation in the "normal" direction, which describes the meaning of $\omega_{rm} > 0$, and should be obvious in a specific application. "Reverse" is defined to mean rotation in the direction opposite to "forward," and corresponds to $\omega_{rm} < 0$.

Consider the more general issue of what positive and negative mean in any technical situation. The critical point is that "positive" must be clearly and unambiguously defined for all variables to be used in an engineering analysis. "Negative" then means "opposite to what was defined as positive." Sometimes this can be done verbally. More typically, however, a diagram is necessary to define positive senses for all variables. Since our emphasis has been on motor operation, we have elected to use motor convention, as shown in Figure 3.1a, which the reader should review at this point, noting in particular the positive definitions for T_{dev} and ω_{rm}.

Refer to Figure 3.8a. Using motor convention, first quadrant operation means that (1) speed is positive (forward) and (2) T_{dev} is positive (also forward), and transferring

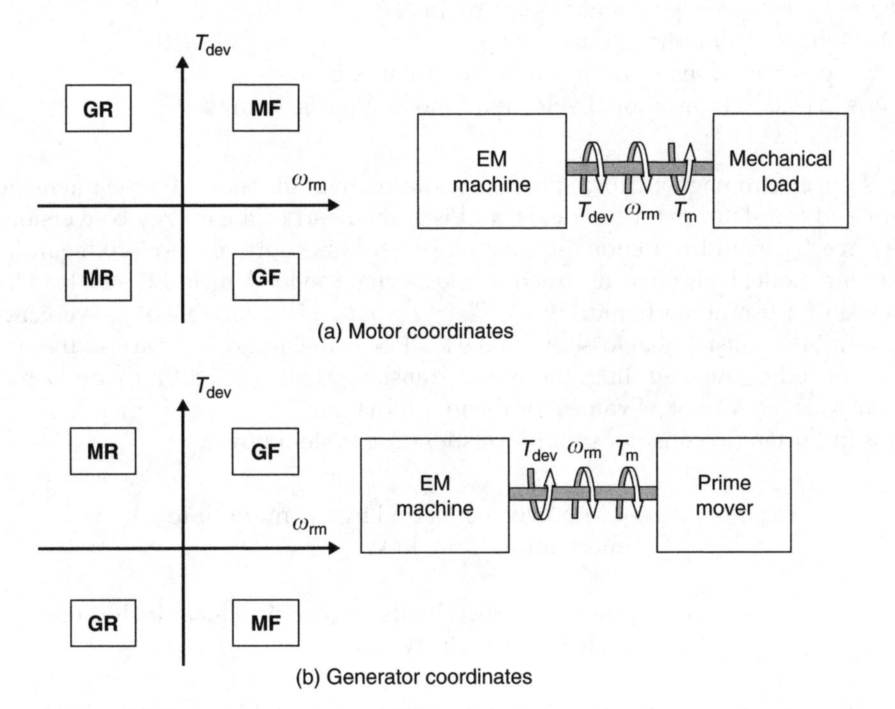

FIGURE 3.8. Operating modes. MF, motor forward; MR, motor reverse; GF, generator forward; GR, generator reverse.

energy from motor to load (motoring). There are four possible operating modes, specific to the four quadrants. In Figure 3.8b, the same physical situation is described using generator convention. In any application, a primary consideration is to determine which of these operating modes will be required.

3.7 Translational Systems

Although a large majority of EM energy conversion systems are rotational, a significant minority are translational. Translational systems are those in which motion is lineal. Consider the general situation in Figure 3.1c. The EM machine has two basic parts: the stator, that cannot move and the translator, that can move in the x direction. Note that the flow of energy through the system is from left to right, or electrical to mechanical. Thus, it is natural to use the terminology, "electrical source" and "mechanical load" to describe the terminations. Also, note the positive definitions of currents, voltages, speed, and forces:

$$M\frac{dv_x}{dt} = F_{dev} - (F_m + F_{TL}) = F_{dev} - F'_m \tag{3.15}$$

where

F_{dev} = EM force, produced by the motor, in N
F_m = Force absorbed by the mechanical load, including the load losses and that used for useful mechanical work, in N
F_{TL} = translational loss force, internal to the motor, in N
$F'_m = F_m + F_{TL}$ = equivalent load force, in N
M = mass of all moving parts, in kg
x = position of the translator relative to stator, in m
$v_x = dx/dt$ = translational velocity of moving parts, in m/s.

F_{dev} is the electromagnetic force produced internally in the motor by the interaction of the stator and translator magnetic fields, and is at the heart of the energy conversion process. The force F_{TL} includes frictional-type motor losses and opposes motion, regardless of the operating mode. F_m is force absorbed by the mechanical load, including the load losses and that used for useful mechanical work. F'_m $(= F_m + F_{TL})$ is a variable of convenience, lumping the motor translational losses with the load. M is the mass of all parts of the system that move, including the coupling, the motor translator, and the load moving elements. The overall M is the sum of M values for the individual parts.

It is important to consider some corresponding system powers:

$$P_{dev} = F_{dev}v_x = \text{EM power, converted by the motor into mechanical form, in W} \tag{3.16a}$$

$$P_m = F_m v_x = \text{power absorbed by the mechanical load, including the load losses, in W} \tag{3.16b}$$

$$P_{RL} = F_{TL}v_x = \text{translational power loss, internal to the motor, in W} \tag{3.16c}$$

3.8 A Comprehensive Example: The Elevator

This section was designed to bring together the issues discussed in this chapter in one specific application. An elevator application was selected because it is familiar to all readers; a minimum of physics system is required to analyze the system; and the system operates in all four modes. Study the system shown in Figure 3.9. Consistent with the mission of this chapter, we will focus on the mechanical issues. In later chapters, we will use this application to

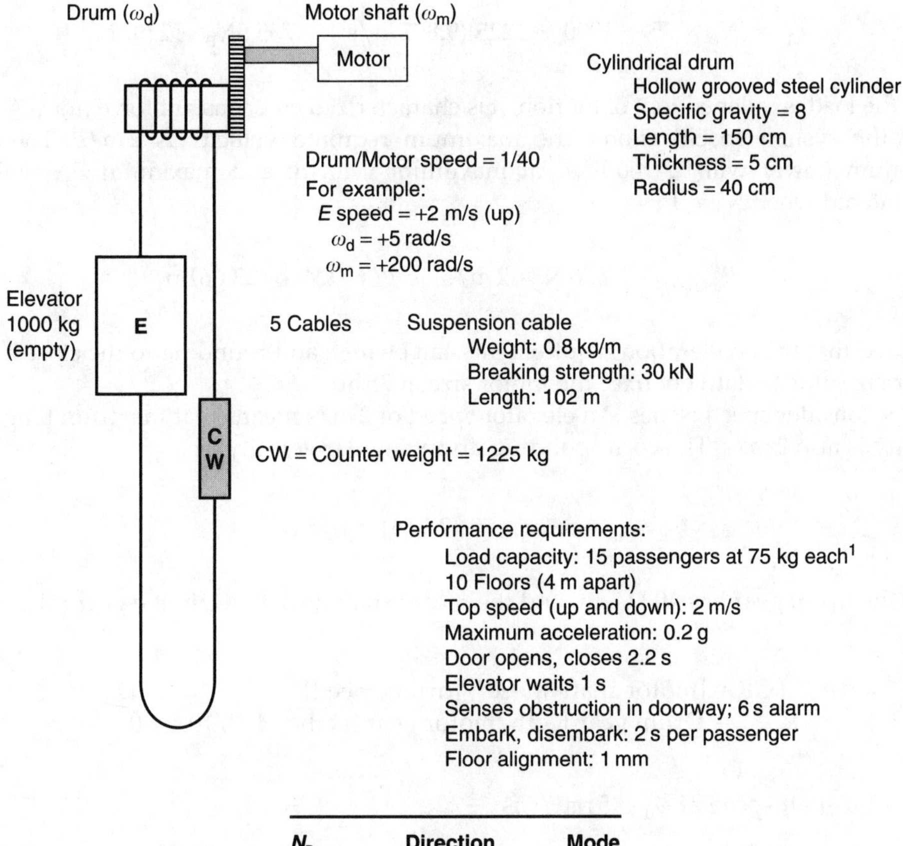

N_P	Direction	Mode
1,2	Up	GF
3	Up	—
4–15	Up	MF
1,2	Down	MR
3	Down	—
4–15	Down	GR

FIGURE 3.9. An elevator application.

[1] There is a wide disparity between a large 150 kg NFL football player and my 23 kg granddaughter Natalie. An assumed average of 75 kg per person is reasonable, considering 15 large people probably would not fit on our elevator anyway. Most elevators have safety interlocks that sense overload and will not run if overloaded.

demonstrate motor control principles. Although the system data are realistic, it is somewhat oversimplified in some details for purposes of clarity.

Our first priority is to establish a coordinate system. We define x to be the vertical position of the elevator relative to the building, with $x = 0$ corresponding to the elevator located on the first floor (up is positive).

Next, we consider the requisite operating modes. Consider the number of onboard passengers to be N_P. Note that when $N_P = 3$, the mass of the loaded elevator ($1000 + 3 \times 75 = 1225$ kg) is counterbalanced by the mass of the counterweight (1225 kg), so that the gravitational forces cancel. Then,

The load force is due to gravity, and is

$$F_L = (N_P \times 75 + 1000 - 1225)(9.807\,\text{m/s}^2) = 735.5N_P - 2207\,\text{N}$$

Since the load is independent of motion, it is characterized as a constant force (torque) load. From the system specifications, the maximum required velocity is 2 m/s. Therefore, maximum power will be required at maximum velocity and maximum F_m, which is encountered when $N_p = 15$:

$$P_{m(max)} = (8826\,\text{N} \times 2\,\text{m/s}) = 17.65\,\text{kW or } 23.66\,\text{hp}$$

Provided that the acceleration requirements can be met, and rounding to the closest stock motor size, this tentatively fixes the motor size at 25 hp.

Now consider speed issues. An elevator speed of 2 m/s means that the drum tangential velocity is also 2 m/s. The corresponding drum angular velocity is

$$\omega_d = v_x/r_{drum} = (2/0.4) = 5\,\text{rad/s}$$

Since the drum gear has 400 teeth, and the motor shaft gear 10 teeth, the gear ratio is

$$\begin{aligned} \text{GR} &= [\text{motor shaft speed}]/[\text{drum speed}] \\ &= \text{Drum gear teeth/motor gear teeth} = 400/10 = 40 \end{aligned}$$

The motor shaft speed at $\omega_d = 5$ rad/s is

$$\omega_m = 40\omega_{drum} = 40(5) = 200\,\text{rad/s} = 1910\,\text{rpm}$$

The maximum load torque on the drum is

$$T_{L(max)} = 8826(0.4\,\text{m}) = 3531\,\text{N m} = 3.531\,\text{kN m}$$

The maximum load torque on the motor is

$$T_{m(max)} = T_{L(max)}/\text{GR} = 3531/40 = 88.26\,\text{N m, or equivalently}$$

$$T_{m(max)} = P_{m(max)}/\omega_{rm} = (17.65/200) = 88.26\,\text{N m}$$

The load masses subject to acceleration includes

$$
\begin{aligned}
M_{\mathrm{E}} &= \text{mass of elevator} = 1000\,\text{kg} \\
M_{\mathrm{P}} &= \text{(maximum) mass of passengers} = 15 \times 75 = 1125\,\text{kg} \\
M_{\mathrm{CW}} &= \text{mass of counter weight} = 1225\,\text{kg} \\
M_{\mathrm{CABLE}} &= \text{mass of cable} = 0.8 \times 102 \times 5 = 408\,\text{kg} \\
M_{\mathrm{DRUM}} &= \text{mass of drum} = \pi\rho(\text{length})(r_{\text{outside}}^2 - r_{\text{inside}}^2) \\
&= \pi(8000)(1.5)[(0.40)^2 - (0.35)^2] = 1414\,\text{kg} \\
\sum M &= M_{\mathrm{E}} + M_{\mathrm{P}} + M_{\mathrm{CW}} + M_{\mathrm{CABLE}} \\
&= 3758\,\text{kg m}^2
\end{aligned}
$$

Next, consider the load inertia. The mass (ΣM) that produces force tangential to the drum is equivalent to point mass located at the drum radius rotating about the drum axis. Therefore,

$$
\begin{aligned}
J_{\Sigma M} &= \sum M\, r_{\mathrm{DRUM}}^2 \\
&= 3758(0.4)^2 = 601.3\,\text{kg m}^2
\end{aligned}
$$

The drum is an annular cylinder.

$$
\begin{aligned}
J_{\mathrm{DRUM}} &= \tfrac{1}{2} M_{\mathrm{DRUM}}(r_{\text{outside}}^2 + r_{\text{inside}}^2) \\
&= 0.5(1414)[(0.40)^2 + (0.35)^2] \\
&= 199.7\,\text{kg m}^2
\end{aligned}
$$

The load inertia is, therefore,

$$
\begin{aligned}
J_{\mathrm{L}} &= J_{\Sigma M} + J_{\mathrm{DRUM}} \\
&= 601.3 + 199.7 = 801.0\,\text{kg m}^2
\end{aligned}
$$

Reflecting the inertia to the motor shaft,

$$
\begin{aligned}
J_{\mathrm{m}} &= \left(\frac{1}{\mathrm{GR}^2}\right) J_{\mathrm{L}} \\
&= \left(\frac{1}{40^2}\right) 801.0 = 0.5006\,\text{kg m}^2
\end{aligned}
$$

Estimating the inertia of a 25 hp motor at 0.25 kg m^2, the total system "J" on the motor shaft is about

$$
J = 0.25 + 0.50 = 0.75\,\text{kg m}^2
$$

Now, the maximum acceleration on the elevator is specified to be $0.2g = 0.2(9.807) = 1.9614\,\text{m/s}^2$, which translates to an angular drum, and motor, acceleration of

$$
\begin{aligned}
\alpha_{\mathrm{DRUM}} &= \left(\frac{\alpha_{\mathrm{DRUM}}}{r}\right) = \frac{1.9614}{0.4} = 4.9035\,\text{rad/s}^2 \\
\alpha_{\mathrm{MOTOR}} &= 40\,\alpha_{\mathrm{DRUM}} = 196.14\,\text{rad/s}^2
\end{aligned}
$$

Converting to torque:

$$T_{ACCEL} = J\alpha_{MOTOR}$$
$$= (0.75)(196.14) = 147\,\text{Nm}$$

Adding the full load steady-state torque:

$$T = T_{ACCEL} + T_{FL}$$
$$= 147 + 88 = 235\,\text{Nm}$$

But full load torque (at 25 hp) was only 88 Nm! Clearly, we need a larger motor! Say, 267% larger! But a larger motor means a larger "J" which requires even more accelerating torque. After a few tries, consider 75 hp, with an equivalent "J_{MOTOR}" of 0.6 kg m^2, resulting in

$$T_{ACCEL} = J\alpha_{MOTOR}$$
$$= (1.1)(196.14) = 216\,\text{Nm}$$
$$T = T_{ACCEL} + T_{FL}$$
$$= 294 + 88 = 304\,\text{Nm}$$

A 75-hp 60-Hz four-pole 1800 rpm motor has a rated torque of about 296 Nm. We intend to operate at $(196/188.5)(60) = 63\,\text{Hz}$ to reach top speed; hence such a motor has enough capacity to meet the maximum steady-state load and dynamic performance specifications. Note that consideration of the dynamic requirements required a motor three times the size of the maximum steady-state load!

Consider a typical load cycle (Table 3.1).

Consider the first step in the load cycle. The time to reach our terminal velocity of 2 m/s with $a = 0.2g$ is 1.02 s. To keep the numbers simple, let us accelerate a bit faster ($a = 0.204g$), which results in a starting time of 1 s.

$$T_1 = \text{starting time at } a = 0.204g = v/a = 2/2 = 1.0\,\text{s}$$

$$D_1 = \text{distance traveled in } 1\,\text{s} = 1.0\,\text{m}$$

Table 3.1

Elevator Load Cycle

From (Floor)	To	N_P	ΔN_P	Dir	Mode	ΔW (kJ)	ΔLoss (10% of $\lvert\Delta W\rvert$) (kJ)
1	2	12	9	up	MF	26.48	2.648
2	4	11	8	up	MF	47.07	4.707
4	5	7	4	up	MF	11.77	1.177
5	7	3	0	up	—	0	0
7	10	2	−1	up	GF	−8.826	0.883
10	7	1	−2	dn	MR	17.65	1.765
7	5	3	0	dn	—	0	0
5	4	7	4	dn	GR	−11.77	1.177
4	2	11	8	dn	GR	−47.07	4.707
2	1	12	9	dn	GR	−26.48	2.648
Totals						8.826	19.71

Note: ΔW = Energy input to the load (change in potential energy), ignoring losses, for the indicated load cycle step.

The same time and distance traveled (D_3, T_3) applies to stopping from 2 m/s at $a = -2\,\text{m/s}^2$. Therefore, we have $D_2 = 4 - (1 + 1) = 2\,\text{m}$ to cover at 2 m/s. The required time traveled (T_2) at 2 m/s is

$$T_2 = (D_2/2) = 1.0\,\text{s}$$

In the starting interval, the motor had to develop enough torque to overcome gravity and accelerate the system at $1.9614\,\text{m/s}^2$. The load torque at $N_P = 12$

$$T_{LOAD} = (9 \times 75)(9.807)(0.4\,\text{m/s}) = 2648\,\text{Nm}$$
$$T_{ACCEL} = (9 \times 75)(9.807)(2\,\text{m/s}^2)(0.4\,\text{m}) = 5296\,\text{Nm}$$
$$T_{TOT} = T_{LOAD} + T_{ACCEL} = 7944\,\text{Nm}$$

Reflecting the torques to the motor shaft, and neglecting T_{RL}

$$T_{LOAD} = 66.2\,\text{Nm} \qquad T_{ACCEL} = 132.4\,\text{Nm}$$
$$T_{DEV} = 198.6\,\text{Nm}$$

In the interval T_2

$$T_{LOAD} = 66.2\,\text{Nm} \qquad T_{ACCEL} = 0\,\text{Nm}$$
$$T_{DEV} = 66.2\,\text{Nm}$$

In the interval T_3

$$T_{LOAD} = 66.2\,\text{Nm} \qquad T_{ACCEL} = -132.4\,\text{Nm}$$
$$T_{DEV} = -66.2\,\text{Nm}$$

We define the first load step duration to begin with an empty elevator on the first floor, with 12 persons to board, and ends when we arrive at the second floor, after the door has opened. Hence,

$$D_1 = 12 \times 2 + 2.2 + 1.0 + 3.0 + 2.2 = 32.4\,\text{s}$$

We assume mechanical locks automatically engage when we are stopped at each floor. The step one results are summarized in Figure 3.10. To evaluate the design, we continue the analysis over the full load cycle. Clearly, this is a complicated situation. We will use these results later as we consider the details of the motor and electrical source.

It is interesting to consider the energy required from the system over one load cycle. To raise one person one floor (4 m)

$$\text{change in potential energy (per person)} = mgh = 75 \times 9.807 \times 4 = 2.942\,\text{kJ}$$

Note that three passengers put the system in perfect balance. Therefore, only passengers in excess of three will change the potential energy.

$$\Delta N_P = N_P - 3$$

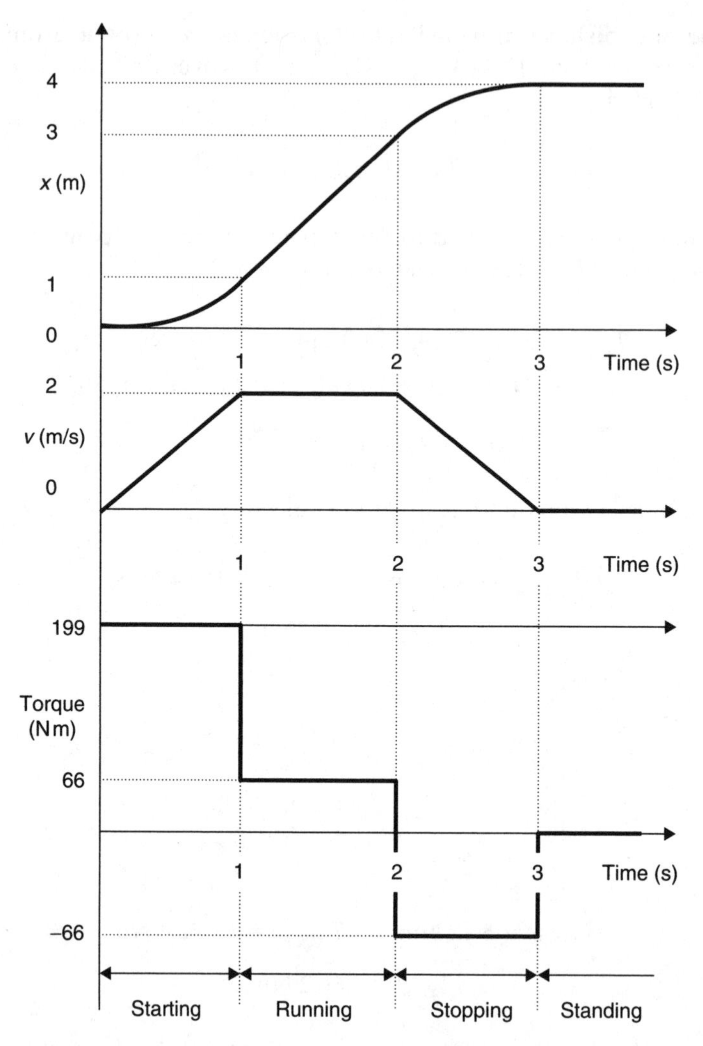

FIGURE 3.10. Step 1 of elevator load cycle.

The change in potential energy (ΔW) for each load step. We note several points from examination of the data. The application requires operation in all four modes, with full reversing. ΔW was sometimes negative, indicating energy recovery when operating in the generator mode. If all passengers made one round trip from the first floor to any other floor, ΔW would have been zero, since the system is conservative. Since this did not happen, some passengers must have used the stairs and departed from a floor other than that at which they arrived (ΔW could have actually been negative). ΔLoss was always positive. System losses are incurred in all operating modes.

The energy ΔW (and ΔLoss) considers only steady-state operation. When the system starts, we must transfer a kinetic energy of $0.5M(v_x)^2$ (or $0.5J(\omega_m)^2$, from the motor shaft perspective) into the moving parts. We are not yet in a position to compute this dynamic loss (we will be before our study is complete), but it is considerable, and in fact exceeds the steady-state losses.

3.9 Prime Movers

To this point, we have focused on motor mode operation. EM machines are also commonly used as generators. Considering generator mode operation, and using generator sign convention, Equation 3.1 is rewritten as

$$J\frac{d\omega_{rm}}{dt} = T_m - (T_{dev} - T_{RL}) \tag{3.17}$$

where

T_{dev} = EM torque, produced by the generator, in N·m
T_m = torque applied by the external mechanical driver, or the "prime mover," in N·m
T_{dev} = EM torque, produced by the generator, in N·m
T_{RL} = rotational loss torque, internal to the machine, in N·m
$T'_m = T_m + T_{RL}$ = equivalent load torque, in N·m
J = mass polar moment of inertia of all rotating parts, in kg·m^2
ω_{rm} = angular velocity of rotating parts, in rad/s.

Note that the role of "source" and "load" has reversed, the former being the mechanical driving device, and the latter being the electrical termination. In Figure 3.1b is the flow of energy through the system shown now from right to left, or from mechanical to electrical. Since the mechanical driving device is the cause of rotation, it is natural to use the descriptive term "prime mover." A particularly important class of prime mover is the turbine, of which there are various types.

3.9.1 Hydraulic Turbines

At hydroelectric generation plants, water flowing from an upper reservoir (pool) is routed through a hydraulic turbine into a lower pool (see Figure 7.18a for details). The turbine design is a function of the "head" (H = head = the vertical distance through which the water drops). The head determines the turbine speed for optimum efficiency. Modern turbine designs can produce efficiencies as high as the low 90s.

Low head plants ($H < 50$ m)
Medium head plants ($50 < H < 250$ m)
High head plants ($H > 250$ m)

In general, the higher the head, the higher the speed. The generator design is chosen to match the turbine speed. A typical value for a 300 m application is about 277 rpm, which corresponds to a 26-pole generator at 60 Hz.

The hydro power available is

$$P = \frac{dW}{dt} = \frac{d(MgH)}{dt} = (gH)\frac{dM}{dt} = (gH\rho)\frac{d(vol)}{dt} \tag{3.18}$$

Thus, for a 300 m head and 100 MW application,

$$\frac{d(vol)}{dt} = \frac{P}{gH\rho} = \frac{10^8}{(9.807)(300)(1000)} = 34\ m^3/s$$

Clearly, the lower the head, the greater the volume of water flow for a given power level.

3.9.2 Steam Turbines

The most common method for generating bulk electrical energy is to use a thermal source to generate steam, which passes through a steam turbine, converting the steam thermal energy into mechanical form, in terms of a torque (T_m) applied to the rotating EM generator shaft (see Figure 7.18b for details). We need a source of heat to produce the steam. The main options are

- Fossil fuel technologies
  - coal-fired boilers
  - oil-fired boilers
  - natural gas-fired boilers
- Nuclear fission
- Geothermal steam generation
- Solar-thermal steam generation.

The efficiencies of such processes are limited by the highest feasible operating temperatures, which in turn are limited by the materials that must operate at these temperatures. State-of-the-art technologies permit maximum thermal efficiencies in the low 40s. Typical heating values of some fossil fuels are

Coal	14,000 Btu/lb$_m$	= 32 MJ/kg
Oil	19,000 Btu/lb$_m$	= 44 MJ/kg
Natural gas	1,000 Btu/scf	= 37 MJ/m^3

Thus, to provide 100 MW electric, a 40% efficient coal-fired plant would consume fuel at about

$$\text{Combustion rate} = \frac{100}{(32)(0.4)} = 8\ \text{kg/s}$$
$$= 28\ \text{metric tons/h} = 675\ \text{metric tons/day}$$

Observe that in a nuclear reaction the amount of matter converted into energy to provide 100 MW electric per day, assuming 40% efficient thermal cycle efficiency, would be

$$m = \frac{(100 \times 10^6\ \text{W}) \times (3600\ \text{s/h}) \times (24\ \text{h/day})}{(3 \times 10^8)^2 (0.4)}$$
$$= \frac{(3600\ \text{s/h}) \times (24\ \text{h/day})}{(9 \times 10)^8 (0.4)} = 0.00024\ \text{kg/day} = 0.24\ \text{g/day}$$

3.9.3 Gas Turbines

A smaller, but more versatile, technology for generating electrical energy is to use a gas turbine to directly drive the generator. The advantages are a simpler overall system, higher efficiencies, and less space required, permitting location nearer load centers, and in developed areas. Disadvantages include public opposition to locating facilities in highly populated areas, and logistical problems in terms of fuel and maintenance.

A common practice used in describing thermal efficiency is the concept of "heat rate"

$$\text{Heat rate} = \frac{\text{thermal power input (Btu/h)}}{\text{electric power output (kW)}} = \frac{3412}{\eta} \tag{3.19}$$

where η is the efficiency.

Therefore, a heat rate of 10,000 Btu/kW h corresponds to

$$\text{Efficiency} = \eta = \frac{3412\ \text{Btu/kW h}}{10{,}000\ \text{Btu/kW h}} = 0.3412 = 34.12\%$$

Note that the higher the heat rate, the lower the efficiency.

3.9.4 Wind Turbines

Since early recorded history, people have used wind energy for boat propulsion, grinding grain, and pumping water. As early as the late 19th century, wind energy was considered for electric power generation. The technology gradually waned until the 1970s when once again it became cost-effective during the oil embargo of the 1970s. At present, wind energy is already one of the most cost-competitive renewable energy technologies at less than 5 cents per kilowatt-hour, and wind energy production is the fastest-growing energy technology in the world, with annual markets in excess of $1.5 billion. The technology is environmentally friendly, with no undesirable air polluting emissions. However, noise, aesthetic (visual) impact, hazards to flying animals, and land usage, remain as significant environmental impact issues.

Wind is moving air, and is caused by irregular solar heating of the Earth's surface, which in turn is caused by the Earth's terrain, bodies of water, man-made structures, and vegetative cover. The mass flow rate of air through a cross-sectional area A is

$$\frac{dm}{dt} = Au\rho = \text{mass flow rate of air, in kg/s} \tag{3.20}$$

where

A = cross-section of wind intercepted by the turbine, m^2 [7]
u = wind velocity, perpendicular to A, m/s
ρ = density of air = 1.2 kg/m^3, as an estimated typical value.

The wind power (Table 3.2) is

$$P_W = \frac{dW}{dt} = \frac{1}{2}u^2\frac{dm}{dt} = \tfrac{1}{2}A\rho u^3,\ \text{in W} \tag{3.21}$$

However, natural wind does not blow at constant speed. At a given site, the statistical distribution of wind speeds must be determined over time. Such studies have been made and data are available.[8]

Wind turbines can capture only a fraction of this energy, depending on their design. The output wind turbine power is therefore

$$P_{WT} = \eta P_W = \tfrac{1}{2}\eta A\rho u^3,\ \text{in W}$$

where η is the turbine efficiency.

---

[7] Air density depends on temperature, pressure, and humidity. ρ = 1.225 kg/m^3 at 15°C, 760 mmHg, 0.013 kg/m^3 water content.

[8] See, for example, D. L. Elliott et al. *Wind Energy Resource Atlas of the United States* (Solar Energy Research Institute, 1986) http://rredc.nrel.gov/wind/pubs/atlas/titlepg.html (accessed August 16, 2006).

TABLE 3.2
Power Density for Some Typical Wind Speeds

Wind Speed			
Knots[a]	m/s	Description	Power Density (W/m^2)
1	0.5	Calm	0.08
3	1.5	Light	2.1
15	7.6	Moderate	265
20	10.2	Fresh	629
25	12.7	Strong	1229
40	20.3	Gale	5034
70	35.6	Hurricane	26,980

[a] knot = nautical mile per hour = 0.5080 m/s.

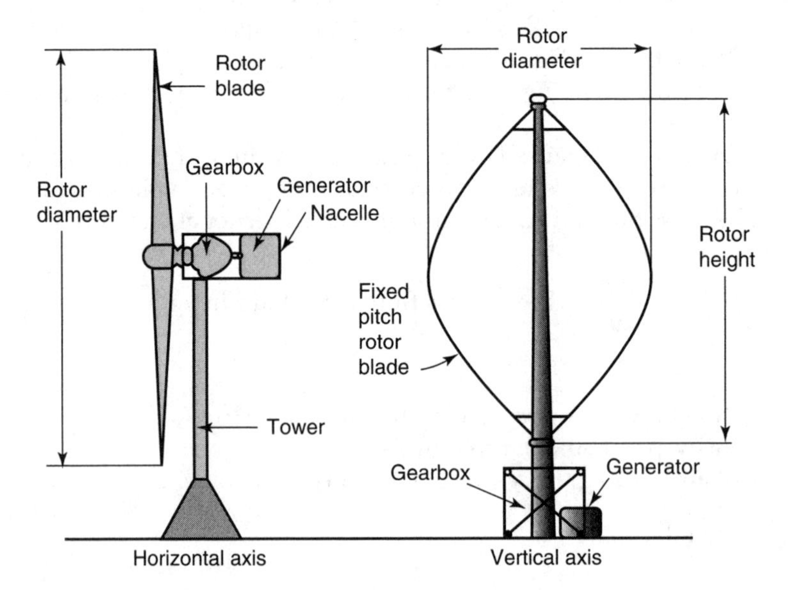

FIGURE 3.11. Wind turbine configurations. Used by permission of the American Wind Energy Association.

The turbine efficiency η depends on turbine design, and is a function of wind speed, with a theoretical maximum value of just below 60%.[9] Modern wind turbines have one of the two basic designs: the horizontal and the vertical axis design, shown in Figure 3.11.

Horizontal-axis wind turbines are most common because of lower costs and somewhat higher efficiency. They are equipped with yaw control, so that they are always facing the wind direction. Vertical-axis wind turbines are almost exclusively of the Darrieus type.[10] Its basic advantages are

- the turbine is omnidirectional, eliminating the need for yaw control;
- the generator, and ancillary equipment, may be at ground level.

[9] The German physicist Albert Betz proved that the theoretical upper limit was 16/27 = 0.5926 in his book on wind energy in 1926, a value which has become known as the "Betz limit."

[10] Named after the French engineer Georges Darrieus, who patented the design in 1931.

Disadvantages of vertical axis wind turbines include

- efficiencies are lower;
- there may be a significant wind velocity gradient over the length of the blades, with low wind speeds near the ground;
- the turbine is not self-starting, developing no torque at zero rotational speed;
- the tower normally must be guyed;
- replacing the main rotor bearing requires disassembling the entire unit.

Modern wind turbine design exploits advances in materials, engineering, electronics, and aerodynamics. The output turbine power must be used immediately or stored. For electric power grid-connected installations, the power is injected into grid, and may be viewed as a negative load, and accommodated as part of the grid operational control problem. For stand-alone installations, the system requires an energy storage facility, such as a battery bank.

Example 3.6
A wind turbine whose horizontal axis is located at 50 m above ground wind turbine is located at remote site at which a constant prevailing wind of 22 knots blows for 15 h/day. The turbine has three blades turning at 20 rpm, with an efficiency of 50%. The turbine drives an 1800 rpm generator through a gear box, which have efficiencies of 92% and 97%. The generator is connected to the local utility grid.

(a) Determine an appropriate turbine rating.
(b) Determine the torques on the turbine and generator shafts.
(c) Assume that the energy may be sold to the utility at 0.02$/kW h. Determine the net annual revenue, after deducting an annual allotment of 10k$ for maintenance and repair.

SOLUTION

(a) 22 knots = 11.18 m/s.

$$P_{WT} = \eta P_W = \frac{1}{2}\eta A\rho u^3 = 0.5(0.5)(\pi(20)^2)(1.2)(11.18)^3 = 526.8\,\text{kW}$$

A reasonable turbine rating would be 600 kW.

(b) 20 rpm = 2.094 rad/s.

$$T_m = \frac{P_m}{\omega_{rm}} = 251.5\,\text{kN m} \quad \text{(on the turbine shaft)}$$

$$T_g = \frac{T_m}{1800/20}(0.97) = 2.711\,\text{kN m} \quad \text{(on the generator shaft)}$$

(c) $P_{OUT} = (0.92)(0.97)(526.8) = 470.1\,\text{kW}$
$W = P_{OUT} \times 15 \times 365 = 2574\,\text{MW} - \text{h}$
gross annual revenue $= W \times 20 = 51.48$ k$
net annual revenue $= 51.48 - 10 = 41.48$ k$

3.9.5 M–G Sets

An EM motor can be used as a prime mover for an EM generator. Such applications are normally for

- AC to DC conversion
- DC to AC conversion
- DC to DC conversion (different voltage levels)
- AC to AC conversion (different voltage levels and/or frequencies).

Although such applications were common in the past, they are rare today because the cost of using two machines, and the widespread availability of smaller, cheaper, and more reliable solid-state converter circuits.

3.9.6 Standalone Emergency Power Supplies

Emergency Power Supplies (or "EPS" systems) are standalone units consisting of a prime mover, normally a diesel or gasoline reciprocating engine, coupled to an electric generator. Such units range in size from <1 kW to hundreds of kW. The smaller units are for individual customers and special applications, whereas the larger units serve to supply standby power for commercial customers, such as hospitals, TV/Radio stations, and telecommunication centers.

Example 3.7

A 5 kW EPS is designed for residential use, and is driven by a 4-cycle 10 hp gasoline engine, with a fuel tank capacity of 8 gallons. Assume that the energy content of gasoline is about 20,000 Btu/lb_m. If the engine efficiency (thermal to mechanical) is 10%, and the generator efficiency (mechanical to electrical) is 85%, how long can the EPS run, delivering (a) 5 kW; (b) 1 kW, without refueling. (c) If gasoline costs $1.50 per gallon, what was the electricity cost?

SOLUTION

$$W = 20{,}000(1055\,\text{J/Btu}) = 21.1\,\text{MJ/lb}_m$$

A gallon weighs about 8 lbs. Therefore, the fuel energy = 1350 MJ = 1350 MW sec = 375.1 kW h at

- 10% efficiency, the engine mechanical energy output is 37.51 kW h
- 85% efficiency, the generator energy output is 31.88 kW h
- 5 kW output, run time is 6.377 h
- 1 kW output, run time is 31.88 h
- The fuel cost for a full tank was $8 \times 1.50 = \$12.00$. The energy cost was therefore $12/31.88 = \$0.38$ per kW h.

3.9.7 Vehicular Electrical Systems

Automobiles, aircraft, boats, and other vehicles have onboard electrical systems, needed to power lights, onboard electronics, and mechanical systems, such as power windows and door locks. For such applications, the prime mover is normally the engine, which provides the thrust to propel the vehicle.

Example 3.8

An automobile travels along an interstate highway at 70 mph, such that its engine turns at 2100 rpm, developing 35 hp. A belt-driven alternator (AC generator) supplies a rectifier, which in turn, supplies a 12 V DC electrical system. The 12 V DC system delivers 50 A. Assuming that the alternator–rectifier is 85% efficient, determine (a) the fraction of engine power taken by the electrical system. (b) Given that the engine and alternator pulley diameters are 20 and 10 cm, respectively, determine the speed and torque of the alternator shaft.

SOLUTION

(a) 35 hp = 26.11 kW
$P_{OUT} = 12 \times 50 = 600\,\text{W} = 0.60\,\text{kW}$
$P_{IN} = 0.7059\,\text{kW}$, or 2.7%

(b) $$\text{Alternator speed} = \frac{20}{10}(2100) = 4200\,\text{rpm} = 439.8\,\text{rad/s}$$

$$T_{IN} = \frac{705.9}{439.8} = 1.605\,\text{N m}$$

3.10 Summary

The principle subject of this book is the study of EM machines. Therefore, we must concern ourselves with the mechanical attributes of such devices. We noted that such systems can operate in four modes: MF, MR, GF, and GR. Machines operating in the motor and generator modes are called "motors" and "generators," respectively. We have considered several mechanical concepts, including force, torque, power, rotational and translational speed, inertia, and energy. Also, since the purpose of EM machines is to convert energy from, or into, mechanical form, we must consider what is on "the other end" of the shaft.

To analyze motor performance, one must consider the load. Although there are hundreds of load applications, it is possible to generalize on many of the most important issues, such as torque, power, speed, and inertia, and variation of load with speed. We have considered constant torque loads, and load that vary with speed. We have studied the elevator application in some detail, including steady state and dynamic performance.

For generators, we must consider prime movers. We examined turbines, including hydraulic, steam, gas, and wind-driven designs. We briefly noted M–G sets, EPS, UPS, and onboard vehicular systems.

We are now prepared to begin our study of the EM machine in detail. There are many types of EM machines and we will discuss the most important. As we proceed, we will need to apply these mechanical and magnetic concepts to understand these remarkable devices.[11]

---

[11] We are most fortunate indeed. We deal with truly remarkable devices, devices which bring together electrical, magnetic, mechanical, and thermal phenomena, mixed into a delicious intellectual stew. It is no wonder that EM generators intrigued Albert Einstein in his youth.

Problems

3.1 Consider a constant torque motor (200 N m) driving a linear torque load defined by

$$T_m = 2\omega_{rm}; \quad 0 < \omega_{rm} < 150\,\text{rad/s}$$
$$J = 2.5\,\text{kg m}^2$$

(a) Accurately sketch the motor and load torque–speed characteristics.
(b) Clearly mark thereon the system steady-state operating speed and torque point.
(c) Solve for the system steady-state operating speed and torque.

3.2 Consider the system of Problem 3.1 at zero speed.

(a) Will the system accelerate, decelerate, or remain at zero speed?
(b) Determine, and sketch, the starting speed transient.
(c) Estimate the starting time.

3.3 Consider the system of Problem 3.1 running at 100 rad/s, when the motor is disconnected, dropping the motor torque to zero.

(a) Will the system accelerate, decelerate, or remain at 100 rad/s?
(b) Determine, and sketch, the stopping speed transient.
(c) Estimate the stopping time.

3.4 Consider a motor, whose torque–speed characteristics are defined by

$$T_{dev} = 300\,\text{N·m}; \quad 0 < \omega_{rm} < 100\,\text{rad/s}$$
$$T_{dev} = 300 - 6(\omega_{rm} - 100); \quad 100 < \omega_{rm} < 150\,\text{rad/s}$$
$$J = 2.5\,\text{kg m}^2$$

Accurately sketch the motor and load torque–speed characteristics, and clearly mark thereon the system steady state operating speed and torque (provide values) for

(a) a constant torque load.
(b) a linear torque load.
(c) a parabolic torque load.

All loads require 200 N m to run at 150 rad/s.

3.5 For the linear torque load case described in Problem 3.4, determine the system starting speed transient, starting from zero speed. Estimate the starting time.

3.6 For the linear torque load case described in Problem 3.4, starting from the run speed determined in Problem 3.2, suppose the motor is suddenly turned off, forcing the motor torque to zero. Determine the system stopping speed transient. Estimate the stopping time.

3.7 A motor supplies a constant 15 hp to a mechanical load for one 30-day month, for which the electric utility bill is $420. The energy rate is $0.05 per kW h.

(a) Find the motor efficiency.
(b) If the motor is replaced with one with an efficiency of 93%, determined the monthly savings.

3.8 A fan requires a mechanical input power of 500 W at 700 rpm.

(a) Find the input torque (N·m).
(b) Find the input power and torque at 1000 rpm.

3.9 A pump requires a mechanical input torque of 5 N m at zero speed, and 30 N m at 500 rpm. Find the input power (W) at

(a) 0 rpm.
(b) 500 rpm.
(c) 800 rpm.

3.10 An application in a paper mill requires a motor to drive a take-up reel, whose radius varies from 40 cm (empty) to 200 cm (full). A requirement of the application is that the paper must be wound on the reel at a constant tangential velocity of 2 m/s and under a uniform tension of 200 N. Determine the corresponding torque, power, and angular velocity of the reel when the effective radius (reel plus paper) is

(a) 40 cm.
(b) 100 cm.
(c) 200 cm.

3.11 Derive M and J for a uniform cylinder of length l, radius r, and mass density ρ.

3.12 Derive M and J for an annular uniform cylinder of length l, inner radius r_1, outer radius r_2, and mass density ρ.

3.13 A given load has a value of $Wk^2 = 150\,\text{lb ft}^2$. Find J in kg m^2.

3.14 A given load has $J = 5\,\text{kg m}^2$. Find Wk^2.

3.15 Given a uniform steel cylinder of radius 20 cm and length 40 cm. The specific gravity of steel is 8. Find J.

3.16 Continuing Problem 3.15, consider the volume of steel to be constant. If the radius is doubled, find J.

3.17 Consider a carousel (merry-go-round). The platform structure may be approximated by a uniform disk of thickness 50 cm, $\rho = 800\,\text{kg/m}^3$, and diameter 12 m. There are two rings of equally spaced fiber glass horses (outer ring: 16 horses, located at radius 5 m; inner ring: 12 horses, located at radius 3 m). Each horse weighs 25 kg, and is mounted by a rider that weighs 75 kg. Each horse + rider may be considered to be a point mass. Calculate J.

3.18 The carousel of Problem 3.17 ride turns at a speed of 4 rpm and is to be driven by a 1800 rpm motor. Find the gear ratio.

3.19 Calculate J for the carousel of Problem 3.17 as seen from the motor shaft.

3.20 Consider the elevator of Figure 3.9. Define the operating mode for the EM machine if the elevator operates as indicated:

Case	Direction	Passengers
1	Up	Empty
2	Down	Empty
3	Up	10
4	Down	10

3.21 In Case 3, Problem 3.20, the elevator travels nonstop from level 1 to level 10. Plot graphs of the motion, as shown in Figure 3.10, for this case.

3.22 In Problem 3.21, how much energy was required for the case considered?

3.23 What is the rate of coal consumption (tons/h) in a coal-fired power plant of 35% efficiency to produce 500 MW electric?

3.24 Suppose the plant of Problem 3.23 operates for a year at constant output. How many tons of coal were burned?

3.25 Determine the volumetric flow rate of water in m^3/sec required in a hydroelectric installation to produce an output of 80 MW. The plant efficiency is 80% and the head is 50 m.

3.26 In a nuclear power plant, how much mass must be converted into energy daily to produce an electric output of 1000 MW? Assume the thermal to electric efficiency is 35%.

3.27 A wind turbine, whose horizontal axis is located at 50 m above ground wind turbine, is located at remote site at which a constant prevailing wind of 25 knots blows for 12 h/day. The turbine has three blades turning at 20 rpm with an efficiency of 50%. The turbine drives an 1800 rpm generator through a gear box, which have efficiencies of 90% and 96%. The generator is connected to the local utility grid.

(a) Determine the torques on the turbine and generator shafts.
(b) Assume that the energy may be sold to the utility at 0.025$ per kW h. Determine the net annual revenue, after deducting an annual allotment of 10k$ for maintenance and repair.

3.28 An M–G set is employed to provide a 250 V DC source from a three-phase 208/120 V, 60 Hz AC source. If each machine has an efficiency of 91%, what is the overall efficiency?

3.29 A 5 kW EPS is designed for residential use, and is driven by a 4-cycle 10 hp gasoline engine, with a fuel tank capacity of 10 gallons. Assume that the energy content of gasoline is about 20,000 Btu/lb_m. If the engine efficiency (thermal to mechanical) is 20%, and the generator efficiency (mechanical to electrical) is 82%, how long can the EPS run, delivering (a) 5 kW. (b) 2 kW, without refueling. (c) If gasoline costs $3.00 per gallon, what was electricity cost?

3.30 An automobile travels along an interstate highway, such that its engine turns at 2100 rpm, developing 38 hp. A belt-driven alternator (AC generator) supplies a rectifier, which supplies the 12 V DC electrical system. Assuming that the alternator–rectifier is 85% efficient, the alternator input is 2 hp, and the engine and alternator pulley diameters are 20 and 10 cm, respectively, determine the speed and torque on the alternator shaft, and the current delivered to the 12 V DC electrical system.

4

The Polyphase Induction Machine: Balanced Operation

As noted in Chapter 3, the basic purpose of EM machines is to convert energy from electrical into mechanical form, or vice versa. We now turn our attention to practical devices designed for this purpose. There are literally hundreds of possibilities, so part of our problem is to decide just where to begin our study. Recall from Chapter 1 that such devices have two basic parts: the stator and rotor.[1] A large and important subset of all EM devices are those which transfer energy between stator and rotor by induction processes, and are therefore logically called "induction machines." We start our study by examining the details of induction machine construction.

4.1 Machine Construction

It is not our objective to provide comprehensive material on how to design EM machines. However, to understand the principles of operation and control, and to provide the fundamentals for deriving valid mathematical models, one does need a basic understanding of how machines are physically constructed. A cutaway diagram of an induction machine is shown in Figure 4.1a.

4.1.1 Stator Design

Examine the structure of Figure 4.1b. It is cylindrical (annular in cross-section, as shown in Figure 4.1c) and composed of laminated ferromagnetic material with equally spaced slots positioned around the inner "hole," parallel to the machine axis. Insulated structures called "coils" will be placed in the slots, and interconnected to form "windings." A coil typically has a few turns (N_{TURN}) of insulated conductor, rectangular in cross-section, which are shaped in an irregular hexagon shape and wrapped into one assembly, as shown in Figure 4.1d and Figure 4.1e. The coil parts that are placed in the slots are called "coil sides;" the parts outside the slots are called "end (front and back) connections." It is clear that there are then two coil sides per coil. Likewise, there are N_{TURN} conductors per coil side (three, as shown in the

[1] For translational machines, the moveable part is the translator. Since the large majority of EM machines are rotational, we concentrate on these devices.

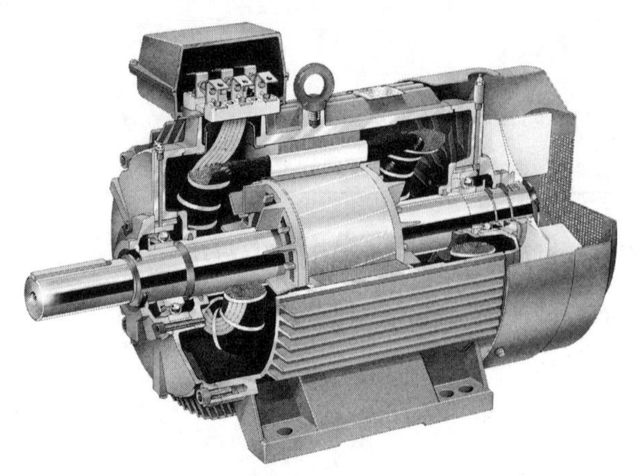

(a) Cutaway diagram of an induction machine. (Courtesy of Baldor Corporation.)

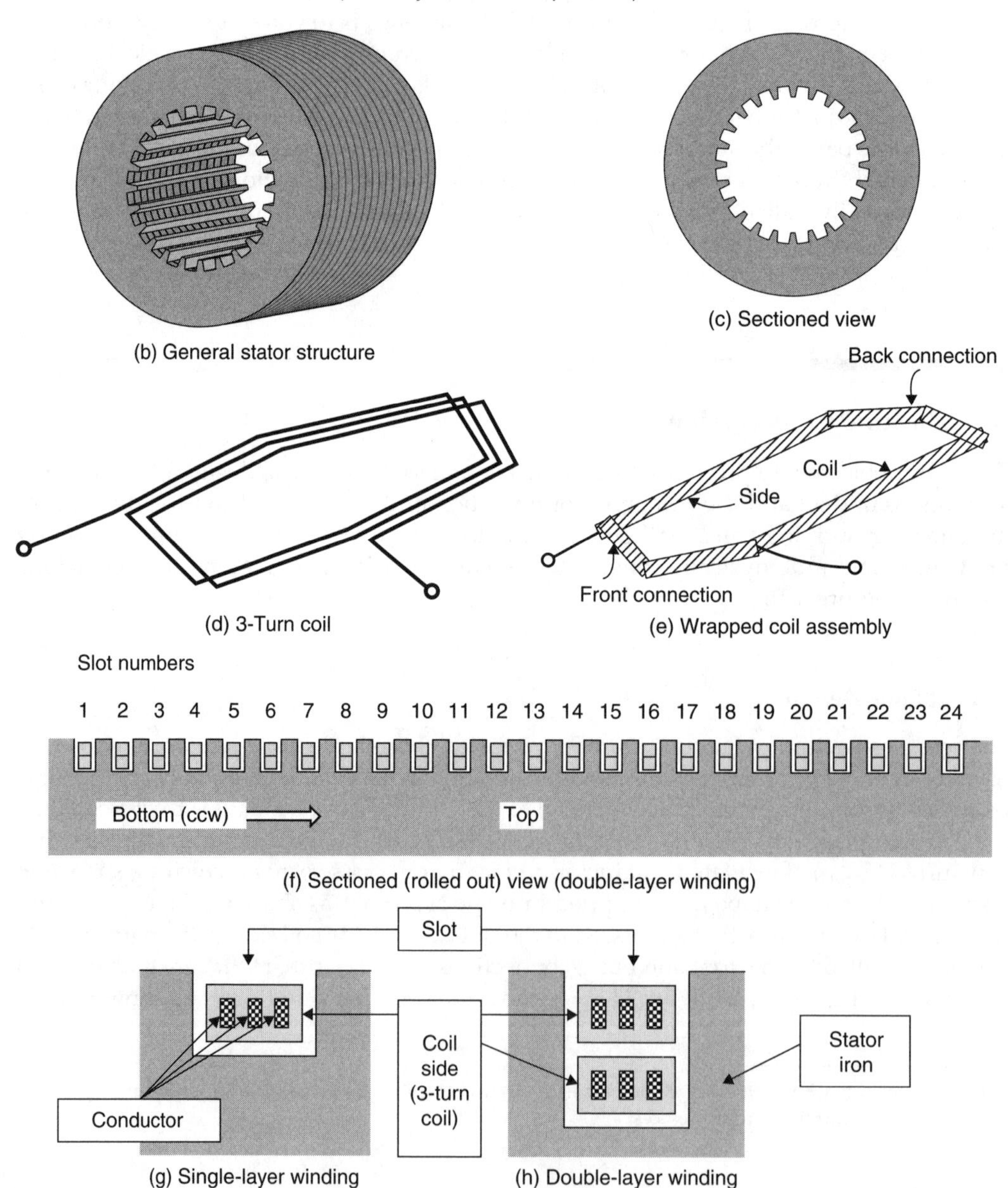

(b) General stator structure

(c) Sectioned view

(d) 3-Turn coil

(e) Wrapped coil assembly

(f) Sectioned (rolled out) view (double-layer winding)

(g) Single-layer winding

(h) Double-layer winding

FIGURE 4.1. Polyphase induction machine stator construction.

example of Figure 4.1). It follows that between coil terminals there are $2 \times N_{TURN}$ (or 6) conductors that are automatically electrically in series.

It is more convenient to represent the stator in "sectioned, rolled out" form: imagine making an axially parallel cut through the stator wall, and "rolling out," or flattening, the structure as shown in Figure 4.1f. To understand how coil sides are placed in the slots, it will be necessary to number the coil sides and the slots. We will demonstrate the necessary insights in the context of a 24-slot example: extension to any number of slots is straightforward.

There are two basic arrangements: single layer (one coil side per slot), and double layer (two coil sides per slot) (see Figure 4.1g and Figure 4.1h). In the double-layer case, it is clear that the number of coils is the same as the number of slots, since each coil has two coil sides: $N_{COIL} = N_{SLOT}$. The slot location of the coil sides, and coil interconnections, are keys to understanding how the stator functions. We will deal with this issue, but first consider the rotor construction.

4.1.2 Rotor Design

Examine the structure in Figure 4.2a. It is cylindrical (circular in cross-section, as shown in Figure 4.2b), and composed of either laminated of solid ferromagnetic material. This construction is the so-called cage rotor design, where conducting bars are placed into slots, and connected on each end with conducting circular rings (called "end rings"). The slots may be parallel, or skewed, to the rotor axis, the latter arrangement used to reduce machine acoustical noise. Note that no external electrical access into the electrical conducting paths

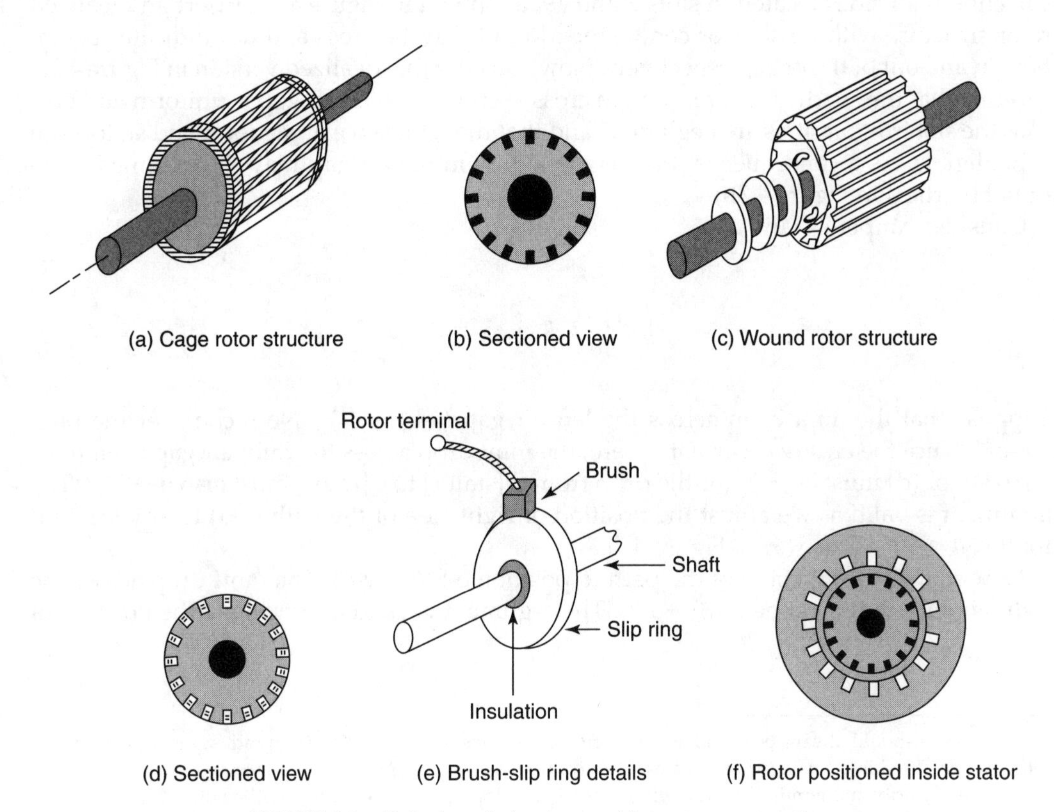

FIGURE 4.2. Polyphase induction machine rotor construction.

is possible. The details (material, shape, etc.) of the rotor bar design are important since they affect the performance of the device.

The second arrangement is the so-called wound rotor design. Examine the structure in Figure 4.2c. It is cylindrical (circular in cross-section, as shown in Figure 4.2d), and composed of either laminated of solid ferromagnetic material. Insulated coils are placed into slots, and interconnected in patterns that form a replica of the stator windings, connected on each end with conducting circular rings (called "end rings"). The bars may be parallel, or skewed, to the rotor axis, the latter arrangement used to reduce acoustical noise in the machine. External electrical access into the electrical conducting paths is possible through terminals connected to brushes which make sliding contact with slip rings (conducting rings which are insulated from the rotor structure and provide termination points for the rotor windings). "Brushes" are copper-impregnated blocks of carbon, providing a low-friction sliding electrical contact with the slip rings (see Figure 4.2e for clarification).

The rotor is placed in the stator "hole," supported by bearings, free to rotate, and positioned such that the air gap is uniform, as shown in Figure 4.2f. The brushes are stationary and mounted on brush riggings, which are attached to the stator.

4.2 Stator Winding Layout

Locate points on the stator in angular measure (θ) from the center of slot 1, in a ccw direction. Since there are 24 slots, they are 15° apart. Consider a single coil of N turns, carrying current i, with sides located in slots 7 and 19, as shown in Figure 4.3a. Insert an idealized rotor structure with no slots or conductors. Recall that the cross and dot indicate current flow in and out of the page, respectively. Now consider the idealized version in Figure 4.3b. Assume that the conductors are small in cross-section, that the air gap is uniform and narrow, the slot irregularities are negligible, and that the reluctance of the rotor and stator iron is negligible. As a consequence, the magnetic flux in the air gap may be assumed to be radial (vertical in Figure 4.3b).

Consider Ampere's law

$$\oint \hat{H}\, dl = i_{\text{enclosed}} = \mathfrak{F}$$

Suppose that the mmf drop across the left air gap (a-b) is $+\mathfrak{F}_o$. Now consider the path a-b-c-d. Since the enclosed current is zero, the mmf drop across the right air gap from rotor (c) to stator (d) must be $-\mathfrak{F}_o$ (or the drop from d [stator] to c [rotor] must also be $+\mathfrak{F}_o$). The argument is valid as we adjust the position of right side of the path (c-d) to any location for $0 < \theta < 90°$.[2] Plot $\mathfrak{F}(\theta)$ in Figure 4.3c.

Now move the right side of the path to position e-f ($\theta > 90°$). The mmf drop across the right air gap (d-c) must be $(+\mathfrak{F}_o - Ni)$. The argument is valid as we adjust the position of

[2] Technically, θ should always be in radians. Indeed, expressions like "$\omega t + 90°$" (ω in rad/s; t in s) require that "90°" be converted into "$\pi/2$ radians" before the addition is executed. Nonetheless, the use of degrees has the advantage of clarity and familiarity, and will be used here. Hopefully, no confusion will result.

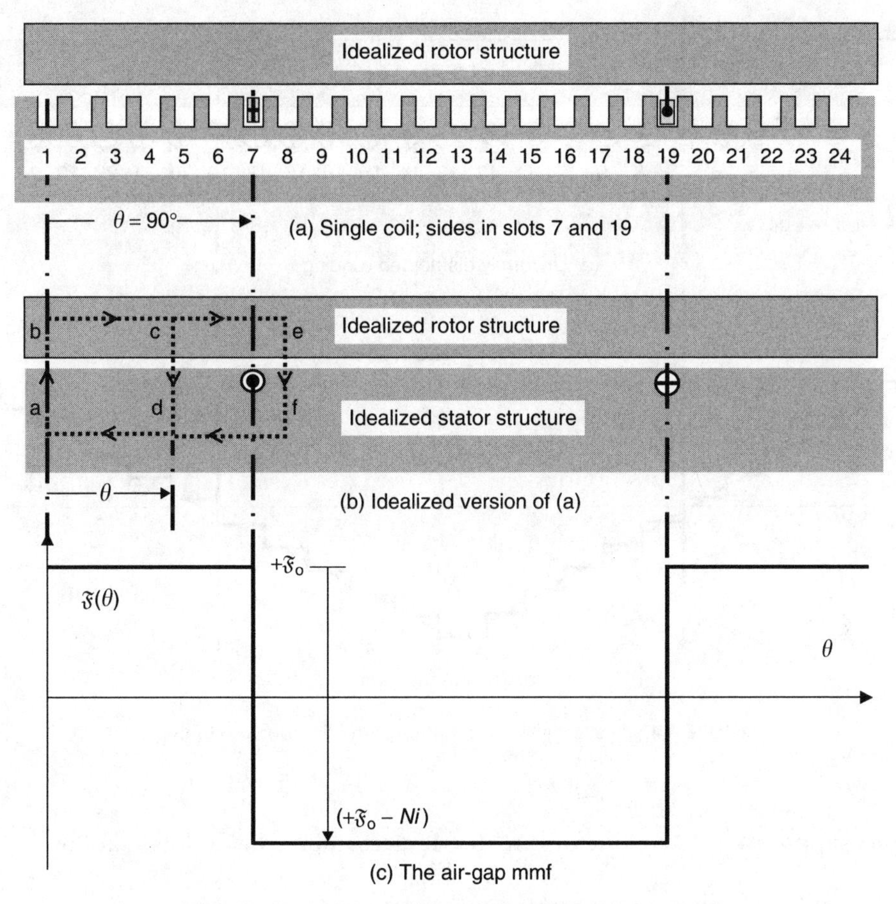

FIGURE 4.3. The air-gap mmf in concentrated windings.

right side of the path (c-d) for $90° < \theta < 270°$, which fixes the air-gap mmf $\mathfrak{F}(\theta)$ at $(+\mathfrak{F}_o - Ni)$. When $270° < \theta < 360°$, the air-gap mmf $\mathfrak{F}(\theta)$ is $(+\mathfrak{F}_o - Ni + Ni) = +\mathfrak{F}_o$.

Integrating over the air-gap surface, Gauss's law requires that:

$$\oint \hat{B} \cdot d\hat{s} = 0$$

Hence, $\mathfrak{F}_o$ must be $0.5Ni$.

The situation just investigated is the so-called concentrated winding case, since the winding consisted of a single coil concentrated in a single slot-pair. Now consider distributing the winding over all the 24 slots, as shown in Figure 4.4a. The mmf waveform is now stepped, as shown in Figure 4.4b. As the number of slots is increased, the granularity of the plot decreases, and becomes triangular in the limit (the dashed plot). This results in the uniformly distributed winding case.

Now if more conductors are placed in some slots than others, the space waveform can be tailored to any particular shape, if enough slots and conductors are available. For

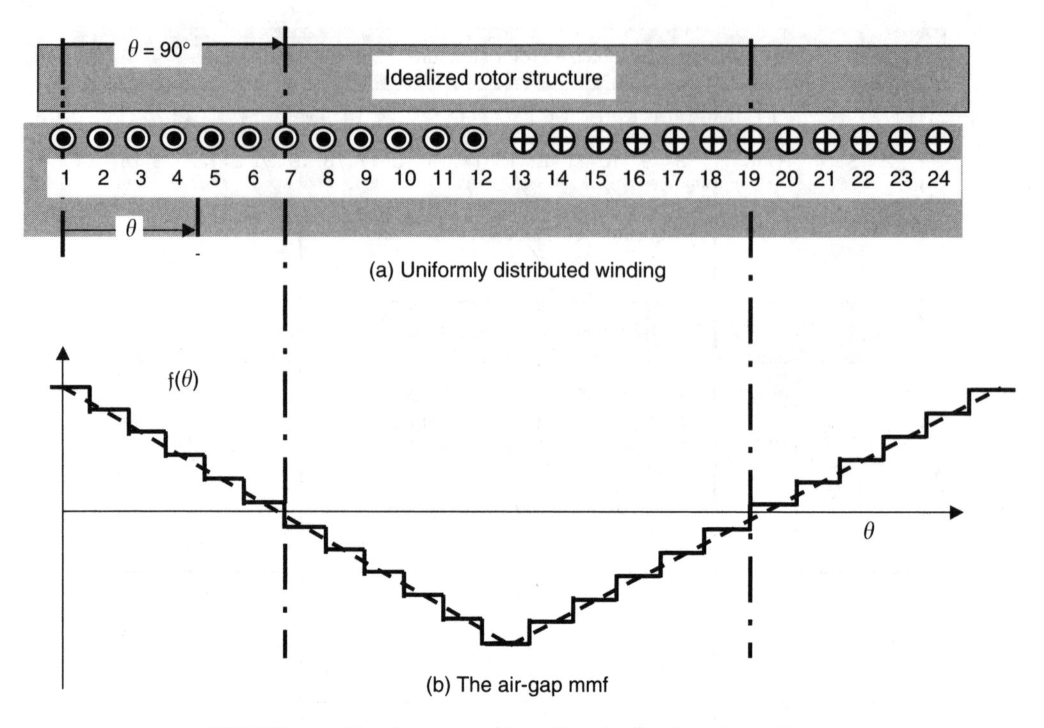

FIGURE 4.4. The air-gap mmf in uniformly distributed windings.

example, suppose 4×10 conductors are distributed among 4×6 slots as follows:

Slot	No. of Conductors
1, 12, 13, 24	1
2, 11, 14, 23	2
3, 10, 15, 22	2
4, 9, 16, 21	3
5, 8, 17, 20	3
6, 7, 18, 19	3

This case is plotted in Figure 4.5. As with any periodic function, it is possible to compute the Fourier series for $\mathfrak{F}(\theta)$:

$$\mathfrak{F}(\theta) = 9.3\cos\theta \;+\; 0.85\cos(3\theta) + \cdots$$

Plotting the fundamental $\mathfrak{F}_1(\theta)$ on the same graph, observe that it is a reasonable approximation.

We can extend the concept with an idealization: the sinusoidally distributed winding. For such a situation, we resort to the same notation as the concentrated case, and rely on the reader's sophistication to understand the difference (see Figure 4.6a). Note that the center of the coil, located halfway between slots 7 and 19, or in the center of slot 13, locates the magnetic axis, the position of maximum field strength. The flux lines flow from stator to rotor in the region among slots 7, 8, ..., 18, 19; therefore the stator surface appears as a

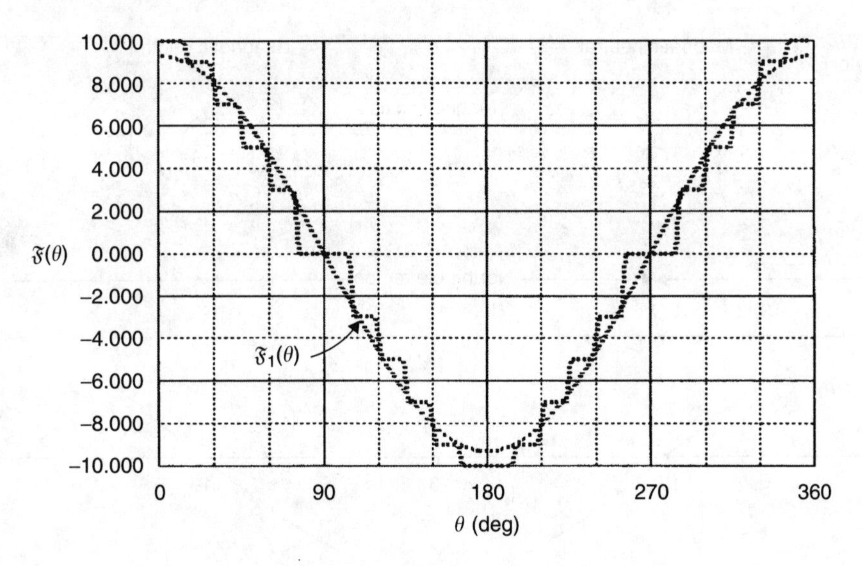

FIGURE 4.5. Normalized $\mathfrak{F}(\theta)$.

north pole. Likewise, the flux returns from rotor to stator in the region among slots 19, 20, …, 6, 7, making this stator surface functionally a south pole. The return flux maximizes at slot 1. The air-gap field strength is zero at slots 7 and 19, locations called "magnetic neutrals."

Now consider installing two sinusoidally distributed series coils, with sides placed in slots 1–7, and 13–19, respectively, shown in Figure 4.6b (slot locations are not shown, but are the same as in Figure 4.4). Note that $\mathfrak{F}(\theta)$ now executes two full cycles as we traverse the full length of the air gap. This inspires the definition of two kinds of angle measure:

θ_m = spatial angular position along the stator, referenced from the center of slot 1 (or the magnetic axis of the coil), ccw positive, in (mechanical) radians or degrees.
θ_e = angular position within $\mathfrak{F}(\theta)$, referenced from the center (or magnetic axis) of the coil, ccw positive, in (electrical) radians or degrees.

There is a fixed relationship between the two angles:

$$\theta_e = (N_P/2)\theta_m \tag{4.1}$$

where N_P is the number of poles formed by the winding (an even integer). In the situation of Figure 4.6c and Figure 4.6d, $\theta_e = 2\theta_m$. The distance between coil sides is defined as the "coil pitch;" "full pitch" is defined as 180 electrical degrees. For our two-pole 24-slot example, full pitch corresponds to

$$24/2 = 12 \text{ slots or } 180 \text{ mechanical degrees}$$

For the four-pole 24-slot example, full pitch corresponds to

$$24/4 = 6 \text{ slots or } 90 \text{ mechanical degrees}$$

Sometimes the coil pitch is expressed in ratio form (actual coil pitch to full pitch). For example, if opposite coil sides were placed in slots 1 and 6 in the four-pole case, the pitch might be described as "5/6."

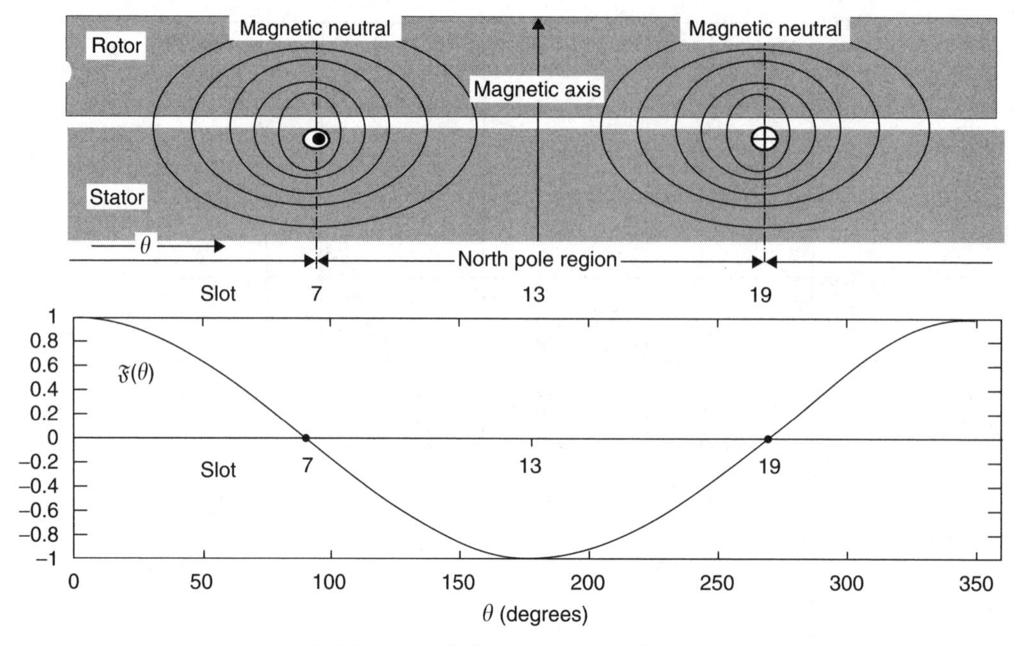

(a) Two-pole winding, showing field pattern

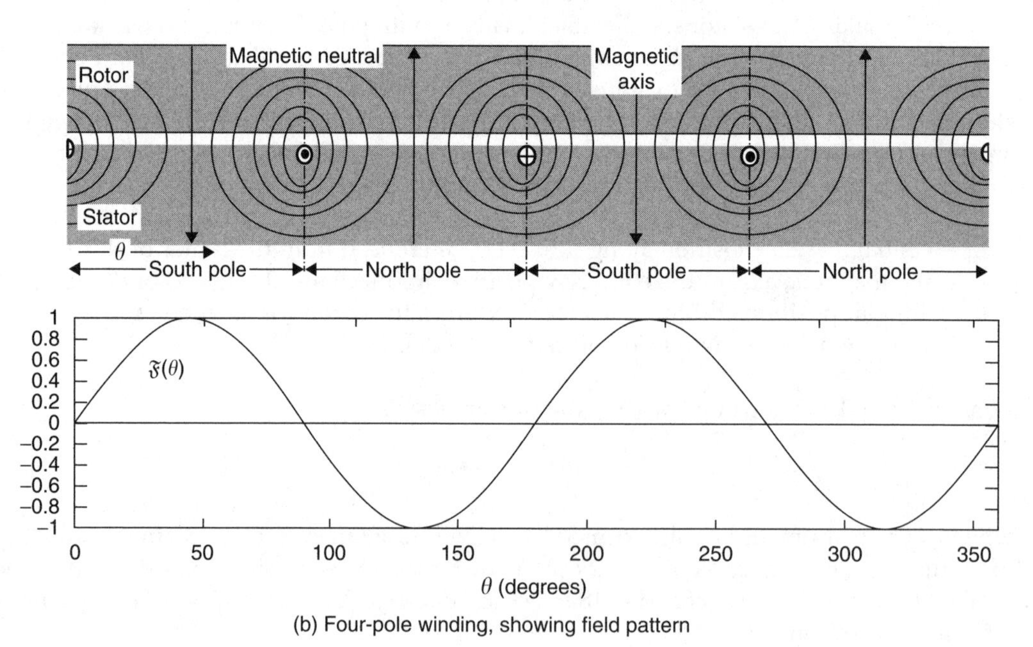

(b) Four-pole winding, showing field pattern

FIGURE 4.6. Sinusoidally distributed air gap field patterns.

4.3 The Rotating Magnetic Field

We will now consider one of the most remarkable phenomena in all of electrical engineering. Consider three sinusoidally distributed windings (a-a′,b-b′,c-c′), positioned so that their magnetic axes are separated 120° as shown in Figure 4.7a, an arrangement which we

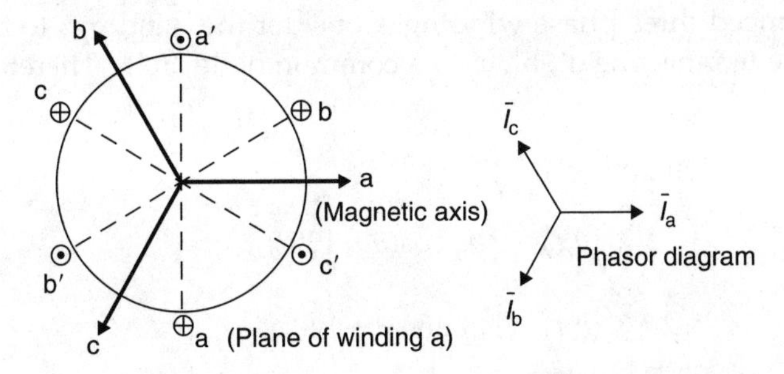

(a) Three sinusoidally distributed windings, spaced displaced 120°

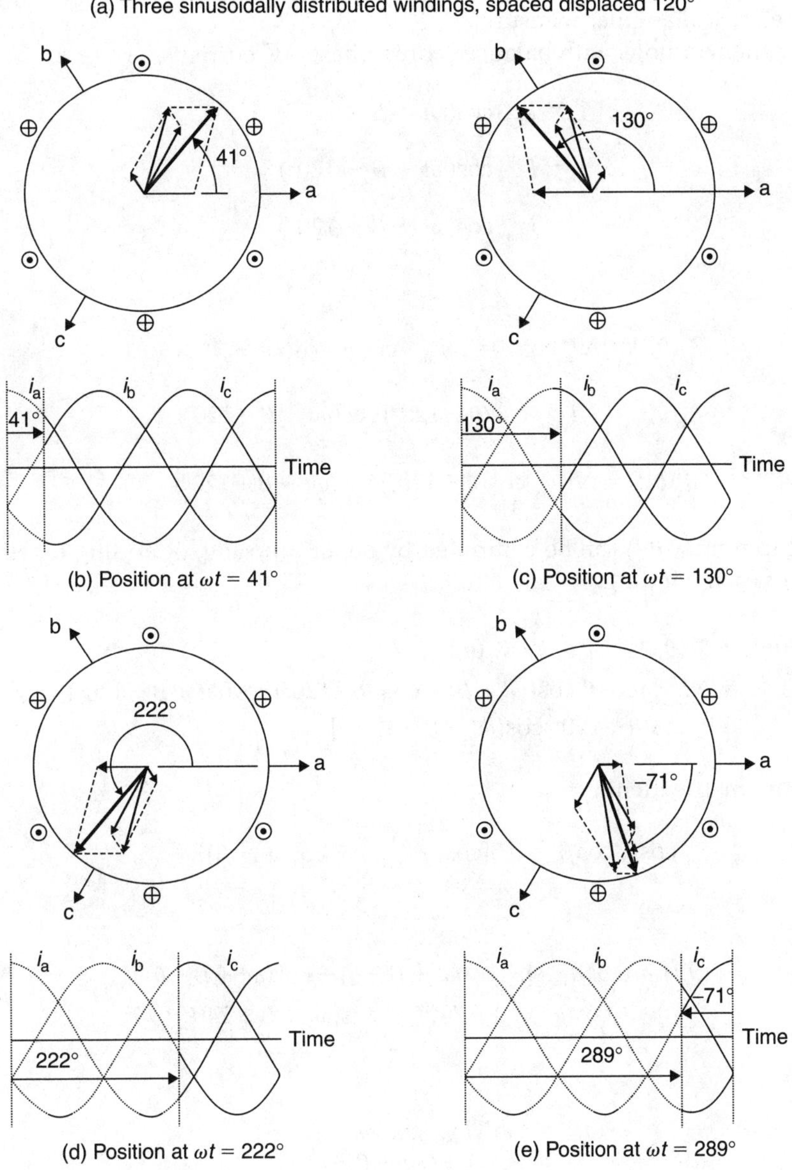

(b) Position at $\omega t = 41°$

(c) Position at $\omega t = 130°$

(d) Position at $\omega t = 222°$

(e) Position at $\omega t = 289°$

FIGURE 4.7. The rotating field.

will classify as a balanced three-phase winding. Consider the windings to be electrically interconnected in wye (connecting a′, b′, c′ to a common node "n").[3] Therefore,

$$\mathfrak{F}_a(\theta,t) = Ni_a \cos\theta \tag{4.2a}$$

$$\mathfrak{F}_b(\theta,t) = Ni_b \cos(\theta - 120°) \tag{4.2b}$$

$$\mathfrak{F}_c(\theta,t) = Ni_c \cos(\theta + 120°) \tag{4.2c}$$

where $\theta = \theta_e$ (electrical angular measure).

Now supply the windings with balanced three-phase AC currents, or

$$i_a = I_{max}\cos(\omega t + \beta) \tag{4.3a}$$

$$i_b = I_{max}\cos(\omega t + \beta - 120°) \tag{4.3b}$$

$$i_c = I_{max}\cos(\omega t + \beta + 120°) \tag{4.3c}$$

Hence,

$$\mathfrak{F}_a(\theta,t) = Ni_a \cos\theta = NI_{max} \cos\theta \, \cos(\omega t + \beta) \tag{4.4a}$$

$$\mathfrak{F}_b(\theta,t) = NI_{max} \cos(\theta - 120°)\cos(\omega t + \beta - 120°) \tag{4.4b}$$

$$\mathfrak{F}_c(\theta,t) = NI_{max} \cos(\theta + 120°)\cos(\omega t + \beta + 120°) \tag{4.4c}$$

The total air-gap mmf $\mathfrak{F}(\theta,t)$ can be computed by superimposing, or adding together, the contributions of each winding:

$$\begin{aligned}\mathfrak{F}(\theta,t) &= \mathfrak{F}_a(\theta,t) + \mathfrak{F}_b(\theta,t) + \mathfrak{F}_c(\theta,t) \\ &= NI_{max}[\cos\theta \cos(\omega t + \beta) + \cos(\theta - 120°)\cos(\omega t + \beta - 120°) \\ &\quad + \cos(\theta + 120°)\cos(\omega t + \beta + 120°)]\end{aligned} \tag{4.5}$$

Recall the trigonometric identity:

$$\cos\alpha \cos\beta = 0.5[\cos(\alpha + \beta) + \cos(\alpha - \beta)]$$

Simplifying,

$$\begin{aligned}\mathfrak{F}(\theta,t) = 0.5NI_{max}[&3\cos(\omega t + \beta - \theta) + \cos(\omega t + \beta + \theta) \\ &+ \cos(\omega t + \beta - 240°) + \cos(\omega t + \beta + 240°)]\end{aligned}$$

or

$$\begin{aligned}\mathfrak{F}(\theta,t) &= 1.5NI_{max} \cos(\omega t + \beta - \theta) \\ &= F_{max} \cos(\omega t + \beta - \theta)\end{aligned} \tag{4.6a}$$

[3] Actually the windings may be wye or delta connected. However, for every delta there is an equivalent wye, which we may convert to before analysis. For readers unfamiliar with three-phase circuits, it may be appropriate to digress to Appendix B at this point.

Consider the meaning of Equation 4.6. Assume θ as the location of an observer on the stator air-gap surface. If $\theta = \omega t + \beta$, the observer is in motion, rotating around the stator at angular velocity ω electrical rad/s, and positioned at $\theta = \beta$ at $t = 0$. Such an observer would see a constant DC mmf of F_{max}. A stationary observer located at $\theta = 0$ would see a time-varying AC mmf, peaking at $\omega t = -\beta$. If the observer moved forward (ccw) to a position $\theta = \beta$, a time-varying AC mmf, peaking at $t = 0$, or *later* in time will be observed.

The nature of $\mathfrak{F}(\theta, t)$ is now revealed. It appears as a sinusoidally distributed waveform in space, and rotating ccw ("forwards") in time at velocity ω, positioned at $\theta = \beta$ at $t = 0$. The location *in time* of the currents is clearly related to the location *in space* of $\mathfrak{F}(\theta,t)$. This is the famous rotating magnetic field, a key to understanding the operation of all AC rotating machines.[4] The corresponding air-gap flux density is

$$B(\theta, t) = (\mu_0/g)\, \mathfrak{F}(\theta,t) = B_{max} \cos(\omega t + \beta + \theta) \tag{4.7}$$

ω represents both the radian frequency of the stator currents and the angular velocity of rotating magnetic field, defined as

$\omega = 2\pi f$ = radian frequency of the stator currents, in electrical rad/s.
$\omega_{se} = \omega$ = angular velocity of rotating magnetic field, or "synchronous speed," in electrical rad/s.

The corresponding mechanical or spatial angular velocity, also called "synchronous speed," is

$$\omega_{sm} = \frac{\omega_{se}}{N_P/2} = \frac{2\omega_{se}}{N_P} = \frac{4\pi f}{N_P} \tag{4.8}$$

where N_P is the number of poles formed by the winding (an even integer).

In the situation of Figure 4.7a, if the current phase sequence is reversed by exchanging the phases of i_b and i_c, Equation 4.6a is altered to

$$\mathfrak{F}(\theta, t) = F_{max} \cos(\omega t + \beta + \theta) \tag{4.6b}$$

$\mathfrak{F}(\theta,t)$ is now rotating cw ("backwards") in time at velocity ω, positioned at $\theta = -\beta$ at $t = 0$.

An example will be useful.

Example 4.1

The two-pole stator of Figure 4.7a is supplied with balanced three-phase currents, sequence a-b-c, $\beta = 0$.

(a) If the current frequency is 60 Hz, determine synchronous speed in rad/s and rev/min and direction of field rotation.
(b) Locate the rotating field maximum position at four arbitrary instants of time ($\omega t = 41°, 130°, 222°$, and $289°$).
(c) Repeat (a) for a four-pole wound stator, $f = 50$ Hz, sequence a-c-b. Where is the field maximum at $\omega t = 90°$?

---

[4] Among the first to grasp this concept was the brilliant and eccentric electrical engineer Nikola Tesla (1856–1943), which he described as a "magnetic whirlwind." According to his own account, Tesla first had the vision of the rotating field while walking in a park at sunset, reciting passages from Goethe's *Faust*. Tesla is generally credited with the invention of the polyphase induction machine.

SOLUTION

(a) Synchronous speed in rad/s and rev/min:

$$\omega = 2\pi f = 2\pi(60) = 377\,\text{rad/s}$$

$$\omega_{sm} = \frac{2\omega}{N_P} = 2(377)/2 = 377\,\text{rad/s, or } 3600\,\text{rev/min, ccw}$$

(b) The field maximum is positioned at $\theta = 41°$, 130°, 222°, and 289°, for $\omega t = 41°$, 130°, 222°, and 289° (see Figures 4.7b–e).

(c) Synchronous speed in rad/s and rev/min:

$$\omega = 2\pi f = 2\pi(50) = 314.2\,\text{rad/s}$$

$$\omega_{sm} = \frac{2\omega}{N_P} = 2(314.2)/4 = 157.1\,\text{rad/s, or } 1500\,\text{rev/min, cw}$$

$$\mathfrak{F}(\theta,t) = F_{max}\cos(\omega t + 30° + \theta)$$

The cosine maximizes when $(\omega t + 30° + \theta) = 0$

Therefore, at $\omega t = 90°$,

$$90° + 30° + \theta = 0 \quad \text{or} \quad \theta = -120° = +240°$$

4.4 Stator–Rotor Interactions

We now insert the rotor into the stator "hole" and consider what happens. First, consider only a wound rotor, with windings that replicate the stator winding configuration, i.e., both stator and rotor have N_P pole balanced three-phase wye-connected windings. Stator and rotor terminals are a,b,c,n and A,B,C,N, respectively. Likewise, consider the rotor windings to be open-circuited and stationary. Assume that the stator and rotor iron have negligible reluctance, and the path reluctance is composed entirely of the air gap, which is uniform. Assume the stator windings are supplied by balanced three-phase currents, creating the air-gap flux density of Equation 4.7, and leading to the following expressions for stator (a) and rotor (A) Phase a, A flux linkages:

$$\lambda_a(\theta,t) = N_S\,\phi_{max}\cos(\omega_{se}t + \beta) \tag{4.9a}$$

$$\lambda_A(\theta,t) = N_R\,\phi_{max}\cos(\omega_{se}t + \beta - \theta) \tag{4.9b}$$

where

θ = angle from stator Phase a magnetic axis to rotor Phase A axis (electrical degrees).

N_S, N_R = the effective number of stator (rotor turns per phase).

If the rotor is rotating at angular velocity ω_{re},

$$\theta = \omega_{re}t - \gamma \tag{4.10a}$$

$$\lambda_A(\theta,t) = N_R\,\phi_{max}\cos((\omega_{se} - \omega_{re})t + \beta + \gamma) \tag{4.10b}$$

The quantity $(\omega_{se} - \omega_{re})$ has a double meaning: it is the velocity of the rotating field relative to the rotor, and simultaneously the radian frequency of the rotor flux linkages. It is sometimes called the "slip speed," and inspires the definition of slip:

$$s = \text{slip} = \frac{\omega_{se} - \omega_{re}}{\omega_{se}} = \frac{\omega_{sm} - \omega_{rm}}{\omega_{sm}} \tag{4.11}$$

Using s in Equation 4.10b,

$$\lambda_A(\theta,t) = N_R\,\phi_{max}\cos(s\omega_{se}t + \beta + \gamma) \tag{4.10c}$$

Now consider the voltages induced into stator and rotor Phases a and A:

$$e_a = \frac{d\lambda_a(\theta,t)}{dt} = -N_S\,\phi_{max}\,(\omega_{se})\sin(\omega_{se}t + \beta + \gamma) \tag{4.12a}$$

$$e_A = \frac{d\lambda_A(\theta,t)}{dt} = -N_R\,\phi_{max}\,(s\omega_{se})\sin(s\omega_{se}t + \beta + \gamma) \tag{4.12b}$$

Consider the ratio of the corresponding rms values of these voltages:

$$\frac{E_a}{E_A} = \frac{N_S}{sN_R}$$

or

$$E_a = \frac{N_S}{N_R}\left(\frac{E_A}{s}\right) \tag{4.13}$$

Because of three-phase symmetry, the abc, and ABC, voltages are balanced three-phase. Consider the stator to be externally terminated in a balanced three-phase source (v_{an}, v_{bn}, v_{cn}). For the moment, assume that all stator windings have negligible resistance, and leakage flux (that is, all stator flux crosses the air gap, and links the rotor windings). Thus the externally applied voltage must be counterbalanced by that voltage induced by the rotating field, or

$$v_{an} = e_a \tag{4.14a}$$

For sinusoidal steady-state operation, we convert to phasor values:

$$\bar{V}_{an} = \bar{E}_a \tag{4.14b}$$

Now suppose that the rotor windings are terminated in a balanced wye-connected set of resistances (R_x). Since there is balanced three-phase rotor voltage, balanced three-phase currents, with radian frequency $s\omega_{se}$, must flow. But we now know that balanced three-phase currents flowing in sinusoidally distributed equally spaced windings will create a (second) rotating magnetic field, rotating at $s\omega_{se}$, *relative to the rotor.* But the rotor is rotating at ω_{re} *relative to the stator*. Therefore, the absolute velocity of the rotor field is $s\omega_{se} + \omega_{re} = (\omega_{se} - \omega_{re} + \omega_{re}) = \omega_{se}$, or synchronous speed. That is, the stator and rotor fields are synchronized at all rotor speeds. Since the two fields are locked in synchronism, the production of EM torque is possible. Consider an example.

Example 4.2

Consider a four-pole wound rotor induction machine excited from a 60 Hz three-phase source. Determine synchronous speed, slip, and speed of rotor with respect to the rotating stator field ($\omega_{R/SF}$), and rotor frequency for rotor speeds of 0, 1200, 2400, and −1200 rev/min.

SOLUTION

At all speeds,

$$\omega = \omega_{se} = 2\pi f = 2\pi(60) = 377.0\ \text{rad/s, or 3600 rev/min}$$

$$\omega_{sm} = 2\omega/N_P = 2(377)/4 = 188.5\ \text{rad/s, or 1800 rev/min, ccw}$$

At zero speed,

$$s = \text{slip} = \frac{\omega_{se} - \omega_{re}}{\omega_{se}} = \frac{1800 - 0}{1800} = 1.0$$

speed of rotor with respect to the rotating field ($\omega_{R/SF}$) = −1800 rev/min

$$f_R = sf_S = 1.0(60) = 60\,\text{Hz}$$

At 1200 rev/min,

$$s = (1800 - 1200)/1800 = 0.3333$$

$$\omega_{R/SF} = 1200 - 1800 = -600\,\text{rev/min}$$

$$f_R = sf_S = 0.3333(60) = 20\,\text{Hz}$$

At 2400 rev/min,

$$s = (1800 - 2400)/1800 = -0.3333$$

$$\omega_{R/SF} = 2400 - 1800 = +600\,\text{rev/min}$$

$f_R = sf_S = -0.3333(60) = -20\,\text{Hz}$ (negative frequency has no special meaning here)

At −1200 rev/min,

$$s = (1800 + 1200)/1800 = 1.667$$

$$\omega_{R/SF} = -1200 - 1800 = -3000\,\text{rev/min}$$

$$f_R = sf_S = 1.667(60) = 100\,\text{Hz}$$

Consider the machine currents. The Phase A rotor current must be

$$\bar{I}_A = \frac{\bar{E}_A}{R_2 + R_x + js\omega_{se}L_2} \tag{4.15a}$$

where
R_2 = internal rotor phase resistance (in Ω).
L_2 = rotor phase leakage inductance (in H).
R_x = external rotor phase resistance (in Ω).

Dividing the numerator and denominator by s leaves $\bar{I}_A$ unaffected:

$$\bar{I}_A = \frac{\bar{E}_A/s}{\frac{R_2 + R_x}{s} + j\omega_{se}L_2} \tag{4.15b}$$

The rotor field will cause an equal and opposite field reaction in the stator, which in turn requires a compensating stator current,[5] according to

$$N_S\,\bar{I}'_A = N_R\,\bar{I}_A \tag{4.16}$$

[5] If this were not the case, the original field would be distorted and the corresponding induced voltage would be altered. But this cannot be, since the phase voltages are fixed due to the external source termination.

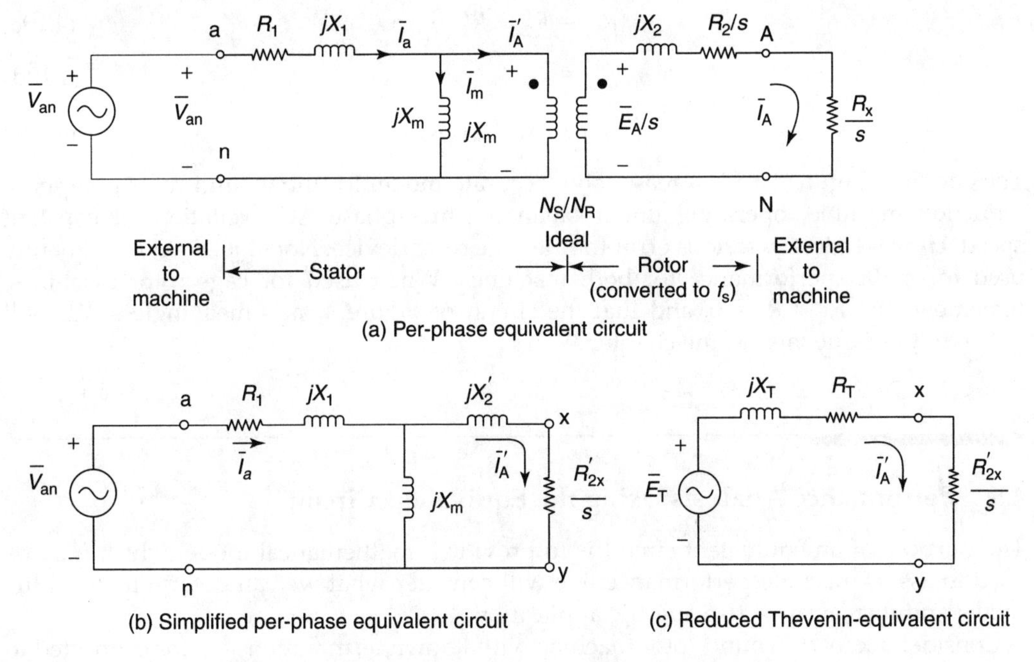

FIGURE 4.8. Polyphase induction machine equivalent circuits.

Here, $\bar{I}_A$ represents the rotor Phase A current, and $\bar{I}_a$, the stator phase "a" current. There are two components to the stator current: the part which establishes the original field ($\bar{I}_m$), and the part which corresponds to the rotor current ($\bar{I}'_A$):

$$\bar{I}_a = \bar{I}_m + \bar{I}'_A \tag{4.17}$$

It would be a more realistic situation to include the stator winding resistance and leakage flux:

$$v_{an} = R_1 i_a + L_1 \frac{di_a}{dt} + e_a \tag{4.18a}$$

where
R_1 = Phase a winding resistance (in Ω).
L_1 = Phase a leakage inductance (in H).

For sinusoidal steady-state operation, we convert into phasor values:

$$\bar{V}_{an} = R_1 \bar{I}_a + jX_1 \bar{I}_a + \bar{E}_a \tag{4.18b}$$

A circuit that accounts for all of the points just discussed appears in Figure 4.8a. The ideal transformer insures that the proper phase and magnitude relationships are maintained between rotor and stator voltages and currents. The rotor section of the circuit has been converted into the stator frequency (f_S). When actual rotor quantities are of no interest, they may be reflected through the ideal transformer, simplifying the circuit to the form of Equation 4.8b:

$$R'_2 = (N_S/N_R)^2 R_2 \tag{4.19a}$$

$$R'_x = (N_S/N_R)^2 R_x \tag{4.19b}$$

$$R'_{2x} = R'_2 + R'_x \tag{4.19c}$$

$$X_2 = \omega_{se} L_2 \tag{4.19d}$$

$$X'_2 = (N_S/N_R)^2 X_2 \tag{4.19e}$$

The circuit of Figure 4.8b is a reasonably accurate model for the wound rotor polyphase induction machine, operating under balanced three-phase AC excitation, at constant speed. However, it is in serious error for the cage rotor device. Nonetheless, it is frequently used to predict performance for both machines. When used for cage rotor machines, remember that $R'_x = R_x = 0$, and that the circuit of Figure 4.8a is meaningless. We will return to this issue later in the chapter.

4.5 Performance Analysis Using the Equivalent Circuit

The purpose of an equivalent circuit is to provide a mathematical model which may be used to assess machine performance. We will consider what we can determine from the model and then apply it to a specific application.

Consider a known wound rotor machine with known terminations (stator connected to a balanced three-phase source; rotor terminated in a balanced wye-connected resistor bank of R_x) running at a known speed. The following terminology is specific to operation in the motor mode. We start by computing the slip:

$$s = \frac{\omega_{sm} - \omega_{rm}}{\omega_{sm}} \tag{4.11}$$

Determination of the machine currents is a key problem. Writing loop equations,

$$\begin{bmatrix} \bar{V}_{an} \\ 0 \end{bmatrix} = \begin{bmatrix} (R_1 + j(X_1 + X_m)) & -j(X_m) \\ -j(X_m) & (R'_{2x}/s + j(X'_2 + X_m)) \end{bmatrix} \begin{bmatrix} \bar{I}_a \\ \bar{I}'_A \end{bmatrix} \tag{4.20}$$

which can be solved for the currents.

From the currents, the machine powers can be computed. Before we proceed with this task, let us digress to a comprehensive loss analysis of the three-phase induction machine.[6] Traditionally, machine power losses have been divided into five categories:

1. *Stator winding loss (SWL).* Since the stator windings have ohmic resistance, currents flowing in these paths will generate thermal power (I^2R) losses.
2. *Rotor winding loss (RWL).* The rotor windings (or rotor bars, in the case of cage rotor machines) also have ohmic resistance, and thermal power (I^2R) losses.
3. *Core loss.* All ferromagnetic parts of the machine have AC magnetic fields inside them, and therefore experience core (hysteresis and eddy current) losses. Since the frequency is normally much higher in the stator iron, most of the core loss occurs there.
4. *Friction and windage loss.* Since parts of the machine are in motion, frictional forces must occur. The corresponding losses are localized at the bearing surfaces. In the

[6] The loss analysis of three-phase induction machines, with minor adjustments, will be applicable to all rotating machines.

wound rotor machine, additional friction loss is caused by the brushes sliding on the slip rings. "Windage" is basically "wind friction" and occurs since machine motion occurs in an atmosphere. Machines with fans mounted on the shaft for forced ventilation cooling have greatly increased windage losses.

5. *Stray load loss (SLL)*. If the machine is tested under load conditions, the measured total loss is consistently greater than the sum of the above losses, even using the most accurate computational and experimental methods available to determine these losses individually. To account for this additional loss, a fifth loss (called the stray load loss) is defined and has been a matter of interest and concern for machine designers from the 1930s through the present.

SLL is thought to be related to a number of factors, including leakage current through lamination due to slot skewing, nonequal division of current through parallel winding paths, irregularities in the magnetic paths due to the slots, nonuniform air gaps due to bearing deformation, the load level, and the particulars of the machine current and voltage waveforms.

It is approximately 1% of the load level, or from 10 to 20% of the total machine losses. An accurate computational accounting for the SLL is extremely difficult, and would require a considerable amount of detailed information about a specific machine, well in excess of that which is available to an applications engineer. Therefore, if SLL is to be considered at all, an engineering approximation is reasonable.

We now return to our mission; analyzing the machine powers under normal running conditions.

$$\text{Power input (motor mode)} = P_{\text{in}} = 3V_{\text{an}}I_{\text{a}}\cos\theta \tag{4.21}$$

$$\text{Stator winding loss} = \text{SWL} = 3I_{\text{a}}^{2}R_{1} \tag{4.22}$$

$$\text{Rotor winding loss} = \text{RWL} = 3I_{\text{A}}'^{2}R_{2\text{x}}' \tag{4.23}$$

Note that the resistance that represents the rotor in Figure 4.8b is R_{2x}'/s and not R_{2x}'. Consider the residual resistance without R_{2x}'

$$\frac{R_{2\text{x}}'}{s} - R_{2\text{x}}' = \left(\frac{1-s}{s}\right)R_{2\text{x}}' \tag{4.24}$$

What about the power absorbed in $R_{2x}'\,[(1-s)/s]$? This is the power that is converted from electrical into mechanical form:

$$\text{Developed power} = P_{\text{dev}} = 3I_{\text{A}}'^{2}\,R_{2\text{x}}'\,((1-s)/s) \tag{4.25}$$

We have accounted for two of the motor's five losses. Consider the following test. Suppose we measure the motor input power running at no load. What have we measured? Since there is no output, all power must have been consumed by the losses. If the small SWL and RWL are deducted, the balance must be core loss, friction and windage loss, and SLL (at no load). Define this combination of losses to be the *rotational loss* (P_{RL}).[7] Experimentation

[7] Frictional loss is actually more complicated than this, as is windage. Likewise, core loss does not depend solely on frequency (speed), but on field strength as well. SLL is even more complex. Yet Equation 4.26 has the advantage of giving us a loss that (1) increases with speed, (2) is easy to obtain data for, and (3) is straightforward to include in analysis. Thus, it is a reasonable approximation, and clearly superior to neglecting these losses entirely, which is sometimes done.

shows that P_{RL} increases with speed. As an approximation, define

$$\text{Rotational loss} = P_{RL} = K_{RL}\,[\omega_{rm}]^2 \tag{4.26}$$

The mechanical power associated with the load is

$$P_m = P_{dev} - P_{RL} \tag{4.27}$$

For motor operation, the output power is

$$P_{out} = P_m \tag{4.28}$$

We can include a least part of the SLL if we use a slightly higher value for R'_{2x}. Therefore, for modeling purposes, the total losses of machine are

$$P_{LOSS} = P_{RL} + \text{SWL} + \text{RWL} \tag{4.29}$$

An important index of performance is the concept of efficiency (η). By definition,

$$\text{Efficiency} = \eta = \frac{P_{out}}{P_{in}} \text{ (expressed as a \%)} \tag{4.30}$$

Clearly,

$$P_{in} = P_{out} + P_{LOSS}$$

The stator input power factor for balanced operation is

$$\text{pf} = \cos\theta \tag{4.31}$$

There are several torques of interest.

$$\text{Developed torque} = T_{dev} = \frac{P_{dev}}{\omega_{rm}} \tag{4.32a}$$

$$= 3I'^2_A R'_{2x} \frac{1-s}{(1-s)s\omega_{sm}} = \frac{3I'^2_A R'_{2x}/s}{\omega_{sm}} \tag{4.32b}$$

Equation 4.32a is intuitive, but Equation 4.32b is not. At first, it would appear that we are using the "wrong" power. But closer examination reveals that we are also dividing by the "wrong" speed, to compensate. Because of its simplicity, Equation 4.32b is preferable. Likewise,

$$\text{Rotational loss torque} = T_{RL} = \frac{P_{RL}}{\omega_{rm}} = K_{RL}\ \omega_{rm} \tag{4.33a}$$

$$\text{Mechanical load torque} = T_m = \frac{P_m}{\omega_{rm}} \tag{4.33b}$$

For motor operation, the output torque is

$$\text{Output torque} = T_{\text{out}} = T_{\text{m}} \tag{4.33c}$$

Sometimes it is convenient to lump the rotational loss with the externally applied load, creating an "equivalent" load:

$$T'_{\text{m}} = T_{\text{RL}} + T_{\text{m}} \tag{4.33d}$$

Example 4.3 will demonstrate the use of these relations.

Example 4.3
Data for a 460 V 100 Hp wound rotor induction motor is provided below.

Three-Phase Induction Motor Data	
Ratings	
Line voltage = 460 V;	Horsepower = 100 hp
Stator frequency = 60 Hz;	Number of poles = 4
Design class = W;	Synchronous speed = 1800.0 rpm
$R'_x = 0.0000\,\Omega$;	$N_s/N_r = 2.5$
Test Data	
Blocked rotor (at 60 Hz)	No-load values
Line voltage = 84.82	460.0 V
Line current = 125.5	35.09 A
3-Phase Power = 4000.2	2394.2 W

Equivalent Circuit Values (in Stator Ω at Rated Frequency)

$R_1 = 0.04232$;	$X_1 = 0.19044$;	$X_m = 7.406$
$R'_2 = 0.04232$;	$X'_2 = 0.19044$;	$J_m = 0.9299\text{ kg m}^2$

Rotational loss torque = $T_{\text{RL}} = 0.06299 \times$ (speed in rad/s) N m

Operating at balanced rated stator voltage at 60 Hz, and running at 1770 rpm with the rotor shorted, find all currents, powers, torques, the slip, the power factor, and the efficiency.

SOLUTION

ω_{rm} = rotor speed = 1770 rpm = 185.4 rad/s

Synchronous speed = $\omega_{\text{sm}} = 4\pi f/N_{\text{P}} = 4\pi(60)/4 = 188.5$ rad/s = 1800 rev/min

$$s = \text{slip} = \frac{1800 - 1770}{1800} = 0.01667$$

Stator line voltage 460 V; frequency 60 Hz; operating mode: MOTOR.

$R'_{2x} = R'_2 + R'_x = 0.04232 + 0 = 0.04232\,\Omega$;

$R'_{2x}/s = 0.04232/0.01667 = 2.539\,\Omega$

$$\begin{bmatrix} \bar{V}_{\text{an}} \\ 0 \end{bmatrix} = \begin{bmatrix} R_1 + j(X_1 + X_{\text{m}}) & -jX_{\text{m}} \\ -jX_{\text{m}} & R'_{2x}/s + j(X'_2 + X_{\text{m}}) \end{bmatrix} \begin{bmatrix} \bar{I}_{\text{a}} \\ \bar{I}'_{\text{A}} \end{bmatrix}$$

$$\begin{bmatrix} 265.6 \\ 0 \end{bmatrix} = \begin{bmatrix} 0.042 + j\,7.596 & -j\,7.406 \\ -j\,7.406 & 2.539 + j\,7.596 \end{bmatrix} \begin{bmatrix} \bar{I}_{\text{a}} \\ \bar{I}'_{\text{A}} \end{bmatrix}$$

Currents

$$\bar{I}_a = 107.4\,\text{A}\ \angle{-26.46°}; \quad \bar{I}'_A = 99.33\,\text{A}\ \angle{-7.98°}$$

Power factor = pf = $\cos\theta$ = cos(26.46°) = 0.8952 lagging

Power input = $P_{in} = 3V_{an}I_a \cos\theta$
= 3(265.6)(107.4) cos(26.46) = 76.619 kW

SWL = $3I_a^2R_1$
= $3(107.4)^2(0.042)$ = 1.465 kW

RWL = $3I_A'^2R'_{2x}$
= $3(99.33)^2(0.042)$ = 1.253 kW

Rotational loss = $P_{RL} = K_{RL}\omega_{rm}$
= (0.06299)(185.4) = 2.164 kW

Total machine loss = $P_{LOSS} = P_{RL}$ + SWL + RWL
= 1.465 + 1.253 + 2.164 = 4.882 kW

Developed power = $P_{dev} = 3I_A'^2R'_{2x}((1 - s)/s)$
= $3(99.33)^2(2.539)(1 - 0.01667)$ = 73.902 kW

Mechanical power = $P_m = P_{out} = P_{dev} - P_{RL}$
= 73.902 − 2.164 = 71.738 (96.163 hp)

Efficiency = $\eta = P_{out}/P_{in}$ = 71.738/76.619 = 93.63%

Developed torque = $T_{dev} = 3I_A'^2R'_{2x}/(s\omega_{sm})$
= $3(107.4)^2(2.539)/(188.5)$ = 398.7 N m

Rotational loss torque = $T_{RL} = P_{RL}/\omega_{rm}$ = 2164/185.4 = 11.7 N m

Mechanical load torque = $T_m = T_{out} = T_{dev} - T_{RL}$ = 387.0 N m

In some cases it is more convenient to use a simpler form of the circuit in Figure 4.8b. Consider reducing the circuit to the left of port x-y to its Thevenin equivalent form:

$$\bar{E}_T = \frac{jX_m\bar{V}_{an}}{R_1 + j(X_1 + X_m)} \tag{4.34a}$$

$$\bar{Z}_T = jX_2 + \frac{jX_m(R_1 + jX_1)}{R_1 + j(X_1 + X_m)} \tag{4.34b}$$

Recall Equation 4.32b,

$$T_{dev} = \frac{3I_A'^2R'_{2x}}{s\omega_{sm}}$$

Observe that torque is maximized when the *power* delivered to R'_{2x}/s is maximized. R'_{2x}/s varies with s (and R'_x, for the wound rotor case). According to the maximum power transfer theorem,

$$\frac{R'_{2x}}{s_{MT}} = Z_T \tag{4.35a}$$

$$s_{MT}\ \text{(slip for maximum developed torque)} = \frac{R'_{2x}}{Z_T} \tag{4.35b}$$

$$\omega_{MT} = \text{speed for maximum torque} = (1 - s_{MT})\,\omega_{sm} \tag{4.35c}$$

The reduced Thevenin circuit is shown in Figure 4.8c. Note the identity of the current $\bar{I}'_A$ is preserved, but not of $\bar{I}_a$. The circuit can be solved for the current and the torque.

Example 4.4
For the machine in Example 4.3, determine the Thevenin-equivalent circuit of Figure 4.8c, the slip and speed for maximum developed torque, and the maximum developed torque.

SOLUTION

$$\begin{aligned}\bar{E}_T &= (jX_m)\,V_{an}/[R_1 + j(X_1 + X_m)]\\ &= (j\,7.406)265.6/[0.04232 + j(0.19044 + 7.406)]\\ &= 258.7\angle 0.32^\circ\end{aligned}$$

$$\begin{aligned}\bar{Z}_T &= jX_2 + (jX_m)(R_1 + jX_1)/[R_1 + j(X_1 + X_m)]\\ &= j\,0.19044 + (j\,7.406)(0.04232 + j\,0.19044)/[0.04232 + j(0.19044 + 7.406)]\\ &= 0.04022 + j\,0.37633 = 0.37847\angle 83.9^\circ\ \Omega\end{aligned}$$

$$\begin{aligned}s_{MT} &= R'_{2x}/Z_T = (0.04232/0.37847) = 0.1118\\ \omega_{MT} &= (1 - s_{MT})\omega_{sm} = (0.8882)1800 = 1599\ \text{rev/min}\ (167.4\ \text{rad/s})\end{aligned}$$

$$\begin{aligned}\bar{I}'_A &= E_T/[R_T + (R'_{2x}/S_{MT}) + jX_T]\\ &= (258.7\ \angle 0.32^\circ)/[0.04022 + (0.04232/0.1118) + j\,0.37633]\\ &= 459.6\angle{-41.6^\circ}\ \text{A}\end{aligned}$$

$$\begin{aligned}T_{dev} &= 3I'^2_A R'_{2x}/(s\omega_{sm})\\ &= 3(459.6)^2(0.04232/0.1118)/(188.5)\\ &= 1274\ \text{N m}\end{aligned}$$

4.6 Equivalent Circuit Constants from Tests

Equivalent circuits have little practical value unless it is possible to obtain accurate values for the required element constants. We have the following four options:

- Request and obtain constants from machine manufacturers
- Calculate constants from design values (dimensions, material constants, etc.)
- Estimate constants from similar known designs
- Design and perform experimental tests on the machine, from which constants can be computed.

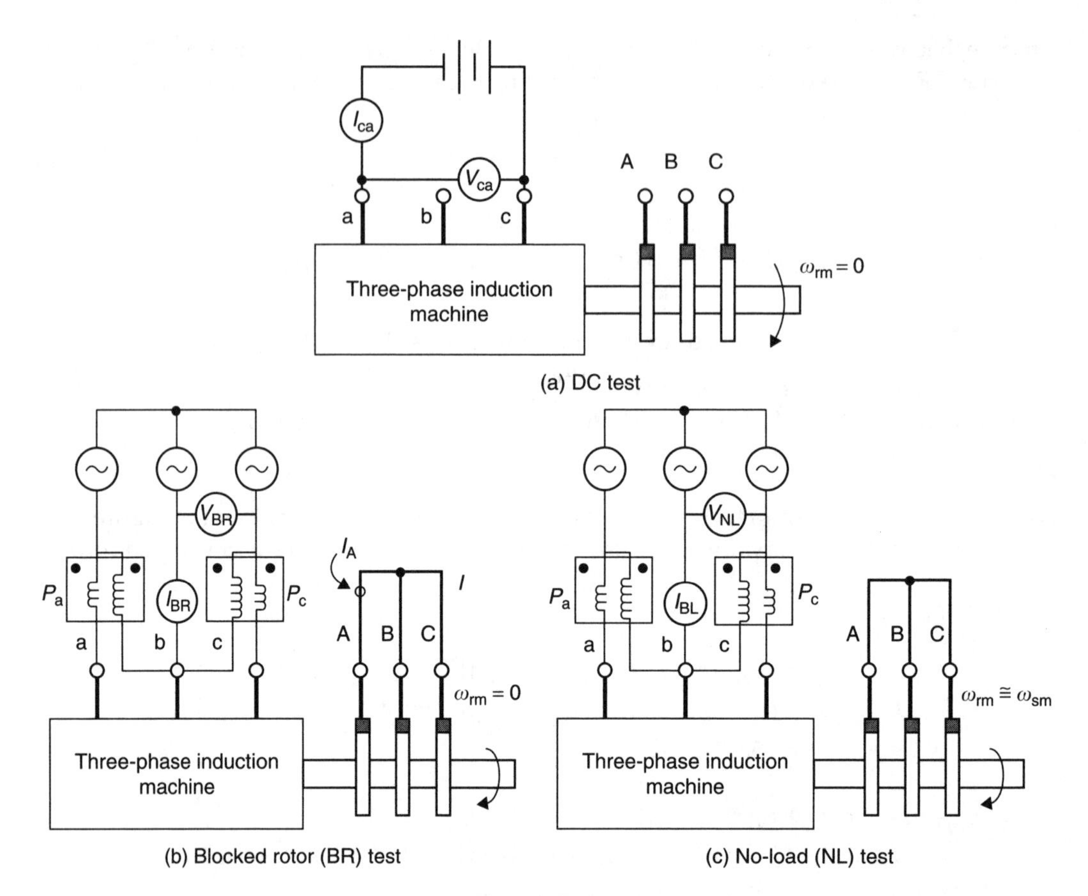

FIGURE 4.9. Test setups.

Fortunately, it is relatively straightforward to design and perform such tests for induction machines from data acquired from certain fairly simple tests. The constants to be determined are the five element values ($R_1, X_1, X_m, R'_2, X'_2$), the constant K_{RL}, and the turns ratio N_S/N_R. Figure 4.9 summarizes the laboratory setups for the required tests.

4.6.1 The DC Test

Consider Figure 4.9a for the laboratory setup. By applying a DC source between terminals (a-b, b-c, c-a; the neutral is usually not available) and measuring the DC voltage and current, the DC stator resistance can be determined. Considering a wye connection, the resistance per phase is half of that measured. Multiplication by two correction factors for temperature and for AC skin effect give R_1:

$$R_{ab} = \frac{V_{ab}}{I_{ab}}; \quad R_{bc} = \frac{V_{bc}}{I_{bc}}; \quad R_{ab} = \frac{V_{ca}}{I_{ca}}; \quad R_{dc} = \frac{R_{ab} + R_{bc} + R_{ca}}{6};$$

$$R_1 = K_{TEMP} K_{ac} R_{dc}$$

$K_{TEMP}K_{ac}$ can vary from about 1.2 to 2.5, depending on the temperature and geometry of the conductor. For rough work, $K_{TEMP}K_{ac} = 1.5$ is a reasonable guess.[8] For wound rotor machines, a similar test at the rotor terminals can be performed to determine R_2.

[8] For precision work, IEEE/ANSI Standard 112 (Standard Test Procedures for Polyphase Induction Machines) should be consulted and followed.

4.6.2 The Blocked Rotor (BR) Test

Consider Figure 4.9b for the laboratory setup. Applying a balanced reduced adjustable three-phase AC source at the stator terminals and measure the voltage, current, and power. At blocked rotor conditions, $s = 1$, so that $Z_2' \ll X_m$, and X_m may be neglected (treated as an open circuit).[9]

The test is normally performed at rated current.

$$P_{BR} = P_a + P_c$$

$$R_{BR} = \frac{P_{BR}}{3I_{BR}^2} = R_1 + R_2'$$

Using the value of R_1 from the DC test,

$$R_2' = R_{BR} - R_1$$

$$Z_{BR} = \frac{V_{BR}}{\sqrt{3}I_{BR}}$$

$$X_{BR} = \sqrt{Z_{BR}^2 - R_{BR}^2} = X_1 + X_2'$$

For wound rotor machines, $X_1 \approx X_2'$. Therefore

$$X_1 = X_2' = 0.5X_{BR}$$

4.6.3 The No-Load (NL) Test

Consider Figure 4.9c for the laboratory setup. Apply a balanced three-phase AC source at the stator terminals and measure the voltage, current, and power. At no-load running conditions, $s \approx 0$. Therefore the resistor R_{2x}'/s is large. Since the output is zero, the input power consists of only the winding losses plus the rotational loss. Most of the no-load winding loss is due to stator current:

$$P_{NL} = P_a + P_c$$

$$P_{RL} = P_{NL} - 3I_{NL}^2 R_1$$

and

$$K_{RL} = \frac{P_{RL}}{\omega_{rm}^2} \simeq \frac{P_{RL}}{\omega_{sm}^2}$$

Assume that I_{NL} has two components: the in-phase loss component (I_{LOSS}) and the out-of-phase magnetizing component (I_m):

$$I_{LOSS} = \frac{P_{NL}/3}{V_{NL}/\sqrt{3}}$$

$$I_m = \sqrt{I_{NL}^2 - I_{LOSS}^2}$$

[9] For precision work, it is recommended that the BR test be executed at low frequency (typically about 5 Hz). At this frequency, the assumption that X_m may be treated as an open circuit is not reasonable. Again, see IEEE/ANSI Standard 112 for a more accurate approach. Another possibility is to use the methods in this section as a first approximation for circuit constants, and iterate between BR and NL test data, using the circuit of Figure 4.8b with no approximations.

Compute X_{NL} and X_m from

$$X_{NL} = \frac{V_{NL}/\sqrt{3}}{I_{NL}}$$
$$X_m = X_{NL} - X_1$$

Example 4.5 will demonstrate these methods.

Example 4.5
Test data is available for the machine discussed in Example 4.3. Consider Figure 4.9 for the laboratory setups.
DC Test:

$$\text{At } I_{ab} = I_{bc} = I_{ca} = 100\text{ A};\quad V_{ab} = 5.640\text{ V},\quad V_{bc} = 5.643\text{ V},\quad V_{ca} = 5.646\text{ V}$$

$$\text{Use } (K_{TEMP})(K_{ac}) = 1.5$$

BR Test:

$$P_{BR} = 4000\text{ W},\quad I_{BR} = 125.5\text{ A } (I_2 = 313.8\text{ A}),\quad V_{BR} = 84.82\text{ A}$$

NL Test:

$$P_{NL} = 2394\text{ W},\quad I_{NL} = 35.09\text{ A},\quad V_{NL} = 460\text{ V}$$

Determine the five element values (R_1, X_1, X_m, R'_2, X'_2), the constant K_{RL}, and the turns ratio N_S/N_R from the given test data.

SOLUTION
DC Test:

$$\text{At } I_{ab} = I_{bc} = I_{ca} = 100\text{ A};\quad V_{ab} = 5.640\text{ V},\quad V_{bc} = 5.643\text{ V},\quad V_{ca} = 5.646\text{ V}$$

$$R_{ab} = 5.640/100,\quad R_{bc} = 5.643/100,\quad R_{ca} = 5.646/100$$

$$R_{dc} = [0.05640 + 0.05643 + 0.05646]/6 = 0.02821\ \Omega$$

$$R_1 = K_{TEMP}K_{ac}R_{dc} = 1.5(0.02821) = 0.04232\ \Omega$$

BR Test:

$$N_S/N_R = (I_2/I_{BR}) = (313.8/125.5) = 2.5$$

$$R_{BR} = P_{BR}/(3I_{BR}^2) = 4000/[(3)(125.5)^2] = 0.08464\ \Omega$$

Using the value of R_1 from the DC test:

$$R'_2 = R_{BR} - R_1 = 0.08464 - 0.04232 = 0.04232\ \Omega$$
$$Z_{BR} = V_{BR}/(\sqrt{3}I_{BR}) = 84.82/[(3)125.5] = 0.3902\ \Omega$$
$$X_{BR} = \sqrt{Z_{BR}^2 - R_{BR}^2} = 0.3809\ \Omega$$

For wound rotor machines, $X_1 \approx X'_2$.[10] Therefore

$$X_1 = X'_2 = 0.5\,X_{BR} = 0.1904\,\Omega$$

NL Test:

$$\begin{aligned}
P_{NL} &= 2397\text{ W}, \quad I_{NL} = 35.07\text{ A}, \quad V_{NL} = 460\text{ V}\\
P_{RL} &= P_{NL} - (3I_{NL}^2)R_1 = 2397 - (3)(35.07)^2(0.04232) = 2241\text{ W}\\
K_{RL} &= P_{RL}/(\omega_{rm})^2 \approx 2241/(188.5)^2 = 0.063\\
I_{LOSS} &= (P_{NL}/3)/(V_{NL}/\sqrt{3}) = (2397/3)/(460/3) = 3.01\text{ A}\\
I_m &= \sqrt{I_{NL}^2 - I_{LOSS}^2} = \sqrt{(35.09)^2 - (3.01)^2} = 34.96\text{ A}
\end{aligned}$$

Compute X_{NL} and X_m from

$$\begin{aligned}
X_{NL} &= (V_{NL}/\sqrt{3})/I_m = (460/3)/34.96 = 7.596\\
X_m &= X_{NL} - X_1 = 7.596 - 0.1904 = 7.406\,\Omega
\end{aligned}$$

To summarize results,

$$R_1 = R'_2 = 0.04232\,\Omega; \quad X_1 = X'_2 = 0.1904\,\Omega$$

$$N_S/N_R = 2.5; \quad K_{RL} = 0.063; \quad X_m = 7.406\,\Omega;$$

4.7 Operating Modes: Motor, Generator, and Braking

Consider repeating the analysis of Section 4.4 over a range of speeds. For example, consider the machine discussed in Example 4.3 driven from a balanced three-phase 460 V 60 Hz source, and with the rotor shorted. For speeds from −600 to +2400 rev/min:

Speed (rpm)	Current (A)	P_{in} (kW)	P_{out} (kW)	T_{dev} (N m)	PF (%)	EFF (%)
−600.0	693.1	104.482	−14.743	230.7	18.92	—
−450.0	692.4	107.155	−11.711	245.5	19.42	—
−300.0	691.6	110.189	−8.307	262.4	20.00	—
−150.0	690.6	113.659	−4.442	281.8	20.66	—
0.0	689.3	117.667	0.000	304.2	21.42	0.00
150.0	687.8	122.344	5.174	330.4	22.32	4.23
300.0	685.9	127.870	11.293	361.5	23.40	8.83
450.0	683.5	134.491	18.653	398.8	24.70	13.87
600.0	680.4	142.554	27.679	444.5	26.30	19.42
750.0	676.0	152.559	39.001	501.5	28.32	25.56
900.0	669.8	165.241	53.578	574.4	30.96	32.42
1050.0	660.4	181.682	72.916	670.1	34.53	40.13
1200.0	645.0	203.397	99.389	798.8	39.58	48.86
1350.0	616.7	231.809	136.378	973.6	47.17	58.83
1500.0	556.3	263.133	184.979	1187.5	59.36	70.30
1650.0	400.4	251.427	209.940	1225.9	78.82	83.50
1800.0	35.0	0.156	−2.238	0.0	0.56	—
1950.0	443.4	−258.516	−309.732	−1503.9	73.17	83.46
2100.0	614.0	−224.801	−321.156	−1446.5	45.95	70.00
2250.0	667.7	−158.492	−272.365	−1141.1	29.79	58.19
2400.0	687.8	−111.143	−232.244	−908.2	20.28	47.86

[10] For cage rotor machines, the division is different. See Section 4.9.

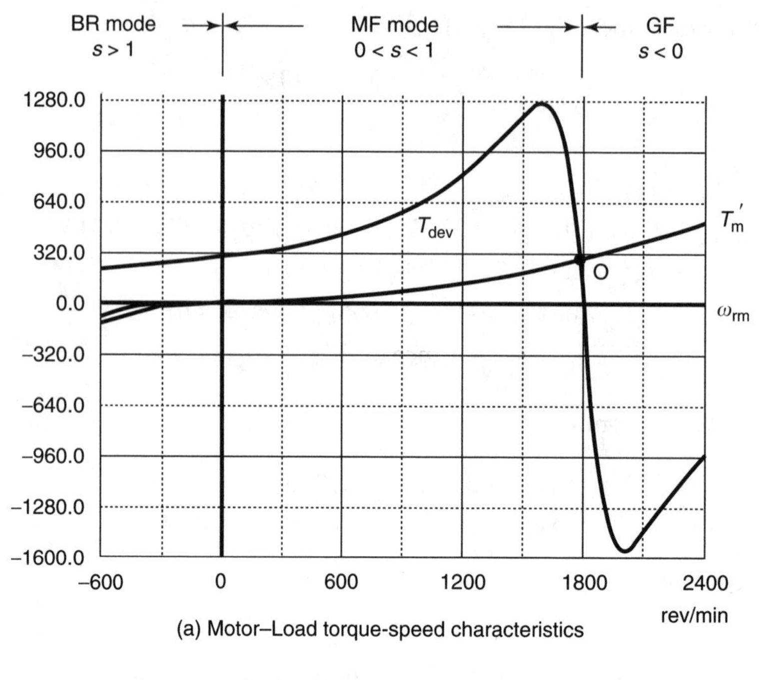

(a) Motor–Load torque-speed characteristics

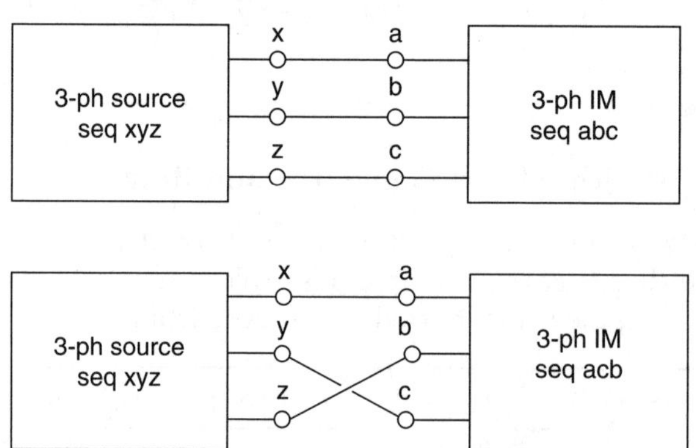

(b) Reverse-phase sequence reverses rotating field.

FIGURE 4.10. Operating modes.

A plot of T_{dev} versus ω_{rm} from -600 to $+2400$ rev/min appears in Figure 4.10a. Consider:

- For $0 < \omega_{rm} < 1800$ rev/min (ω_{rm}) or $1 > s > 0$: Torque and speed are positive, as is the input power. This reveals that we are operating in the *motor forward* (MF) mode.
- For $\omega_{rm} > 1800$ rev/min (ω_{rm}) or $s < 0$: Torque and the input power are negative, and speed remains positive. This reveals that we are operating in the *generator forward* (GF) mode. You might question the sign on T_{dev}. Recall Equation 4.32b,

$$T_{dev} = 3I_A'^2 R_{2x}'/(s\omega_{sm})$$

Hence, if $s < 0$, $T_{dev} < 0$, since all other terms are positive, meaning that T_{dev} opposes rotation.

- For $\omega_{rm} < 0$, or $s > 1$. Torque and the input power are positive and speed is negative, as is the output power. This is a mode that we have not anticipated! The machine has ceased to function as an energy conversion device! Energy flows in to the machine from both the electrical and mechanical sides and is consumed entirely in the machine losses! Let us define this new mode as *braking reverse* (BR) since T_{dev} does oppose rotation and is trying to stop the machine, and $\omega_{rm} < 0$.

Now consider reversing the phase sequence of the applied voltage, which may be done by reversing any two stator connections as indicated in Figure 4.10b. From an analytical perspective the positive sense of speed is reversed, since that was defined as the direction in which the stator field was rotating. However, if the forward sense of speed is not changed, it is clear how we can enter the *motor reverse* (MR), the *generator reverse* (GR), and the *braking forward* (BF) modes.

Let us consider a classic problem in motor applications, in the context of Example 4.6.

Example 4.6

Motor of Example 4.3 driven from a balanced three-phase 460 V 60 Hz source. Motor pulley radius = 15 cm.

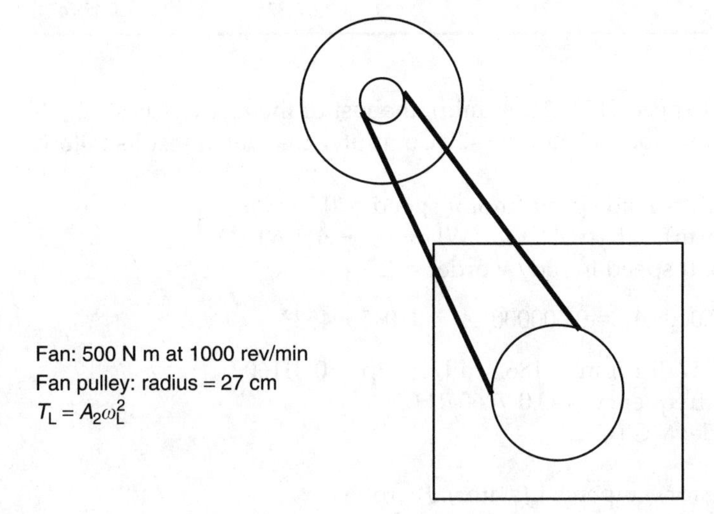

Provide a comprehensive performance analysis for the given system.

SOLUTION

Since 1000 rev/min converts into 104.7 rad/s, the corresponding motor shaft speed is 1800 rev/min (188.5 rad/s)

$$A_2 = 500/(104.7)^2 = 0.04559.$$

$$T_m = (15/27)T_L = 0.02533(\omega_L)^2 = 0.02533(\omega_{rm}/1.8)^2 = 0.007818\omega_{rm}^2$$

Combining the load with the rotational loss:

$$T'_{\text{m}} = T_{\text{m}} + T_{\text{RL}} = 0.007818\omega^2_{\text{rm}} + 0.063\omega_{\text{rm}}$$

The key to the analysis is to determine the system speed. To resolve this, recall an observation from Chapter 3:

The steady-state running speed for any motor–load system occurs at the intersection of the motor and load torque-speed characteristics, i.e., where $T_{\text{dev}} = T'_{\text{m}}$.

A graphical solution appears in Figure 4.10a. Plot T'_{m} on the same graph as T_{dev} and search for the point of intersection (0). For greater accuracy, we can execute a search procedure (implemented by computer) for the speed, by guessing an initial speed, and computing corresponding torques. Then based on the results, refine the speed estimation until the equilibrium point is located. Results of such an analysis are

ω_{rm} (rpm)	ω_{L} (rpm)	($T_{\text{dev}} - T_{\text{RL}}$) (N m)	T_{m} (N m)	T_{L} (N m)
1764.0	980.0	459.61	266.78	480.20
1781.5	989.7	239.78	272.11	489.80
1778.6	988.1	278.18	271.21	488.18
1779.2	988.5	269.95	271.41	488.53
1779.1	988.4	271.67	271.37	488.46
1779.1	988.4	271.31	271.37	488.47

Now that we know the shaft speed (1779.1 rev/min), the rest of the analysis is straightforward and follows the same method as Example 4.3. Computer-calculated results follow:

Gear ratio = GR = load speed/motor speed = 0.5556
Load torque (N m) = $T_{\text{L}} = A_0 + A_1$*WL + ⋯ + A_N*WL**N
WL = Load shaft speed in rad/s; order = 2

$$A_0 = 0.000;\ A_1 = 0.000000;\ A_2 = 0.045594533$$

ω_{rm} = speed = 1779.1 rpm = 186.3 rad/s; slip = 0.01160
Stator voltage, frequency: 460.0 V; 60.0 Hz
Operating mode: MOTOR

Currents in (stator) amperes I_{a}(stator), I'_{A} (rotor)
$I_{\text{a}} = 79.452$ at $-31.16°$; $I'_{\text{A}} = 69.823$ at $-5.51°$

Torques in N m (RL = rotational loss):
$T_{\text{dev}}, T_{\text{RL}}, T_{\text{m}}$ = 283.12; 11.735; 271.37

Powers in kW:
Mechanical output = 50.562 (67.777 hp)
Stator input = 54.168; Developed = 52.748

Losses in kW:
SWL = 0.801; RWL = 0.619; P_{RL} = 2.186; P_{LOSS} = 3.607

Efficiency = 93.34%; Power factor = 85.57%

4.8 Dynamic Performance

There is a need to understand motor–load systems when accelerating (or decelerating). Recall Equation 3.1:

$$T_{dev} - T'_m = T_a = J\frac{d\omega_{rm}}{dt}$$

If $T_{dev} > T'_m$, then T_a and $(d\omega_{rm}/dt)$ are greater than zero, the system accelerates. If $T_{dev} < T'_m$, then T_a and $(d\omega_{rm}/dt)$ are negative, the system slows down.

To analyze the situation, it is necessary to solve Equation 3.1. The solution must be implemented using numerical methods, since the torques are nonlinear functions of speed. Actually, the torques are functions of time, since we are now in a time-varying situation. Indeed, the equivalent circuit of Figure 4.8b is technically inapplicable, since it is restricted to balanced three-phase AC situations. What is required is a more general and powerful model for a rigorously correct analysis. Fortunately, experience shows that for most practical situations, excellent results can be obtained by using "quasi-AC" methods: that is, the circuit model of Figure 4.8b may still be used, and balanced three-phase AC assumed, provided that the rms voltage and speed do not change significantly within the time step selected to compute the dynamic response. For 60 Hz applications, the smallest appropriate time step is about 0.02 s. Most loads and motors will have sufficient inertia to meet this restriction. However, there are high-performance situations where this restriction is not met, and the reader is cautioned to be aware of this issue. Also note that more data are needed, specifically the inertia constants for the motor and load.

Starting the system is an important dynamic situation. A practical concern is that the motor may draw excessive current under starting conditions. The longer it takes to start, the more serious this over-current condition becomes. It not only may thermally damage the windings, but it complicates the motor protection scheme, which is designed to protect against over-current. Furthermore, the electrical source must tolerate this excessive current as well, which can be a problem. One way to reduce the current is to reduce the applied voltage; unfortunately this also reduces T_{dev}, making the starting interval even longer. What is needed is a more sophisticated approach, somehow controlling the current during starting, while maintaining high starting torque.

An ingenious solution to the problem is the temporary insertion of rotor resistance during starting. As we shall see, this will reduce stator current, while actually increasing T_{dev}! We follow our usual approach by studying the problem through an example.

Example 4.7
Consider starting the system of Example 4.6 from a 460 V source from standstill.

$$J_{MOTOR} = 0.9299\,\text{kg m}^2, \qquad J_{LOAD} = 7.5\,\text{kg m}^2$$

(a) Determine and plot the dynamic speed and current for zero rotor resistance ($R_x = 0$).
(b) Determine and plot the dynamic speed and current for nonzero rotor resistance. Design a three-step starter (select three values of R_x).

SOLUTION

(a) Reflecting J_L to the motor shaft:

$$J_m = (GR)^2 J_L = (0.5556)^2 (7.5) = 2.3148\,\text{kg m}^2$$

or

$$J = 0.9299 + 2.3138 = 3.2447\,\text{kg}\,\text{m}^2$$

Thus the equation to be solved is

$$T_{\text{dev}} - T'_{\text{m}} = T_{\text{dev}} - (0.007818\omega_{\text{rm}}^2 + 0.063\omega_{\text{rm}}) = T_{\text{a}} = 3.2447\,(\text{d}\omega_{\text{rm}}/\text{d}t)$$

T_{dev} is a nonlinear function of ω_{rm}. Consider the approximation

$$T_{\text{a}} = 3.2447(\text{d}\omega_{\text{rm}}/\text{d}t) \approx (3.2447)(\Delta\omega_{\text{rm}}/\Delta t)$$

In the first time step, at $t = 0$, $\omega_{\text{rm}} = 0$. For $\Delta t = 0.02$ s:

$$\Delta\omega_{\text{rm}} = (T_{\text{a}})(\Delta t)/(3.2447) = (304 - 0)(0.02)/(3.2447) = 1.874\,\text{rad/s}$$

Slip and speed changed about 1% (instead of holding fixed at 1.00, s varied from 1.00 to 0.99). Consequently, the assumption that T_{a} is constant for Δt is not unreasonable, particularly if we evaluate T_{a} at the midpoint of the step ($s = 0.995$).

Computer-calculated results follow:

Load torque (N m) = T_{L} = 0.045594*WL**2; WL = load speed in rad/s
Gear ratio = GR = load/motor speed = 0.5556; Js in kg m^2
Motor inertia (J_{M}) = 0.9299; load inertia (J_{L}) = 7.5000
$J_{\text{T}} = J_{\text{M}}$ + sqr(GR) × J_{L} = 0.9299 + 2.3148 = 3.2447

Time (s)	Speed (rpm)	Current (A)	T_{dev} (N m)	T_{m} (N m)	T_{a} (N m)	R_{x} (mΩ)	W_{loss} (kJ)
0.00	0	689.3	304	0	304	0.00	0.2
0.19	181	687.5	336	3	332	0.00	23.0
0.37	361	685.1	375	11	362	0.00	43.5
0.53	541	681.8	425	25	396	0.00	62.3
0.68	720	677.1	488	44	439	0.00	79.0
0.81	901	669.9	574	69	498	0.00	93.9
0.92	1082	658.1	691	100	585	0.00	106.5
1.02	1264	636.0	862	136	718	0.00	116.6
1.10	1444	587.3	1099	177	912	0.00	123.8
1.16	1620	453.3	1270	223	1036	0.00	128.1
1.33	1779	70.5	286	271.3	3	0.000	130.2

Examine the results plotted in Figure 4.11a. The starting time was 1.33 s, which is pretty good. However, the maximum current was 689 A, which is 621% of rated! At 50% voltage, the maximum current drops to 345 A (311% of rated), but the starting time extends to 8.1 s (609% of the rated voltage value)! Keep these figures in mind as we consider insertion of rotor resistance.

You may have been curious about the right column in the above table (W_{loss}). This is the total energy loss in the machine, over the starting/stopping interval, or

$$W_{\text{loss}} = 3\int_0^{T_{\text{S}}} [R_1(I_1)^2 + R_2(I_2)^2 + P_{\text{RL}}]\,\text{d}t$$

This energy primarily heats the machine windings and must by dissipated before excessive temperatures are reached. For the full voltage start ($R_{\text{x}} = 0$),

$$W_{\text{loss}} = 130.2\,\text{kJ}$$

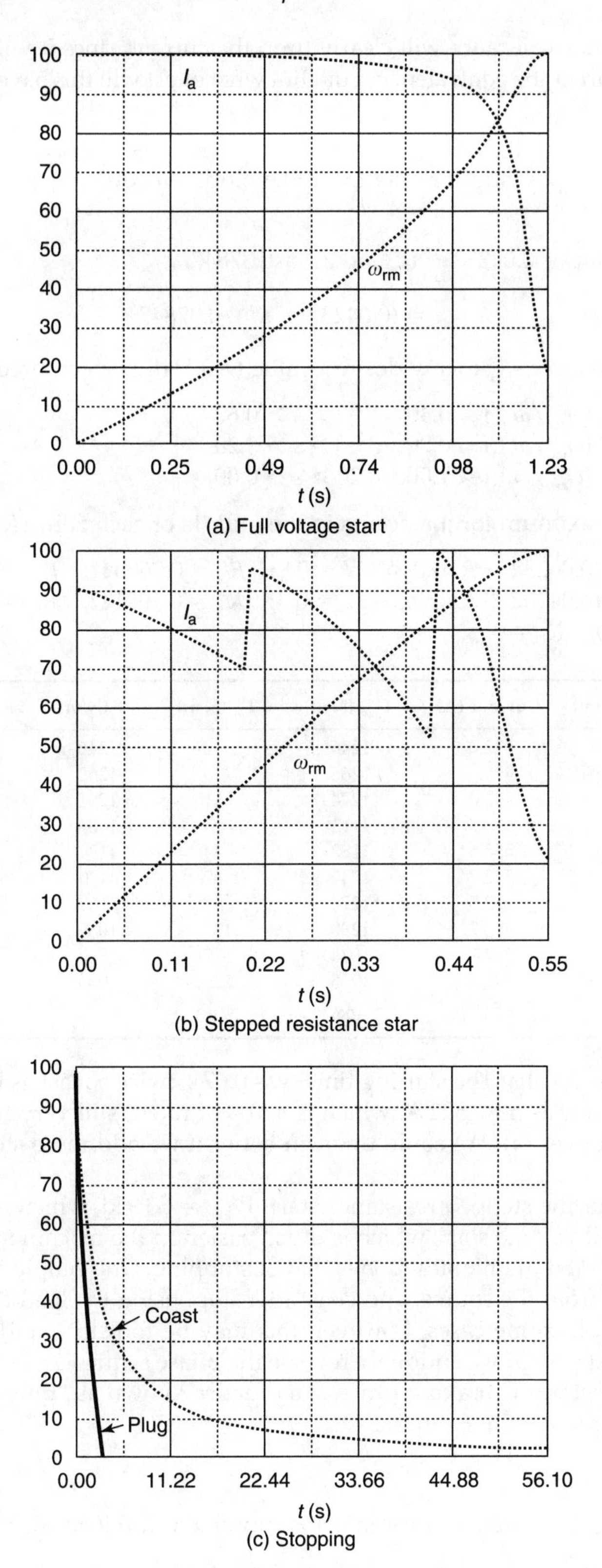

FIGURE 4.11. Dynamic performance.

(b) Insertion of rotor resistance will clearly drop the current, since it will raise the impedance of the rotor part of the equivalent circuit. But what effect will this have on torque? Recall Equation 4.35b

$$s_{MT} = \text{slip for maximum developed torque} = \frac{R'_{2x}}{Z_T}$$

For our machine, $R'_{2x} = 0.04232 + R'_x$ and $Z_T = 0.37847$

$$s_{MT} = (0.04232 + R'_x)/0.37847$$

We can select R'_x to force s_{MT} to any desired value (>0.1118)! Select three speed zones:

Zone 1: $0.00 < (\omega_{rm}/\omega_{sm}) < 0.40 \quad 1.00 > s > 0.60$
Zone 2: $0.40 < (\omega_{rm}/\omega_{sm}) < 0.80 \quad 0.80 > s > 0.20$
Zone 3: $0.80 < (\omega_{rm}/\omega_{sm}) < 1.00 \quad 0.20 > s > 0.00$

Select R'_x so that maximum torque occurs in the middle of each zone ($R'_x = 0$ in Zone 3).

Zone 1: $s_{MT} = (0.04232 + R'_x)/0.37847 = 0.8; \quad R'_x = 0.2605\,\Omega; \quad R_x = 41.68\,\text{m}\Omega$
Zone 2: $s_{MT} = (0.04232 + R'_x)/0.37847 = 0.4; \quad R'_x = 0.1091\,\Omega; \quad R_x = 17.46\,\text{m}\Omega$
Zone 3: $R'_x = 0; \quad R_x = 0.$

Time (s)	Speed (rpm)	Current (A)	T_{dev} (N m)	T_m (N m)	T_a(N m)	R_x (mΩ)	W_{loss} (kJ)
0.00	0	522.0	1246	0	1246	41.68	0.1
0.05	185	499.6	1266	3	1262	41.68	3.4
0.10	363	473.0	1274	11	1261	17.46	6.2
0.16	544	583.7	1112	25	1083	17.46	11.2
0.21	720	558.5	1182	44	1134	17.46	15.6
0.26	903	522.7	1245	69	1170	17.46	19.6
0.32	1083	473.5	1274	99	1168	17.46	22.8
0.37	1263	403.2	1230	135	1086	17.46	25.4
0.43	1443	304.2	1043	177	856	0.00	27.4
0.49	1625	447.4	1268	225	1033	0.00	31.9
0.67	1779	70.6	286	271.3	3	0.000	33.9

Again, examine the results. The starting time was 0.67 s, twice as fast as in (a). Even better, the maximum current is now 522 A, which is 470% of rated, still very high, but a significant improvement over (a)! We could do even better if we add more steps. The plot is in Figure 4.11b.

Also note that for the stepped resistance start, $W_{loss} = 33.9$ kJ, which is only 26% of that absorbed in the full voltage start, which is much easier on the machine thermally.

Another important dynamic situation is that of stopping. One simple approach is to disconnect the stator from the source, and coast to a stop, using the load and rotational loss as braking torque. In some cases, however, this may be too slow, and we may wish to resort to a powered stop. If we suddenly reverse the phase sequence (see Figure 4.10b), we can use T_{dev} to assist in the braking process, a practice known as "plugging." We investigate the stopping problem in Example 4.8.

Example 4.8
Consider stopping the system of Example 4.7, running at full load ($R_x = 0$).

(a) Determine and plot the speed for a coasting stop.
(b) Determine and plot the speed for a plug stop.

SOLUTION

(a) The equation to be solved is

$$T_{\text{dev}} - T_{\text{m}} = 0 - (0.007818\omega_{\text{rm}}^2 + 0.063\omega_{\text{rm}}) = T_{\text{a}} = 3.2447\frac{d\omega_{\text{rm}}}{dt}$$

Observe that T_{dev} is zero. Computer-calculated results follow:

Load torque (N m) = T_L = (0.04559) *WL**2
WL = Load shaft speed in rad/s;

Gear ratio = GR = load/motor speed = 0.5556;
$J_T = J_M + \text{sqr(GR)} \times J_L = 0.9299 + 2.3148 = 3.2447\,\text{kg}\,\text{m}^2$

Time (s)	Speed (rpm)	Current (A)	T_{dev} (N m)	T_{m} (N m)	T_{a} (N m)	R_x (mΩ)	W_{loss} (kJ)
0.00	1800	0.0	0	278	−290	0.00	0.2
0.30	1559	0.0	0	229	−240	0.00	0.8
0.50	1430	0.0	0	191	−201	0.00	1.2
0.90	1228	0.0	0	139	−147	0.00	1.7
1.30	1076	0.0	0	106	−113	0.00	2.0
2.00	884	0.0	0	71	−77	0.00	2.5
2.90	719	0.0	0	46	−51	0.00	2.9
4.50	537	0.0	0	26	−29	0.00	3.4
7.60	356	0.0	0	11	−14	0.00	3.8
15.80	179	0.0	0	3	−4	0.00	4.1
82.80	18	0.0	0	0.0	−0	0.000	4.4

Examine our results. At first, we stop fairly quickly, reaching 25% speed in about 5 s. However, it will take over a minute to reach a complete stop, which may or may not be a problem. W_{loss} is small, since the currents are zero. Plots are in Figure 4.11c.

(b) For the plug stop, the rotor and rotating field are in opposite directions. For example, at $t = 0$,

$$s = \text{slip} = \frac{1800 - (-1799)}{1800} = 1.988$$

But since the rotating field has reversed, so has T_{dev}. T_{dev} and T'_{m} are both negative, and the machine is clearly in the braking mode. As the system slows, T_{dev} will become even greater, since we are moving toward $s = 1$. Computer-calculated results follow:

Time (s)	Speed (rpm)	Current (A)	$-T_{\text{dev}}$ (N m)	T_{m} (N m)	T_{a} (N m)	R_x (mΩ)	W_{loss} (kJ)
0.00	1800	696.5	155	−278	445	0.00	0.2
0.15	1619	696.2	163	−225	399	0.00	18.1
0.31	1438	695.8	172	−178	360	0.00	37.8
0.49	1259	695.4	182	−136	327	0.00	59.3
0.68	1079	694.9	193	−100	300	0.00	83.0
0.89	899	694.3	206	−70	281	0.00	108.2
1.11	720	693.7	220	−45	269	0.00	134.8
1.34	540	692.9	236	−25	265	0.00	162.2
1.57	359	692.0	255	−11	269	0.00	189.6
1.80	178	690.8	278	−3	282	0.00	216.1
2.01	2	672.1	304	−0.0	304	0.000	240.6

Examine our results. The system stops in about 2 s!
Also note that

$$\text{plug stop: } W_{\text{loss}} = 240.6\,\text{kJ}$$

which is 185% of the 130.2 kJ absorbed in the full voltage start, indicating that the plug stop is very hard on the machine thermally!

4.9 Cage Rotor Machines

It was noted in Section 4.4 that the equivalent circuit in Figure 4.8b, which is our basic machine model, is in serious error for cage rotor machines. We will now investigate this issue.

Recall that the rotor frequency varies with speed. In the wound rotor case, the current is confined to a well-defined path within the rotor conductors. Whereas the rotor resistance and leakage inductance vary slightly with frequency, the effect is minor and the rotor parameters (R_2, L_2) are reasonably constant.

However, this is not the case with the cage rotor machine. As the rotor frequency varies, the current distribution within the rotor bars may vary quite a bit, which has a significant effect on the machine characteristics. To appreciate the effect, examine the torque-speed characteristics of Figure 4.12, which are for different cage rotor bar designs. Recall that we added resistance to the wound rotor machine in Section 4.7, and dramatically altered the torque-speed characteristic. Hence, if the rotor bars are made out of materials of different resistivities, or the shape is altered, we would anticipate that this would have a significant impact on the torque and current at a given slip, which indeed it does.

The relation between rotor bar design and machine performance is fairly well understood, so much so that these designs have standard designations. The agency in the United States which has established standards is the National Electrical Manufacturers' Association (NEMA), and the corresponding designs are called "NEMA A," "NEMA B," etc. Torque characteristics and rotor bar designs are shown in Figure 4.12. Five common cage rotor NEMA designs follow. Numbers are in percentage of rated, and typical for a 100 hp machine at rated voltage. Values will vary for other sizes.

NEMA A. For all-purpose applications. Normal starting torque (105%), normal starting current (700%), high breakdown torque (210%), low-full-load slip (2 to 4%).

NEMA B. Similar to A, except lower starting current. For all-purpose applications. Normal starting torque (105%), low starting current (500%), high breakdown torque (200%), low-full-load slip (2 to 4%).

NEMA C. For applications requiring high starting torque. High starting torque (225%), low starting current (500%), medium breakdown torque (190%), low- to medium-full-load slip (3 to 6%).

NEMA D. For applications requiring unusually high starting torque. Very high starting torque (275%), low starting current (400%), low breakdown torque, high-full-load slip (5 to 10%).

NEMA E. For applications requiring unusually low starting torque. Low starting torque (70%), very low starting current (300%), low breakdown torque (160%), low-full-load slip (1 to 3%).

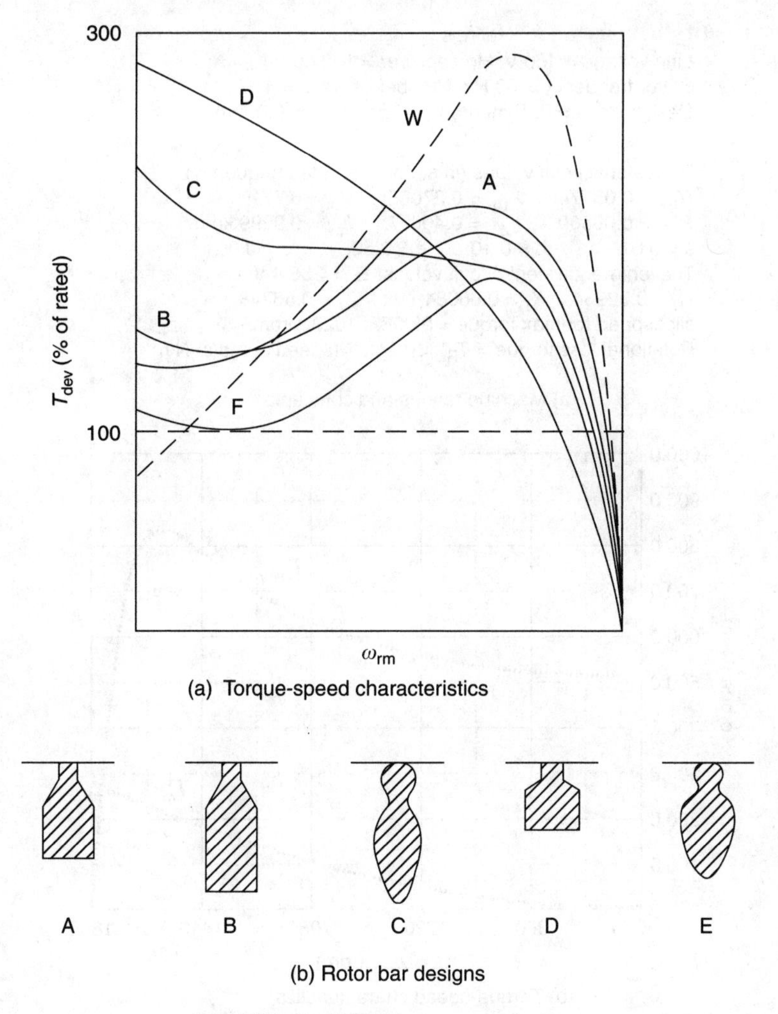

FIGURE 4.12. NEMA cage rotor designs.

Wound Rotor ("Design W"). This is not a NEMA design, but is a useful designation for our purposes. For applications requiring operation over wide speed ranges. At zero resistance: low starting torque, high starting current, very high breakdown torque, medium- to low-full-load slip. Characteristics can be adjusted by insertion of rotor resistance.

Our main interest is to consider what modifications can be made to the circuit in Example 4.8b, so that we might have an accurate simulation of the machine torque and current characteristics for cage rotor machines.

The topology of the circuit is logical from a theoretical perspective; the problem is that R_2 and L_2 vary with rotor frequency, and thus speed, or slip. Hence consider

$$R_2(s) = R_{20}(1 + as + bs^2) \tag{4.36a}$$

$$L_2(s) = L_{20}(1 + cs + ds^2) \tag{4.36b}$$

Ratings
Line voltage = 460 V; Horsepower = 100 hp
Stator frequency = 60 Hz; Number of poles = 4
Design class = B; Synchronous speed = 1800.0 rpm

Equivalent circuit values (in stator Ω at rated frequency)
$R_{10} = 0.03174$; $X_{10} = 0.27085$; $X_m = 6.771$
$R'_{20} = 0.06560$; $X'_{20} = 0.40627$; $J_m = 0.9299$ kg m^2
a = 0.61; b = 0.10; c = 0.52; d = 0.00
Thevenin-equivalent circuit values: $E_T = 255.4$ V
$R_T = 0.02934$; $X_T = 0.66684$; $Z_T = 0.66748$
slip, speed for max torque = 0.0983; 1623.1 rpm
Rotational loss torque = $T_{RL} = 0.042 \times$ (speed in rad/s) N m

(a) Machine ratings and constants

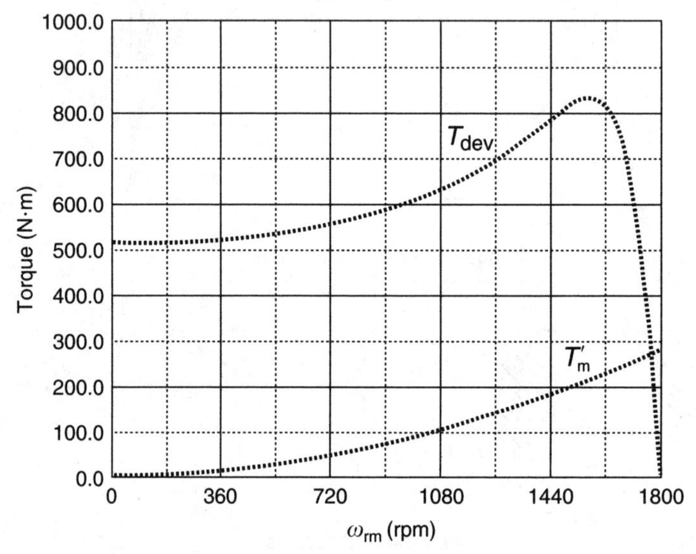

(b) Torque-speed characteristics

at rated output	$\omega_{rm} = 1746.9$ rpm	$I = 119$ A	$T_m = 408$ N m
at $\omega_{rm} = 0$ rpm	$I = 551$ A	$T_m = 512$ N m	
at $\omega_{rm} = 1623$ rpm	$T_{dev} = T_{max} = 833$ N m		

(c) Performance values

FIGURE 4.13. Data for a three-phase Design B cage rotor induction motor.

We could perform a variable frequency blocked rotor test, varying the frequency from 0 to the rated stator frequency, and measuring $V_{BR}(\omega)$, $I_{BR}(\omega)$, $P_{BR}(\omega)$, and $T_{BR}(\omega)$. If the parameters R_1 and L_m are known, it is possible to get a least mean-squared error fit to the measured data by selecting an optimum combination of the seven constants L_1, R_{20}, L_{20}, a, b, c, and d. Data for a typical 460 V 100 hp Design B machine is provided in Figure 4.13.

We return to the example application presented in Example 4.6, namely that of a motor driving a fan load. This time we will use a Design B motor; the details are presented in Example 4.9.

Example 4.9
The Design B motor of Figure 4.13 driven from a balanced three-phase 460 V 60 Hz source. Motor pulley radius = 15 cm.

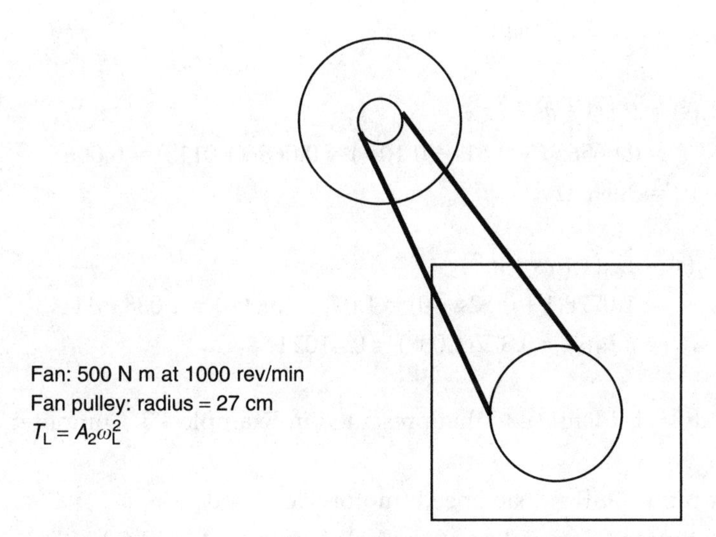

Provide a comprehensive performance analysis for the given system.

SOLUTION
The load analysis is the same as presented in Example 4.6. At the motor shaft,

$$T_m = 0.007818\omega_{rm}^2$$

Combining the load with the rotational loss:

$$T'_m = T_m + T_{RL} = 0.007818\omega_{rm}^2 + 0.042\omega_{rm}$$

Again, recall

> *The steady-state running speed for any motor–load system occurs at the intersection of the motor and load torque-speed characteristics, i.e., where $T_{dev} = T'_m$.*

The load torque characteristic (T'_m) is presented in Figure 4.13b. We can execute a search procedure (implemented by computer) for the intersection speed by guessing an initial speed, adjusting R_2 and L_2, and computing the corresponding torques. Then, based on the results, refine the speed estimation until the equilibrium point is located. Results of such an analysis are:

ω_{rm} (rpm)	ω_L (rpm)	$(T_{dev} - T_{RL})$ (N m)	T_m (N m)	T_L (N m)
1764.0	980.0	288.68	266.78	480.20
1766.0	981.1	273.74	267.38	481.29
1766.6	981.4	269.36	267.56	481.60
1766.7	981.5	268.12	267.61	481.69
1766.8	981.5	267.76	267.62	481.72

Now that we know the shaft speed (1766.8 rev/min), the rest of the analysis is straightforward and follows the same method as in Example 4.3. The significant difference (from the wound rotor case) is that the rotor elements must be adjusted for frequency:

At 1766.8 rpm, the slip $= s = 0.01845$

$$R_2(s) = R_{20}(1 + as + bs^2)$$
$$= 0.0656\ (1 + 0.61s + 0.10s^2) = 0.0656(1.0113) = 0.0663$$
$$R_2(s)/s = 3.596\ \Omega$$

$$L_2(s) = L_{20}(1 + cs + ds^2)$$
$$= 1.0776(1 + 0.52s + 0) = 1.0776(1.0096) = 1.088\ \text{mH}$$
$$X_2 = \omega L_2(s) = 0.377(1.088) = 0.4102\ \Omega$$

The rest of the analysis is identical to that presented in Example 4.3. Computer-calculated results follow:

Gear ratio = GR = load speed/motor speed = 0.5556
Load torque (N m) = $T_L = A_0 + A_1$*WL + ··· + A_N*WL**N
WL = Load shaft speed in rad/s; order = 2

$A_0 = 0.000$; $A_1 = 0.000000$; $A_2 = 0.045594533$

ω_{rm} = speed = 1766.8 rpm = 185.0 rad/s; slip = 0.01845
Stator voltage, frequency: 460.0 V; 60.0 Hz
Operating mode: MOTOR

Currents in (stator) amperes ... I_a(stator), I'_A (rotor)
$I_a = 81.986$ at $-36.43°$; $I'_A = 69.373$ at $-9.73°$

Torques T_{dev}, T_{RL}, T_m: 275.43; 7.769; 267.62 N m
Gear ratio = GR = load speed/motor speed = 0.5556

Powers in kW:
Mechanical output = 49.523 (66.384 hp)
Stator input = 52.558; Developed = 50.960

Losses in kW:
SWL = 0.640; RWL = 0.958; $P_{RL} = 1.437$; $P_{LOSS} = 3.025$

Efficiency = 94.22%; Power factor = 80.46%

The results are comparable to the wound rotor case. The speed is a little slower, the efficiency slightly higher, and the power factor about 5 points lower. Now consider the starting problem.

Example 4.10
The system of Example 4.9 is started from a 460 V source from standstill. The same inertias (as in Example 4.7) were assumed.

$$J_{MOTOR} = 0.9299\ \text{kg m}^2,\quad J_{LOAD} = 7.5\ \text{kg m}^2$$

Determine and plot the dynamic speed and current.

SOLUTION

EMAP-calculated results follow:

$$\text{Load torque (N m)} = T_L = A_0 + A_1\text{*WL} + \cdots + A_N\text{*WL**}N$$

$$\text{WL} = \text{Load shaft speed in rad/s; order} = 2$$

$$A_0 = 0.000; \quad A_1 = 0.000000; \quad A_2 = 0.045594533$$

$$\text{Gear ratio} = \text{GR} = \text{load/motor speed} = 0.5556;\ Js \text{ in kg m}^2$$

$$\text{Motor inertia } (J_M) = 0.9299; \quad \text{load inertia } (J_L) = 7.5000$$

$$J_T = J_M + \text{sqr(GR)} \times J_L = 0.9299 + 2.3148 = 3.2447$$

Time (s)	Speed (rpm)	Current (A)	T_{dev} (N m)	T_m (N m)	T_a (N m)	T_{ap} (%)	W_{loss} (kJ)
0.00	0	550.8	512	0	512	100.0	0.2
0.12	181	537.9	515	3	511	100.0	10.2
0.24	361	524.3	522	11	510	100.0	19.6
0.36	541	510.0	535	25	508	100.0	28.6
0.48	723	494.3	555	44	508	100.0	37.2
0.60	903	476.8	585	69	512	100.0	45.1
0.72	1082	455.9	631	100	526	100.0	52.3
0.83	1260	428.5	696	135	555	100.0	58.5
0.94	1440	384.6	786	177	603	100.0	63.4
1.04	1620	288.8	817	224	586	100.0	66.8
1.32	1766	70.1	278	267.5	3	100.0	68.7

The advantages of the Design B motor are now clearer. There is a significant reduction in the starting current, compared to the full voltage, zero resistance wound rotor machine start, even though the starting times and rotor energy dissipation was almost the same. The other NEMA designs have comparable specialized characteristics that make them particularly suited to certain applications.

4.10 Thermal Considerations

It was noted in Chapter 1 that EM system performance is always limited by waste heat, which raises winding and core temperatures. This is a constant concern with all motors and generators. Higher temperatures can damage the winding insulation, and the core, and in extreme cases, melt the conductor. Hence, it is important to correlate machine performance with temperature. To do this accurately, one will require a thermal model for the machine, which may assume the form of a thermal circuit. It is helpful to make an electrical–thermal analogy, as presented in Figure 4.14a. A simple first-order thermal circuit machine is presented in Figure 4.14b. The corresponding fundamental thermodynamic equation is

$$P = G\theta + C\frac{d\theta}{dt}$$

where P is the net power internally dissipated inside the motor (equal to the total losses of the motor, minus the heat extracted). The losses, of course, depend on motor operating conditions. θ is the machine temperature above ambient. Two common temperature (θ)

Concept	Electrical	Thermal
Fundamental quantity	Charge (Q in coulombs)	Energy (Q in joules)
Flow variable	Current (I in C/s = A)	Power (P in J/s = W)
Nodal variable	Voltage (V in volts)	Temperature (θ in °C)
Conductivity	σ (S/m)	κ (W/m °C)
Flow conductance	$G = \sigma A/l$ (S)	$G = \kappa A/l$ (W/°C)
Conv. Trans. Coef.		h(W/m^2 °C)
Flow convection		G = hA (W/°C)
Specific heat		C_p (J/kg °C)
Storage	Capacitance (C in C/V)	Capacitance (C in J/°C) $C = C_p \times$ mass
"Ohm's law"	$I = GV$	$P = G\theta$
	$i = C\,dv/dt$	$p = C\,d\theta/dt$
Time constant	$\tau = C/G$	$\tau = C/G$
Polarity	+, −	Hot, cold

(a) Thermal–Electrical circuit analogs

θ = Nodal Variable

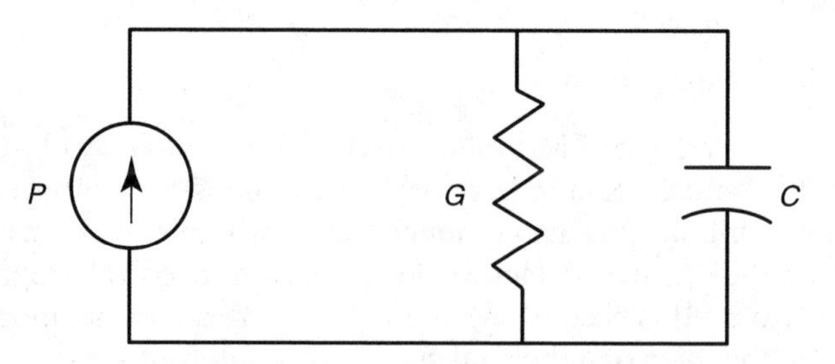

Temperature above ambient = θ in °C
Net thermal power (total losses) injected into machine = P in W
Machine thermal capacity = C in J/°C
Machine thermal conductance = G in W/°C

(b) First-order motor thermal circuit model

FIGURE 4.14. A simple first-order thermal circuit model for EM machines.

ratings are 40 and 55°C. When the machine is "cold" (i.e., at the temperature of its surrounding environment), θ is zero. G represents the motor thermal conductance (from the machine to its operating environment).

When $\theta > 0$, heat flows *out of* the machine and when $\theta < 0$, heat flows *into* the machine. We will explore the situation with an example.

Example 4.11

The 460 V 100 hp wound rotor motor defined in Example 4.3 continuously runs at rated conditions, reaching rated temperature, which is 55°C over ambient. Assume that

(a) The motor is suddenly disconnected from the source, after which it stops rapidly. Measurements show that it cools to essentially room temperature in 20 min after stopping. Determine the thermal circuit constants (G, C), and the thermal time constant.
(b) Assume the motor is started across the line at full voltage, drawing a current of 600 A over the starting period, which takes 1.3 s. Estimate the temperature rise.
(c) Studies show that when the load inertia is increased to 30 kg m^2, the starting time is stretched to 13 s. Again, estimate the temperature rise.

SOLUTION

(a) At rated conditions the motor losses are 4.882 kW. Hence,

$$G = \frac{P}{\theta} = \frac{4.882}{55} = 0.08876 \text{ kW/°C}$$

Assume that motor cools to ambient temperature in five time constants. Therefore,

$$\tau = C/G = (20 \times 60)/5 = 240 \text{ s}$$

and

$$C = \tau G = 21.3 \text{ kJ/°C}$$

(b) At full load,

$$\text{SWL} + \text{RWL} = \text{Total losses} - P_{\text{RL}} = 4.882 - 2.164 = 2.718 \text{ kW}$$

During starting, these losses rise to

$$(600/107.4)^2\,(2.718) = 84.83 \text{ kW}$$

$$84.83 = 0.08876\theta + 21.3\frac{d\theta}{dt}$$

$$240\frac{d\theta}{dt} + \theta = 955.7$$

Solving for θ:

$$\theta = 955.7(1 - e^{-t/240})$$

Since starting lasts for only 1.3 s,

$$\theta = 5.16°\text{C}$$

Hence, the machine can easily start with no excessive temperature rise.

TABLE 4.1

Maximum Allowable Temperatures by Insulation Class

Insulation Class	Maximum Temperature (°C)
A	105
B	130
F	155
H	180

(c) The same temperature relation as was used in (b) is applicable. For a starting interval of 13 s,

$$\theta = 50.4°\text{C}$$

This time we barely made it, coming just under our allowed rise of 55°C. In fact, if the motor had been running at rated conditions, then stopped, and restarted, we would have driven the temperature to 55 + 36.6 = 91.6°C over ambient. In fact, assuming an ambient temperature of 40°C, the temperature would have risen to 91.6 + 40 = 131.6°C, well in excess of the maximum temperature for Class A insulation (105°C) (Table 4.1).

One might pose the question of just how much current, and for how long can a motor tolerate? Using the data presented in Example 4.11, for the motor of Example 4.3, the losses can be approximated using the expression

$$P = P_{\text{RL}} + \text{SWL} + \text{RWL} = 2.164 + 2.356\left(\frac{I}{100}\right)^2$$

Assume the motor has Class B insulation (rated at 130°C), and is operating in a 40°C environment. Also assume operation at rated conditions, such that the initial temperature is 40 + 55 = 95°C, when suddenly the motor current rises to some level I. We wish to solve for the time it takes the motor to reach the maximum allowed temperature. We realize that our thermal model is somewhat simplistic, since it assumes that the machine temperature is uniform throughout. We know that the hottest locations will be at the source of the losses, which is primarily within the windings. Allowing a 20°C margin to account for hot spots, we conclude that the maximum allowable temperature is 130 − 20 = 110°C, or θ_{MAX} = 110 − 40 = 70°C. Hence, our problem shall be to determine how long it will take (T) for the motor to rise from θ_0 (55°C) to θ_{MAX} (70°C) for sustained over-current I.

$$P = G\theta + C\frac{d\theta}{dt}$$

or

$$\tau\frac{d\theta}{dt} + \theta = \frac{P}{G}, \quad \tau = \frac{C}{G}$$

which solves to

$$\theta_{\text{MAX}} = \frac{P}{G} + \left(\theta_0 - \frac{P}{G}\right)e^{-T/\tau}$$

Solving for T,

$$T = \tau \ln\left(\frac{P - G\theta_0}{P - G\theta_{MAX}}\right) = 240 \ln\left(\frac{P - G\theta_0}{P - G\theta_{MAX}}\right)$$

Some typical values appear below.

Current (% of rated)	Current (I) (A)	Losses (P) (kW)	T (s)
175	188.0	10.49	65.0
200	214.8	13.04	42.8
225	241.7	15.92	30.8
250	268.5	19.15	23.5
275	295.4	22.72	18.6
300	322.2	26.63	15.2

Again, the reader is cautioned that the foregoing analysis was based on an approximate first-order thermal model. A more accurate model is presented in Figure 4.15. The problem is in acquiring data to implement the model.

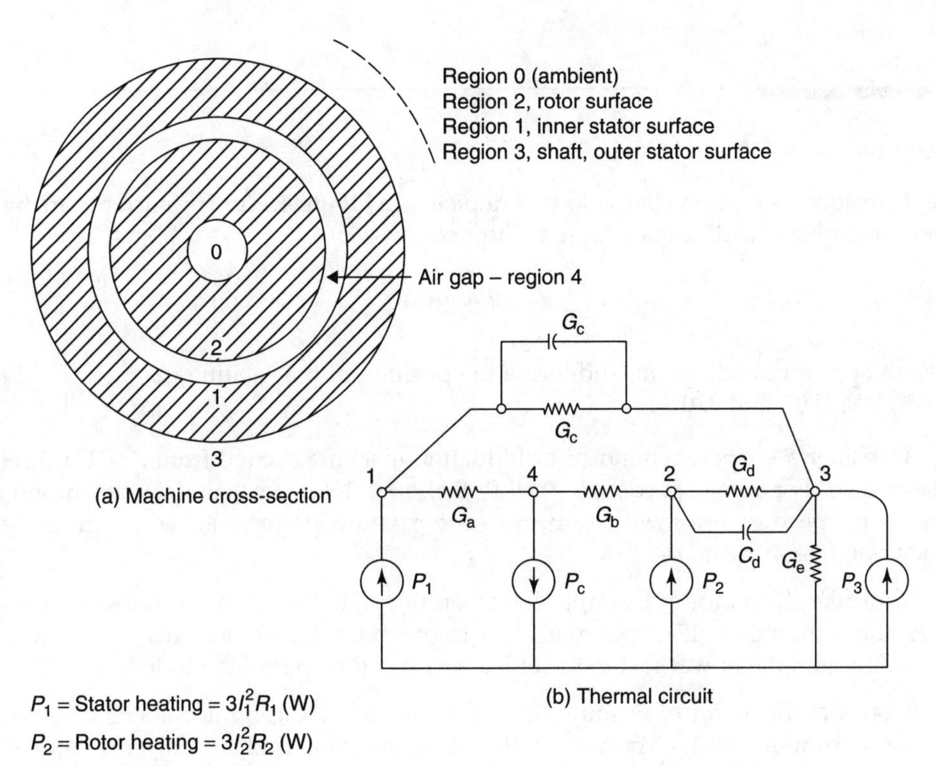

FIGURE 4.15. A more complex thermal model.

4.11 Summary

Arguably, the single most important electric motor is the polyphase induction machine, which is the subject of this chapter. It is elegantly simple and efficient, and one of the triumphs of modern engineering, without which many industrial applications would be severely degraded. We have examined the main structural details of the machine and developed accurate mathematical models, which can be used to accurately assess motor performance for a given application. We have demonstrated how machine loading and excitation is correlated to machine performance, in terms of speed, currents, powers, and torques. We noted that there were two basic rotor designs (wound and cage), and have examined the latter in some detail. We extended our analysis to include an estimate of machine temperature.

The next logical step is to investigate control options, which are necessary for many applications. This important topic is the subject of the next chapter.

We must consider the limitations and restrictions of our work. Our approach can be used to accurately assess motor performance, operating under balanced three-phase sinusoidal excitation, and running at constant speed. The approach can be extended to system dynamic performance, such as starting and stopping, as long as quasi-AC balanced operation can be justified. But what if the applied voltages are not balanced? Indeed, what if the excitation voltages are not sinusoidal? And finally, what if the mechanical transients have time constants comparable to the electrical transients. These important topics are considered in Chapter 6.

Problems

4.1 Consider a situation similar to that depicted in Figure 4.12a. The currents are balanced three-phase, with sequence a-c-b. Suppose

$$i_a = I_{max} \cos(\omega t + \beta) = I_{max} \cos(\omega t - 90°)$$

Draw an appropriate diagram, and locate the position of the rotating magnetic field at $\omega t = 0°, 90°, 180°$, and $270°$.

4.2 Consider a six-pole wound rotor induction machine excited from a 60 Hz three-phase source. For rotor speeds of −900, 0, 900, and 1500 rpm, tabulate synchronous speed, slip, speed of rotor with respect to the rotating stator field ($\omega_{R/SF}$), rotor frequency, and operating mode.

4.3 Consider the motor of Example 4.3. Operating at balanced rated stator voltage at 60 Hz, and running at 1775 rpm with the rotor shorted, find all currents, powers, torques, the slip, the power factor, the efficiency, and the operating mode.

4.4 Consider the motor of Example 4.3. Operating at balanced rated stator voltage at 60 Hz, and running at 1785 rpm with the rotor shorted, find all currents, powers, torques, the slip, the power factor, the efficiency, and the operating mode.

4.5 Consider the motor of Example 4.3. Operating at balanced rated stator voltage at 60 Hz, and running at 1815 rpm with the rotor shorted, find all currents, powers, torques, the slip, the power factor, the efficiency, and the operating mode.

4.6 Consider the motor of Example 4.3. Operating at balanced rated stator voltage at 60 Hz, and running at −1800 rpm with the rotor shorted, find all currents, powers, torques, the slip, the power factor, the efficiency, and the operating mode.

4.7 Test data are available for a 300 hp 2400 V 60 Hz four-pole machine of Example 4.3. See Figure 4.9 for the laboratory setups.
DC Test:

$$\text{At } I_{ab} = I_{bc} = I_{ca} = 70\text{ A}; \quad V_{ab} = 26\text{ V}, \quad V_{bc} = 30\text{ V}, \quad V_{ca} = 28\text{ V}$$

$$\text{Use } K_{TEMP}K_{ac} = 1.5$$

BR Test:

$$P_{BR} = 12{,}000\text{ W}, \quad I_{BR} = 72\text{ A} \quad (I_2 = 144\text{ A}), \quad V_{BR} = 442\text{ V}$$

NL Test:

$$P_{NL} = 7000\text{ W}, \quad I_{NL} = 20\text{ A}, \quad V_{NL} = 2400\text{ V}$$

Determine the five element values (R_1, X_1, X_m, R'_2, X'_2), the constant K_{RL}, and the turns ratio N_S/N_R from the given test data.

4.8 The values for the machine considered in Problem 4.7 are:

Three-Phase Induction Motor Data

Ratings

Line voltage = 2400 V; Horsepower = 300 hp
Stator frequency = 60 Hz; Number of poles = 4
Design class = W; Synchronous speed = 1800 rpm
$R'_x = 0.0000\ \Omega$; $N_s/N_r = 2.0$

Equivalent Circuit Values (in Stator Ω at Rated Frequency)

$R_1 = 0.3000$; $X_1 = 1.73$; $X_m = 67.774$
$R'_2 = 0.4716$; $X'_2 = 1.73$; $J_m = 2.5\text{ kg m}^2$
Rotational loss torque $T_{RL} = 0.18688 \times$ (speed in rad/s) N m

Find the Thevenin-equivalent circuit, the slip for maximum torque, and the maximum torque.

4.9 The values for the machine considered in Problem 4.8 are:

Thevenin-equivalent circuit values: $E_T = 1351.1$ V
$R_T = 0.28525$; $X_T = 3.41747$; $Z_T = 3.42935$
slip, speed for max torque = 0.1375; 1552.5 rpm

The motor drives a load described by

$$T_m = 6.0\,\omega_{rm} \text{ in N m}; \quad T'_m = T_m + T_{RL}; \quad \omega_{rm} \text{ in rad/s}$$

(a) Determine the running speed analytically.
(b) Plot the T_{dev} vs. ω_{rm} and T'_m vs. ω_{rm} torque-speed characteristics on the same plot, and prominently indicate the steady-state operating point.

4.10 Repeat Problem 4.9 for a load of $T_m = 2000\,\text{N·m}$; ω_{rm} in rad/s. Note that there are two solutions.

4.11 For the two speeds determined in Problem 4.10, the system can run at only one. Which speed is constitutes a realistic operating point? Explain.

4.12 Plot T_{dev} and T'_m for $-\omega_{sm} < \omega_{rm} < +\omega_{sm}$. For $T_m = 0.06\,\omega_{rm}^2$ in N·m. Note that there are two solutions.

4.13 For the two speeds determined in Problem 4.12, the system can run at only one. Which speed constitutes a realistic operating point? Explain.

4.14 Consider the machine of Problems 4.8 and 4.9. We wish to move the maximum torque point to zero speed by inserting external resistance into the rotor circuit. Determine

(a) the resistance per phase, if a wye connection is used;
(b) the resistance per phase, if a delta connection is used.

4.15 Examine Figure 4.12, which shows the torque-speed characteristics for various NEMA design motors (A,B,C,D,F,W). Rank these in order from highest to lowest in terms of:

(a) starting torque.
(b) running slip (at 100% torque).

4.16 Design a four-step (three rotor resistances) starter for the motor of Example 4.7 (see Example 4.7 for the procedure). Plan to switch at 30, 55, and 85% speed.

4.17 Design a five-step (four rotor resistances) starter for the motor of Example 4.7 (see Example 4.7 for the procedure). Plan to switch at 25, 50, 75, and 90% speed.

4.18 Consider the motor of Example 4.8 to be decoupled from the load.

(a) Compute the time to coast to a stop from 1800 rpm.
(b) Estimate the time for a plugged stop from 1800 rpm. Make the simplifying assumption that the plugging torque is constant at $-200\,\text{N m}$.

4.19 Consider the motor of Example 4.8. Use MATLAB® (or any other computer-aided analysis) to

(a) compute the time to coast to a stop from 1800 rpm.
(b) compute the time for a plugged stop from 1800 rpm.

4.20 Consider a 75 hp 460 V three-phase four-pole 60 Hz induction motor with full load (i.e., rated) speed, efficiency, and pf of 1764 rpm, 93%, 0.866 lagging, respectively, and negligible rotational loss. The motor has $J = 0.5\,\text{kg m}^2$, and drives a load with a linear torque–speed characteristic, and $J = 2.5\,\text{kg m}^2$.

(a) Determine the rated torque and current.
(b) Suppose during starting, the system is controlled in such a way that the motor torque is clamped at 200% rated. Determine the time to start.
(c) While running at rated conditions, the motor circuit breaker is tripped, and the system coasts to a stop. Determine the stopping time.

4.21 Consider the motor of Example 4.3 with a load of

$$T_m = 1.6\,\omega_{rm}\,\text{N·m};\ \omega_{rm}\ \text{rad/s};\ J_L = 5.0\,\text{kg·m}^2$$

Use MATLAB® (or any other computer-aided analysis) to compute the starting time at rated voltage; $R_x = 0$. In particular, plot the speed versus time.

4.22 Repeat Problem 4.21 using the three-step starter designed in Example 4.7.

4.23 An induction motor has a mass of 600 kg and is primarily constructed of iron, with a specific heat of 440 J/kg K, with full load losses of 4 kW; an ambient temperature of 40°C, and a rated temperature rise of 55°C,

(a) Determine its thermal capacity C_T in kJ/K.
(b) Suppose it takes 30 min for the motor temperature to reach 95°C from a cold start (ambient temperature 40°C), supplying full load. Determine its thermal conductance G_T.

4.24 The motor of Example 4.8 is rated at 55°C over a 40°C ambient, and has Class B insulation. Tests show that it cools to essentially room temperature from 95°C in 20 min after stopping. Determine the thermal circuit constants (G, C), and the thermal time constant. Draw the thermal equivalent circuit.

4.25 Consider a 100 kW 460 V three-phase motor rated at 55°C over a 40°C ambient with Class B insulation, whose losses may be approximated as

$$P_{LOSS} = 2 + 2\left(\frac{P_{OUT}}{100}\right)^2 \qquad \text{(powers in kW)}$$

The thermal circuit constants are $G = 10°\text{C/kW}$ and $C = 250\,\text{kJ/°C}$. The motor serves the following load cycle: 100% load for 10 min; 150% load for T_x min; 50% load for 20 min.

Find the maximum value of T_x if the insulation thermal rating is not to be exceeded. Assume an ambient temperature is 30°C and allow 10°C for hot spots.

4.26 Repeat Problem 4.25 if the machine has Class F insulation.

5

Control of AC Motors

In Chapter 4, we studied the polyphase AC induction machine, with emphasis on its application as a motor. For some applications, the system speed is not critical; however for many others, it is necessary to control the running speed and dynamic performance of the system. We now turn our attention to the issue of speed control of AC motors. Even though we have considered only one type of AC machine, the issues we shall discuss are generally applicable to all AC devices, and we use the polyphase AC induction motor as a major application of these principles.

Recall a basic principle from Chapter 3:

> ***The steady-state running speed for any motor–load system occurs at the intersection of the motor and load torque–speed characteristics, i.e., where $T_{dev} = T'_m$.***

Thus, to control the speed, we must adjust the motor torque–speed characteristic (T_{dev}), the load torque–speed characteristic (T'_m), or both.

Consider an automobile as example of a system whose speed must vary over wide ranges (including reversing). How is this done? The accelerator controls fuel flow to the engine, which in effect controls the engine torque–speed characteristic ("T_{dev}"). Also, the transmission, which is in effect a variable-ratio gear box, controls the effective load torque–speed characteristic ("T'_m"),[1] present to the engine at its shaft. The vehicle must run at the steady-state speed defined by this torque balance condition.

Thus, to control the speed, we must adjust the motor torque–speed characteristic (T_{dev}), the load torque–speed characteristic (T'_m), or both, to force the intersection point to occur at the desired speed.

5.1 Control of the Load Torque–Speed Characteristic

There is no technical reason why the speed of the system could not be controlled by mechanical means, at least over some limited range. We could insert some sort of variable-ratio gear box between the load and the motor, which would vary T_m at a given speed. In fact, there is a design, called a "Reeves drive," which employs a pulley whose side spacing is adjustable, thereby permitting a belt to seek out different depths in the pulley groove, and rendering the effective pulley diameter variable. The blower in my grandson Steven's

[1] The true load torque (T_L) is the torque applied to the drive wheels at the point of contact between the tires and the pavement, which transfers to the axial, reflects through the transmission, and appears as T_m at the engine crankshaft.

home forced air heating and cooling system has a motor pulley with an adjustable groove that can be set wide (for the high-speed air conditioning mode) or narrow (for the low-speed heating mode). However, these are the exceptions, not the rule, and in general it is far simpler and cheaper to control the motor torque–speed characteristic (T_{dev}), and thus the speed, by electrical means. It is this second option that we will explore in the remainder of this chapter.

There is one rather obvious and direct way to control the speed. Recall that the number of stator magnetic poles is related to the conducting pattern of currents in the stator windings. Hence, if we control the paths available for stator current flow, we can control the number of poles and, therefore, the synchronous speed. This approach, called "pole-changing," can be and is used for some applications. However, since the number of poles must always be an even integer, system speeds are confined to discrete values. Additional disadvantages of pole changing are that the necessary pole-switching schemes are complicated, and the system speed will still vary with load. What we really need is a technology that permits full-range continuous speed control, such that machine performance is not seriously degraded over the desired speed range.

5.2 Control of the Motor Torque–Speed Characteristic

Recall the example machine of Chapter 4 (see Example 4.3). Again, suppose it drives the belt-driven fan, defined in Example 4.6. We solved for the running speed in Example 4.6, which was found to be 1779.1 rev/min. Suppose we wish to reduce air flow in the system by slowing the fan and, consequently, the motor.

Let us do some experimentation. For the sake of discussion, let us try for half-speed. We shall reduce the applied voltage to 50% of rated, hoping for a corresponding reduction in the speed. The results are:

- Speed = 1700.0 rpm = 178.0 rad/s; slip = 0.05553; stator voltage, 230.0 V; frequency, 60.0 Hz; operating mode: MOTOR
- Currents (stator) (A): I_a (stator) = $150.596\angle-30.5°$; I'_A(rotor) = $146.088\angle-24.8°$
- Torques (N m): developed = 258.85; TRL = 11.2; T_m = 247.78
- Powers (in kW): mech. output = 44.086 (59.096 hp); stator input = 51.671; developed = 46.082
- Losses (in kW): SWL = 2.879; RWL = 2.710; P_{RL} = 1.996; P_{LOSS} = 7.585; efficiency = 85.32%; power factor = 86.13%

Results are very disappointing. The speed only dropped 4.7%! Worse yet, the efficiency decreased from 93.34 to 85.32%! The torque–speed characteristics are shown in Figure 5.1.

Let us try frequency control. Reduce the applied voltage frequency to 50% of rated (30 Hz), hoping for a corresponding reduction in speed. The results are:

- Speed = 898.7 rpm = 94.1 rad/s; slip = 0.00149; stator voltage, 460.0 V; frequency, 30.0 Hz; operating mode: MOTOR
- Currents: I_a = 70.44 A at −82.1°; I'_A = 9.12 A at 0.3°
- Torques (in N M): developed = 75.08; T_{RL} = 5.928; T_m = 69.24
- Powers (in kW): output = 6.508 (8.724 hp); input = 7.706; developed = 7.066
- Losses (in kW): SWL = 0.630; RWL = 0.011; P_{RL} = 0.558; P_{LOSS} = 1.198; efficiency = 84.45%; power factor = 13.73%

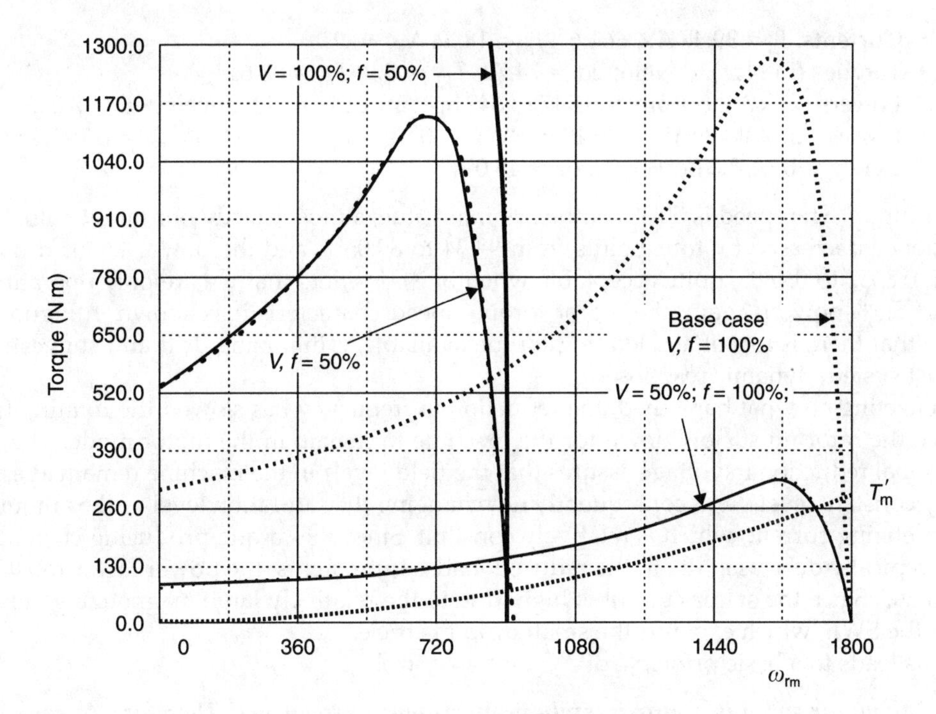

FIGURE 5.1. Torque–speed characteristics.

The results are encouraging. The speed adjusted to 50.5% of the original value, very close to our target value! Unfortunately, the efficiency decreased from 93.34 to 84.45%, even worse than was the case for voltage control! And look at the power factor! It dropped from 0.8557 to only 0.1373! The motor and load torque–speed characteristics are shown in Figure 5.1.

Let us consider our situation. Voltage (magnitude) control will not work because system speed appears to be relatively insensitive to changes in voltage. And if the voltage is changed sufficiently to appreciably change the speed, there is a terrible cost in terms of efficiency. In addition, there is another problem. The available accelerating torque, at speeds other than the equilibrium speed, has been greatly reduced, making the system susceptible to drift away from the desired speed, and making the system dynamic response much more sluggish.

Frequency control seems more promising. However, it is still unsatisfactory because the system efficiency and particularly the power factor are seriously degraded. Actually the situation is worse than it would appear. Remember that even though we have modeled the path for magnetizing current with a linear shunt inductor in the equivalent circuit of Figure 4.8b, the effect is actually nonlinear. If the frequency is reduced, X_m reduces proportionately, increasing the current I_m. But increased magnetizing current will *increase* the internal magnetic field levels, which increases the machine saturation and *decreases* the inductor L_m, further decreasing X_m. In the extreme case, X_m presents a short circuit between the stator and rotor, blocking current flow to the rotor.

Now try simultaneously controlling voltage *and* frequency. Reduce the applied voltage magnitude and frequency to 50% of their respective rated values (230 V, 30 Hz). The results are:

- Speed = 894.6 rpm = 93.7 rad/s; slip = 0.0060; stator voltage, 230.0 V; frequency, 30.0 Hz; operating mode: MOTOR

- Currents: $I_a = 39.47\,\text{A}\angle{-62.6°}$; $I'_A = 18.24\,\text{A}\angle{-0.9°}$
- Torques (in N m): developed = 74.72; $T_{RL} = 5.9$; $T_m = 68.61$
- Powers (in kW): output = 6.447 (8.642 hp); input = 7.240; developed = 7.000
- Losses (in kW): SWL = 0.198; RWL = 0.042; $P_{RL} = 0.553$; $P_{LOSS} = 0.793$; efficiency = 89.05%; power factor = 46.05%

Serendipity! The speed is 50.3% of the original value, almost exactly our target value! The efficiency decreased by four points from 93.34 to 89.05%, and the power factor dropped from 0.8557 to 0.4605, quite acceptable when one considers that we are only operating at about 8.6% power! Again, the motor torque–speed characteristic is shown in Figure 5.1. Note that there is lots of accelerating torque available, minimizing drift and suggesting a robust system dynamic response.

Reflecting on what happened, the reduction in frequency has slowed the rotating field; hence the rotor must slow down for the machine to remain in the motor mode. The proportional reduction in voltage assures that the field levels in the machine remain at essentially constant levels, and consequently not changing the saturation level or the amount of magnetizing current, which is relatively constant. Since the torque-producing current (I'_A) will typically decrease, due to the reduced load requirements, the power factor must also decrease. Since the stator current is high, due to the relatively large magnetizing current, so is the SWL, which explains the small drop in efficiency.

This leads to a basic principle of AC motor control:

> ***Ac motor speed is approximately proportional to frequency. Therefore, to control the speed, control the frequency. As the frequency is varied, the voltage should be likewise varied such that V/f is constant (the "constant volts per hertz" principle).***[2]

5.3 Controlling Voltage and Frequency

We now know that to control the speed of an AC motor–load system, it is necessary to control the (rms) voltage and frequency. Consider the problem of designing a circuit capable of providing an AC voltage with controlled (rms) magnitude and frequency.

It is possible to design such a circuit using switches. Consider the ideal switch, a two-terminal device which has two states, as shown in Figure 5.2a. In the OPEN or OFF state, the device presents a OPEN CIRCUIT between its terminals, and is NONCONDUCTING; in the CLOSED or ON state, the device presents a SHORT CIRCUIT between its terminals, and is CONDUCTING. The following notation will be convenient for our purposes:

open switches ("OFF")	blank bands:
closed switches ("ON"; nonconducting)	cross-hatched bands:
closed switches ("ON"; conducting)	shaded bands:

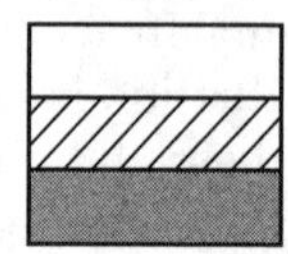

Now consider the circuit of Figure 5.2b. If we synchronously close switches 1,3 and 2,4 as indicated, the load voltage $v_L(t)$ will have the waveform indicated in Figure 5.2c. Note that the circuit has two distinct topologies or states:

Switches 1 to 4 closed; Switches 3 to 2 open
Switches 3 to 2 closed; Switches 1 to 4 open

[2] Like all rules of thumb, this is the basic idea but need not be slavishly followed in all cases. At very low speeds, the circuit resistance becomes predominant and a voltage boost is required to overcome ohmic effects.

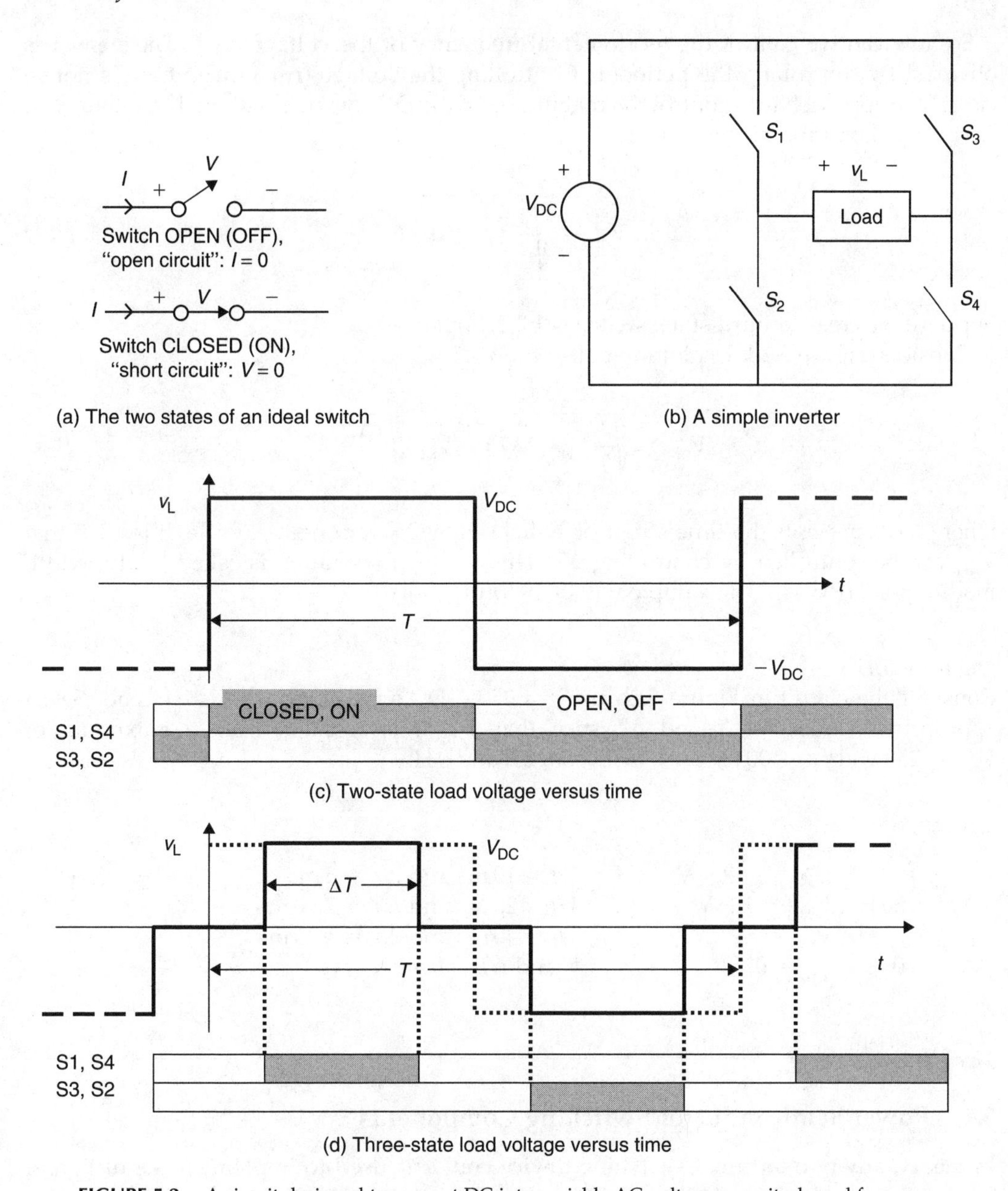

FIGURE 5.2. A circuit designed to convert DC into variable AC voltage magnitude and frequency.

The load voltage $v_L(t)$ could be characterized as "AC" since it is bipolar. However, it is certainly not sinusoidal. If the load is a motor, how will it respond to such a voltage? Any periodic waveform may be thought of as being made up of sinusoidal components, called "harmonics." The lowest frequency harmonic, called the "fundamental," has the same frequency as the original waveform, the remainder of the harmonics having integer-multiple frequencies of the fundamental. We will investigate the response of the machine to non-sinusoidal waveforms later in this chapter; for now it will be sufficient to argue that the machine responds basically to the fundamental.

So how can we control the fundamental frequency of the voltage $v_L(t)$? The answer is obvious! By controlling the period T! Controlling the voltage (rms) magnitude is not so clear. One approach is to control the magnitude of the DC source. But there is another way. Recall the definition of rms voltage,

$$V_{rms} = \sqrt{\frac{1}{T}\int_0^T v(t)^2 dt} \tag{5.1}$$

Suppose we create a third state: switches 1,2,3,4 open.

Consider Figure 5.2d. Exploiting half-wave symmetry,

$$V_{rms} = \sqrt{\frac{1}{T} V_{DC}^2 (2\Delta T)} = V_{DC}\sqrt{\frac{2\Delta T}{T}} \tag{5.2}$$

where ΔT represents the time a pair of switches are ON for one-half cycle. It is clear that V_{rms} can be controlled by controlling ΔT. This mode of operation is called "pulse width modulation" (PWM). An example will be helpful.

Example 5.1

Consider the circuit in Figure 5.2b using a 250 V DC source and a resistive load. Select appropriate values for T and ΔT such that V_{rms}/f is constant from a maximum of $V_{rms} = 250$ V at $f_0 = 100$ Hz for $f = 100, 50, 25$, and 0 Hz.

SOLUTION

At $f_0 = 100$ Hz, $V_{rms} = 250$ V; $T = 1/f = 10$ ms and $\Delta T = 5$ ms
At $f = 50$ Hz, $V_{rms} = 125$ V; $T = 1/f = 20$ ms and $\Delta T = 2.50$ ms
At $f = 25$ Hz, $V_{rms} = 62.5$ V; $T = 1/f = 40$ ms and $\Delta T = 1.25$ ms
At $f = 0$ Hz, $V_{rms} = 0$ V; $T = \infty$ and $\Delta T = 0$ ms

5.4 Power Semiconductor Switching Components

In theory, any two-terminal switching device could be used to implement the design of Section 5.3, provided it had the proper characteristics. A satisfactory switch must:

- Provide a reasonable approximation to an open circuit in the OFF state.
- Provide a reasonable approximation to a short circuit in the ON state.
- Handle the high voltages and currents associated with the high power levels associated with motor applications.
- Be cost effective.
- Have fast response capabilities, switching from one state to the other in microseconds, or faster.
- Be electronically controllable, requiring small amounts of energy for control.

There are several power semiconductor devices available today which can meet all of the above criteria. The study of these devices and the associated circuitry, is a subspecialty within electric power engineering called "power electronics." This is an exciting and dynamic subject in its own right, with many applications other than motor control. There are several excellent books on the subject, with the technology growing at a breath-taking rate. Our purposes will be served by considering only idealized versions of three such devices.

5.4.1 The Power Semiconductor Diode

The power diode meets all of the criteria required for motor control applications, except one. It is "uncontrollable" in that its operating state is determined by its own terminal voltage and current, and not by additional control circuitry. That is, the diode switches automatically from one state to the other, depending on the polarity of the terminal voltage. Its characteristics are defined in Figure 5.3b. The ON, OFF states are sometimes described as forward- and reverse-biased, respectively. Note that the diode symbol inherently indicates the permissible direction of current flow. Diodes are available in current ratings in excess of 1000 A, and blocking voltages in excess of 1000 V, with switching speeds in fractions of microseconds.

Example 5.2
Consider the circuit of Figure 5.3c with a 10 Ω resistive load. Find v_d and i_d for:

(a) $v_s = +100\,\text{V}$
(b) $v_s = -100\,\text{V}$

SOLUTION

(a) The circuit is operating at point "a" in Figure 5.3b (the diode is forward-biased). Therefore

$$v_d = 0\,\text{V}; \qquad i_d = +10\,\text{A}$$

(b) The circuit is operating at point "b" in Figure 5.3b (the diode is reverse-biased). Therefore

$$v_d = -100\,\text{V}; \qquad i_d = 0\,\text{A}$$

5.4.2 The Power Semiconductor Thyristor

The thyristor (sometimes called a silicon controlled rectifier or SCR) is an improvement on diode, in that it is at least partially controllable. Refer to Figure 5.4. There is a third terminal,

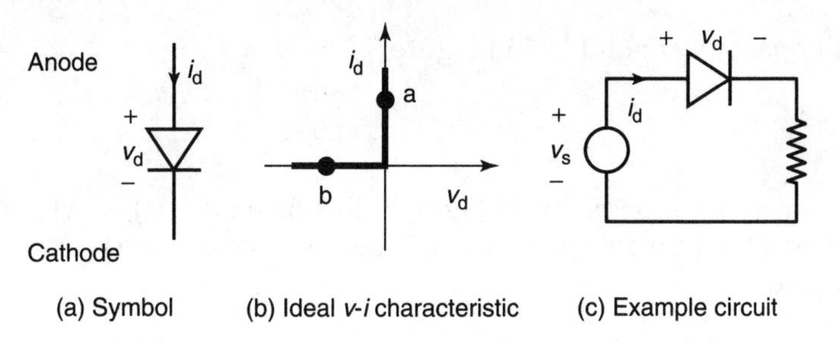

FIGURE 5.3. The ideal semiconductor diode.

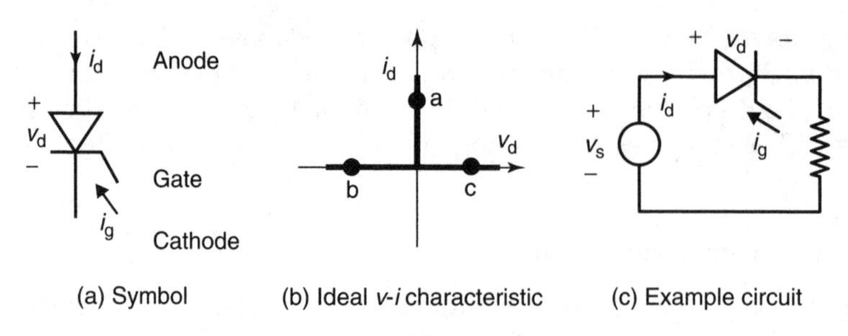

FIGURE 5.4. The ideal semiconductor thyristor.

called the "gate." The SCR turns ON if a pulse of positive gate current is supplied when $v_d > 0$, and remains ON until i_d attempts to reverse: otherwise the SCR is OFF. The gate current is supplied by additional circuitry, not shown on our circuit diagrams. The SCR characteristics are defined in Figure 5.4b. The thyristor symbol also indicates the permissible direction of current flow. Thyristors are available in current ratings in excess of 1000 A, and blocking voltages in excess of 1000 V, with switching speeds in microseconds.

Example 5.3

Consider the circuit of Figure 5.4c with a 10 Ω resistive load. Find v_d and i_d for:

(a) $v_s = +100\,\text{V}$, $i_g \leq 0$.
(b) $v_s = +100\,\text{V}$, $i_g = \delta(t)$.
(c) $v_s = -100\,\text{V}$, $i_g \leq 0$.
(d) $v_s = -100\,\text{V}$, $i_g = \delta(t)$.

SOLUTION

(a) The diode junction is forward-biased. However, conduction cannot begin until charge is injected into the gate. Therefore, the circuit is operating at point "c" in Figure 5.4b:

$$v_d = +100\,\text{V}; \qquad i_d = 0\,\text{A}$$

(b) The circuit is operating at point "c" in Figure 5.4b, when $i_g = \delta(t)$, at which time the SCR turns ON (moving to point "a"):

$$v_d = 0\,\text{V}; \qquad i_d = +10\,\text{A}$$

(c) The circuit operates at point "b" in Figure 5.4b:

$$v_d = -100\,\text{V}; \qquad i_d = 0\,\text{A}$$

(d) The circuit operates at point "b" in Figure 5.4b, when $i_g = \delta(t)$, at which time the SCR remains OFF (remains at point "b") because the diode structure remains reverse-biased:

$$v_d = -100\,\text{V}; \qquad i_d = 0\,\text{A}$$

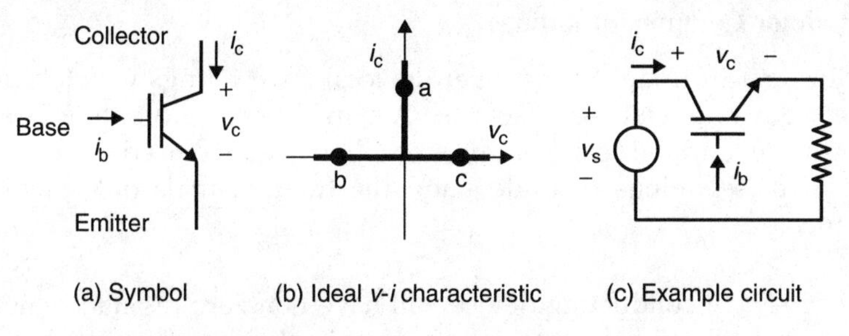

FIGURE 5.5. The ideal insulated gate bipolar junction transistor (IGBT).

5.4.3 The Insulated Gated Bipolar Junction Transistor

The main disadvantage of the thyristor is that while it is easy to turn ON, it turns OFF only when the external power current attempts to reverse. The insulated gate bipolar junction transistor (IGBT) overcomes that problem. It is ON when current flows into the base terminal and OFF when the base current is zero, independent of the collector current. Like the diode and thyristor, it conducts only in one direction (as indicated by the arrow on the emitter lead). Note that the base current is also unidirectional (into the IGBT). Its characteristics are defined in Figure 5.5. IGBTs are available in current ratings in excess of 500 A, and blocking voltages in excess of 1000 V, with switching speeds in microseconds.

Example 5.4

Consider the circuit of Figure 5.5c with a 10 Ω resistive load. Find v_c and i_c for:

(a) $v_s = +100\,\text{V}, i_b = 0$
(b) $v_s = +100\,\text{V}, i_b > 0$
(c) $v_s = -100\,\text{V}, i_b = 0$
(a) $v_s = -100\,\text{V}, i_b > 0$

SOLUTION

(a) The IGBT is forward-biased. However, conduction cannot occur until $i_b > 0$. Therefore, the IGBT remains OFF, and operates at point "c":

$$v_c = +100\,\text{V}; \qquad i_c = 0\,\text{A}$$

(b) The circuit is operating at point "c" in Figure 5.5b, while charge is injected into the gate ($i_b > 0$), at which time the IGBT turns ON (moving to point "a"):

$$v_c = 0\,\text{V}; \qquad i_c = +10\,\text{A}$$

(c) The circuit operates at point "b" in Figure 5.5b

$$v_c = -100\,\text{V}; \qquad i_c = 0\,\text{A}$$

(d) The circuit is operating at point "b" in Figure 5.5b, and remains at point "b".

$$v_c = -100\,\text{V}; \qquad i_c = 0\,\text{A}$$

5.4.4 Semiconductor Component Ratings

All semiconductor devices (indeed, all power devices) have ratings which must be considered if the device is to operate satisfactorily. A comprehensive discussion of all semiconductor ratings is beyond the scope of our work; however, we need to be aware of the most important of these ratings to understand the fundamentals of switching circuit design. Ratings include:

- *Rated rms current.* A conducting device will have non-zero resistance; hence the flow of current will produce heat, which must be dissipated or the device temperature will rise to a point that will damage or destroy the device.
- *Reverse breakdown voltage.* A nonconducting device, like an open switch, will have a voltage across its terminals. Excessive voltage will break down the open circuit properties of the device.
- *TURN-ON time.* It takes a certain finite time for a device to switch from the OFF state to the ON state.
- *TURN-OFF time.* It takes a certain finite time for a device to switch from the ON state to the OFF state.
- *Maximum* di/dt. If currents change too rapidly internal through a device due to the external circuitry, damage can result.
- *Maximum* dv/dt. If voltages change too rapidly across a device due to the external circuitry, damage can result.

Although we will typically assume ideal devices, we will occasionally consider ratings issues. All three components have the disadvantage that they permit current flow in only one direction; this disadvantage is overcome by using multiple components and circuit topologies that permit multiquadrant operation. Also, there are devices that are bipolar. Indeed, the reader is advised that there are many power semiconductor devices that have various advantages and specialized performance characteristics, and the list is growing. However, all are fundamentally switching devices, and a consideration of the circuit design using these three will be sufficient for understanding the basics of motor control. Also, keep in mind that control is implemented by additional circuitry, called "gating" or "control" circuitry, which shall not be considered and is not shown.

5.5 The Single-Phase Inverter

Before we study the single-phase inverter, consider the demonstration circuit of Figure 5.6a. We begin with the IGBT OFF. Turn the IBGT ON at $t = 0$. For $0 < t < T_1$, the current $i_L(t)$ follows the usual transient trajectory, according to

$$i_L(t) = \frac{V_{DC}}{R}(1 - e^{-Rt/L}) \tag{5.3}$$

Now suppose we turn off the IGBT at $t = T_1$. Consider KVL around the circuit:

$$V_{DC} = v_T + Ri_L + L\frac{di_L}{dt} \tag{5.4}$$

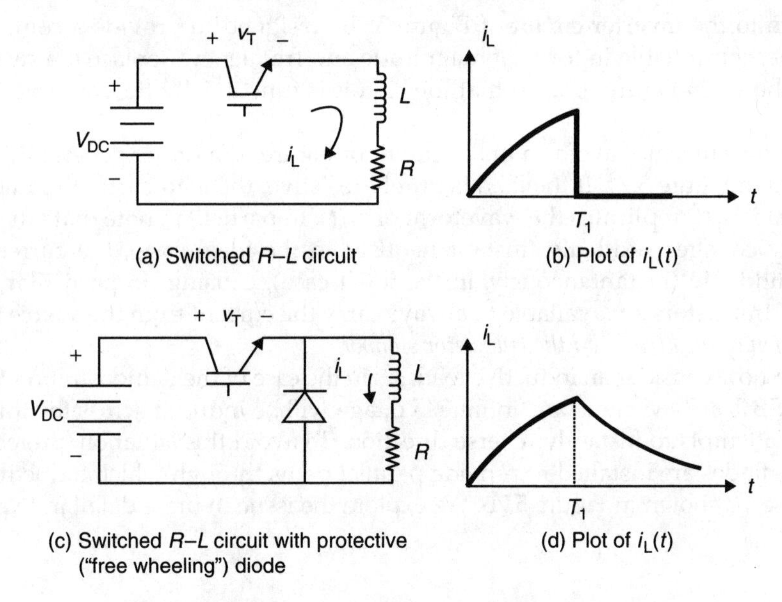

FIGURE 5.6. A demonstration *R–L* circuit.

The current must fall to zero (instantly, for an ideal IGBT), as shown in Figure 5.6b. Therefore

$$\frac{di_L}{dt} = -\infty$$

and

$$L\frac{di_L}{dt} = -\infty$$

Since V_{DC} and $i_L(t)R$ are finite,

$$v_T(t) = -L\frac{di_L}{dt} = +\infty.$$

Such a voltage will destroy the transistor, of course. The conclusion is that the circuit of Figure 5.6a is impractical, in that if the IGBT is switched, it can be destroyed! Now consider placing a diode in parallel with the load, as shown in the circuit of Figure 5.6c and start over. As before, the current $i_L(t)$ rises, according to

$$i_L(t) = \frac{V_{DC}}{R}(1 - e^{-Rt/L})$$

Again, turn off the IGBT at $t = T_1$. As the current collapses, the strong negative voltage across the inductor will match Ri_L, and forward bias the diode, providing a path through which the current can dissipate, as shown in Figure 5.6d. This time, the transistor voltage $v_T(t)$ will never exceed V_{DC}, which should be easily withstood, if the IGBT is properly rated.

We return to the inverter circuit of Figure 5.2b, designed to provide a controllable AC load voltage, controllable in turns of magnitude and frequency. Replace the switches with IGBTs, as shown in Figure 5.7a, so that the circuit is functionally the same as that shown in Figure 5.2b.

Consider the current waveform in the circuit of Figure 5.7a under general PWM control as described in Figure 5.7c. If the load is purely resistive, the load current mimics the load voltage, that is, $i_L(t)$ replicates the waveform of $v_L(t)$. In particular, note that when the voltage polarity changes suddenly (instantaneously, in the ideal case), the current changes direction suddenly (instantaneously, in the ideal case), causing no particular problems, since "ON" transistors are available to always carry the current from the source to the load *in the direction of the arrows on the transistor symbol!*

However, now consider an inductive load. As in the case of the demonstration R–L circuit, when the IGBTs are switched, an "infinite" voltage will be induced across the transistors as the current attempts to instantly reverse direction. To avoid this situation, protective ("free wheeling") diodes are installed to provide parallel paths through which inductive currents may collapse, as shown in Figure 5.7b. We explore the issue in more detail in Example 5.5.

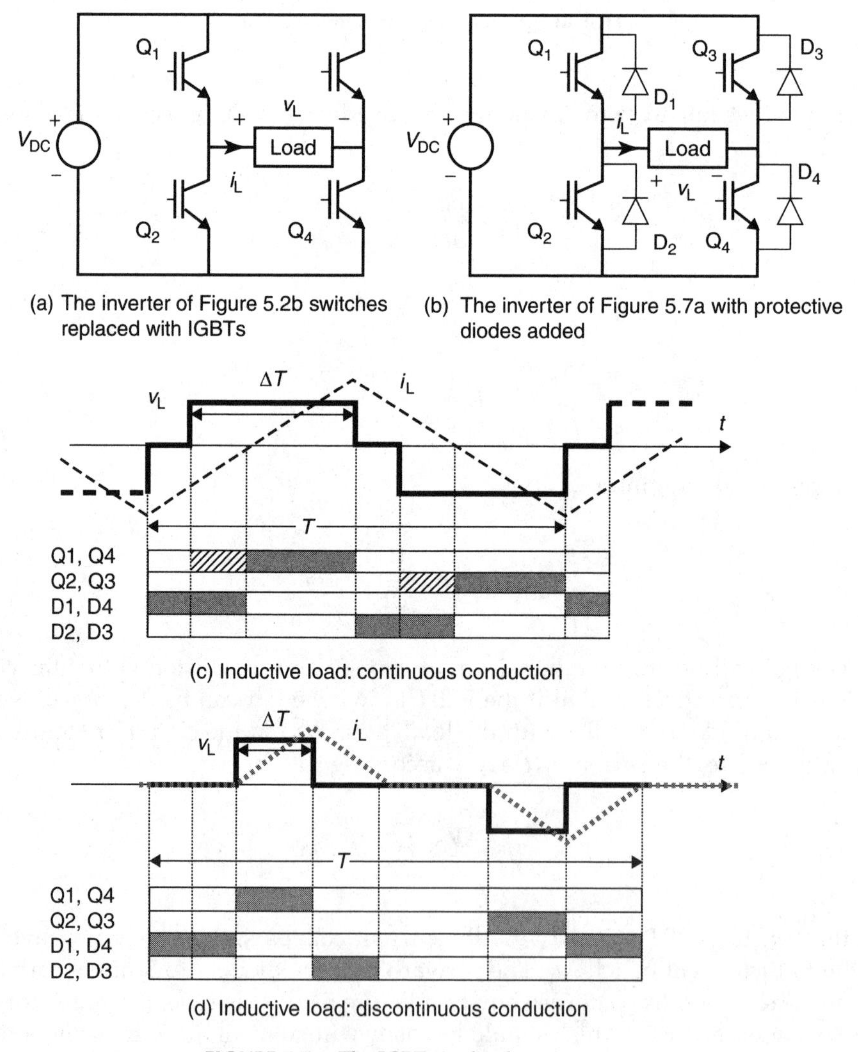

(a) The inverter of Figure 5.2b switches replaced with IGBTs

(b) The inverter of Figure 5.7a with protective diodes added

(c) Inductive load: continuous conduction

(d) Inductive load: discontinuous conduction

FIGURE 5.7. The IGBT single-phase inverter.

Example 5.5
For the circuit of Figure 5.7b, the switching frequency is 62.5 Hz, and the load is a pure inductance of 40 mH, and $V_{DC} = 200$ V. The IGBTs are symmetrically switched to force the given Δt values.

(a) Suppose $\Delta t = 6$ ms. Compute and plot the load current waveform, indicating the conducting state of all diodes and IGBTs.
(b) Suppose $\Delta t = 2$ ms. Compute and plot the load current waveform, indicating the conducting state of all diodes and IGBTs.

SOLUTION

(a) See Figure 5.7c for the relevant plot. $\Delta t = 6$ ms

$$\text{period} = 1/f = 1/62.5 = 16\,\text{ms}; \qquad \text{therefore} \quad T/2 = 8\,\text{ms}$$

Locate the time origin at the falling edge of the $v_L(t)$ waveform, as shown. Suppose the current $i_L(t)$ at $t = T_1^- = 0^-$ is 20 A, with Q1–Q4 ON. Now turn Q1–Q4 OFF. Since there is no "Q" path for the current, and it *must* momentarily remain at 20 A, D2–D3 are forced ON. D2–D3 remain ON for $T_1 = 0 < t < 2$, which applies $v_L = -200$ V to the load.

$$-200 = 40(\Delta i_L/\Delta t) = 40(\Delta i/2)$$

or

$$\Delta i_L = -10\,\text{A}$$

so that the current is

$$i_L = 20 - 10 = +10 \text{ at } t = 2$$

at $t = 2$, Q2–Q3 are turned ON. However, conduction continues through D2–D3, since Q2–Q3 blocks current flow "UP." At $t = 4$, i_L reaches 0, the current transfers to Q2–Q3, and all diodes become reverse biased. Over the next 4 ms ($4 < t < 8$),

$$-200 = 40(\Delta i_L/\Delta t) = 40(\Delta i/4)$$

or

$$\Delta i_L = -20\,\text{A}; \qquad i_L = 0 - 20 = -20\,\text{A at } t = 8$$

We have now executed ½ cycle (8 ms). There is complete half-wave symmetry. Diodes D1–D4 are forced ON, when Q2–Q3 are turned OFF. The current rises +10 A to $(-20 + 10) = -10$ A at $t = 10$, at which point Q1–Q4 are turned ON. However, conduction continues through D1–D4, since Q1–Q4 blocks current flow "DOWN." At $t = 12$, (or $t = -4$, since the situation is periodic), i_L reaches 0, the current transfers to Q1–Q4, and all diodes again become reverse biased. Over the next 4 ms ($-4 < t < 0$) the current rises to the starting value of +20 A.[3]

Since the current waveform was continuous, this mode of operation is called "continuous conduction."

(b) See Figure 5.7d for the relevant plot. $\Delta t = 2$ ms
Relocate the time origin to the rising edge of the $v_L(t)$ waveform, as shown. Also, assume $i_L(0) = 0$. Since Q1–Q4 are ON, the current rises according to

$$40(di_L/dt) = +200; \text{ or } i_L(t) = 5t + 0$$

$$\text{At } t = 2 \qquad i_L = +10\,\text{A}$$

[3] We argue that i_L must have zero average value (half-wave symmetry) since the given applied voltage had zero average value. This only occurs if we start at +20 A. Had we started at +21 A, the current minimum would have been −19 A.

Diodes D2–D3 are forced ON. The current falls to 0 in 2 ms. At this point all diodes and IGBTs are all OFF.

Since the current waveform was discontinuous, this mode of operation is called "discontinuous conduction." The IGBTs were not ON long enough to support conduction for the full cycle.

Although Example 5.5 considered a pure inductor load to simplify the analysis, the principles are relevant for more general *R–L* loads, creating exponential current waveforms, rather than triangular. Current automatically transfers to the diodes to avoid the voltage spikes that must occur when inductive currents are suddenly interrupted. This is important to our work since our load (the polyphase induction machine) is clearly inductive.

The inverter circuit of Figure 5.7b provides a controlled voltage and frequency source; however it only accommodates a single-phase load, whereas our machine is three-phase. We extend the design to the three-phase case in the next section.

5.6 The Three-Phase Inverter

We extend the inverter design to the three-phase case, as shown in Figure 5.8a. The load terminal voltages are plotted at full voltage ($\Delta T = T/2$), and resistive load in Figure 5.8b. The PWM control concepts presented in Section 5.3 can be applied to the circuit of Figure 5.8a. Now we have what is required for AC motor control: a high power, variable voltage, variable frequency balanced three-phase source! We control the frequency by controlling the timing of IGBT switching, the rms values by controlling ΔT. We have converted both magnitude and frequency control into a timing problem, a problem which is particularly amenable to digital electronic solution. Let us examine this circuit in a bit more detail in an example.

Example 5.6

Consider the three-phase inverter of Figure 5.8a operating from a 300 V DC source.

(a) Plot v_{an}, v_{bn}, v_{cn}, and v_{ab}, and indicate status of all IGBTs for one cycle.
(b) Compute the necessary switching times for outputs of
 (i) 100% voltage; 100% frequency (100 Hz)
 (ii) 50 V; 50 Hz.

SOLUTION

(a) The solution appears in Figure 5.8b.
(b) Several strategies for voltage control are possible. Suppose in each step we make pulses of equal width ΔT
 (i) At 100% voltage; 100% frequency:

$$T = 1/f = 1/100 = 10\,\text{ms}$$

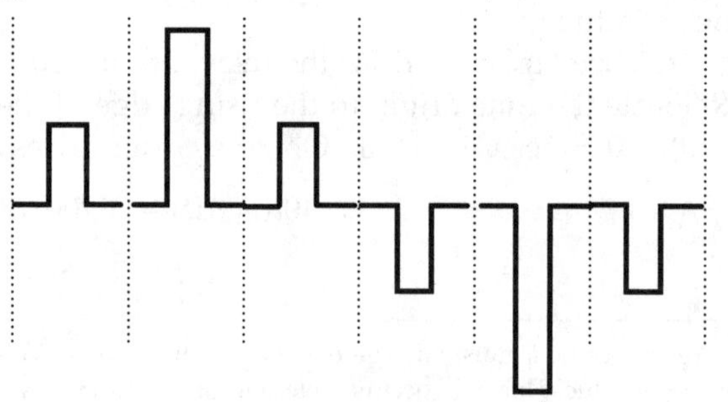

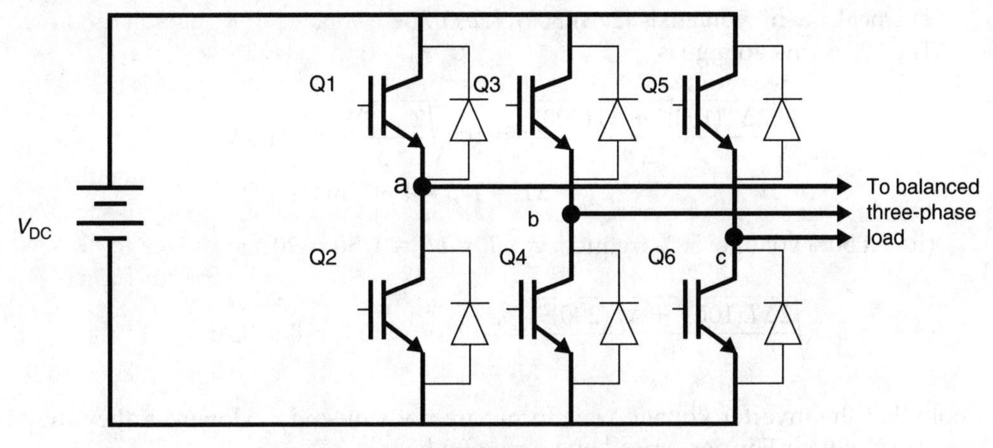

(a) The three-phase IGBT inverter circuit diagram

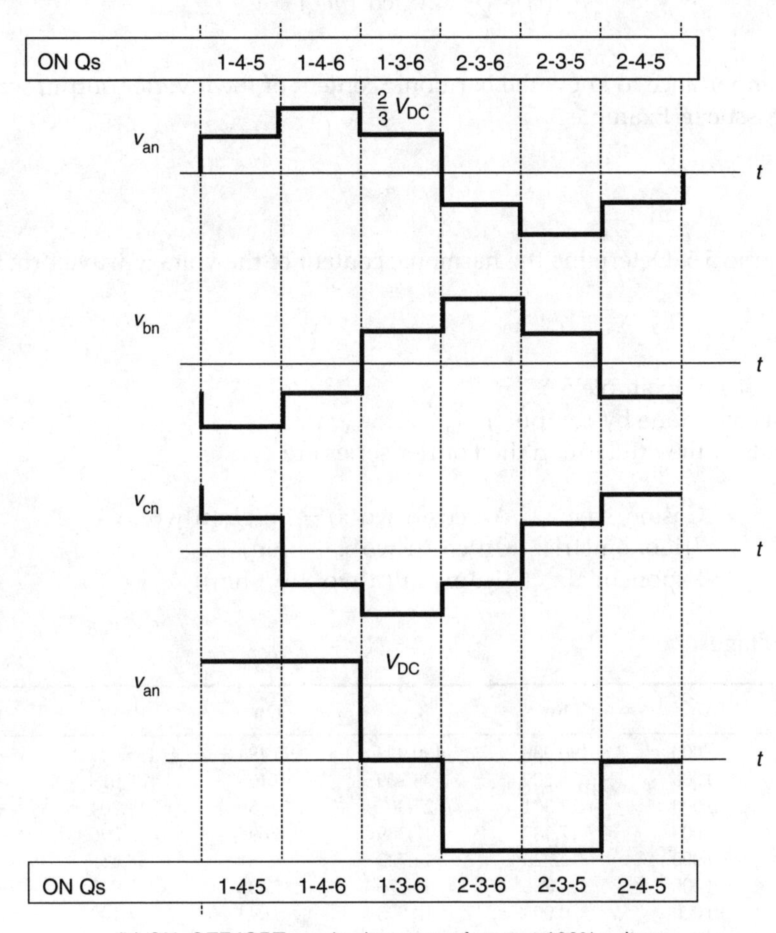

(b) ON–OFF IGBTs and voltage waveforms at 100% voltage

FIGURE 5.8. The IGBT three-phase inverter.

The peak "a–n" voltage is (2/3)(300) = 200 V.
The 100% rms voltage is

$$\sqrt{\frac{2\Delta T(100)^2 + \Delta T(200)^2}{3\Delta T}} = 100\sqrt{\frac{2+(2)^2}{3}} = 141.4\,\text{V}$$

$$\Delta T = T/6 = 1.667\,\text{ms}$$

(ii) At 50% voltage; 50% frequency: $T = 1/f = 1/50 = 20\,\text{ms}$

$$\sqrt{\frac{2\Delta T(100)^2 + \Delta T(200)^2}{10}} = 70.71\,\text{V}; \quad \Delta T = 0.8333\,\text{ms}$$

We note that the inverter voltage waveforms are not sinusoidal. However, they may be represented by their Fourier series,[4] one form of which is

$$v_{\text{an}}(t) = \sum_{n=0}^{N} \sqrt{2}V_{\text{n}} \cos(n\omega_{\text{n}}t + \phi_{\text{n}})$$

It is of some importance to know the harmonic content of the inverter output waveforms. We study the issue in Example 5.7.

Example 5.7
Refer to Example 5.6. Determine the harmonic content of the voltage waveform v_{an}.

SOLUTION
Refer to Solutions of Example 5.6.
The analysis was done by computer.
The three forms of the nth term in the Fourier series are

Cosine, sine: An*cos(n*wo*x) + Bn*sin(n*wo*x)
Trigonometric: Cn*cos(n*wo*x + phin)
Exponential: Dn*exp (j*(n*wo*x + phin))

At 100% voltage

Harm	An	Bn	Cn	Dn	Rms	phin
1	−0.000	190.988	190.988	95.494	135.049	−90.0
5	−0.000	38.209	38.209	19.105	27.018	−90.0
7	−0.000	27.301	27.301	13.650	19.304	−90.0
11	−0.000	17.389	17.389	8.695	12.296	−90.0
13	−0.000	14.723	14.723	7.361	10.411	−90.0
17	−0.000	11.276	11.276	5.638	7.973	−90.0
19	−0.000	10.098	10.098	5.049	7.140	−90.0

rms, 141.4214; sqr(rms), 20000.000; 2 × sum[sqr(Dn)], 19715.042; values: DC, −0.0000; AC, 141.4214; fund, 135.0492; harm, 41.9729; THD, 31.08; TIF factors (%), 7389035.12.

[4] Readers not familiar with Fourier analysis may wish to digress to Appendix C.

At 50% voltage

Harm	An	Bn	Cn	Dn	Rms	phin
1	−0.000	57.385	57.385	28.693	40.577	−90.0
3	0.000	−13.017	13.017	6.508	9.204	90.0
5	−0.000	12.085	12.085	6.042	8.545	−90.0
7	0.000	−49.603	49.603	24.802	35.075	90.0
9	−0.000	44.782	44.782	22.391	31.666	−90.0
11	0.000	−7.197	7.197	3.598	5.089	90.0
13	−0.000	5.169	5.169	2.584	3.655	−90.0
15	0.000	−26.373	26.373	13.186	18.648	90.0
17	−0.000	19.742	19.742	9.871	13.960	−90.0
19	−0.000	0.642	0.642	0.321	0.454	−90.0

rms, 71.4920; sqr(rms), 5111.111; 2 × sum[sqr(Dn)], 4619.301; values: DC, 0.0000; AC, 71.4920; fund, 40.5774; harm, 58.8608; THD, 145.06; TIF factors (%), 11920235.84.

We continue our harmonics study by considering the square waves encountered in Section 5.3. Consider Figure 5.9.

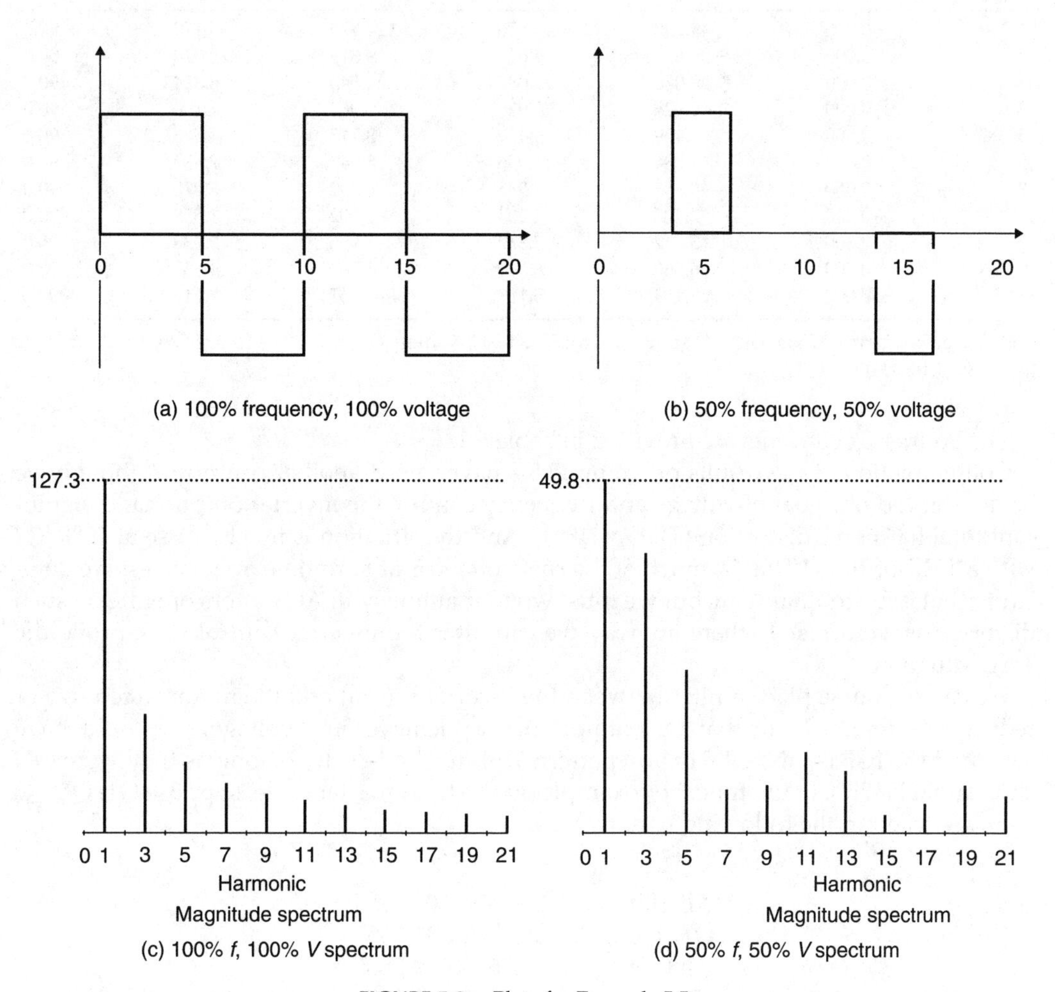

FIGURE 5.9. Plots for Example 5.5.

TABLE 5.1

(a) Fourier Analysis for the 100% V, f Square Wave

Harm	An	Bn	Cn	Dn	Rms	phin
1	−0.000	127.326	127.326	63.663	90.033	−90.0
3	−0.000	42.446	42.446	21.223	30.014	−90.0
5	−0.000	25.473	25.473	12.736	18.012	−90.0
7	−0.000	18.200	18.200	9.100	12.870	−90.0
9	−0.000	14.162	14.162	7.081	10.014	−90.0
11	−0.000	11.593	11.593	5.796	8.197	−90.0
13	−0.000	9.815	9.815	4.908	6.940	−90.0
15	−0.000	8.513	8.513	4.256	6.019	−90.0
17	−0.000	7.517	7.517	3.759	5.315	−90.0
19	−0.000	6.732	6.732	3.366	4.760	−90.0
21	−0.000	6.097	6.097	3.049	4.311	−90.0

rms, 100.0; sqr(rms), 10000.0; 2 × sum[sqr(Dn)], 9818; values: DC, 0.0000; AC, 100.0; fund, 90.03; harm, 43.5; THD = 48.34%.

(b) Fourier Coefficients for the 50% V, f Square Wave

Harm	An	Bn	Cn	Dn	Rms	phin
1	−0.000	49.750	49.750	24.875	35.179	−90.0
3	0.000	−39.627	39.627	19.813	28.020	90.0
5	−0.000	23.086	23.086	11.543	16.324	−90.0
7	0.000	−5.925	5.925	2.963	4.190	90.0
9	0.000	−6.429	6.429	3.215	4.546	90.0
11	−0.000	11.086	11.086	5.543	7.839	−90.0
13	0.000	−8.585	8.585	4.292	6.070	90.0
15	−0.000	2.203	2.203	1.102	1.558	−90.0
17	−0.000	3.872	3.872	1.936	2.738	−90.0
19	0.000	−6.560	6.560	3.280	4.638	90.0
21	−0.000	5.113	5.113	2.557	3.616	−90.0

rms, 50.5525; sqr(rms), 2555.556; 2 × sum[sqr(Dn)], 2470.200; values: DC, 0.0000; AC, 50.5525; fund, 35.1786; harm, 36.3046; THD = 103.20%.

The Fourier Coefficients are provided in Table 5.1.

Contemplation of the results of Example 5.7 have raised another concern. Although we have achieved our goal of voltage and frequency control, the inverter output has a significant total harmonic distortion (THD = 48%). And the situation is much worse at 50% V, f with a THD of 103%! That is, much of the rms voltage is at harmonic frequencies! We defer study that issue to Chapter 6, but we must wonder at how will AC motors operate on such distorted waveforms? Is there anyway we can alter the inverter control to improve the THD situation?

We can of course place a filter between the inverter output and the motor to remove or reduce harmonics. Note that in Example 5.3, to achieve 50% voltage, we could have switched the IGBTs ON, OFF in any pattern within a half-cycle, as long as they were ON 25%, and OFF 75% of the time. For example, in the first-quarter cycle, suppose Q1, Q4, Q6 were switched in the following pattern:

All OFF:	$0 < \theta < 30°$
Q1, 4, 6 ON:	$30 < \theta < 41.25°$
All OFF:	$41.25 < \theta < 78.75°$
Q1, 4, 6 ON:	$78.75 < \theta < 90°$

A Fourier analysis on the output voltage waveform produces:

Harm	An	Bn	Cn	Dn	Rms	phin
1	−0.000	38.468	38.468	19.234	27.201	−90.0
5	−0.000	21.892	21.892	10.946	15.480	−90.0
7	0.000	−38.817	38.817	19.409	27.448	90.0
11	−0.000	0.305	0.305	0.153	0.216	−90.0
13	−0.000	24.149	24.149	12.075	17.076	−90.0
17	0.000	−14.346	14.346	7.173	10.144	90.0
19	0.000	−6.034	6.034	3.017	4.266	90.0

rms, 49.4413; sqr(rms), 2444.444; 2 × sum[sqr(Dn)], 2145.671; values: DC, 0.0000; AC, 49.4413; fund, 27.2011; harm, 41.2861; THD = 152%.

At first, the situation seems even worse since the THD has risen to 152%. However, notice that the odd triplen harmonics (3rd, 9th, 15th, 21st) are now zero, and that the 11th harmonic is negligible. This has greatly simplified the filtering problem. There are many clever algorithms for controlling the IGBT switching, to the end of eliminating or drastically reducing many of the harmonics; this is an important design problem in power electronics. A more typical waveform appears below.

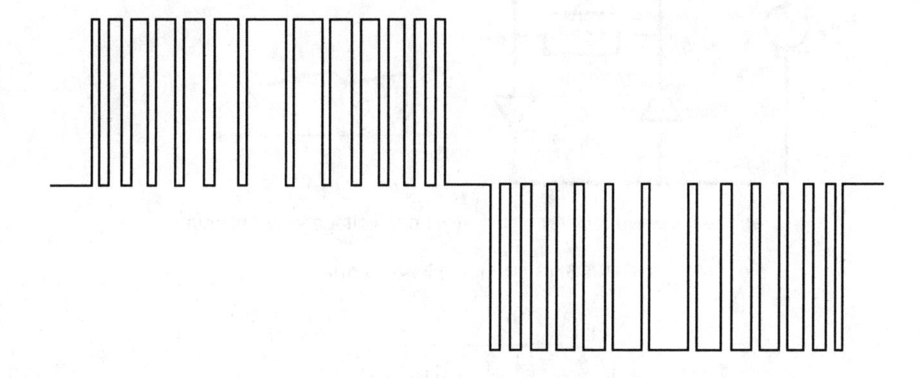

5.7 AC to DC Conversion: Rectifiers

The inverter circuits of Section 5.4 and Section 5.5 require DC electrical sources. The typical available source, as supplied by the local electric utility, is constant AC (rms) voltage and frequency, and single or three phase. What is required is a power converter circuit to convert AC to DC.

Consider the circuit of Figure 5.10a, called a single-phase half-wave rectifier. The load voltage $v_L(t)$, shown in Figure 5.10b, becomes

$$v_L(t) = V_{max}\cos(\omega t);\quad 0 < \omega t < \pi/2;\quad 3\pi/2 < \omega t < 2\pi;\quad v_L(t) = 0;\quad \pi/2 < \omega t < 3\pi/2$$

$$V_{average} = V_{DC} = \frac{1}{T}\int_0^T v(t)\mathrm{d}t = \frac{V_{max}}{\pi} \tag{5.5a}$$

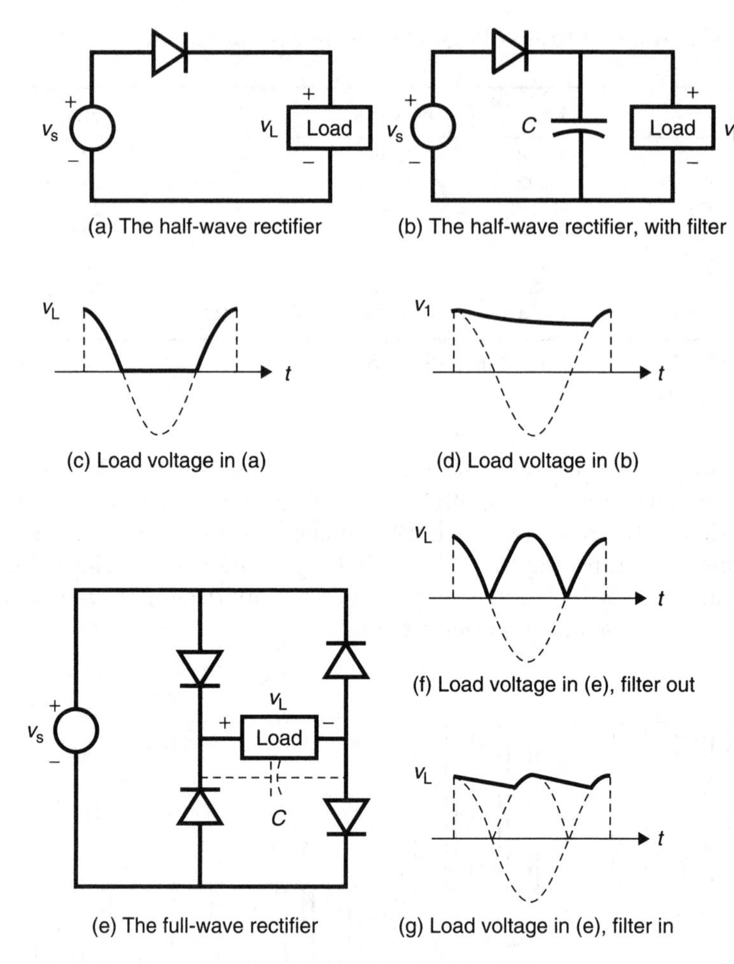

FIGURE 5.10. Single-phase rectifiers.

$$V_{rms} = V_{eff} = \sqrt{\frac{1}{T}\int_0^T v^2(t)\mathrm{d}t} = \frac{V_{max}}{2} \tag{5.5b}$$

The circuit can be modified to enhance its "DC productivity" by adding a "filter" capacitor, as shown in Figure 5.10c. The load voltage $v_L(t)$ is shown in Figure 5.10d.

A better design, called a single-phase full-wave rectifier, is the circuit of Figure 5.10e. The load voltage $v_L(t)$ shown in Figure 5.10f, becomes

$$v_L(t) = |V_{max}\cos(\omega t)|, \qquad 0 < \omega t < 2\pi$$

$$V_{average} = V_{DC} = \frac{2V_{max}}{\pi} \tag{5.6a}$$

$$V_{rms} = V_{eff} = \frac{V_{max}}{\sqrt{2}} \tag{5.6b}$$

Adding a filter capacitor modifies the load voltage $v_L(t)$ to that shown in Figure 5.10g. An example will develop additional details.

Example 5.8

A 10 Ω resistive load is to be supplied with DC voltage. The available source is 120 V rms 60 Hz AC. Determine the rms and DC load voltage and diode current, draw the circuit diagrams, and plot $v_L(t)$, if the load is served from a:

(a) Half-wave rectifier, no filter capacitor.
(b) Half-wave rectifier, with filter capacitor. Determine C if the voltage should drop no more than 10% from its peak value.
(c) Full-wave rectifier, no filter capacitor.
(d) Full-wave rectifier, with filter capacitor. Determine C if the voltage should drop no more than 10% from its peak value.

SOLUTION

(a) The circuit diagram and $v_L(t)$ plot appears in Figure 5.10a and Figure 5.10c respectively,

$$v_s = 120\sqrt{2}\sin(377t) = 169.7\sin(377t)$$

$$V_{average} = V_{DC} = V_{max}/\pi = 54.02\,\text{V}$$

$$V_{rms} = 169.7/2 = 84.86\,\text{V}$$

$$I_{average} = I_{DC} = V_{DC}/R_L = 5.402\,\text{A}$$

$$I_{rms} = V_{rms}/R_L = 8.486\,\text{A}$$

(b) The circuit diagram and $v_L(t)$ plot appear in Figure 5.10b and Figure 5.10d, respectively. As the voltage passes its crest value, the capacitor will tend to support the voltage, reverse biasing the diode, and become the source of load current.[5] Assuming a 10% drop in the voltage, the exponentially decaying $v_L(t)$ intersects the source $v_s(t)$ at 90%(169.7) = 152.7 V. Solving for the intersection time,

$$169.7\sin(\theta) = 152.7; \quad \theta = 1.1198\,\text{rad}; \quad \omega t = \theta + (3/2)\pi = 1.1198 + 4.7124$$

$$\omega t = 5.8322 = 377\,T_1, \quad T_1 = 15.47\,\text{ms}$$

$$169.7\exp(T/\tau) = 152.7; \quad T_1/\tau = 0.1054$$

$$\tau = 146.8\,\text{ms} = RC, \quad \text{for} \quad R = 10\,\Omega, \quad C = 14.68\,\text{mF}$$

Approximating the waveform as triangular

$$V_{DC} = \frac{169.7 + 152.7}{2} = 161.2\,\text{V}$$

The RMS value is almost the same (since the variation in v_L is small).

$$V_{rms} \cong 161.2\,\text{V}$$

$$I_{DC} = I_{rms} = \frac{161.2}{10} = 16.12\,\text{A}$$

[5] The location of this point is not precisely at the voltage maximum. However, the error incurred by assuming this to be so has a minor effect on the value of C.

(c) The circuit diagram and $v_L(t)$ plot appear in Figure 5.10e and Figure 5.10f, respectively.

$$V_S = 120\sqrt{2}\sin(377t) = 169.7\sin(377t)$$

$$V_{\text{average}} = V_{DC} = 2V_{\max}/\pi = 108.0\,\text{V}$$

$$V_{\text{rms}} = V_{\max}/\sqrt{2} = 120\,\text{V}$$

(d) The circuit diagram and $v_L(t)$ plot appear in Figure 5.8e and Figure 5.8g respectively. Approximating the waveform as triangular,

$$V_{DC} = V_{\text{rms}} = 161.2\,\text{V}$$

$$I_{DC} = I_{\text{rms}} = V_{\text{rms}}/R_L = 16.12\,\text{A}$$

As before, the capacitor will support the voltage, reverse biasing the diode and become the source of load current. Assuming a 10% drop in the voltage, the exponentially decaying $v_L(t)$ intersects the source $v_s(t)$ at 90%(169.7) = 152.7 V.

Solving for the intersection time

$$\sin(\theta) = 0.9, \quad \theta = 1.1198\,\text{rad}; \quad \omega t = \theta + \pi/2 = 1.1198 + 1.5702 = 2.6906$$

$$377\,T_1 = 2.6906, \qquad T_1 = 7.137\,\text{ms}$$

$$169.7\exp(T/\tau) = 152.7; \qquad T_1/\tau = 0.1054$$

$$\tau = 67.74\,\text{ms} = RC, \qquad \text{for} \quad R = 10\,\Omega, \quad C = 6.774\,\text{mF}$$

5.8 Three-Phase Rectifiers

For applications larger than a few kVA, it is preferable to use a three-phase source whenever possible. The single-phase rectifier circuits of Figure 5.10 can be extended to the three-phase rectifier circuit design of Figure 5.11a. The load voltage $v_L(t)$ in Figure 5.11b, becomes

$$v_L(t) = V_{\max}\cos(\omega t - 30°), \qquad 0 < \omega t < \pi/3 \quad \text{(waveform has 1/6 period symmetry)}$$

$$V_{\text{average}} = V_{DC} = \frac{6}{T}\int_0^{T/6} V(t)\,dt = 0.9549V_{\max} \tag{5.7a}$$

$$V_{\text{rms}} = V_{\text{eff}}\sqrt{\frac{6}{T}\int_0^{T/6} v^2(t)\,dt} = 0.9558V_{\max} \tag{5.7b}$$

A filter capacitor in parallel with the load will smooth the load voltage $v_L(t)$.

Example 5.9

A 30 Ω resistive load is to be supplied with DC voltage. The available source is three-phase 208/120 V rms 60 Hz AC. Determine the rms and DC load voltage and diode current, draw the circuit diagrams, and plot $v_L(t)$, if the load is served from a three-phase rectifier, with and without a filter capacitor.

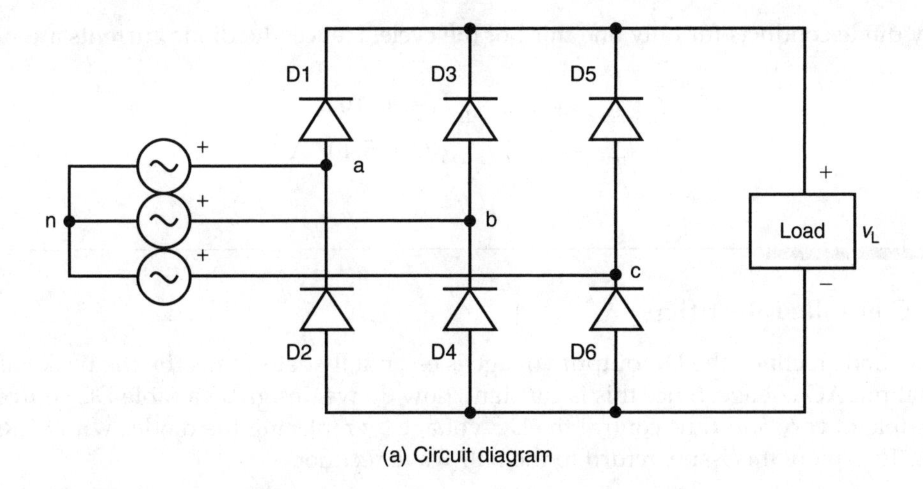

(a) Circuit diagram

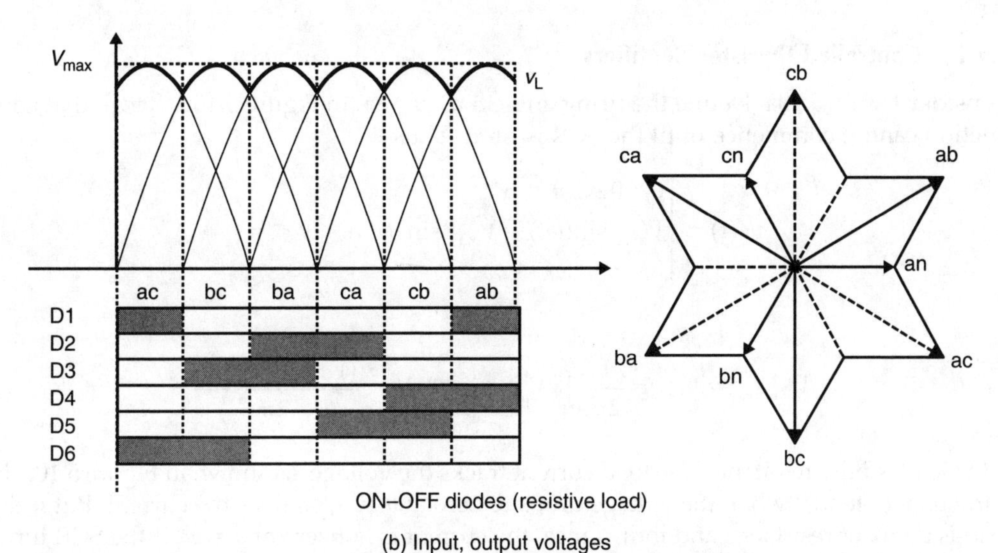

ON–OFF diodes (resistive load)

(b) Input, output voltages

FIGURE 5.11. Three-phase diode rectifier.

SOLUTION

The circuit diagram and $v_L(t)$ plot appear in Figure 5.11. Without C,

$$V_{max} = 120\left(\sqrt{3}\sqrt{2}\right) = 293.9\,\text{V}$$

$$V_{average} = V_{DC} = 0.9549V_{max} = 0.9549(293.9) = 280.7\,\text{V}$$

$$V_{rms} = 0.9558V_{max} = 0.9558(293.9) = 280.9\,\text{V}$$

The load currents

$$I_{DC} = V_{DC}/R_L = 9.357\,\text{A}; \qquad I_{rms} = V_{rms}/R_L = 9.365\,\text{A}$$

The load voltage averages 95.49% of the maximum even with no filter capacitor.

Any diode conducts for only one third of the cycle. Hence, the diode currents are

$$I_{\text{average}} = (I_{\text{DC}})_{\text{LOAD}}/3 = 3.119\text{ A}$$
$$I_{\text{rms}} = (I_{\text{rms}})_{\text{LOAD}}/\sqrt{3} = 5.407\text{ A}$$

5.9 Controlled Rectifiers

For the diode rectifier, the DC output voltage is essentially determined by the peak values of the input AC voltage. Since this is constant, how do we design a variable DC source? It is possible to vary and thus control the DC voltage by replacing the diodes with SCRs or IGBTs. To explore this issue, return to the half-wave rectifier.

5.9.1 Controlled Thyristor Rectifiers

Consider Figure 5.10a. Define the firing angle α as shown in Figure 5.10b. Recall that conduction cannot commence until the SCR is fired. Thus

$$v_{\text{L}}(t) = \begin{cases} 0, & 0 < \omega t < \alpha \\ V_{\text{max}} \sin(\omega t) = V_{\text{max}} \sin\theta, & \alpha < \theta < \pi \\ 0, & \pi < \omega t < 2\pi \end{cases}$$

$$V_{\text{average}} = V_{\text{DC}} = \frac{1}{2\pi}\int_{\alpha}^{\pi} V_{\text{max}} \sin(\theta)\,\text{d}\theta = \frac{(1+\cos\alpha)V_{\text{max}}}{2\pi}$$

If the load is pure resistance, the load current tracks the voltage, as shown in Figure 5.10c. In particular, note that when the voltage drops to zero at $\theta = \pi$, so does the current. But if the load is a mix of resistance and inductance, the current is nonzero at $\theta = \pi$. If the SCR turns off at this point, $\text{d}i/\text{d}t$ must be $-\infty$. This large negative voltage will force the SCR to remain on and permit the current to follow its natural trajectory to zero at $\theta = \beta$, according to

$$v_{\text{L}}(t) = V_{\text{max}} \sin(\omega t) = Ri_{\text{L}} + L\frac{\text{d}i_{\text{L}}}{\text{d}t}$$

For this more general situation

$$V_{\text{average}} = V_{\text{DC}} = \frac{1}{2\pi}\int_{\alpha}^{\beta} V_{\text{max}} \sin(\theta)\,\text{d}\theta = \frac{(\cos\alpha - \cos\beta)V_{\text{max}}}{2\pi}$$

Considering the full-wave controlled rectifier is Figure 5.10e. The output voltage is

$$V_{\text{average}} = V_{\text{DC}} = \frac{2}{2\pi}\int_{\alpha}^{\beta} V_{\text{max}} \sin(\theta)\,\text{d}\theta = \frac{(\cos\alpha - \cos\beta)V_{\text{max}}}{\pi}$$

There appears to be a problem if $\beta > \alpha + \pi$. If the SCRs 1 to 4 are on when 2 to 3 are fired, it would appear that we have caused a direct short circuit across the source! However, as the short-circuit currents surge up through SCRs 2 to 1 and 4 to 3, 1 and 4 are forced off, and the conducting path 2-load-3 is restored. Hence, continuous conduction is permitted. The output DC voltage is

$$V_{\text{average}} = V_{\text{DC}} = \begin{cases} \dfrac{(\cos\alpha - \cos\beta)V_{\max}}{\pi}; & \text{discontinuous conduction} \\ \dfrac{2\cos\alpha\ V_{\max}}{\pi}; & \text{continuous conduction} \end{cases}$$

The analysis can be extended to the six-step three-phase rectifier shown in Figure 5.11, with the diodes replaced with SCRs. The output DC voltage is

$$V_{\text{average}} = V_{\text{DC}} = \begin{cases} \dfrac{3(\cos\alpha - \cos\beta)V_{\max}}{2\pi}; & \text{discontinuous conduction} \\ \dfrac{3\cos\alpha\ V_{\max}}{\pi}; & \text{continuous conduction} \end{cases}$$

5.9.2 Controlled IGBT Rectifiers

The circuits discussed in Section 5.9.1 can be modified by replacing the SCRs with IBGTs. The difference is that we now have direct control over the extinction angle β. To provide a path for load current if all IGBTs are off, we add a free-wheeling diode in parallel with the load. The outputs are

$$V_{\text{average}} = V_{\text{DC}} = \begin{cases} \dfrac{(\cos\alpha - \cos\beta)V_{\max}}{2\pi} & \text{half-wave} \\ \dfrac{(\cos\alpha - \cos\beta)V_{\max}}{\pi} & \text{full-wave} \\ \dfrac{3(\cos\alpha - \cos\beta)V_{\max}}{2\pi} & \text{six-step} \end{cases}$$

Example 5.10

A load is to be supplied with adjustable DC voltage from a single-phase 120 V 60 Hz AC. source. Using a full-wave controlled thyristor rectifier, determine the rms and DC load voltage and current, when α is set at 60° for:

(a) A 14.14 Ω resistive load
(b) A 10 Ω, 53.06 mH *R–L* load

SOLUTION

The circuit diagram appears in Figure 5.12e:

$$V_{\max} = 120\sqrt{2} = 169.7\ \text{V}$$

(a) The situation is discontinuous, with $\beta = 180°$:

$$V_{\text{average}} = V_{\text{DC}} = \frac{(\cos\alpha - \cos\beta)V_{\max}}{\pi} = \frac{(\cos 60° + 1)169.7}{\pi} = 81.03\ \text{V}$$

$$V_{\text{rms}} = 70.84\ \text{V}$$

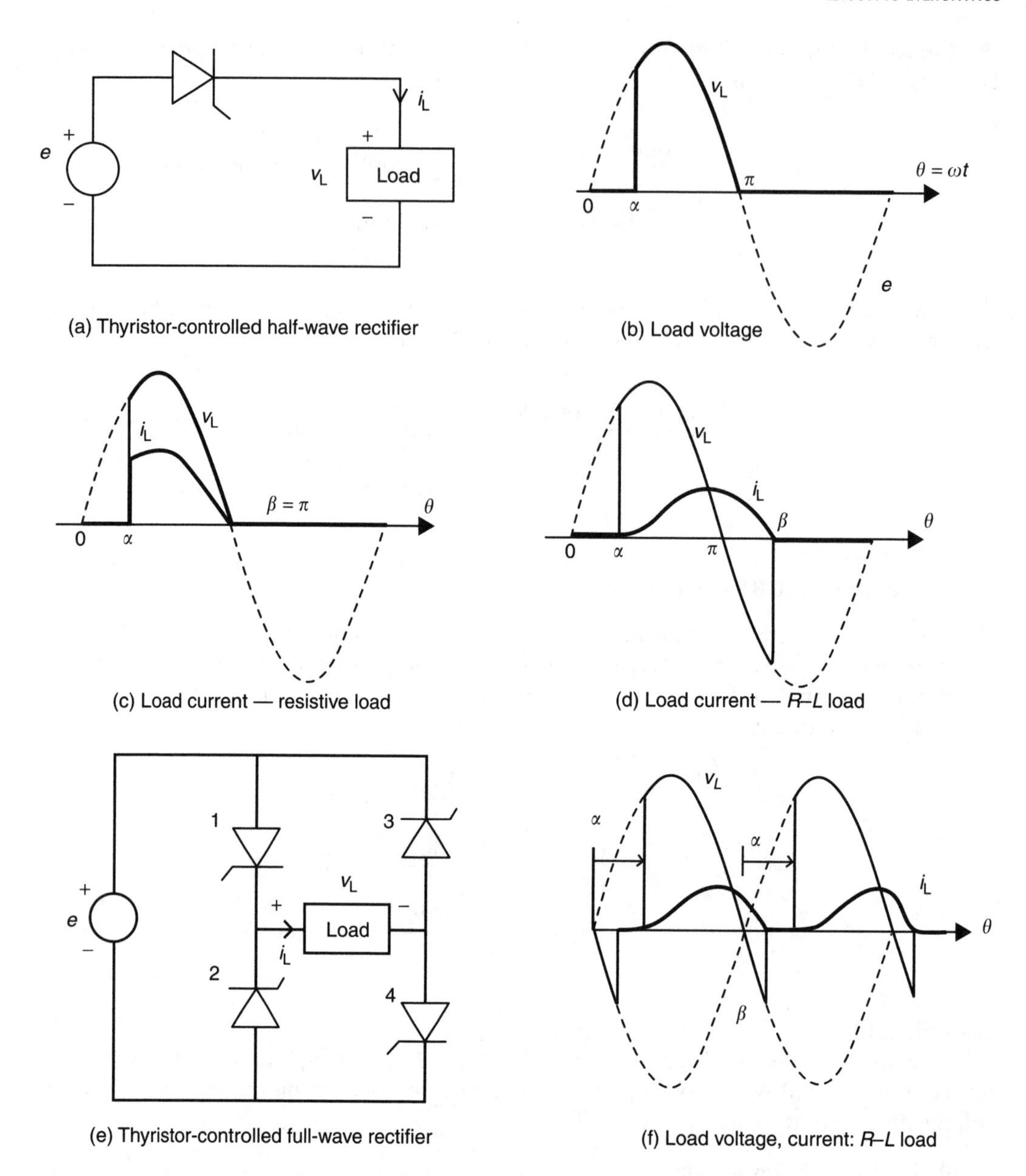

FIGURE 5.12. Controlled single-phase rectifiers.

The load current values are proportional for the resistive load:

$$I_{DC} = V_{DC}/R_L = 5.731\,\text{A}; \qquad I_{rms} = V_{rms}/R_L = 5.010\,\text{A}$$

(b) The situation is continuous

$$V_{average} = V_{DC} = \frac{2\cos\alpha\, V_{max}}{\pi} = \frac{(2)169.7\cos(60°)}{\pi} = 54.02\,\text{V}$$

$$V_{rms} = 120\,\text{V}$$

$$I_{DC} = V_{DC}/R_L = 5.402\,\text{A}; \qquad I_{rms} = 5.864\,\text{A}$$

5.10 AC Motor Drives

We now return to our original objective, that is to control the speed of a three-phase induction motor. We assume that a constant voltage, constant frequency three-phase AC source is available. The fundamental problem then is to convert the constant voltage, constant frequency three-phase source to variable voltage, variable frequency three-phase, from which we intend to supply our motor. Such a converter circuit is called an "AC motor drive" or "adjustable speed drive" (ASD). There are several possible design approaches. We shall mention four.

1. We could convert the constant voltage, constant frequency three-phase AC source to a controllable DC voltage output, using a three-phase SCR, or IGBT, rectifier, which in turn supplies a variable frequency three-phase inverter for frequency control. A parallel capacitor filter is placed between the rectifier and inverter, which tends to maintain constant voltage at the inverter input. Such a drive is sometimes called a "variable voltage" (VV) drive.
2. We could convert the constant voltage, constant frequency three-phase AC source to a controllable DC current output, using a three-phase SCR, or IGBT, rectifier, which in turn supplies a variable frequency three-phase inverter for frequency control. A series inductor filter is placed between the rectifier and inverter, which tends to maintain constant current at the inverter input. Such a drive is sometimes called a "variable current" (VC) drive.
3. We could convert the constant voltage, constant frequency three-phase AC source to a constant DC voltage output, using a three-phase diode rectifier, which in turn, supplies a three-phase inverter, such that voltage and frequency control is simultaneously implemented in the inverter section. This is achieved by a strategy called "pulse width modulation" (PWM), and uses IGBTs (or SCRs) in the inverter stage. The IGBTs are switched in such a way as to produce variable frequency, variable voltage. Such a drive is sometimes called a "PWM drive."
4. The PWM drive can be supplied with sensors to feed back speed and position information to provide more precise speed and position control. Such a drive is sometimes called an "AC vector drive."

Consider the PWM drive, which is summarized in Figure 5.13.

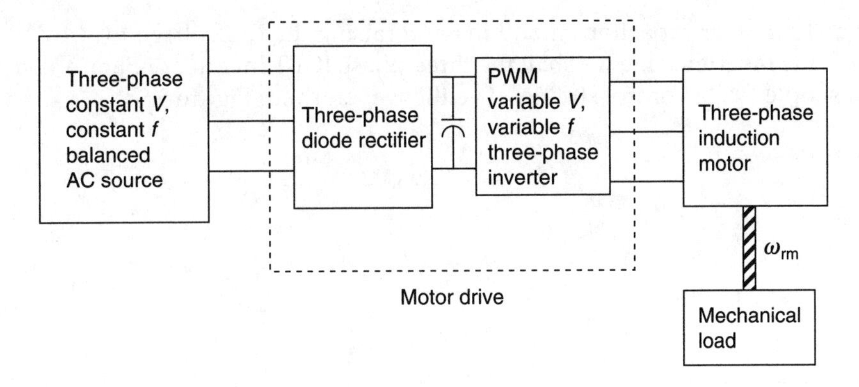

FIGURE 5.13. General configuration of a PWM motor drive.

Example 5.11
Recall the system of Example 4.6. We now wish to drive the load at 50% speed, using a PWM drive. A 480 V 60 Hz source is available.

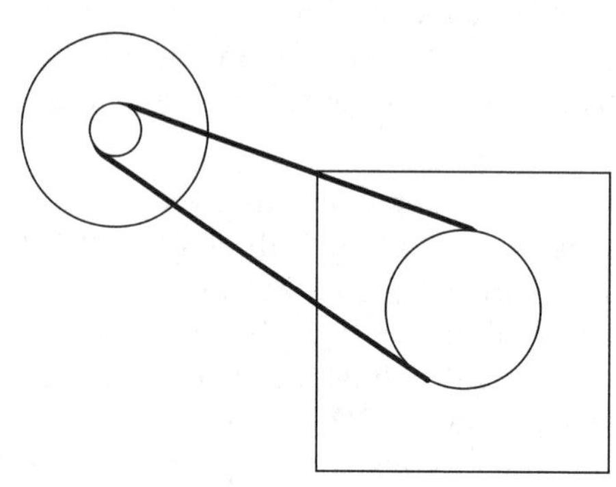

(a) Determine the rectifier design and output DC voltage.
(b) Determine the inverter design and output voltage.
(c) With the drive set for 50% V, 50% f provide a comprehensive performance analysis for the given system, assuming that the actual voltage can be replaced by a sinusoidal voltage of the proper magnitude and frequency.

SOLUTION

(a) For the rectifier stage we shall use the six-step diode rectifier of Figure 5.11a. For a 480 V source,

$$V_{max} = 480\sqrt{2} = 678.8\text{ V}$$

$$V_{average} = V_{DC} = 0.9549\ V_{max} = 648.2\text{ V}$$

We add a 11 mF filter capacitor, smoothing and raising the DC voltage to 664 V.[6]

(b) For the inverter stage we shall the three-phase IGBT inverter of Figure 5.8a, with a simple PWM control strategy. See the waveform of Figure 5.8b.

$$V_{rms} = 230\text{ V}$$

$$f_0 = 30\text{ Hz}$$

[6] We skip the details of capacitor sizing. See Example 5.3 for the methodology.

For $v_L(t)$ below

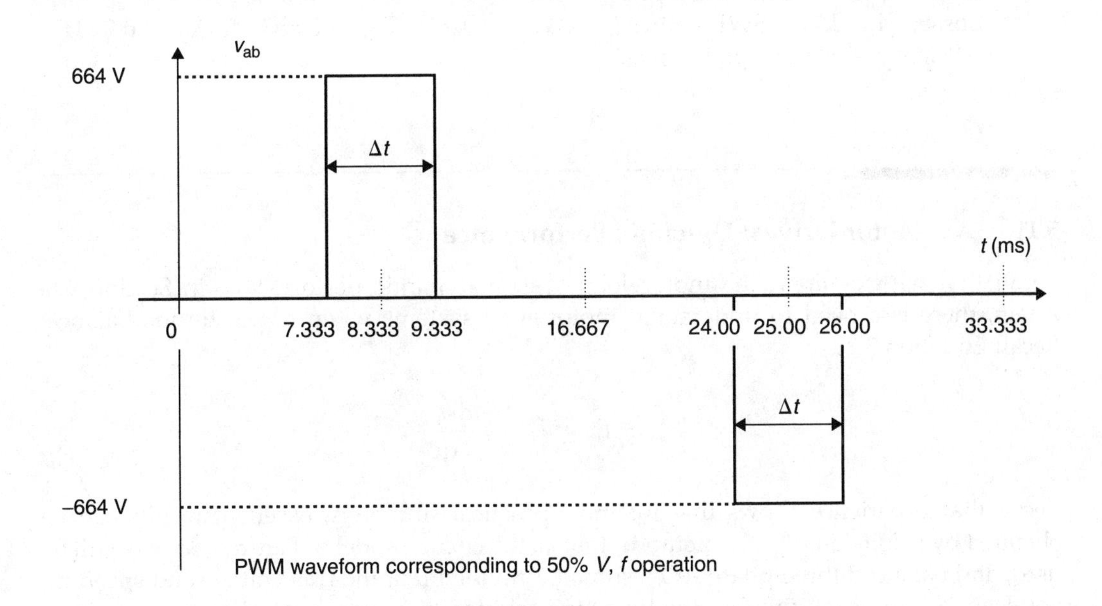

PWM waveform corresponding to 50% V, f operation

$$V_{rms} = \sqrt{\frac{(664)^2 \Delta t}{16.67}} = 230\,\text{V}$$

$$\Delta t = 2.0\,\text{ms}$$

$$T = \frac{1}{30} = 33.33\,\text{ms}$$

(c) Considering a gear ratio of 1000/1800

$$T'_m = T_m + T_{RL} = 0.007818\omega_{rm}^2 + 0.063\omega_{rm}$$

Again, recall our main observation from Chapter 3:

> ***The steady-state running speed for any motor–load system occurs at the intersection of the motor and load torque–speed characteristics, i.e., where $T_{dev} = T'_m$.***

We execute a search procedure (implemented by EMAP) for the speed, by guessing an initial speed and computing corresponding torques, and based on the results, refine the speed estimation until the equilibrium point is located. The result of such an analysis are

- ω_{rm} = speed = 894.6 rpm = 93.7 rad/s; slip = 0.00600; ω_L = 497.0 rpm; stator voltage, 230.0 V; frequency, 30.0 Hz
  Operating mode: MOTOR
- Stator–rotor equivalent circuit mesh [Z] matrix at 30 Hz: $Z_{11} = 0.042 + j3.798$; $Z_{12} = 0.000 - j3.703$; $Z_{21} = 0.000 - j3.703$; $Z_{22} = 7.053 + j3.798$
- Currents (stator) (A), I_a (stator) = 39.467 at −62.58°; I'_A (rotor) = 18.243 at −0.88°
- Torques in newton-meters (RL = rotational loss): developed = 74.72; T_{RL} = 5.901; T_m = 68.61

- Powers in kW: mech. output = 6.447 (8.642 hp)
  Stator input = 7.240; developed = 7.000
- Losses in kW: SWL = 0.198; RWL = 0.042; P_{RL} = 0.553; P_{LOSS} = 0.793; efficiency = 89.05%; power factor = 46.05%

5.11 AC Motor Drives: Dynamic Performance

Recall that we examined the motor–load system dynamic performance in Section 4.8. Again, there is a need to understand motor–load systems when not in torque balance. Recall Equation 3.1:

$$T_{dev} - T'_{m} = T_{a} = J\frac{d\omega_{rm}}{dt}$$

Recall that experience shows that for most practical situations, excellent results can be obtained by using "quasi-AC" methods, that is, the circuit model of Figure 4.8b may still be used, and balanced three-phase AC assumed, provided that the rms voltage and speed do not change significantly within the time step selected to compute the dynamic response. Most loads and motors will have sufficient inertia to meet this restriction. The addition of the AC motor drive adds the additional requirement that the voltage magnitude and frequency will now vary in time. However, there are high-performance situations where this restriction is not met, and the reader is cautioned to be aware of this issue. Also note that more data are needed; specifically, the inertias for the motor and load.

Starting the system is a particularly important dynamic situation. Whereas we were previously limited to changing the voltage in a few discrete steps, or switching rotor resistance in a few discrete steps, the AC motor drive permits changing voltage magnitude and frequency continuously over the starting cycle. Hence, we have a very flexible tool to achieve our object of a high-torque low-current fast and smooth start. The drive can be set to ramp up k, where $V = kV_{rated}$ and $f = kf_{rated}$, and/or limit current to a user-defined value. Since the problem is complicated and nonlinear, it does not lend itself to a simple closed-form solution. The problem is best solved by computer, and even then, one must "play" with the adjustable drive parameters to obtain a desired solution.

We consider starting the system of Example 4.8 from an AC drive from standstill. A drive simulation program is available in EMAP, and we will use it to demonstrate the analysis. The EMAP dynamic AC drive simulation automatically ramps up "k," where $V = kV_{rated}$ and $f = kf_{rated}$. The user can define

- k_{max} (= 1.0)
- k_{min} (= 0.05)
- T_S = desired starting time (s), which determines the ramp rate of k.
- I_{max}/I_{rated}

We set k_{max} = 1.0, because we have no desire to go faster than the speed corresponding to 60 Hz. We set k_{min} = 0.05, because we require some torque to get started, and experience shows 5% is a reasonable value. The choice of I_{max}/I_{rated} depends on the realization that high current means high torque and fast starts, where at the same time too much current could damage the motor or the drive (actually the drive is self-protected, and will not permit dangerous current for any operating condition). We select I_{max}/I_{rated} = 2.

TABLE 5.2
Motor Starting Using an AC Drive

Time (s)	*kf*	*kv*	Speed (rpm)	Current (A)	T_{dev} (N m)	T_a (N m)
Setting $T_s = 1.5$ s						
0.00	0.05	0.08	0	249.1	785	785
0.12	0.13	0.13	182	142.6	457	453
0.27	0.22	0.22	360	116.7	410	397
0.43	0.32	0.32	542	115.5	416	387
0.58	0.42	0.42	721	119.0	434	385
0.74	0.52	0.52	902	124.7	459	383
0.90	0.62	0.62	1082	132.2	489	382
1.06	0.72	0.72	1262	141.3	525	381
1.22	0.83	0.83	1441	152.1	567	379
1.39	0.93	0.93	1621	164.8	614	378
1.62	1.00	1.00	1779	80.2	286	3
Setting $T_s = 0.9$ s						
0.00	0.05	0.08	0	249.1	785	785
0.10	0.16	0.16	182	251.0	682	678
0.19	0.25	0.25	364	242.4	767	754
0.27	0.34	0.34	542	208.5	716	687
0.37	0.44	0.44	723	198.1	702	652
0.46	0.54	0.54	901	200.1	718	642
0.56	0.64	0.64	1082	206.5	745	637
0.65	0.74	0.74	1262	215.5	779	634
0.75	0.84	0.84	1440	226.6	818	631
0.85	0.94	0.94	1622	240.2	864	628
1.04	1.00	1.00	1779	80.2	286	3
Setting $T_s = 0.88$ s						
0.00	0.05	0.08	0	249.1	785	785
0.10	0.16	0.16	182	251.0	653	649
0.20	0.27	0.22	362	251.0	600	586
0.33	0.40	0.24	542	251.0	391	362
0.77	0.88	0.33	720	251.0	85	35
6.57	1.00	0.37	900	251.0	81	5
8.92	1.00	0.38	929	251.0	83	3

Now select three different values of T_s: examine Table 5.2 and the plots in Figure 5.14. In (a), we set the drive for a ramp-up time $T_s = 1.5$ s. The system easily makes a smooth start in about 1.6 s. However, the current is only around 60% its limit during the start. This indicates that a faster start is possible. In (b) we try $T_s = 0.9$ s. Again, the system makes a smooth start, starting in about 0.94 s. The current is around 90% for most of the start, hitting the limit at some points. Finally in (c), we try $T_s = 0.88$ s. What a difference! After 8 s, we are only running a little above 900 rpm! What has happened? The drive has "outrun" the system! That is, we commanded the drive to change the voltage and frequency so fast that the motor could not keep up. For the given load torque and inertia, the motor could not develop enough torque to remain anywhere close to its maximum torque point. The current is at its limit throughout the start. We will probably never come up to full speed, finally reaching an equilibrium point somewhere on the back side of the torque–speed characteristic.

Our conclusions are that (1) a drive must be tuned to the load for optimum performance, and (2) when properly tuned, drive dynamics can easily outperform older technologies. Drives can do this and can be programed to do much more. They can also provide diagnostics, which are extremely useful for troubleshooting and other evaluation functions.

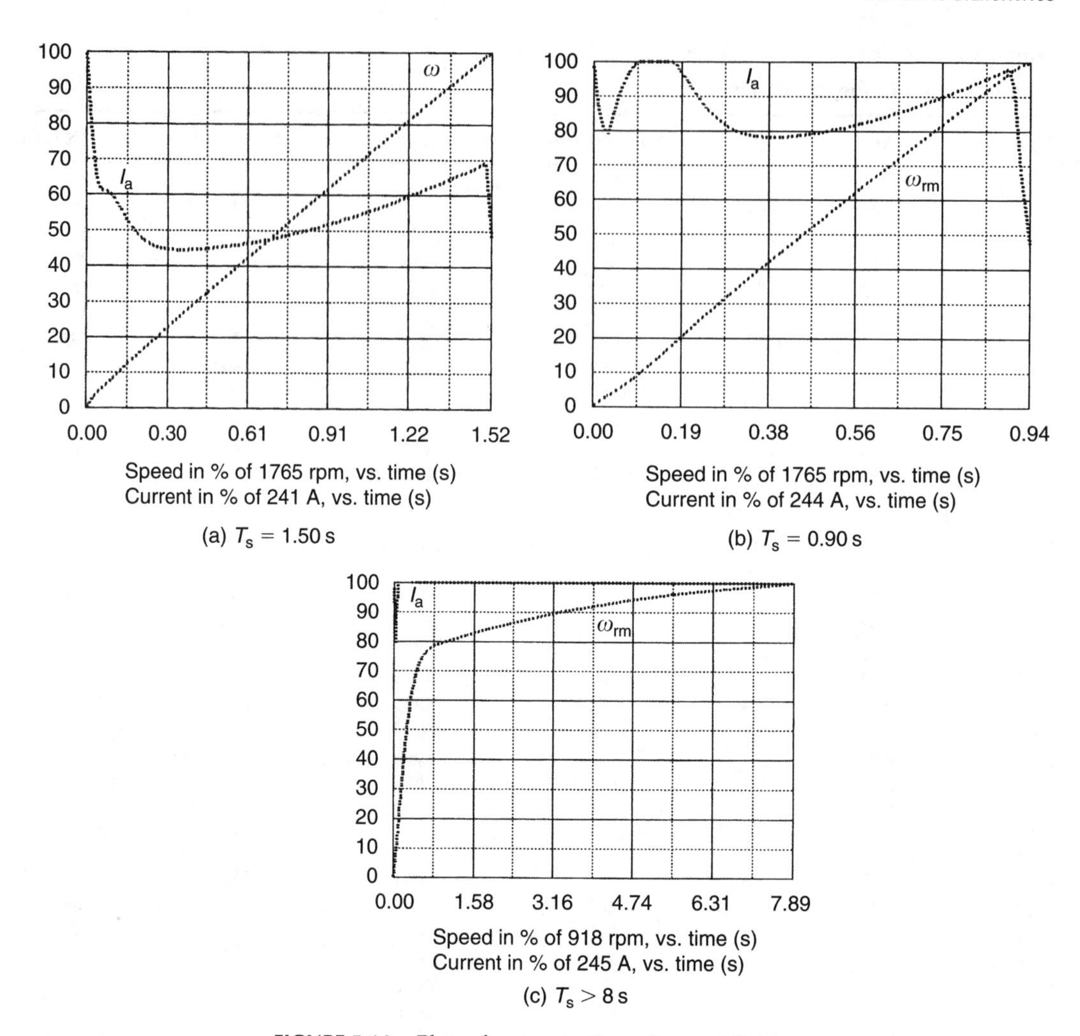

FIGURE 5.14. Plots of motor starting using an AC drive.

5.12 Motor Reverse Performance

Our discussion to this point has essentially been confined to first quadrant (motor-forward) operation. But what if an application requires running in both directions? How can motor drives accommodate these conditions?

Recall that the direction of the rotating stator field determined the rotor direction of rotation for motor operation. Also, recall that the stator-phase sequence established the stator-field direction of rotation. We realize that all that is required to change the phase sequence is reverse the firing order of the IGBTs! Refer to Figure 5.8, which shows sequence abc.

Sequence abc	Sequence cba
Q1–Q4–Q6	Q5–Q4–Q2
Q1–Q3–Q6	Q5–Q3–Q2
Q2–Q3–Q6	Q6–Q3–Q2
Q2–Q3–Q5	Q6–Q3–Q1
Q2–Q4–Q5	Q6–Q4–Q1
Q1–Q4–Q5	Q5–Q4–Q1

Hence, we can operate in the motor mode in either direction with equal facility simply by properly controlling the IGBTs!

5.13 The Cycloconverter

If steady-state and dynamic performance is required in all four quadrants, a different approach may be justified. Consider the arrangement in Figure 5.15a. Although the design, called a cycloconverter, requires two complete 6-SCR banks per phase (36 SCRs total), the DC section has been eliminated. Note that each phase winding is electrically isolated. Now consider Figure 5.15b. If the SCRs are fired in the sequence indicated, the voltage (v) applied to phase "a" is as indicated. Specifically, at time t_1, the ON SCRs for motor-mode operation are those which are circled (Q4–Q5). For generator operation, the ON SCRs are Q10–Q11. By inspection, the waveshape has a strong 20 Hz component, and its amplitude is about one third the full voltage, complying with the constant volts/hertz principle.

Again, the direction of the rotating stator field can be determined by the fundamental phase sequence, and hence the firing order of the SCRs, so that operation in both directions is equally possible. Since we can operate in either motor or generator mode and in both directions, the cycloconverter permits full four-quadrant operation.

5.14 Summary

An AC motor directly connected to a load runs at a speed determined by torque balance: that is, where the motor and load torque–speed characteristics intersect. Unfortunately, there are many applications that require the system to operate at some desired speed. Thus, it is important to understand what are the basic physical principles that determine speed, and to design systems that permit speed control. We learned that the key to AC motor control was to vary voltage magnitude and frequency in such a way that their ratio is constant. In order to achieve this goal, it was necessary to design appropriate power processing circuits.

The rectifier, along with its filter, serves as an approximation of an "ideal" DC source, a pure source of voltage and current, from which we can in effect "start over," and create an AC voltage of any rms value and frequency, using a circuit configuration, called an "inverter." Such circuits are practical only because of the availability of high-power low-loss fast switching components. Today dozens of appropriate semiconductor switch exist of which we examined: the diode, the thyristor, and the IGBT.

A common design utilizes PWM, which permits translation of voltage magnitude and frequency control to a timing problem, and hence is ideally suited to microprocessor control. The drive easily accommodates reversing. If full four-quadrant operation is necessary, the cycloconverter is suitable. Whereas PWM control is excellent for many applications, there are applications that require precision speed and position control. For these, a technology called "AC vector control" is appropriate. We defer this topic until Chapter 7, first using the synchronous motor, and then consider AC induction motor applications.

In analyzing such systems, we realize that our machine model is inadequate. The primary defect was that the model was limited to balanced three-phase sinusoidal constant speed operation. We continue our study in Chapter 6 to remove these restrictions from our analysis.

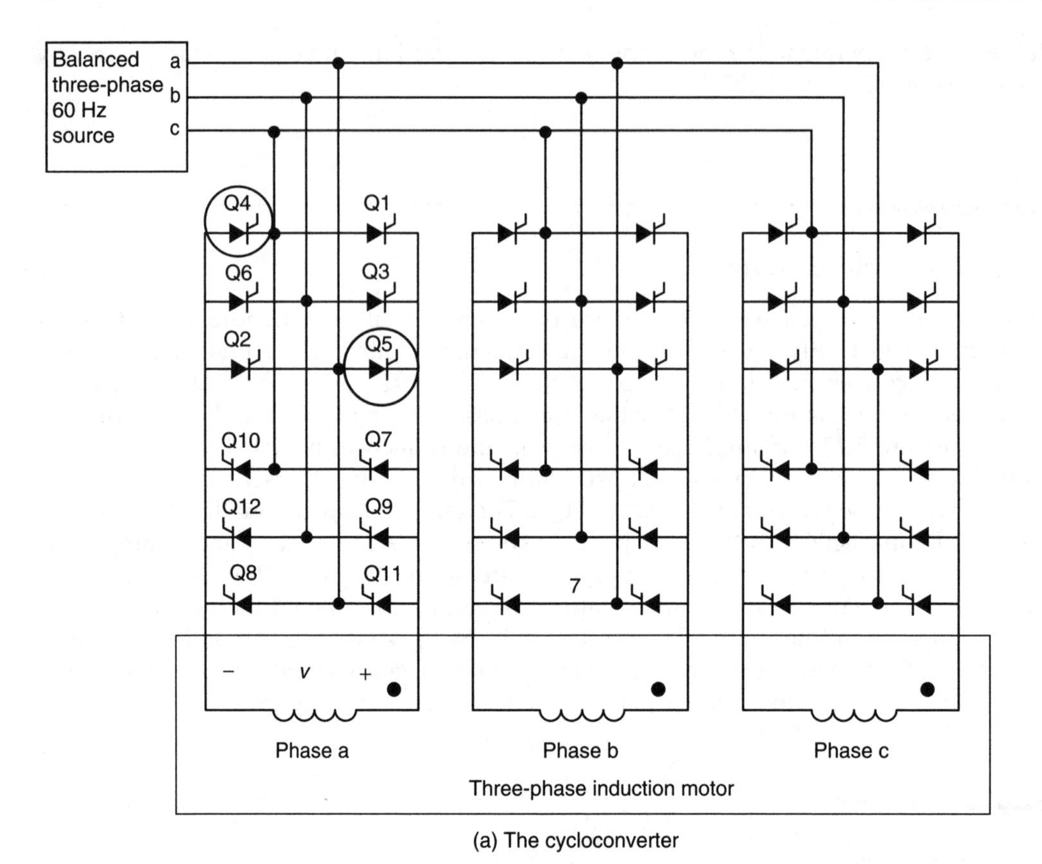

(a) The cycloconverter

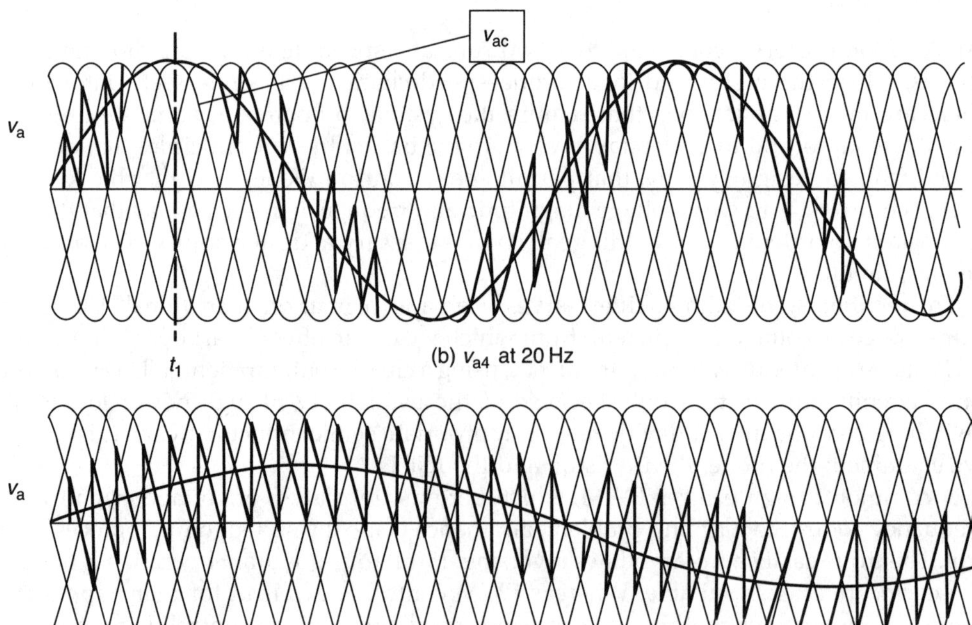

(b) v_{a4} at 20 Hz

(b) v_{a4} at 10 Hz

FIGURE 5.15. The cycloconverter and associated output waveforms.

Problems

5.1 Consider the motor of Example 4.3 operating with shorted rotor. Find all currents, powers, torques, the speed, the power factor, and the efficiency at

(a) 30% stator voltage; 100% frequency, slip = 1.5%.
(b) 100% stator voltage; 30% frequency, slip = 1.5%.
(c) 30% stator voltage; 30% frequency, slip = 1.5%.

5.2 Refer to Problem 5.1. Discuss the differences between results of (a), (b), and (c), explaining why operation at (c) is preferable at reduced speed.

5.3 Consider the circuit of Figure 5.2b using a 300 V DC source and a 10 Ω resistive load.

(a) Select appropriate values for T and ΔT such that V_{rms}/f is constant from a maximum of $V_{rms} = 300\,\text{V}$ at $f_0 = 100\,\text{Hz}$ for $f = 80$ and 50 Hz.
(b) Sketch the waveforms of load voltage and current for the settings of (a).

5.4 Consider the circuit of Figure 5.2b using a 150 V DC source and a 10 Ω resistive load.

(a) Select appropriate values for T and ΔT such that V_{rms}/f is constant from a maximum of $V_{rms} = 150\,\text{V}$ at $f_0 = 100\,\text{Hz}$ for $f = 70$ and 30 Hz.
(b) Sketch the waveforms of load voltage and current for the settings of (a).

5.5 Consider the diode circuit of Figure 5.3c, with a 10 Ω resistive load

(a) With $v_s = 100\,\text{V}$
  1. Draw the circuit diagram. Show all voltages and currents thereon.
  2. Draw a plot like Figure 5.3b. Locate the operating point.

(b) With $v_s = -100\,\text{V}$
  1. Draw the circuit diagram. Show all voltages and currents thereon.
  2. Draw a plot like Figure 5.3b. Locate the operating point.

5.6 Consider the thyristor circuit of Figure 5.4c, with a 10 Ω resistive load; $v_s = 100\,\text{V}$; and $i_g(t) = \delta(t-2)$.

(a) Draw the circuit diagram.
(b) Draw a plot like Figure 5.4b. Locate the operating points (2).
(c) Plot $v_d(t)$ and $i_d(t)$; $0 \le t \le 4\text{s}$.

5.7 Consider the IGBT circuit of Figure 5.5c, with a 10 Ω resistive load, $v_s = 100\,\text{V}$; and $i_b(t) = u(t-1) - u(t-3)$.

(a) Draw the circuit diagram.
(b) Draw a plot like Figure 5.5b. Locate the operating points (2).
(c) Plot $v_d(t)$ and $i_d(t)$; $0 \le t \le 4\text{s}$.

5.8 The circuit of Figure 5.6c has the following values: $V_{DC} = 100\,\text{V}$, $R = 10\,\Omega$, $L = 10\,\text{mH}$. The IGBT turns ON at $t = 0$, and OFF at $t = 1\,\text{ms}$. Compute and plot the load current, the IGBT current, and the diode current, all on the same graph. Let "t" be in ms (not seconds).

5.9 Repeat Problem 5.8, except the IGBT is turned on and off **periodically** (ON at $t = 0$; OFF at $t = 1\,\text{ms}$; ON at $t = 2\,\text{ms}$; etc.).

5.10 In Problem 5.9, compute **rms** values for the

(a) load current
(b) IGBT current
(c) diode current

5.11 In Problem 5.9, compute **maximum** values for the

(a) load voltage
(b) IGBT voltage
(c) diode voltage

5.12 Rework Example 5.5 with revised data
Switching frequency = 100 Hz; Load: $L = 50$ mH; $V_{dc} = 300$ V
IGBTs are symmetrically switched to force the given ΔT values.

Given $\Delta t = 6$ ms, compute and plot the load current waveform, indicating the conducting state of all diodes and IGBTs.

5.13 Repeat Problem 5.12, given $\Delta t = 2$ ms.

5.14 Consider the three-phase inverter of Figure 5.8a operating from a 300 V DC source.

(a) Determine the maximum rms phase-to-phase voltage possible.
(b) Determine the maximum rms phase-to-neutral voltage possible.

5.15 Consider the three-phase inverter of Figure 5.8a operating from a 300 V DC source. Draw the circuit diagram. Plot $v_{an}(t)$ and $v_{ab}(t)$ versus t, and compute the necessary switching times for outputs of 100% voltage at 60 Hz.

5.16 Consider the three-phase inverter of Figure 5.8a operating from a 300 V DC source. Draw the circuit diagram. Plot $v_{an}(t)$ and $v_{ab}(t)$ versus t, and compute the necessary switching times for outputs of 33% voltage at 20 Hz.

5.17 Consider the situation described in Problem 5.16. Assume a load of three 10 Ω wye-connected resistors. Draw the circuit diagram. The cycle starts with Q1, Q4, and Q6 ON at $t = 0$. At $t = 12.5$ ms, determine the current in all parts of the circuit and write the values onto the diagram.

5.18 Consider the half-wave rectifier of Figure 5.10a operating from a 120 V 60 Hz single-phase source, with a load resistor of 20 Ω. Accurately sketch the load voltage and current with

(a) no filter capacitor
(b) an 8 mF filter capacitor

5.19 Continuing Problem 5.18, compute the rms and DC load voltage and current with

(a) no filter capacitor
(b) an 8 mF filter capacitor

5.20 Consider the full-wave rectifier of Figure 5.10b operating from a 120 V 60 Hz single-phase source, with a load resistor of 20 Ω. Accurately sketch the load voltage and current with

(a) no filter capacitor
(b) an 4 mF filter capacitor

5.21 Continuing Problem 5.20, compute the rms and DC load voltage and current with

(a) no filter capacitor
(b) an 4 mF filter capacitor

5.22 Consider the 6-diode bridge rectifier of Figure 5.11 operating from a 120/208 V 60 Hz three-phase source, with a 10 Ω load resistor. Accurately sketch the load voltage and current with

(a) no filter capacitor
(b) an 4 mF filter capacitor

5.23 Continuing Problem 5.22, compute the rms and DC load voltage and current with

(a) no filter capacitor
(b) an 4 mF filter capacitor

5.24 Consider the half-wave thyristor-controlled rectifier of Figure 5.12 operating from a 120 V 60 Hz single-phase source. Draw the circuit diagram. Find β and compute the DC load voltage, for $\alpha = 60°$

(a) for a resistive load of 20 Ω
(b) for a *R–L* load of 10 Ω, 45.94 mH

5.25 Consider the full-wave thyristor-controlled rectifier of Figure 5.12e operating from a 120 V 60 Hz single-phase source. Draw the circuit diagram. Find β and compute the DC load voltage, for $\alpha = 60°$

(a) for a resistive load of 20 Ω
(b) for a *R–L* load of 10 Ω, 45.94 mH

5.26 Consider a six-step thyristor-controlled rectifier operating from a 120/208 V 60 Hz three-phase source. Draw the circuit diagram. Find β and compute the DC load voltage, for $\alpha = 60°$

(a) for a wye-connected load of $R = 20\,\Omega$, per phase
(b) for a wye-connected load *R–L* load of 10 Ω, 45.94 mH, per phase

5.27 Consider the half-wave IGBT-controlled rectifier operating from a 120 V 60 Hz single-phase source. Draw the circuit diagram with a free-wheeling diode added across the load. Compute and plot the DC load voltage, for $\alpha = 60°$; $\beta = 150°$

(a) for a resistive load of 20 Ω
(b) for a *R–L* load of 10 Ω, 45.94 mH

5.28 Consider the full-wave IGBT-controlled rectifier operating from a 120 V 60 Hz single-phase source. Draw the circuit diagram with a free-wheeling diode added across the load. Compute and plot the DC load voltage, for $\alpha = 60°$; $\beta = 150°$ for a resistive load of 20 Ω.

5.29 Repeat Problem 5.28 for a *R–L* load of 10 Ω, 45.94 mH.

5.30 Recall the system of Example 5.7. We now wish to drive the load at 70% speed, using a PWM drive, supplied from a 277/480 V 60 Hz three-phase source. Assume the rectifier output is filtered so as to produce a 664 V DC bus voltage. Plot the inverter PWM output voltage waveform.

6

The Polyphase Induction Machine: Unbalanced Operation

Recall that we have investigated the polyphase induction machine in Chapter 4. We examined the machine's main structural details, and developed accurate mathematical models. We also examined motor performance, operating under balanced three-phase sinusoidal excitation, and running at constant speed. The approach was extended to system dynamic performance, such as starting and stopping, assuming quasi-AC balanced operation. But what if the applied voltages are not balanced? Indeed, what if the excitation voltages are not sinusoidal? And finally, what if the mechanical transients have time constants comparable to the electrical transients? We shall extend our work to these important topics now.

6.1 Unbalanced Operation

The circuits of Figure 4.8 are valid models for the polyphase induction machine, operating under balanced three-phase AC excitation at constant speed. They can even be used to assess dynamic performance, if the excitation voltage remains balanced three-phase AC, as long as the speed changes are slow relative to the power frequency. But what if the applied voltages are not balanced?

Recall that a set of three-phase voltages may be transformed into three so-called sequence voltages by means of the symmetrical component transformation,[1] according to

$$\begin{bmatrix} \bar{V}_0 \\ \bar{V}_1 \\ \bar{V}_2 \end{bmatrix} = \frac{1}{3}\begin{bmatrix} 1 & 1 & 1 \\ 1 & a & a^2 \\ 1 & a^2 & a \end{bmatrix}\begin{bmatrix} \bar{V}_{\text{an}} \\ \bar{V}_{\text{bn}} \\ \bar{V}_{\text{cn}} \end{bmatrix}$$

or

$$\hat{V}_{012} = [T]^{-1}\hat{V}_{abc}$$

where $a = 1\angle 120°$, and $\bar{V}_0$, $\bar{V}_1$, and $\bar{V}_2$ are called the zero, positive, and negative sequence voltages, respectively. The transformation in effect decomposes an arbitrary set of phase

[1] Presumably, the reader is familiar with the symmetrical component transformation. If not, it will be necessary to consider this subject to appreciate the ideas discussed in this section. Appendix B is provided for that purpose. Specifically see Section B.4.

(a, b, c) values into a set of *sequence* ($0, 1, 2$) values. To understand sequence values, we shall work a demonstration example.

Example 6.1
Work out the sequence voltages for the following cases:

(a) The phase voltages are balanced three-phase with phase sequence abc.
(b) The phase voltages are balanced three-phase with phase sequence acb.
(c) The phase voltages are equal.

SOLUTION

(a) Assume that Phase a voltage is $\bar{V}_{an} = \bar{E}$. If the phase voltages are balanced three-phase, with phase sequence abc, $\bar{V}_{bn}$ must lag $\bar{V}_{an}$ by 120°, and $\bar{V}_{cn}$ must lead $\bar{V}_{an}$ by 120°. Hence, $\bar{V}_{bn} = a^2 \bar{V}_{an} = a^2 \bar{E}$ and $\bar{V}_{cn} = a\bar{V}_{an} = a\bar{E}$. Substituting into Equation B-17a:

$$\begin{bmatrix} \bar{V}_0 \\ \bar{V}_1 \\ \bar{V}_2 \end{bmatrix} \begin{matrix} = \\ = \\ = \end{matrix} \frac{1}{3}\begin{bmatrix} 1 & 1 & 1 \\ 1 & a & a^2 \\ 1 & a^2 & a \end{bmatrix}\begin{bmatrix} \bar{E} \\ a^2\bar{E} \\ a\bar{E} \end{bmatrix} = \begin{bmatrix} 0 \\ \bar{E} \\ 0 \end{bmatrix}$$

(b) Again, set $\bar{V}_{an} = \bar{E}$. For sequence acb, $\bar{V}_{bn} = a\bar{E}$ and $V_{cn} = a^2\bar{E}$. Substituting into Equation B-17a:

$$\begin{bmatrix} \bar{V}_0 \\ \bar{V}_1 \\ \bar{V}_2 \end{bmatrix} \begin{matrix} = \\ = \\ = \end{matrix} \frac{1}{3}\begin{bmatrix} 1 & 1 & 1 \\ 1 & a & a^2 \\ 1 & a^2 & a \end{bmatrix}\begin{bmatrix} \bar{E} \\ a\bar{E} \\ a^2\bar{E} \end{bmatrix} = \begin{bmatrix} 0 \\ 0 \\ \bar{E} \end{bmatrix}$$

(c) Now set $\bar{V}_{an} = \bar{E} = \bar{V}_{bn} = \bar{V}_{cn}$. Substituting into Equation B-17a:

$$\begin{bmatrix} \bar{V}_0 \\ \bar{V}_1 \\ \bar{V}_2 \end{bmatrix} \begin{matrix} = \\ = \\ = \end{matrix} \frac{1}{3}\begin{bmatrix} 1 & 1 & 1 \\ 1 & a & a^2 \\ 1 & a^2 & a \end{bmatrix}\begin{bmatrix} \bar{E} \\ \bar{E} \\ \bar{E} \end{bmatrix} = \begin{bmatrix} \bar{E} \\ 0 \\ 0 \end{bmatrix}$$

Reflect on the results of Example 6.1.

- In (a) we discovered that for a three-phase balanced set of phase voltages, sequence abc, the zero and negative sequence voltages were zero, and that the positive sequence voltage was identical to $\bar{V}_{an}$.

$$\bar{V}_1 = \bar{E} = \bar{V}_{an} \tag{6.1a}$$

- In (b) we discovered that for a three-phase balanced set of phase voltages, sequence acb, the zero and positive sequence voltages were zero, and that the negative sequence voltage was identical to $\bar{V}_{an}$.

$$\bar{V}_2 = \bar{E} = \bar{V}_{an} \tag{6.1b}$$

- In (c) we discovered that for an equal set of phase voltages, the positive and negative sequence voltages were zero, and that the zero sequence voltage was identical to $\bar{V}_{an}$.

$$\bar{V}_0 = \bar{E} = \bar{V}_{an} \tag{6.1c}$$

Now extend these perceptions to the general case. For general unbalanced phase voltages, all three sequence voltages will be present. We can replace the phase values with their symmetrical components, calculate the machine sequence currents, and transform the sequence currents back to phase values, *provided we can derive appropriate sequence circuit models.* Consider calculating the machine currents in response to each sequence acting alone.

6.1.1 Positive Sequence Response

What would be an appropriate circuit model for the machine reacting to positive sequence excitation? This is equivalent to the response to the application of a three-phase balanced set of phase voltages, sequence abc, with $\bar{V}_{an} = \bar{V}_1$. We realize we have already solved this problem in Chapter 4. The proper circuit model is presented in Figure 4.8b. The procedure is as follows. Set $\bar{V}_{an} = \bar{V}_1$. The slip is

$$\text{positive sequence slip} = s = s_+ = \frac{\omega_{sm} - \omega_{rm}}{\omega_{sm}} \tag{6.2a}$$

The circuit of Figure 4.8b is applicable. Solve for the stator and rotor currents, and symbolize as

$$\bar{I}_{S1} \quad \text{and} \quad \bar{I}'_{R1}$$

6.1.2 Negative Sequence Response

What would be an appropriate circuit model for the machine reacting to negative sequence excitation? This is equivalent to the response to the application of a three-phase balanced set of phase voltages, sequence acb, with $\bar{V}_{an} = \bar{V}_2$. Again, the circuit model of Figure 4.8b is applicable with one important modification. Realize that because of the reversal in phase sequence, the rotor, and the rotating stator field now have opposite directions of rotation. This will require a re-computation of the slip. The procedure is as follows: Set $\bar{V}_{an} = \bar{V}_2$. The slip is now:

$$\text{negative sequence slip} = s_- = \frac{-\omega_{sm} - \omega_{rm}}{-\omega_{sm}} = \frac{\omega_{sm} + \omega_{rm}}{\omega_{sm}} = 2 - s \tag{6.2b}$$

The circuit of Figure 4.8b is applicable, using the negative slip. Solve for the currents, and symbolize as

$$\bar{I}_{S2} \quad \text{and} \quad \bar{I}'_{R2}$$

6.1.3 Zero Sequence Response

Now $\bar{V}_{an} = \bar{V}_{bn} = \bar{V}_{cn} = \bar{V}_0$. Whatever currents flow, they must be equal in phase and magnitude: $\bar{I}_a = \bar{I}_b = \bar{I}_c = \bar{I}_0$. Since the corresponding mmf's cancel, the air-gap flux is essentially zero (except for some leakage), and there are no interactions with the rotor windings. Hence;

$$\bar{I}_0 = \frac{\bar{V}_0}{R_0 + jX_0} = \frac{\bar{V}_0}{\bar{Z}_0} \tag{6.3}$$

where

$\bar{Z}_0$ = complex zero sequence impedance.

$\bar{Z}_0$ is available from the manufacturer or from test (connect the windings in series, apply a single-phase AC source, and measure the applied voltage ($3\bar{V}_0$), current ($\bar{I}_0$), and power). If the connection is three-wire, or no path to the system neutral is available, $\bar{Z}_0$ is open.

Symbolize the zero sequence currents as

$$\bar{I}_{S0} = \bar{I}_0, \qquad \bar{I}'_{R0} = 0$$

We can use superposition[2] to compute the machine stator currents:

$$\begin{bmatrix} \bar{I}_a \\ \bar{I}_b \\ \bar{I}_c \end{bmatrix} = \begin{bmatrix} 1 & 1 & 1 \\ 1 & a^2 & a \\ 1 & a & a^2 \end{bmatrix} \begin{bmatrix} \bar{I}_{S0} \\ \bar{I}_{S1} \\ \bar{I}_{S2} \end{bmatrix}$$

Consider some powers:[3]

$$P_{in} = \bar{V}_{an}\,\bar{I}_a \cos\theta_a + \bar{V}_{bn}\,\bar{I}_b \cos\theta + \bar{V}_{cn}\,\bar{I}_c \cos\theta_c \tag{6.4a}$$

$$Q_{in} = \bar{V}_{an}\,\bar{I}_a \sin\theta_a + \bar{V}_{bn}\,\bar{I}_b \sin\theta + \bar{V}_{cn}\,\bar{I}_c \sin\theta_c \tag{6.4b}$$

$$\text{SWL} = \bar{I}_a^2 R_1 + \bar{I}_b^2 R_1 + \bar{I}_c^2 R_1 \tag{6.5}$$

Rotor quantities require special consideration. Even though the current $\bar{I}'_A$ was at the same stator frequency for both positive and negative sequence excitation, the "real" rotor current ($\bar{I}_A$) was not, since its frequency was $s(f)$. Therefore, it is possible to compute

$$P_f = 3(\bar{I}'_{R1})^2(R'_{2x}/s_+) \tag{6.6a}$$

$$P_b = 3(\bar{I}'_{R2})^2(R'_{2x}/s_-) \tag{6.6b}$$

$$\text{RWL} = s_+P_f + s_-P_b \tag{6.7}$$

$$P_{dev} = (1 - s_+)P_f + (1 - s_-)P_b \tag{6.8}$$

---

[2] At least at constant speed, which implies constant slips, resulting in linear circuit models.

[3] Be careful here. Note that it would NOT be correct to compute SWL = $\bar{I}_{S0}^2 R_1 + \bar{I}_{S1}^2 R_1 + \bar{I}_{S2}^2 R_1$. That is, superposition in general does not apply to power calculations.

The developed torque may be computed as follows:

$$T_{dev} = P_{dev}/\omega_{rm} = \frac{P_{dev}}{(1-s)\omega_{rm}} \tag{4.32a}$$

Substituting Equation 6.8 into Equation 4.32a:

$$T_{dev} = \frac{(1-s)P_f + (1-s_{-})P_b}{(1-s)\omega_{sm}} \tag{6.9}$$

Defining forward and backward torques:

$$T_f = \frac{P_f}{\omega_{sm}} \tag{6.10a}$$

$$T_b = \frac{P_b}{\omega_{sm}} \tag{6.10b}$$

and finally

$$T_{dev} = T_f - T_b \tag{6.11}$$

That is, the developed electromagnetic torque is composed of two counteracting components, the forward torque, acting on the rotor in the direction of rotor rotation, and the backward component, acting in the reverse direction. The rest of the analysis is the same as for the balanced case:

$$P_{RL} = K_{RL}[\omega_{rm}]^2 \tag{4.26}$$

$$P_m = P_{dev} - P_{RL} \tag{4.27}$$

$$\text{Output power} = P_{out} = P_m \tag{4.28}$$

$$P_{LOSS} = P_{RL} + \text{SWL} + \text{RWL}$$

$$\eta = \frac{P_{out}}{P_{in}}, \text{ expressed as a \%} \tag{4.30}$$

$$T_{RL} = \frac{P_{RL}}{\omega_{rm}} = K_{RL}\omega_{rm}$$

$$T_m = \frac{P_m}{\omega_{rm}}$$

For motor operation, the output torque is

$$T_{out} = T_m$$

$$T_{dev} = T_{RL} + T_m = T'_m$$

There is a special case of unbalance that deserves our attention. We consider a three-wire case where only magnitudes for $\bar{V}_{ab}$, $\bar{V}_{bc}$, and $\bar{V}_{ca}$ are known. $\bar{V}_0$ is zero. Hence:

$$\bar{V}_{ab} = \bar{V}_{an} - \bar{V}_{bn} = (1-a^2)\bar{V}_1 + (1-a)\bar{V}_2 \tag{B.23a}$$

$$\bar{V}_{bc} = \bar{V}_{bn} - \bar{V}_{cn} = (a^2 - a)\bar{V}_1 + (a - a^2)\bar{V}_2 \qquad \text{(B.23b)}$$

which can be solved simultaneously for $\bar{V}_1$ and $\bar{V}_2$. An example would be useful.

Example 6.2

The 460 V 100 hp wound-rotor induction motor of Example 4.3 is operating from an unbalanced three-phase three-wire 60 Hz source, such that

$$\bar{V}_{ab} = 450\ \text{V}, \quad \bar{V}_{bc} = 470\ \text{V}, \quad \bar{V}_{ca} = 440\ \text{V}$$

Running at 1770 rpm with the rotor shorted, find all currents, powers, torques, the slips, and the efficiency.

SOLUTION

$$\omega_{rm} = \text{Rotor speed} = 1770\ \text{rpm} = 185.4\ \text{rad/s}$$

$$\begin{aligned}\text{Synchronous speed} &= \omega_{sm} = 4\pi f/N_P \\ &= 4\pi(60)/4 = 188.5\ \text{rad/s} = 1800\ \text{rev/min}\end{aligned}$$

$$s = (1800 - 1770)/1800 = 0.01667 = s_+$$

$$s_- = 2 - s = 1.98333$$

For the details of the voltage calculations, see Section B.4, Example B.3. The results are

$$\begin{aligned}\bar{V}_{ab} &= 450.0\angle 31.5^\circ & \bar{V}_{an} &= 251.9\angle 0^\circ \\ \bar{V}_{bc} &= 470.0\angle -91.4^\circ & \bar{V}_{bn} &= 269.4\angle -119.3^\circ \\ \bar{V}_{ca} &= 440.0\angle 147.7^\circ & \bar{V}_{cn} &= 263.8\angle 117.1^\circ\end{aligned}$$

$$\begin{aligned}\bar{V}_0 &= 0\angle 0^\circ \\ \bar{V}_1 &= 261.6\angle -0.7^\circ \\ \bar{V}_2 &= 10.2\angle 160.6^\circ\end{aligned}$$

Use the equivalent circuit of Figure 4.8b with $\bar{V}_{an} = \bar{V}_1 = 261.6\angle -0.7^\circ$ at $s = 0.01667$. Repeating the analysis of Example 4.3, the results are

$$\begin{aligned}\bar{I}_{S1} &= 105.8\angle -27.16^\circ, & T_{dev} &= T_f = 386.9\ \text{Nm} \\ \bar{I}'_{R1} &= 97.84\angle -8.68^\circ, & (\text{RWL})_f &= 1.215\ \text{kW}\end{aligned}$$

Again, use the equivalent circuit of Figure 4.8b with $\bar{V}_{an} = \bar{V}_2 = 10.2\angle 160.6^\circ$ at $s = 1.98333$. Repeating the analysis of Example 4.3, the results are

$$\begin{aligned}\bar{I}_{S2} &= 26.74\angle 80.04^\circ, & T_{dev} &= T_b = 0.23\ \text{N m} \\ \bar{I}'_{R2} &= 26.07\angle 80.2^\circ, & (\text{RWL})_b &= 0.086\ \text{kW}\end{aligned}$$

Since the situation is three-wire: $\bar{V}_{S0} = 0, \quad \bar{I}_{S0} = 0$

The phase currents are

$$\begin{aligned}\hat{I}_{abc} &= [T]\hat{I}_{012} = [T][0 \quad 105.8\angle{-27.16°} \quad 26.74\angle 80.04°]^T \\ &= [101.2\angle{-12.5°} \quad 132.2\angle{-149.8°} \quad 89.8\angle 80.1°]^T\end{aligned}$$

The machine powers and torques are

$$\begin{aligned}P_{in} &= \bar{V}_{an}\bar{I}_a \cos\theta_a + \bar{V}_{bn}\bar{I}_b \cos\theta + \bar{V}_{cn}\bar{I}_c \cos\theta_c \\ &= (251.9)(101.2)\cos(-12.5°) + (269.4)(132.2)\cos(30.5°) \\ &\quad + (263.8)(89.8)\cos(37°) = 24.879 + 30.686 + 18.927 = 74.492\,\text{kW}\end{aligned}$$

$$\begin{aligned}Q_{in} &= \bar{V}_{an}\bar{I}_a \sin\theta_a + \bar{V}_{bn}\bar{I}_b \sin\theta + \bar{V}_{cn}\bar{I}_c \sin\theta_c \\ &= (251.9)(101.2)\sin(-12.5°) + (269.4)(132.2)\sin(30.5°) \\ &\quad + (263.8)(89.8)\sin(37°) = 5.526 + 18.053 + 14.250 = 37.829\,\text{kvar}\end{aligned}$$

$$\begin{aligned}\text{SWL} &= \bar{I}_a^2 R_1 + \bar{I}_b^2 R_1 + \bar{I}_c^2 R_1 = [(101.2)^2 + (132.2)^2 + (89.8)^2]\,(0.04232) \\ &= 1.513\,\text{kW}\end{aligned}$$

$$\text{RWL} = (\text{RWL})_f + (\text{RWL})_b = 1.215 + 0.086 = 1.301\,\text{kW}$$

$$P_{RL} = 2.164\,\text{kW}; \qquad T_{RL} = P_{RL}/\omega_{rm} = 2164/185.4 = 11.68\,\text{N m}$$

$$T_{dev} = T_f - T_b = 386.9 - 0.23 = 386.7\,\text{N m}$$

$$P_{dev} = T_{dev}\,\omega_{rm} = (386.7)(185.4) = 71.68\,\text{kW}$$

$$T_{out} = T_m = T_{dev} - T_{RL} = 386.7 - 11.7 = 375.0\,\text{N m}$$

$$P_{out} = P_m = T_{dev}\,\omega_{rm} = (375)(185.4) = 69.51\,\text{kW (93.18 hp)}$$

$$\text{Efficiency} = \eta = P_{out}/P_{in} = 69.51/74.492 = 93.31\%$$

$$\begin{aligned}P_{LOSS} &= \text{SWL} + \text{RWL} + P_{RL} \\ &= 1.513 + 1.301 + 2.164 = 4.980\,\text{kW}\end{aligned}$$

6.2 Single Phasing

A common unbalanced condition occurs if one phase inadvertently opens, a condition referred to as "single phasing" (see Figure 6.1a). We write

$$\bar{I}_a = 0 = \bar{I}_0 + \bar{I}_1 + \bar{I}_2 = 3\bar{I}_0 \tag{6.12a}$$

or

$$\bar{I}_0 = 0 \tag{6.12b}$$

Also, $\bar{I}_b + \bar{I}_c = 0$ so that

$$\bar{I}_b + \bar{I}_c = (\bar{I}_0 + a^2\bar{I}_1 + a\bar{I}_2) + (\bar{I}_0 + a\bar{I}_1 + a^2\bar{I}_2) = 0 \tag{6.13a}$$

simplifying to

$$2\bar{I}_0 = \bar{I}_1 + \bar{I}_2 = 0 \tag{6.13b}$$

Therefore,

$$\bar{I}_1 = -\bar{I}_2 \tag{6.14}$$

We also note

$$\begin{aligned} \bar{V}_{bc} &= \bar{V}_b - \bar{V}_c \\ &= \bar{V}_0 + a^2\bar{V}_1 + a\bar{V}_2 - (\bar{V}_0 + a\bar{V}_1 + a^2\bar{V}_2) \\ &= (a^2 - a)(\bar{V}_1 - \bar{V}_2) \\ &= -j\sqrt{3}(\bar{V}_1 - \bar{V}_2) \end{aligned} \tag{6.15a}$$

or

$$\frac{\bar{V}_{bc}}{-j\sqrt{3}} = \bar{V}_1 - \bar{V}_2 \tag{6.15b}$$

For balanced three-phase, $\dfrac{\bar{V}_{bc}}{-j\sqrt{3}} = \bar{V}_{an}$. Therefore,

$$\bar{V}_{an} = \bar{V}_1 - \bar{V}_2 \tag{6.15c}$$

All conditions are met with the terminations shown in Figure 6.1b. As usual, the situation will be analyzed in an example.

Example 6.3

The 460 V 100 hp wound-rotor induction motor of Example 4.3 is operating from a three-phase three-wire 60 Hz source, with Phase a open, as shown in Figure 6.1a. Running at 1770 rpm with the rotor shorted, find all currents, powers, torques, the slips, and the efficiency.

SOLUTION

ω_{rm} = Rotor speed = 1770 rpm = 185.4 rad/s

Synchronous speed = $\omega_{sm} = 4\pi f/N_P$ = 188.5 rad/s = 1800 rev/min

$s = (1800 - 1770)/1800 = 0.01667$

$s_- = 2 - s = 1.9833$

Reducing the positive and negative sequence circuits to impedances:

$$\bar{Z}_1 = 2.2132 + j1.1018, \qquad \bar{Z}_2 = 0.06260 + j0.3762$$

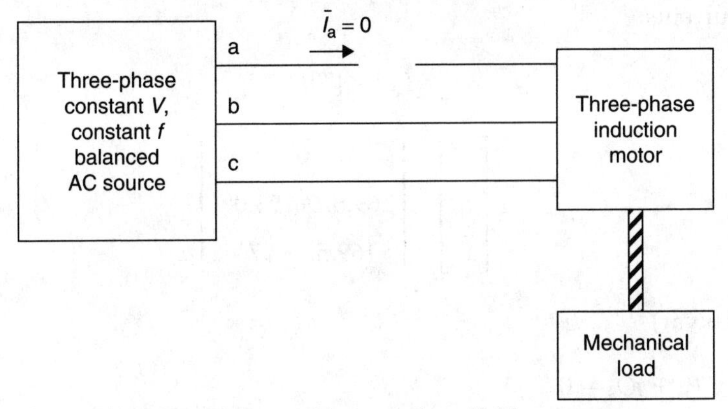

(a) A three-phase machine operating with Phase a open

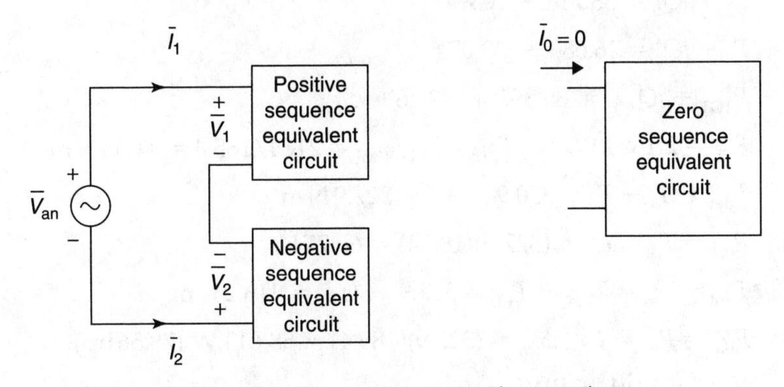

(b) Corresponding sequence network connections

FIGURE 6.1. A three-phase machine operating in a single-phase mode.

Solving for the sequence currents,

$$\bar{I}_1 = -\bar{I}_2 = (\bar{V}_{bc}/-j\sqrt{3})/(\bar{Z}_1 + \bar{Z}_2) = 97.9\angle{-33.0°}$$

The sequence voltages:

$$\bar{V}_1 = \bar{Z}_1\bar{I}_1 = 242.0\angle{-6.5°}\text{ V}$$
$$\bar{V}_2 = \bar{Z}_2\bar{I}_2 = 37.3\angle 227.6°\text{ V}$$
$$\bar{V}_0 = \bar{I}_0 = 0$$

The phase voltages:

$$\bar{V}_{an} = 222.1\angle{-14.4°}\text{ V}$$
$$\bar{V}_{bn} = 229.3\angle{-118.0°}\text{ V}$$
$$\bar{V}_{cn} = 279.1\angle{-112.7°}\text{ V}$$

The line voltages:

$$\bar{V}_{ab} = 354.9\angle 24.5°\text{ V}$$
$$\bar{V}_{bc} = 460.0\angle{-90°}\text{ V}$$
$$\bar{V}_{ca} = 449.4\angle 135.9°\text{ V}$$

The phase currents:

$$\hat{I}_{abc} = [T]\hat{I}_{012}$$

$$\begin{bmatrix} \bar{I}_a \\ \bar{I}_b \\ \bar{I}_c \end{bmatrix} = \begin{bmatrix} 0 \\ 169.5\angle -123.0° \\ 169.5\angle +57° \end{bmatrix}$$

Powers (kW, kvar)

$$P_a + jQ_a = 0 + j0$$

$$P_b + jQ_b = 38.718 + j3.394$$

$$P_c + jQ_c = 26.680 + j39.074$$

$$P_{TOT} + jQ_{TOT} = 65.397 + j42.469$$

$$P_{RL} = 2.164\,\text{kW}; \qquad T_{RL} = P_{RL}/\omega_{rm} = 2164/185.4 = 11.68\,\text{N m}$$

$$T_{dev} = T_f - T_b = 330.9 - 3.1 = 327.9\,\text{N m}$$

$$P_{dev} = T_{dev}\,\omega_{rm} = (327.9)(185.4) = 60.77\,\text{kW}$$

$$T_{out} = T_m = T_{dev} - T_{RL} = 327.9 - 11.7 = 316.2\,\text{N m}$$

$$P_{out} = P_m = T_{dev}\,\omega_{rm} = (327.9)(185.4) = 58.61\,\text{kW (78.56 hp)}$$

$$\begin{aligned} P_{LOSS} &= \text{SWL} + \text{RWL} + P_{RL} \\ &= 2.432 + 2.196 + 2.164 = 6.792\,\text{kW} \end{aligned}$$

$$\text{Efficiency} = \eta = P_{out}/P_{in} = 58.61/65.40 = 89.61\%$$

$$\text{Power factor} = 83.87\%$$

6.3 Running Three-Phase Motors from Single-Phase Sources

The foregoing analysis illustrates that although three-phase motors will run on single phase (one phase open), the currents are severely unbalanced, and motor performance is degraded. Is there some way we can operate a three-phase motor from a single-phase source without this degradation in performance? A straightforward solution of course is to produce DC from the AC source, and convert the DC back into balanced three-phase AC, which is applied to the motor. However, this is complicated and expensive, costing almost as much as the motor itself. It is possible to supply the open motor phase from a capacitor bridge, as shown in Figure 6.2a. If appropriate values for the capacitors are selected, the motor voltages can be restored to balanced three-phase conditions.[4] This circuit was designed by Otto J.M. Smith,[5] and is sometimes called the "Smith connection."

[4] Actually, appropriate capacitor values can be found only for operating power factors in the range from 0.5 to 0.866 lagging. Outside of this range, one or more of the Z's must be inductive.

[5] See, for example, *A Novel Method of Operating a Single-Phase Motor with Three-Phase Motor Efficiency* by Ali Shaban and David B. Mar, Paper No. 2 in NAPS-5-C Proceedings of the 25th North American Power Symposium, Howard University, Washington, DC, October 11, 1993.

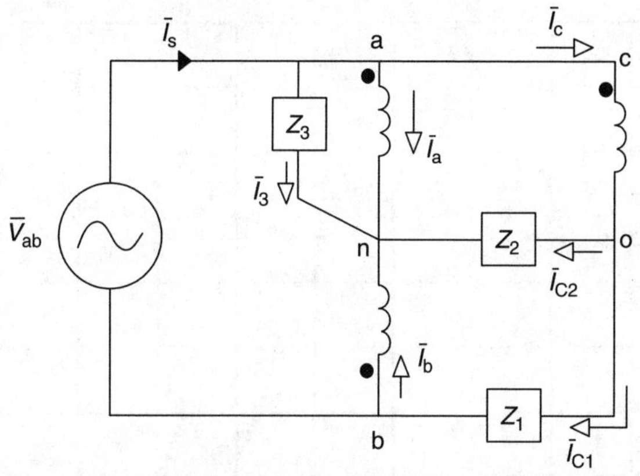

(a) Operating a three-phase motor in the Smith connection
The Z's are capacitors

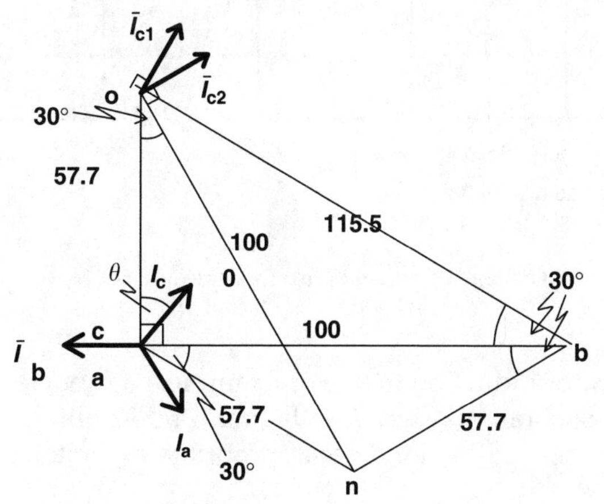

(b) The phasor diagram in closed form. Note that the reference phasor is V_{ab}

FIGURE 6.2. Operating in a single-phase mode using the Smith connection.

Under balanced conditions, the phasor diagram of Figure 6.2b is applicable. The reference voltage phasor is $\bar{V}_{ab}$. Note that phasor magnitudes are in percent of $\bar{V}_{ab}$. The motor speed will define the motor currents, the power factor, and θ. The capacitor currents required for balanced operation can be shown to be

$$0.500\bar{I}_{c1} + 0.866\bar{I}_{c2} = \bar{I}_c \sin\theta$$

$$0.866\bar{I}_{c1} + 0.500\bar{I}_{c2} = \bar{I}_c \cos\theta$$

which may be solved for $\bar{I}_{c1}$ and $\bar{I}_{c2}$. Finally

$$\bar{Z}_1 = 1.155\frac{\bar{V}_{ab}}{\bar{I}_{c1}}$$

$$\bar{Z}_2 = \frac{\bar{V}_{ab}}{\bar{I}_{c2}}$$

$$\bar{Z}_3 = 0.5\bar{Z}_1$$

An example will clarify matters.

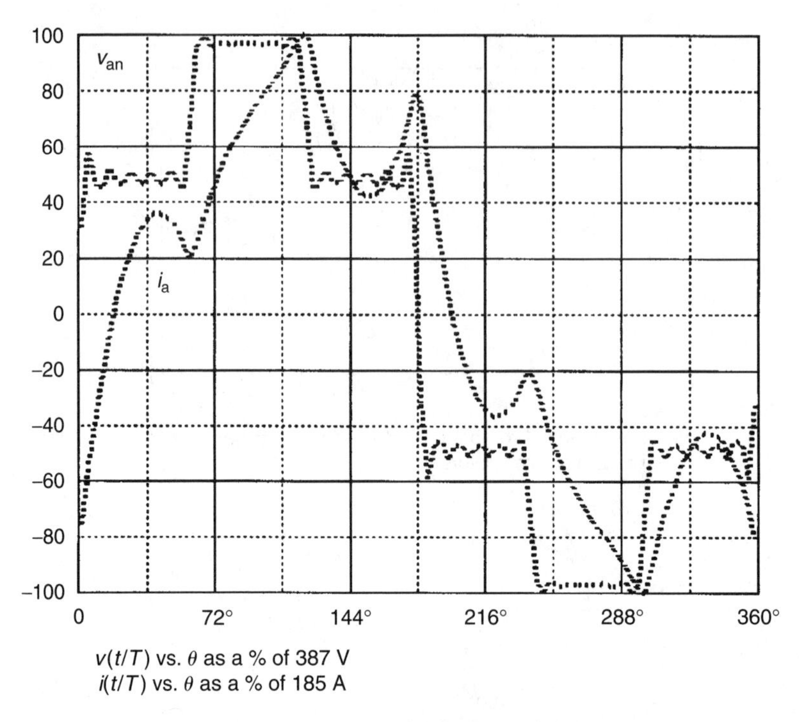

FIGURE 6.3. v_{an} and i_a for Example 6.4.

Example 6.4
The 460 V 100 hp wound-rotor induction motor of Example 4.3 is operating from a single-phase 460 V two-wire 60 Hz source as shown in Figure 6.1a. Running at 1784.5 rpm with the rotor shorted, the pf = 0.8 lagging. Find the appropriate capacitor values.

SOLUTION

At 1784.5 rpm,

$$\bar{I}_a = 63.77\angle{-36.9°}\ \text{A}$$

Adjusting the phase of $\bar{V}_{an}$ to $-30°$ (to move $\bar{V}_{ab}$ to 0°),

$$\bar{I}_a = 63.77\angle{-66.9°}\ \text{A}$$

The phase "c" current is

$$\bar{I}_c = 63.77\angle{+53.1°}\ \text{A}$$

Since "θ" is 36.9°,

$$0.500\bar{I}_{c1} + 0.866\bar{I}_{c2} = 63.77\sin 36.9° = 38.26$$

$$0.866\bar{I}_{c1} + 0.500\bar{I}_{c2} = 63.77\cos 36.9° = 51.02$$

$$\bar{I}_{c1} = 50.11$$

$$\bar{I}_{c2} = 15.24$$

$$\bar{Z}_1 = 1.155V_{ab}/\bar{I}_{c1} = 10.60\,\Omega \quad (C_1 = 250\,\mu\text{F})$$

$$\bar{Z}_2 = \bar{V}_{ab}/\bar{I}_{c2} = 30.18\,\Omega \quad (C_2 = 87.89\,\mu\text{F})$$

$$\bar{Z}_3 = 0.5\bar{Z}_1 = 5.301\,\Omega \quad (C_3 = 500\,\mu\text{F})$$

6.4 Operation on Nonsinusoidal Voltage

Recall that in Chapter 5, to achieve speed control, it was necessary to supply the motor from an AC motor drive, the output of which was nonsinusoidal. A way of dealing with nonsinusoidal signals is to break them into a sum of sinusoidal components at integer-multiple frequencies, a technique called "harmonic" or "Fourier" analysis.[6] Hence, each phase voltage may be represented by an expression of the form

$$\bar{V}(t) = \bar{V}_0 + \sum_{n=1}^{n=N} \sqrt{2}\,\bar{V}_n \cos(n\omega_0 t + \phi_n) \tag{6.16a}$$

where

$$T = \text{period of } v(t) \tag{6.16b}$$

$$f_0 = 1/T = \text{fundamental frequency} \tag{6.16c}$$

$$\omega_0 = 2\pi f_0 = \text{fundamental radian frequency} \tag{6.16d}$$

$$nf_0 = n\text{th harmonic frequency} \tag{6.16e}$$

$$n\omega_0 = n\text{th harmonic radian frequency} \tag{6.16f}$$

Also

$$V_0 = \text{DC term of } v(t) \tag{6.16g}$$

$$\bar{V}_n = \text{rms value of } n\text{th harmonic of } v(t) \tag{6.16h}$$

$$\phi_n = \text{phase angle of } n\text{th harmonic of } v(t) \tag{6.16i}$$

Consider the three phase voltages ($v_{an}(t)$, $v_{bn}(t)$, $v_{cn}(t)$). Collecting the set at each harmonic frequency, we can compute the corresponding sequence (symmetrical component) set using Equation B.17. Then, use the approach explained in detail in Section 6.1 to compute the response of the machine to each harmonic, one at a time. However, we must account for the change in frequency.

Consider the three inductors (X_1, X_m, X_2') in our basic equivalent circuit model (Figure 4.8b). Recall that inductive reactance is directly proportional to frequency:

$$X = \omega L$$

[6] Readers unfamiliar with this subject are advised to study Appendix C at this point.

Thus, at the nth harmonic,

$$X = n\omega_0 L = nX_0$$

where X_0 represents reactance at the fundamental frequency.

The next issue is to consider slip at each harmonic. Define

$\Omega_0 = \omega_{sm}$ = synchronous mechanical speed; the speed of the rotating magnetic field caused by positive sequence stator currents at the fundamental frequency, in mechanical radians per second[7]

$\Omega_n = n\Omega_0$ = synchronous speed, caused by stator currents at the nth harmonic frequency

$\Omega_r = \omega_{rm}$ = rotor mechanical speed; mechanical radians per second

Care must be exercised in comparing phase angles of harmonics in stator phases a, b, and c. Consider a nonsinusoidal voltage consisting only of a fundamental, a second, and a third harmonic, with fundamental frequency 60 Hz ($\omega_0 = 377$ rad/s).

$$v_a(t) = 100\cos(377t) + 20\cos(754t) + 30\cos(1131t)$$

Suppose that this is the "a" phase voltage applied to an induction motor, and that the voltages are balanced; that is, the voltages applied to the b and c phase windings have the same nonsinusoidal waveform, but occur $T/3 = 16.67/3 = 5.556$ ms later in the cycle ($v_b(t)$ is delayed by 5.556 ms and $v_c(t)$ is delayed by 11.11 ms). This requires a 120° phase lag for each fundamental component, but 2 × 120° = 240° phase lag for each second harmonic, and a 3 × 120° = 360° phase lag for each third harmonic. Hence,

$$v_b(t) = 100\cos(377t - 120°) + 20\cos(754t - 240°) + 20\cos(1131t - 360°)$$

$$v_c(t) = 100\cos(377t - 240°) + 20\cos(754t - 480°) + 20\cos(1131t - 720°)$$

Note that the fundamentals are balanced three-phase, sequence abc; the second harmonics are balanced three-phase, sequence cba; and the third harmonics are in phase. The point is that the nth harmonic phase angle must be adjusted by $120(n)°$. The pattern for all harmonics is presented in Table 6.1.

Positive sequence harmonic slip: The positive sequence quantities will always create a field rotating in the *same* direction as the rotor. Hence,

$$s_+ = \frac{\Omega_n - \Omega_r}{\Omega_n} \tag{6.17a}$$

Negative sequence harmonic slip: The negative sequence quantities will always create a field rotating in the *opposite* direction as the rotor. Hence,

$$s_- = \frac{-\Omega_n - \Omega_r}{-\Omega_n} = \frac{\Omega_n + \Omega_r}{\Omega_n} \tag{6.17b}$$

To demonstrate the use of Equation 6.17, consider a four-pole 60 Hz machine with the rotor turning at 1620 rpm, as presented in Table 6.2. An example analysis would be useful.

[7]As previously discussed, we may use "rpm" instead of "rad/s" in certain situations, if we are careful.

TABLE 6.1

Harmonics of Balanced Three-Phase Nonsinusoidal Voltages

Harmonic	Phase Sequence
1	+ (abc)
2	− (cba)
3	0 (in phase)
4	+
5	−
6	0
7	+
8	−
9	0
etc.	

TABLE 6.2

Harmonic Slips for a Four-Pole 60 Hz Motor Running at 1620 rpm

Harmonic	Ω_n (rpm)	s_+	s_-
1	1800	0.100	1.900
2	3600	0.550	1.450
3	5400	0.700	1.300
4	7200	0.775	1.225
5	9000	0.820	1.180
etc.			

Example 6.5

The 460 V 100 hp wound-rotor induction motor of Example 4.3 is operating from a six-step inverter. The inverter is at 100% voltage ($\bar{V}_{an} = 265.6$ V; see Figure 5.8 for voltage waveshapes). Using the first 32 harmonics, running at 1770 rpm with the rotor shorted, find all currents, powers, torques, the slips, and the efficiency.

SOLUTION

Consider the fundamental components only. Since the phase voltages are balanced, the fundamentals are "pure" positive seqence. Results are:

Harm	Rot	Slip	R	X	V_{rms}	V_{ang}	I_{rms}	I_{ang}	T_{dev}
1	+	0.0167	2.2132	1.1018	253.65	−90.0	102.60	−116.5	363.68
2	−	1.4917	0.0000	0.0000	0.00	−85.6	0.00	−85.6	0.00
3	0	—	0.5729	0.0000	0.00	−8.0	0.00	−93.8	—
4	+	0.7542	0.0000	0.0000	0.00	87.2	0.00	87.2	0.00
5	−	1.1967	0.0759	1.8806	50.87	−90.0	27.03	−177.7	−0.08
6	0	—	1.1434	1.8806	0.00	43.0	0.00	−44.9	—
7	+	0.8595	0.0891	2.6328	36.43	−90.0	13.83	−178.1	0.02
8	−	1.1229	0.0000	0.0000	0.00	−87.6	0.00	−87.6	0.00
9	0	—	1.7145	0.0000	0.00	113.3	0.00	24.7	—
10	+	0.9017	0.0000	0.0000	0.00	87.2	0.00	87.2	0.00
11	−	1.0894	0.0792	4.1372	23.38	−90.0	5.65	−178.9	−0.00
12	0	—	2.2857	4.1372	0.00	−22.5	0.00	−111.5	—
13	+	0.9244	0.0858	4.8894	19.89	−90.0	4.07	−179.0	0.00
14	−	1.0702	0.0000	0.0000	0.00	−85.9	0.00	−85.9	0.00
15	0	—	2.8569	0.0000	0.00	2.3	0.00	−86.8	—
16	+	0.9385	0.0000	0.0000	0.00	86.5	0.00	86.5	0.00
17	−	1.0578	0.0803	6.3938	15.42	−90.0	2.41	−179.3	−0.00
18	0	—	3.4282	6.3938	0.00	−35.8	0.00	−125.1	—
19	+	0.9482	0.0847	7.1460	13.92	−90.0	1.95	−179.3	0.00
20	−	1.0492	0.0000	0.0000	0.00	−84.3	0.00	−84.3	0.00
21	0	—	3.9995	0.0000	0.00	98.1	0.00	8.7	—
22	+	0.9553	0.0000	0.0000	0.00	87.0	0.00	87.0	0.00
23	−	1.0428	0.0809	8.6504	11.72	−90.0	1.36	−179.5	−0.00
24	0	—	4.5708	8.6504	0.00	−6.3	0.00	−95.8	—
25	+	0.9607	0.0842	9.4027	10.91	−90.0	1.16	−179.5	0.00
26	−	1.0378	0.0000	0.0000	0.00	−91.2	0.00	−91.2	0.00
27	0	—	5.1421	0.0000	0.00	47.3	0.00	−42.2	—
28	+	0.9649	0.0000	0.0000	0.00	92.2	0.00	92.2	0.00
29	−	1.0339	0.0812	10.9071	9.65	−90.0	0.88	−179.6	−0.00

(*Continued*)

(Continued)

Harm	Rot	Slip	R	X	V_{rms}	V_{ang}	I_{rms}	I_{ang}	T_{dev}
30	0	—	5.7134	10.9071	0.00	−21.3	0.00	−110.9	—
31	+	0.9683	0.0839	11.6593	9.15	−90.0	0.79	−179.6	0.00
32	−	1.0307	0.0000	0.0000	0.00	−90.3	0.00	−90.3	0.00

$\omega_{rm} = 185.4$ rad/s;
$T_{dev} = 363.62$ N m; $T_m = 351.95$ N m
$P_{in} = 70.22$ kW; $P_{dev} = 67.40$ kW; $P_{out} = 65.25$ kW
Efficiency = 92.92%;
rms stator current = 107.29 A

For the unbalanced nonsinusoidal case, the approach would be to compute the Fourier harmonics for each phase voltage, convert these into sequence quantities at each frequency, and to compute the current response of each sequence network to its appropriate sequence voltage at each harmonic. Phase currents could then be computed from Equation B.17a (Figure 6.3).

6.5 The Two-Phase Induction Motor

The rotating magnetic field, discussed in Section 4.3, can be produced by arrangements other than three-phase. A more general statement of the concept could be stated as follows:

> *Given any integer $K \geq 3$, "K" identical sinusoidally distributed stator windings, whose axes are located (360 electrical degrees/K) apart, and supplied with equal sinusoidal currents (360 electrical degrees/K) apart in phase, with radian frequency ω, collectively produce a rotating magnetic field $\mathfrak{F}(\theta,t)$, which is sinusoidal waveform in space, and rotating at velocity $\omega_{se} = \omega$, in electrical rad/s ("synchronous speed").*

The field may be described analytically with a more general version of Equation 4.6a:

$$\mathfrak{F}(\theta, t) = (KN/2)I_{max}\cos(\omega t + \beta - \theta) \tag{6.18a}$$

A machine could be designed for any value of $K \geq 3$.

$K = 4$ is of particular interest. Consider Figure 6.4a. The four windings are spaced $360°/4 = 90°$ apart, and carry currents 90° apart in phase. Specifically, $\overline{I}_1$ and $\overline{I}_3$ are 180° phase-separated, as are $\overline{I}_2$ and $\overline{I}_4$. But observe that since $\overline{I}_3$ is 180° out of phase with $\overline{I}_1$, there need not be a separate winding! $\overline{I}_3$ could simply be the return value of $\overline{I}_1$! Likewise, $\overline{I}_2$ and $\overline{I}_4$ could be the same current!

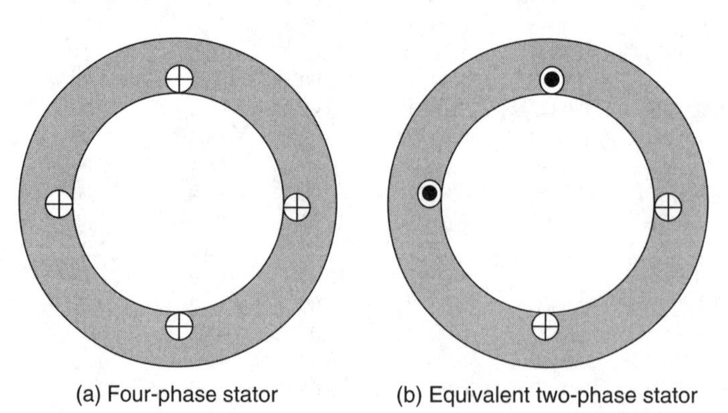

FIGURE 6.4. Stator winding layout for the two-phase machine.

Thus, a rotating field could be created from two currents [$(\bar{I}_1-\bar{I}_3)$ and $(\bar{I}_2-\bar{I}_4)$] and there need not be four, but only two separate windings, spaced 90 electrical degrees apart. Such a machine is called a "two-phase" induction machine. For this case, Equation 6.18a becomes

$$\mathfrak{F}(\theta,t) = NI_{\max}\cos(\omega t+\beta-\theta) \tag{6.18b}$$

The corresponding symmetrical component transformation[8] for this case becomes

$$\begin{bmatrix}\bar{V}_m\\ \bar{V}_a\end{bmatrix} = \begin{bmatrix}1 & 1\\ j & -j\end{bmatrix}\begin{bmatrix}\bar{V}_f\\ \bar{V}_b\end{bmatrix} \tag{6.19a}$$

We use "f" and "b" for the positive and negative sequence quantities because they relate to the forward and backward rotating fields, respectively. Solving for the f,b voltages,

$$\begin{bmatrix}\bar{V}_f\\ \bar{V}_b\end{bmatrix} = \frac{1}{2}\begin{bmatrix}1 & -j\\ 1 & +j\end{bmatrix}\begin{bmatrix}\bar{V}_m\\ \bar{V}_a\end{bmatrix} \tag{6.19b}$$

The approach for computing the machine's performance is basically the same as that outlined in Section 6.1 for three-phase machines. The two-phase machine, with some modifications in design, is frequently operated from a single-phase AC source, in which case it is described as a "single-phase" motor. When this is the situation, two-phase operation and the "pure" two-phase machine become a special case of the single-phase motor, which we shall consider next.

6.6 The Single-Phase Induction Motor

Consider the two-phase machine of Section 6.4 with the following modified design. The "main winding" (which we will designate as "m") is intended to be permanently in the circuit, and must be designed for continuous duty. The "auxiliary winding" (designated as "a") is frequently designed for intermittent duty and may be energized only during starting. Because the stator windings are no longer of identical design, this creates the need for more complex circuit models, as depicted in Figure 6.5.

$$\bar{V}_m = \bar{Z}_m\bar{I}_m + \bar{E}_m \tag{6.20a}$$

$$\bar{V}'_a = \bar{Z}_C\bar{I}_a + \bar{V}_a \tag{6.20b}$$

$$\bar{V}_a = \bar{Z}_a\bar{I}_a + \bar{E}_a \tag{6.20c}$$

where

$\bar{V}_m, \bar{V}_a$ = main, auxiliary winding terminal voltage (in V)

$\bar{I}_m, \bar{I}_a$ = main, auxiliary winding current (in A)

---

[8] The reader must surely be curious as to why we elected to identify the stator windings as "m" and "a," rather than the more logical choice of "a" and "b." "m" and "a" stands for "main" and "auxiliary," respectively. The terms selected shall become clear in the next section.

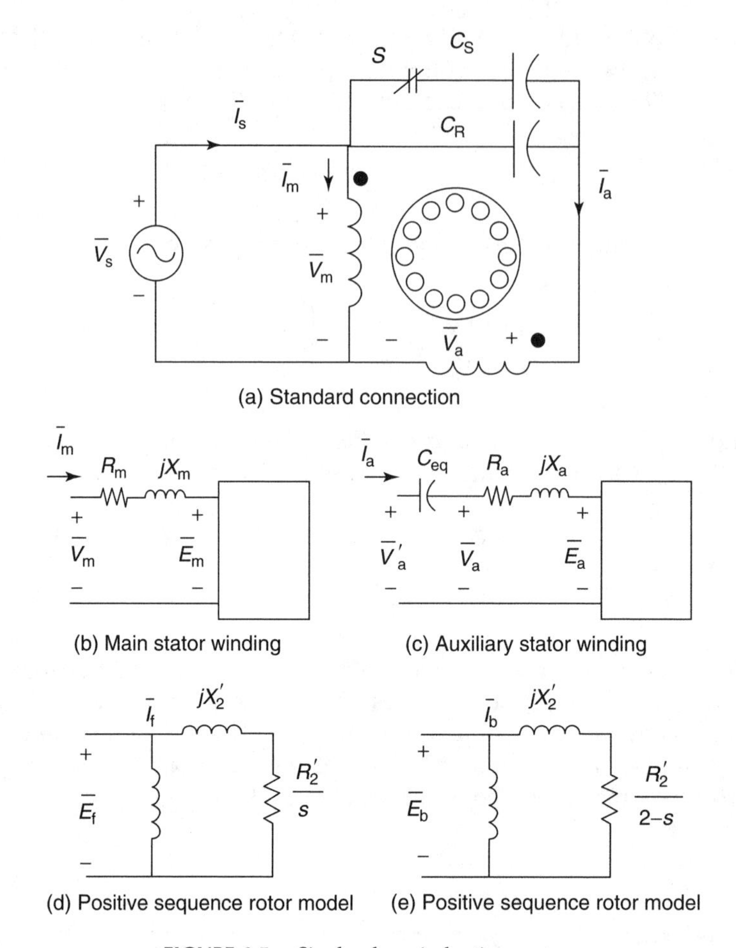

FIGURE 6.5. Single-phase induction motor.

$\bar{E}_m, \bar{E}_a$ = main, auxiliary winding internal induced voltage, due to the air gap rotating fields (in V)

R_m, R_a = main, auxiliary winding resistance (in Ω)

X_m, X_a = main, auxiliary winding leakage reactance (in Ω)

$\bar{Z}_m = R_m + jX_m$

$\bar{Z}_a = R_a + jX_a$

$\bar{Z}_C$ = External impedance (typically capacitance), placed in series with the auxiliary winding

$\bar{Z}'_a = \bar{Z}_a + \bar{Z}_C$

Converting to matrix notation,

$$\begin{bmatrix} \bar{V}_m \\ \bar{V}_a \end{bmatrix} = \begin{bmatrix} \bar{Z}_m & 0 \\ 0 & \bar{Z}'_a \end{bmatrix} \begin{bmatrix} \bar{I}_m \\ \bar{I}_a \end{bmatrix} + \begin{bmatrix} \bar{E}_m \\ \bar{E}_a \end{bmatrix} \tag{6.20d}$$

Now the main and auxiliary windings have different numbers of turns:

N_m, N_a = main, auxiliary effective number of winding turns

$$n = N_a/N_m = \text{auxiliary to main winding turns ratio} \tag{6.21}$$

The induced voltage in the main winding due to the forward and backward fields is

$$\bar{E}_{m} = \bar{E}_{fm} + \bar{E}_{bm} = \bar{E}_{f} + \bar{E}_{b} \tag{6.22a}$$

The induced voltage in the auxiliary winding due to the forward and backward fields is

$$\bar{E}_{a} = \bar{E}_{fa} + \bar{E}_{ba} = jn\bar{E}_{f} - jn\bar{E}_{b} \tag{6.22b}$$

Converting to matrix notation:

$$\begin{bmatrix} \bar{E}_{m} \\ \bar{E}_{a} \end{bmatrix} = \begin{bmatrix} 1 & 1 \\ jn & -jn \end{bmatrix} \begin{bmatrix} \bar{E}_{f} \\ \bar{E}_{b} \end{bmatrix} \tag{6.22c}$$

It can also be shown that

$$\begin{bmatrix} \bar{I}_{m} \\ \bar{I}_{a} \end{bmatrix} = \begin{bmatrix} 1 & 1 \\ \dfrac{j}{n} & \dfrac{-j}{n} \end{bmatrix} \begin{bmatrix} \bar{I}_{f} \\ \bar{I}_{b} \end{bmatrix} \tag{6.22d}$$

or

$$\begin{bmatrix} \bar{I}_{f} \\ \bar{I}_{b} \end{bmatrix} = \left(\frac{1}{2}\right) \begin{bmatrix} 1 & -j\dfrac{1}{n} \\ 1 & +j\dfrac{1}{n} \end{bmatrix} \begin{bmatrix} \bar{I}_{m} \\ \bar{I}_{a} \end{bmatrix} \tag{6.22e}$$

Returning to Equation 6.20d

$$\begin{bmatrix} \bar{V}_{m} \\ \bar{V}_{a} \end{bmatrix} = \begin{bmatrix} \bar{Z}_{m} & 0 \\ 0 & \bar{Z}'_{a} \end{bmatrix} \begin{bmatrix} \bar{I}_{m} \\ \bar{I}_{a} \end{bmatrix} + \begin{bmatrix} \bar{E}_{m} \\ \bar{E}_{a} \end{bmatrix} \tag{6.20d}$$

Substituting Equation 6.22c into Equation 6.20d

$$\begin{bmatrix} \bar{V}_{m} \\ \bar{V}_{a} \end{bmatrix} = \begin{bmatrix} \bar{Z}_{m} & 0 \\ 0 & \bar{Z}'_{a} \end{bmatrix} \begin{bmatrix} \bar{I}_{m} \\ \bar{I}_{a} \end{bmatrix} + \begin{bmatrix} 1 & 1 \\ jn & -jn \end{bmatrix} \begin{bmatrix} \bar{E}_{f} \\ \bar{E}_{b} \end{bmatrix} \tag{6.20e}$$

But from the circuits of Figure 6.5d and Figure 6.5e:

$$\bar{Z}_{f} = \frac{jX_{\phi}\left(\dfrac{R'_{2}}{s} + jX'_{2}\right)}{\dfrac{R'_{2}}{s} + j(X'_{2} + X_{\phi})} = R_{f} + jX_{f}$$

$$\bar{Z}_{b} = \frac{jX_{\phi}\left(\dfrac{R'_{2}}{2-s} + jX'_{2}\right)}{\dfrac{R'_{2}}{2-s} + j(X'_{2} + X_{\phi})} = R_{b} + jX_{b}$$

and

$$\begin{bmatrix} \bar{E}_f \\ \bar{E}_b \end{bmatrix} = \begin{bmatrix} \bar{Z}_f & 0 \\ 0 & \bar{Z}_b \end{bmatrix} \begin{bmatrix} \bar{I}_f \\ \bar{I}_b \end{bmatrix} \tag{6.23}$$

If the speed is known, so is the slip, and $\bar{Z}_f$ and $\bar{Z}_b$ can be evaluated. Substituting Equation 6.23 into Equation 6.20e,

$$\begin{bmatrix} \bar{V}_m \\ \bar{V}_a \end{bmatrix} = \begin{bmatrix} \bar{Z}_m & 0 \\ 0 & \bar{Z}'_a \end{bmatrix} \begin{bmatrix} \bar{I}_m \\ \bar{I}_a \end{bmatrix} + \begin{bmatrix} 1 & 1 \\ jn & -jn \end{bmatrix} \begin{bmatrix} \bar{Z}_f & 0 \\ 0 & \bar{Z}_b \end{bmatrix} \begin{bmatrix} \bar{I}_f \\ \bar{I}_b \end{bmatrix}$$

$$\begin{bmatrix} \bar{V}_m \\ \bar{V}_a \end{bmatrix} = \begin{bmatrix} \bar{Z}_m & 0 \\ 0 & \bar{Z}'_a \end{bmatrix} \begin{bmatrix} \bar{I}_m \\ \bar{I}_a \end{bmatrix} + \begin{bmatrix} \bar{Z}_f & \bar{Z}_b \\ jn\bar{Z}_f & -jn\bar{Z}_b \end{bmatrix} \begin{bmatrix} \bar{I}_f \\ \bar{I}_b \end{bmatrix} \tag{6.20f}$$

Finally, eliminating the f,b currents using 6.22e:

$$\begin{bmatrix} \bar{V}_m \\ \bar{V}_a \end{bmatrix} = \begin{bmatrix} \bar{Z}_m & 0 \\ 0 & \bar{Z}'_a \end{bmatrix} \begin{bmatrix} \bar{I}_m \\ \bar{I}_a \end{bmatrix} + \begin{bmatrix} \bar{Z}_f & \bar{Z}_b \\ jn\bar{Z}_f & -jn\bar{Z}_b \end{bmatrix} \left(\frac{1}{2}\right) \begin{bmatrix} 1 & -j\frac{1}{n} \\ 1 & +j\frac{1}{n} \end{bmatrix} \begin{bmatrix} \bar{I}_m \\ \bar{I}_a \end{bmatrix}$$

$$\begin{bmatrix} \bar{V}_m \\ \bar{V}_a \end{bmatrix} = \begin{bmatrix} \bar{Z}_m + \dfrac{\bar{Z}_f + \bar{Z}_b}{2} & -jn\dfrac{\bar{Z}_f - \bar{Z}_b}{2} \\ +jn\dfrac{\bar{Z}_f - \bar{Z}_b}{2} & \bar{Z}'_a + n^2\dfrac{\bar{Z}_f + \bar{Z}_b}{2} \end{bmatrix} \begin{bmatrix} \bar{I}_m \\ \bar{I}_a \end{bmatrix} \tag{6.20g}$$

Equation 6.20g is the key to single-phase motor analysis. Given the applied voltages to each winding ($\bar{V}_m$, $\bar{V}'_a$), Equation 6.20g can be solved for the m,a currents. The f, b currents can then be determined using Equation 6.22e. Once all the currents are known, the calculation of all powers and torques is straightforward.

Consider the torques

$$T_f = \frac{2I_f^2 R_f}{\omega_s} \tag{6.24a}$$

$$T_b = \frac{2I_b^2 R_b}{\omega_s} \tag{6.24b}$$

$$T_{dev} = T_f - T_b \tag{6.24c}$$

$$T_{RL} = K_{RL}\omega_{rm} \tag{6.24d}$$

$$T_m = T_{dev} - T_{RL} \tag{6.24e}$$

Consider the powers,

$$\text{SWL} = I_m^2 R_m + I_a^2 R_a \tag{6.25a}$$

$$P_{in} = V_m I_m \cos\theta_m + V_a' I_a \cos\theta_a \tag{6.25b}$$

The power crossing the air gap from stator to rotor is

$$P_{ag} = 2I_f^2 R_f + 2I_b^2 R_b \tag{6.25c}$$

This power divides into RWL and P_{dev} according to

$$\text{RWL} = (s)(2I_f^2 R_f) + (2-s)(2I_b^2 R_b) \tag{6.25d}$$

and

$$\begin{aligned} P_{dev} &= P_{ag} - \text{RWL} \\ &= (1-s)(2I_f^2 R_f) + (s+1)(2I_b^2 R_b) \end{aligned} \tag{6.25e}$$

As was the case for the three-phase motor,

$$P_m = P_{dev} - P_{RL} \tag{4.27}$$

$$\text{Output power} = P_{out} = P_m \tag{4.28}$$

$$P_{LOSS} = P_{RL} + \text{SWL} + \text{RWL} \tag{4.29}$$

$$\eta = \frac{P_{out}}{P_{in}}, \text{ expressed as a \%} \tag{4.30}$$

Before we work an example, the centrifugal switch (S) in Figure 6.5 requires an explanation. As we shall see, the capacitors in the circuit are there to improve motor performance, specifically to increase developed torque. The optimum values of capacitance are a function of speed, more capacitance being required at lower speeds. Thus at slow speeds, S is closed. S can be set to open at any speed ω_C (commonly, ω_C, = $0.75\omega_{sm}$). Thus,

$$C_{eq} = C_R + C_S \qquad \omega_{rm} < \omega_C \tag{6.26a}$$

$$C_{eq} = C_R \qquad \omega_{rm} \geq \omega_C \tag{6.26b}$$

$$\bar{Z}_C = \frac{1}{j\omega C_{eq}} \tag{6.26c}$$

An example will demonstrate the application of these equations.

Example 6.6

Given the following single-phase induction motor data:

Single-Phase Induction Motor Data

Ratings

Voltage = 120.0 V; $a = N_a/N_m = 1.50$

Horsepower = 0.250 hp; $J_{Motor} = 0.00330\ \text{kg m}^2$

Synchronous speed = 1800.0 rpm; 4 poles; Frequency = 60 Hz

Equivalent Circuit Values (in ohms)

$R_m = 3.456$; $X_m = 4.320$; $X_\phi = 115.200$

$R_a = 12.442$; $X_a = 9.720$; $R_2' = 6.912$; $X_2' = 4.320$

Rotational loss (RL) torque = $T_{RL} = K_{RL} * \omega_{rm}$

$T_{RL} = 0.000525 \times$ (speed in rad/s) N m

Auxiliary Winding

```
                     Cr
---,---------| (----,---UUUUUU-----
   |                |  Switch CLOSED (Lo w) ... C = Cr + Cs
   |                |     C = 61.40 µF (43.20 Ω)
   |    Switch   Cs |  Switch OPEN (Hi w) .... C = Cr
   |---/----| (----|     C = 8.19 µF (324.00 Ω)
```

Switching speed = (75.0%) × 1800.0 rpm = 1350.0 rpm

Calculate all currents, torques, and powers at a speed = 1681 rpm when operated from a 120 V single-phase source.

SOLUTION

(a) Note that

$$s = (\omega_{sm} - \omega_{rm})/\omega_{sm} = (1800 - 1681)/1800 = 0.06611$$

$$\bar{Z}_f = \frac{(R_2'/s + jX_2')(jX_\phi)}{R_2'/s + j(X_2' + jX_\phi)} = 55.03 + j52.30 = R_f + jX_f$$

$$\bar{Z}_b = \frac{(R_2'/(2-s) + jX_2')(jX_\phi)}{(R_2'/(2-s) + j(X_2' + jX_\phi)} = 3.317 + j4.263 = R_b + jX_b$$

It follows that

$$\begin{bmatrix} \bar{Z}_m + \dfrac{\bar{Z}_f + \bar{Z}_b}{2} & -jn\dfrac{\bar{Z}_f - \bar{Z}_b}{2} \\ +jn\dfrac{\bar{Z}_f - \bar{Z}_b}{2} & \bar{Z}_a' + n^2\dfrac{\bar{Z}_f + \bar{Z}_b}{2} \end{bmatrix} = \begin{bmatrix} 32.63 + j32.60 & +36.03 - j38.78 \\ -36.03 + j38.78 & 78.77 - j250 \end{bmatrix}$$

Now $\bar{V}_m = \bar{V}_a' = 120\angle 0°$. Therefore, solving for the currents from Equation 6.20g:

$$\begin{bmatrix} \bar{I}_m \\ \bar{I}_a \end{bmatrix} = \begin{bmatrix} 1.924\angle -40.1° \\ 0.641\angle 33.6° \end{bmatrix}$$

The source current is $2.178\angle{-24.4°}$ A. Transforming to f, b values,

$$\begin{bmatrix} \bar{I}_f \\ \bar{I}_b \end{bmatrix} = \left(\frac{1}{2}\right) \begin{bmatrix} 1 & -j\dfrac{1}{n} \\ 1 & +j\dfrac{1}{n} \end{bmatrix} \begin{bmatrix} \bar{I}_m \\ \bar{I}_a \end{bmatrix} = \begin{bmatrix} 1.410\angle{-45.3°} \\ 0.536\angle{-26.1°} \end{bmatrix}$$

The torques

$$T_f = \frac{2\bar{I}_f^2 R_f}{\omega_s} = 1.160 \text{ N m}$$

$$T_b = \frac{2\bar{I}_b^2 R_b}{\omega_s} = 0.0101 \text{ N m}$$

$$T_{dev} = T_f - T_b = 1.150 \text{ N m}$$

$$T_{RL} = K_{RL}\omega_{rm} = 0.0925 \text{ N m}$$

$$T_m = T_{dev} - T_{RL} = 1.057 \text{ N m}$$

The powers:

$$\text{SWL} = I_m^2 R_m + I_a^2 R_a = 17.48 \text{ W}$$

$$P_{in} = V_m I_m \cos\theta_m + V_a' I_a \cos\theta_a = 238.1 \text{ W}$$

$$\text{RWL} = (s)2I_f^2 R_f + (2-s)2I_b^2 R_b = 18.15 \text{ W}$$

$$P_{dev} = 202.5 \text{ W}$$

$$\text{Output power} = P_{out} = P_m = P_{dev} - P_{RL} = 186.2 \text{ W } (0.25 \text{ hp})$$

$$P_{LOSS} = P_{RL} + \text{SWL} + \text{RWL} = 51.90 \text{ W}$$

$$\eta = P_{out}/P_{in} = 78.2\%$$

$$\text{Power factor} = \cos(-24.4°) = 0.9109 \text{ lagging}$$

We return to consideration of the external capacitance. The best possible situation would be to have the m,a mmf's to be equal and in phase quadrature. This would result in zero backward voltage, current, and torque components, and would be equivalent to balanced two-phase operation. Whereas we could design the m,a windings to achieve this optimum condition at one specific speed, and with one corresponding capacitance value, performance would be compromised at other speeds. Experience with the problem shows that more capacitance is needed at low speeds than at higher speeds. Experience also shows that reasonable performance can be achieved over the entire speed range with two values of capacitance. Thus the circuit of Figure 6.8 is designed to insert a maximum of capacitance ($C_{eq} = C_R + C_S$) at starting, and remove the surplus capacitance (C_S) at speeds approaching run conditions. The capacitors are called the "Run" capacitor (C_R), and the "Start" capacitor (C_S), for obvious reasons.

The issue will be developed in the context of an example.

Example 6.7

Given the single-phase induction motor of Example 6.6, (a) find the value of C_R that nearly approximates operation as a balanced two-phase motor running at 1681 rpm and (b) Find the value of C_S that (nearly) maximizes starting torque.

SOLUTION

(a) As in Example 6.6,

$$s = 0.06611$$

$$\bar{Z}_f = 55.03 + j52.30$$

$$\bar{Z}_b = 3.317 + j4.263$$

Assume a balanced two-phase operation such that

$$\bar{I}_a = \frac{j}{n}\bar{I}_m$$

Therefore, from Equation 6.20g,

$$\bar{V}_m = (\bar{Z}_m + \bar{Z}_f)\bar{I}_m$$

and

$$\bar{I}_m = 1.474\angle{-44.1°}$$

Thus the desired $\bar{I}_a$

$$= 0.9828\angle{+45.9°}$$

Again from Equation 6.20g

$$\bar{V}'_a = (\bar{Z}'_a + n^2\bar{Z}_f)\bar{I}_a = (\bar{Z}_a + \bar{Z}_C + n^2\bar{Z}_f)\bar{I}_a$$

$$(12.44 + j9.72 + \bar{Z}_C + 2.25(55.03 + j52.30)) = 84.97 - j87.69$$

$$\bar{Z}_C = -51.28 - j215.1$$

Ignoring the negative resistance[9]

$$\bar{Z}_C = -j215.1\,\Omega$$

[9] Negative resistance is required to achieve a true balanced two-phase operation. However, since our intent is to use a capacitor for $\bar{Z}_C$, negative resistance is not a realizable option. Note that though the backward current is a substantial 25% of the forward current, the backward torque is only 0.4% of the forward torque.

Using this value for C_R (12.33 μF) in EMAP:

$\bar{V}_m = \bar{V}'_a = 120.0\,\text{V at } 0°$;
Speed = 1681.0 rpm = 176.0 rad/s; $s = 0.06611$
Running $C = 12.34\,\mu\text{F}$ (215.00 Ω)
$\bar{Z}_f = 55.025 + j(52.297)\Omega$; $E_f = 111.8$ at $-2.4ø$
$\bar{Z}_b = 3.317 + j(4.263)\Omega$; $E_b = 2.0$ at $87.2ø$

Currents in amperes
$\bar{I}_m = 1.570$ at $-32.78°$; $\bar{I}_a = 0.974$ at $29.84°$
$\bar{I}_f = 1.472$ at $-45.97°$; $\bar{I}_b = 0.363$ at $35.11°$
$\bar{I}_s = 2.195$ at $-9.58°$

Torques in N m (forward, backward, developed)
$T_f = 1.2657$; $T_b = 0.0046$; $T_{dev} = 1.2610$
$T_{RL} = 0.0925$; $T_m = 1.1686$

Powers in watts
SWL = 20.32; RWL = 17.46; $P_{RL} = 16.28$ Total = 54.05
$P_{in} = 259.8$; $P_{dev} = 222.0$; $P_{out} = 205.7$ (0.2757 hp)

$\text{HP}_{out} = 0.276$ hp; pf = 98.60%; Eff = 79.19%

Note that, compared with the values used in Example 6.5, overall performance was enhanced; greater torque and power, less main winding current; and higher efficiency. The problem is that this is the near-optimum value of C_R only at 1681 rpm. Still, if it were known that the motor was to operate at some given speed and power, this procedure could be used to determine a near-optimum C_R.

(b) The problem now is to select C_S. The criterion to meet is to maximize the starting torque. Recall at starting, $\bar{Z}_f = \bar{Z}_b$. Hence,

$$\begin{bmatrix} \bar{V}_m \\ \bar{V}_a \end{bmatrix} = \begin{bmatrix} \bar{Z}_m + \bar{Z}_f & 0 \\ 0 & \bar{Z}'_a + n^2\bar{Z}_f \end{bmatrix} \begin{bmatrix} \bar{I}_m \\ \bar{I}_a \end{bmatrix}$$

Solving for I_m,

$$\bar{I}_m = 9.057\angle{-41.9°}\ \text{A}$$

Also, again for balanced two-phase operation,

$$\bar{I}_a = \frac{j}{n}\bar{I}_m = 6.038\angle{+48.1°}$$

$$(\bar{Z}'_a + n^2\bar{Z}_f)\,\bar{I}_a = \bar{V}_a = 120\angle 0°$$

Therefore

$$\bar{Z}'_a + n^2\,\bar{Z}_f = \bar{Z}_C + 12.44 + j9.72 + 14.40 + j10.20 = 13.28 - j14.79$$

$$\bar{Z}_C = -13.56 - j34.71$$

Ignoring the negative resistance,[10]

$$\bar{Z}_C = -j34.71\,\Omega$$

Therefore $C_R + C_S = 76.43\,\mu F$, or $C_S = 64.1\,\mu F$. Using this value for C_S in EMAP.

$\bar{V}_m = \bar{V}'_a = 120.0\,V$ at 0°; Switch speed = 1350.0 rpm
Speed = 0.0 rpm = 0.0 rad/s; $s = 1.00000$
Starting C = 76.43 μF (34.71 Ω)
$\bar{Z}_f = 6.400 + j(4.534)\Omega$; $E_f = 57.8$ at −14.2ø
$\bar{Z}_b = 6.400 + j(4.534)\Omega$; $E_b = 15.7$ at 22.2ø

$\bar{Z}_{11} = 9.856 + j(8.854)$; $\bar{Z}_{12} = 0.000 + j(0.000)$
$\bar{Z}_{21} = 0.000 + j(0.000)$; $\bar{Z}_{22} = 26.842 + j(-14.785)$

Currents in amperes
$\bar{I}_m = 9.057$ at −41.93°; $\bar{I}_a = 3.916$ at 28.85°
$\bar{I}_f = 7.366$ at −49.48°; $\bar{I}_b = 2.004$ at −13.09°
$\bar{I}_s = 10.987$ at −22.27°

Torques in N m (forward, backward, developed)
$T_f = 3.6841$; $T_b = 0.2727$; $T_{dev} = 3.4113$
$T_{RL} = 0.0000$; $T_m = 3.4113$

Powers in watts
SWL = 474.31; RWL = 745.84; $P_{RL} = 0.00$ Total = 1220.14
$P_{in} = 1220.1$; $P_{dev} = 0.0$; $P_{out} = 0.0$ (0.0000 hp)

$HP_{out} = 0.000$ hp; pf = 92.54%; Eff = 0.00%

Compared with the values used in Example 6.5, starting (zero speed) torque was enhanced (3.411 N m, compared with 3.092 N m). Trial and error reveals that the true maximum starting torque (3.4404 N m) is achieved with 83 μF. However, the original value of 61.4 μF is not necessarily the worse choice because only one speed was considered. Still, this procedure can be used to determine a near-optimum C_S.

6.7 The Single-Phase Induction Motor Operating on One Winding

It is interesting to consider the single-phase motor running on one winding (m). Let us disconnect the auxiliary winding (a), creating the situation illustrated in Figure 6.6a.

$$\bar{I}_a = \frac{j}{n}\bar{I}_f - \frac{j}{n}\bar{I}_b = 0$$

or

$$\bar{I}_f = \bar{I}_b$$

Therefore,

$$\bar{I}_m = \bar{I}_f + \bar{I}_b = 2\bar{I}_f$$

[10] Refer to the footnote commentary in (a) of this example. The consequences of the approximation are more serious at starting because $\bar{Z}_b$ is much larger relative to $\bar{Z}_f$.

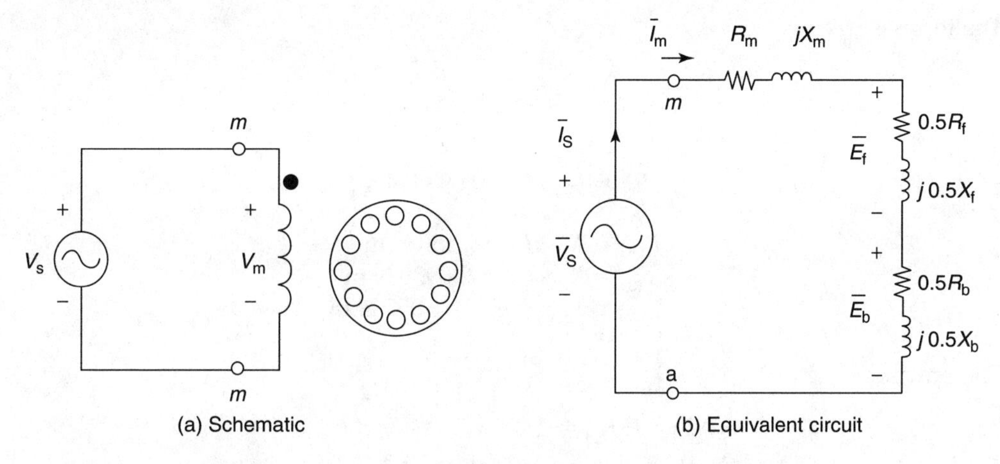

FIGURE 6.6. The single-phase single-winding induction motor.

Now

$$\bar{E}_m = \bar{E}_f + \bar{E}_b = \bar{I}_f\bar{Z}_f + \bar{I}_b\bar{Z}_b$$

$$\bar{E}_m = (2\bar{I}_f)(0.5\bar{Z}_f) + (2\bar{I}_b)(0.5\bar{Z}_b) = \bar{I}_m(0.5\bar{Z}_f + 0.5\bar{Z}_b)$$

which inspires the circuit of Figure 6.6b. It is now simple to solve for all currents.

$$\bar{I}_m = \frac{\bar{V}_m}{\bar{Z}_m + 0.5\bar{Z}_f + 0.5\bar{Z}_b}$$

An example will demonstrate this.

Example 6.8

Given the single-phase induction motor of Example 6.6, provide a comprehensive analysis of the motor operating on the main winding only while running at 1681 rpm.

SOLUTION

(a) As in Example 6.6, $s = 0.06611$:

$$0.5\bar{Z}_f = 27.52 + j26.15$$

$$0.5\bar{Z}_b = 1.659 + j2.132$$

Now

$$\bar{I}_m = \frac{\bar{V}_m}{\bar{Z}_m + 0.5\bar{Z}_f + 0.5\bar{Z}_b} = \frac{120}{3.456 + j4.320 + 27.52 + j26.15 + 1.659 + j2.132}$$

$$\bar{I}_m = 2.602\angle{-45^\circ}\text{ A}$$

$$\bar{I}_a = 0$$

$$\bar{I}_f = \bar{I}_b = 0.5\bar{I}_m = 1.301\angle{-45.0^\circ}\text{ A}$$

The torques

$$T_f = \frac{2I_f^2 R_f}{\omega_s} = 0.9880\,\text{N m}$$

$$T_b = \frac{2I_b^2 R_b}{\omega_s} = 0.0925\,\text{N m}$$

$$T_{dev} = T_f - T_b = 1.150\,\text{N m}$$
$$T_{RL} = K_{RL}\omega_{rm} = 0.0925\,\text{N m}$$
$$T_m = T_{dev} - T_{RL} = 1.057\,\text{N m}$$

The powers

$$\text{SWL} = I_m^2 R_m = 23.39\,\text{W}$$

$$P_{in} = V_m I_m \cos\theta_m = 220.9\,\text{W}$$

$$\text{RWL} = (s)2I_f^2 R_f + (2-s)2I_b^2 R_b = 34.03\,\text{W}$$

$$P_{RL} = 16.28\,\text{W}$$

$$P_{LOSS} = 73.70\,\text{W}$$

$$P_{dev} = 163.4\,\text{W}$$

$$P_{out} = 147.2\,\text{W}\ (0.1973\,\text{hp})$$

$$\eta = \frac{P_{out}}{P_{in}} = \frac{147.2}{220.9} = 70.74\%$$

$$\text{pf} = 0.7074 \text{ lagging}$$

It is instructive to compare single-phase motor performance in several modes. Consider Example 6.9.

Example 6.9
Given the single-phase induction motor of Example 6.6, defined by

Single-Phase Induction Motor Data

Ratings

Voltage = 120.0 V;	$a = N_a/N_m = 1.50$
Horsepower = 0.250 hp;	$J_{Motor} = 0.00330\,\text{kg m}^2$
Synchronous speed = 1800.0 rpm;	4 poles; Frequency = 60 Hz

Equivalent Circuit Values (in ohms)

$R_m = 3.456$; $X_m = 4.320$; $X_\phi = 115.200$
$R_a = 12.442$; $X_a = 9.720$; $R_2' = 6.912$; $X_2' = 4.320$
Rotational loss (RL) torque = $T_{RL} = K_{RL} * \omega_{rm}$
$T_{RL} = 0.000525 \times$ (speed in rad/s) N m

Auxiliary Winding

```
                 Cr
--- .---------| (----.---UUUUUU-----
    |                |  Switch CLOSED (Lo w) ... C = Cr + Cs
    |                |     C = 61.40 μF (43.20 Ω)
    | Switch    Cs   |  Switch OPEN (Hi w) ... C = Cr
    '--/ ------| (----'     C =  8.19 μF (324.00 Ω)
```

Switching speed = (75.0%) × 1800.0 rpm = 1350.0 rpm

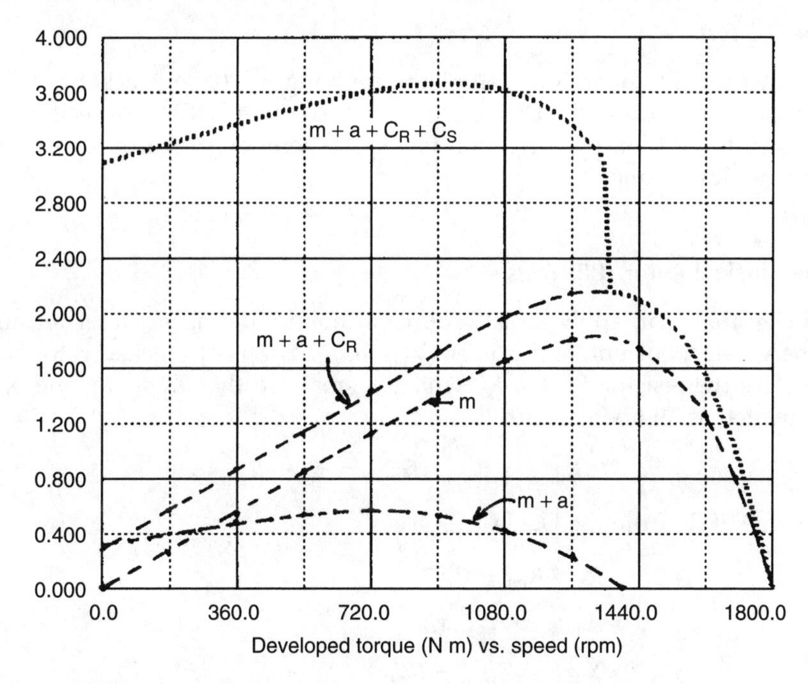

FIGURE 6.7. Comparative single-phase motor torque-speed characteristics.

Determine the torque-speed characteristic for the following operating modes.

(a) Operating on the m winding only.
(b) Operating on the m and a windings.
(c) Operating on the m and a windings, with C_R added.
(d) Operating on the m and a windings:
- with $C_R + C_S$ added for $\omega_{rm} < 75\%\omega_{sm}$
- with C_R added for $\omega_{rm} > 75\%\omega_{sm}$

Comment on the results.

SOLUTION

Examine Figure 6.7. The torque-speed characteristics were computed with computer assistance. (a) Observe that we get no starting torque, but respectable running torque when operating on the m winding only. (b) We get a small starting torque, but at the expense of seriously degraded running torque, when operating on the m and a windings. (c) The addition of the capacitor C_R improves operation at all speeds, but particularly at running speed. (d) Finally, the addition of the capacitor C_S dramatically improves operation at starting, but would degrade operation at running speed, if left in the circuit.

The mode for best overall performance is clear. We add the capacitor C_S to maximize starting torque, and switch to a smaller capacitor C_R for running conditions.

6.8 Equivalent Circuit Constants from Tests

The constants for equivalent circuits for single-phase machines may be determined from tests similar to those described in Section 4.4. The required constants are the seven element values (R_m, X_m, R_a, X_a, X_ϕ, R_2', X_2'), the constant K_{RL}, and the turns ratio $n = N_a/N_m$.

6.8.1 The DC Test

See Figure 4.9a for the laboratory setup. By applying a DC source between terminals (m–m′, a–a′), and measuring the DC voltage and current, the DC stator resistances can be determined. Multiplication by correction factors for temperature and for AC skin effect produces values for R_m and R_a.

6.8.2 The Blocked Rotor (BR) Tests

We immobilize the rotor, apply an AC source at main winding m stator terminals, and measure the voltage, current, and power ($\overline{V}_{BRm}$, $\overline{I}_{BRm}$, P_{BRm}). At blocked rotor conditions, $s = 1$. Therefore the resistor $R_2'/s = R_2'$, which is small, so that $Z_2' \ll X_\phi$ and X_ϕ may be neglected (treated as an open circuit).[11]

$$R_{BRm} = P_{BRm}/(I_{BRm}^2) = R_m + R_2'$$

Using the value of R_m from the DC test,

$$R_2' = R_{BR} - R_m$$

$$Z_{BRm} = V_{BRm}/(I_{BRm})$$

$$X_{BRm} = Z_{BRm}^2 - R_{BRm}^2 = X_m + X_2'$$

Repeat the test for the auxiliary (a) winding, measuring the voltage, current, and power ($\overline{V}_{BRa}$, $\overline{I}_{BRa}$, P_{BRa}):

$$Z_{BRa} = V_{BRa}/(I_{BRa})$$

Assume[12]

$$n = \sqrt{\frac{Z_{BRa}}{Z_{BRm}}}$$

At blocked rotor conditions, it can be shown that

$$\overline{V}_a = (\overline{Z}_a' + n^2\overline{Z}_f)\overline{I}_a$$

$$R_{BRa} = P_{BRa}/(\overline{I}_{BRa}^2) = R_a + n^2R_2'$$

Using the value of R_a from the DC test

$$n^2R_2' = R_{BRa} - R_a$$

If the two R_2''s, as determined in the m,a tests, differ (and they usually do), refine the R_2' value by averaging the two. Now

---

[11] For precision work, it is recommended that the BR test be executed at low frequency (typically about 5 Hz). At this frequency, the assumption that X_ϕ may be treated as an open circuit is not reasonable. Again, see IEEE/ANSI Standard 112 for a more accurate approach. Another possibility is to use the methods in this section as a first approximation for circuit constants, and iterate between BR and NL test data, using the circuit of Figure 4.8b with no approximations.

[12] The assumption is not quite as arbitrary as it appears. First, the rotor contribution ($\overline{Z}_f$) is in fact in the assumed n^2 ratio. Second, since the m,a windings are designed to contribute roughly equal mmf's, high turns implies low current. At roughly equal flux linkage, high turns implies high voltage. The high voltage, low current combination would suggest that winding resistance and leakage reactance would be in the approximate n^2 ratio, compared to the other winding. More sophisticated methods are available if greater accuracy is needed.

$$X_{\text{BRm}} = \sqrt{Z_{\text{BRm}}^2 - R_{\text{BRm}}^2} = X_{\text{m}} + X_2'$$
$$X_{\text{BRa}} = \sqrt{Z_{\text{BRa}}^2 - R_{\text{BRa}}^2} = X_{\text{a}} + n^2 X_2'$$

which can be solved simultaneously for X_2'. Once X_2' is known, the same equations provide values for X_{m} and X_{a}.

6.8.3 The No-Load Test

Run the machine as a motor on one winding (m) terminals at rated voltage with no external mechanical load, and measure the voltage, current, power, and speed ($\bar{V}_{\text{NL}}\ \bar{I}_{\text{NL}}\ P_{\text{NL}}\ \omega_{\text{NL}}$).

$$s_{\text{NL}} = \frac{\omega_{\text{sm}} - \omega_{\text{NL}}}{\omega_{\text{sm}}}$$

The circuit of Figure 6.6b applies. X_ϕ' is negligible (in $\bar{Z}_{\text{b}}$, but not in $\bar{Z}_{\text{f}}$), therefore $\bar{Z}_{\text{b}}$ can be evaluated. Neglecting the RWL due to the forward current,

$$P_{\text{RL}} = P_{\text{NL}} - [R_{\text{m}} + (2 - s)R_{\text{b}}]I_{\text{NL}}^2$$
$$K_{\text{RL}} = P_{\text{RL}}/(\omega_{\text{NL}})^2$$

Solve for E_{f} (by subtracting the $\bar{Z}_{\text{m}}\bar{I}_{\text{NL}}$ and $0.5\,\bar{Z}_{\text{b}}\bar{I}_{\text{NL}}$ phasor voltage drops from $\bar{V}_{\text{NL}}$). Assume $\bar{I}_{\text{NL}}$ as having two components, the in-phase loss component ($\bar{I}_{\text{LOSS}}$) and the out-of-phase magnetizing component ($\bar{I}_\phi$):

$$I_{\text{LOSS}} = P_{\text{RL}}/E_{\text{f}}$$
$$I_\phi = \sqrt{I_{\text{NL}}^2 - I_{\text{LOSS}}^2}$$

Finally

$$X_\phi = 2E_{\text{f}}/I_\phi$$

The required constants for the machine model (R_{m}, X_{m}, R_{a}, X_{a}, X_ϕ, R_2', X_2', K_{RL}, and n) are now determined.

6.9 Dynamic Performance

Dynamic performance may be analyzed using the same approach as was used for polyphase motors (see Section 4.8). Recall Equation 3.1:

$$T_{\text{dev}} - T_{\text{m}}' = T_a = J\frac{\text{d}\omega_{\text{rm}}}{\text{d}t}$$

To analyze the situation, it is necessary to solve Equation 3.1. The solution must be implemented using numerical methods, since the torques are nonlinear functions of speed. Note that the torques are functions of time, since we are now in a time-varying situation. Experience shows that for most practical situations, excellent results can be obtained by using "quasi-AC" methods: that is, the circuit models of Figure 6.5 and Figure 6.6 may still be used, provided that the rms voltage and speed do not change significantly within the time step selected to compute the dynamic response. For 60 Hz applications, the smallest appropriate time step is about 0.02 s. Most loads and motors will have sufficient inertia to meet this restriction. However, there are high-performance situations where this restriction is not met, and the reader is cautioned to be aware of this issue. Also note that more data are needed; specifically, the inertia constants for the motor and load.

Starting the system is arguably the most important dynamic situation. Consider the following example.

Example 6.10

The system illustrated represents a residential heating/air-conditioning blower. The motor is that of Example 6.6.

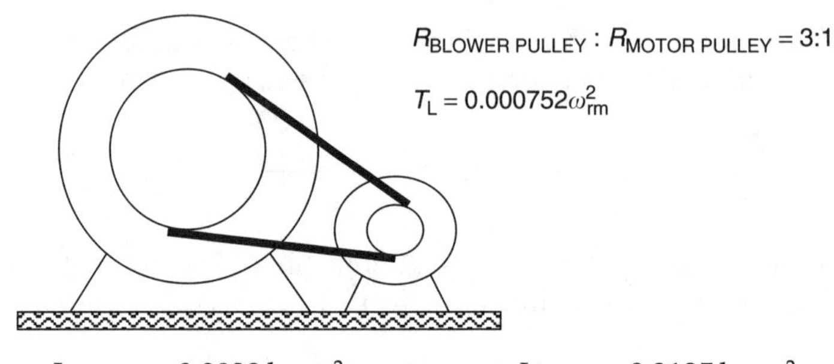

$J_{MOTOR} = 0.0033\,kg\,m^2$ $\qquad$ $J_{LOAD} = 0.3125\,kg\,m^2$

SPMOT.OUT
Gear ratio = GR = Load speed/motor speed = 0.3333
Load torque (N m) = $T_L = A_0 + A_1$*WL + ⋯ + A_N*WL**N
WL = LOAD shaft speed in rad/s; Order = 2

A_0 = 0.000; A_1 = 0.000000; A_2 = 0.000751864
$V_m = V'_a$ = 120.0 V at 0°; Switch speed = 1350.0 rpm
Speed = 1702.9 rpm = 178.3 rad/s; s = 0.05395
Running C = 8.19 μF (324.00 Ω)
$\bar{Z}_f$ = 55.384 + j(63.536)Ω; E_f = 109.1 at −0.8ø
$\bar{Z}_b$ = 3.297 + j(4.262)Ω; E_b = 2.0 at 19.6ø

$\bar{Z}_{11}$ = 32.796 + j(38.219); $\bar{Z}_{12}$ = 44.455 + j(−39.066)
$\bar{Z}_{21}$ = −44.455 + j(39.066); $\bar{Z}_{22}$ = 78.458 + j(−238.008)

Currents in amperes
$\bar{I}_m$ = 1.650 at −45.96°; $\bar{I}_a$ = 0.632 at 33.71°
$\bar{I}_f$ = 1.294 at −49.73°; $\bar{I}_b$ = 0.369 at −32.65°
$\bar{I}_s$ = 1.870 at −26.55°

Torques in N m (forward, backward, developed)
T_f = 0.9837; T_b = 0.0048; T_{dev} = 0.9789
T_{RL} = 0.0937; T_m = 0.8855

Powers in watts
SWL = 14.37; RWL = 11.75; P_{RL} = 16.70 Total = 42.82
P_{in} = 200.7; P_{dev} = 174.6; P_{out} = 157.9 (0.2116 hp)
HP_{out} = 0.212 hp; pf = 89.46%; Eff = 78.66%

Determine and plot the speed and current as the motor starts under load.

SOLUTION

(a) Reflecting J_L to the motor shaft:

$$J_m = (GR)^2 J_L = (1/3)^2 (0.3125) = 0.0125\,kg\,m^2$$

$$\text{or } J = 0.0033 + 0.0125 = 0.0158\,\text{kg m}^2$$

Also

$$T_{\text{m}} = (\text{GR})T_{\text{L}} = (1/3)(0.000752)\omega_{\text{rm}}^2 = 0.0002507\,\omega_{\text{rm}}^2\,\text{N m}$$

and

$$T'_{\text{m}} = T_{\text{m}} - T_{\text{RL}} = 0.0002507\,\omega_{\text{rm}}^2 + 0.063\,\omega_{\text{rm}}\,\text{N m}$$

Thus the equation to be solved is

$$T_{\text{dev}} - (0.0002507\,\omega_{\text{rm}}^2 + 0.063\,\omega_{\text{rm}}) = 0.0158\,(\text{d}\omega_{\text{rm}}/\text{d}t)$$

The equation is solved numerically, and the solution appears in Figure 6.8.

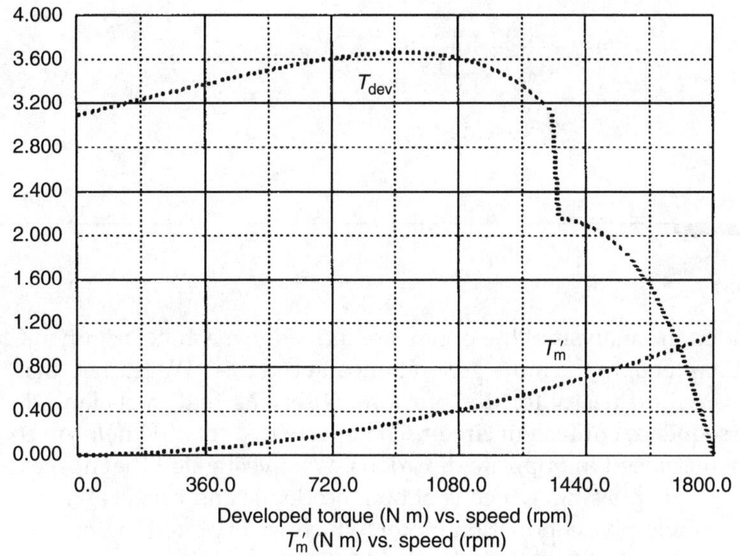

(a) Torque-speed characteristic for system of Example 6.10

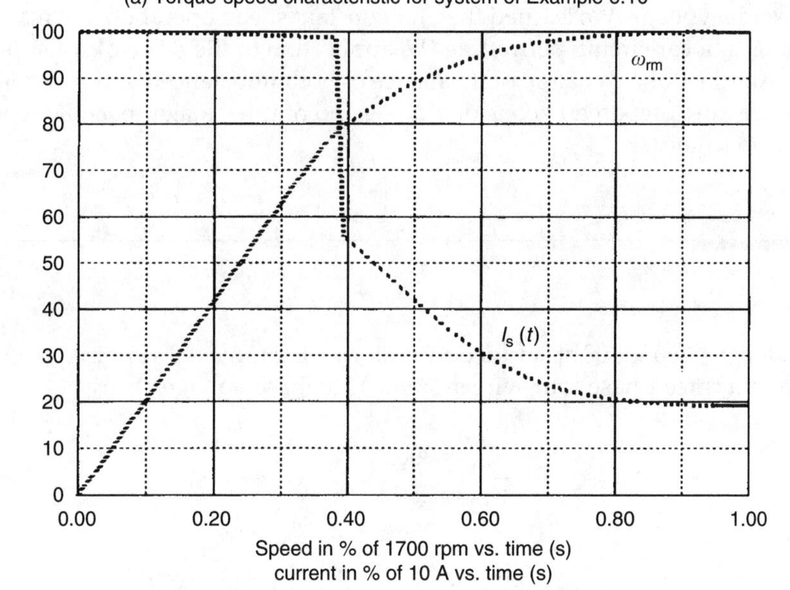

(b) Current-speed starting transient for system of Example 6.10

FIGURE 6.8. Plots for system of Example 6.10.

The powers

$$\text{SWL} = I_m^2 R_m = 23.39\,\text{W}$$

$$P_{in} = V_m I_m \cos\theta_m = 220.9\,\text{W}$$

$$\text{RWL} = (s)2I_f^2 R_f + (2 - s)2\bar{I}_b^2 R_b = 34.03\,\text{W}$$

$$P_{RL} = 16.28\,\text{W}$$

$$P_{LOSS} = 73.70\,\text{W}$$

$$P_{dev} = 163.4\,\text{W}$$

$$P_{out} = 147.2\,\text{W}\ (0.1973\,\text{hp})$$

$$\eta = \frac{P_{out}}{P_{in}} = \frac{147.2}{220.9} = 70.74\%$$

$$\text{pf} = 0.7074 \text{ lagging}$$

6.10 Summary

We have extended the analysis of the polyphase induction machine, begun in Chapter 4 and continued in Chapter 5, to the more general unbalanced case. We learned that symmetrical components were most useful for this purpose, observing that the balanced case reduced to the positive sequence equivalent circuit. There were two rotating fields to consider, rotating at synchronous speed in opposite directions. We investigated the most extreme case of unbalance; i.e., single-phasing, which is of both academic and practical interest. This leads naturally to the single-phase induction motor, which we explored in detail.

Solid-state speed control forced us to consider the proposition of operating the machine on nonsinusoidal voltage. We learned that, for constant speed operation, the machine could be modeled as a linear circuit, permitting the application of the principle of superposition. Finally, we examined the dynamic performance of both unbalanced and nonsinusoidal situations. We are now prepared to consider the second of three major machine types, i.e., the synchronous machine.

Problems

6.1 Consider the 460 V 100 hp induction motor of Example 4.3 running at 1785 rpm, supplied from a three-phase four-wire system. The phase voltages are:

$$\bar{V}_{an} = 300\angle 0^\circ\,\text{V}$$
$$\bar{V}_{bn} = 250\angle -90^\circ\,\text{V}$$
$$\bar{V}_{cn} = 250\angle +90^\circ\,\text{V}$$

(a) Find the line (ab, bc, ca) and sequence (0,1,2) voltages.
(b) Draw all relevant circuit diagrams.

(c) Find all currents.
(d) Find all powers.
(e) Find all torques.
(f) Find the efficiency.

6.2 The line voltages in a three-phase three-wire system are:

$$V_{ab} = 500\,\text{V}$$
$$V_{bc} = 420\,\text{V}$$
$$V_{ca} = 460\,\text{V}$$

(a) Find the sequence (0,1,2) and phase (an, bn, cn) voltages.
Apply the voltages of (a) to the motor of Example 4.3 running at a speed of 1785 rpm.
(b) Draw all relevant circuit diagrams.
(c) Find all currents.
(d) Find all powers.
(e) Find all torques.
(f) Find the efficiency.

6.3 Consider the 460 V 100 hp induction motor of Example 4.3 running at 1785 rpm, operating from a balanced 460 V three-phase three-wire source, when a fuse blows in Phase a. The load is such that the system slows to 1775 rpm.

(a) Draw the connection diagram.
(b) Find all voltages.
(c) Draw all relevant circuit diagrams.
(d) Find all currents.
(e) Find all powers.
(f) Find all torques.
(g) Find the efficiency.

6.4 Consider the 460 V 100 hp induction motor of Example 4.3 running at 1775 rpm, operating from a 460 V single-phase two-wire source in a Smith connection.

(a) Draw the connection diagram.
(b) Find all external elements. Add to the diagram of (a).
(c) Find all external currents.

6.5 Consider Example 6.4.

(a) Plot $\overline{V}_{an}(t)$ over one cycle. The fundamental frequency is 60 Hz.
(b) Compute the fundamental (positive sequence) voltage
(c) Compute the fifth harmonic (negative sequence) voltage (check results in the table)

Check results in the table of Example 6.4.

6.6 Continuing Problem 6.5,

(a) Find the speed, direction, and slip of the fundamental rotating field.
(b) Find the speed, direction, and slip of the fifth harmonic rotating field.

Check results in the table of Example 6.4.

6.7 Continuing Problem 6.5, at the fundamental frequency,

(a) Determine the frequency.
(b) Draw the appropriate equivalent circuit.
(c) Compute the stator current and forward torque.

Check results in the table of Example 6.4.

6.8 Continuing Problem 6.5, at the fifth harmonic frequency,

(a) Determine the frequency.
(b) Draw the appropriate equivalent circuit.
(c) Compute the stator current and backward torque.

Check results in the table of Example 6.4.

6.9 Consider the single-phase induction motor of Example 6.6. Compute all currents, powers, and torques at a running speed of:

(a) 1300 rpm.
(b) 1700 rpm.
(c) 1700 rpm, assuming the centrifugal switch failed to open, keeping the starting capacitor in the circuit. Compare answers with (b) and comment (does it make sense to remove the starting capacitor?).

6.10 Consider the single-phase induction motor of Example 6.6. Compute all currents, powers, the efficiency, and torques, running at a speed of

(a) 1300 rpm.
(b) 1720 rpm.
(c) 1720 rpm, assuming the centrifugal switch failed to open, keeping the starting capacitor in the circuit. Compare answers with (b) and comment (does it make sense to remove the starting capacitor?).

6.11 Consider the single-phase induction motor of Example 6.6, running at a speed of 1720 rpm.

(a) Determine the near-optimum value of C_R for this speed, as demonstrated in Example 6.7.
(b) Compute all currents, powers, and torques, and the efficiency with the capacitance of (a). Compare answers in Problems 6.8(b) with 6.9(b) and comment.

6.12 Consider the single-phase induction motor of Example 6.6, with all capacitors removed and the auxiliary winding permanently disconnected (on open circuit). Compute all currents, powers, and torques, running at a speed of

(a) 0 rpm.
(b) +1700 rpm.
(c) −1700 rpm.

Comment on your results.

7

The Polyphase Synchronous Machine: Balanced Operation

The polyphase induction machine has one obvious characteristic: there must be relative motion between the rotating stator field and the rotor if induction is to occur. If the rotor rotates at synchronous speed, the induction process breaks down, and rotor voltages and currents drop to zero. Also, if rotor currents were to exist, their frequency would have to be zero to produce nonzero average torque: But zero frequency AC is DC. It is possible to modify the design of the machine to supply the rotor with DC current, not by *induction* processes but rather by *conduction* means. The resultant machine is characterized as a *synchronous* machine since the EM energy conversion process can only occur when the rotating stator field and the rotor are *synchronized*; that is, they rotate at the same speed.[1] Hence, the speed of the machine is

$$\omega_{rm} = \omega_{sm} = \frac{2\pi f}{N_P/2} \tag{7.1}$$

Observe that for motor operation, the speed control problem is greatly simplified in that the rotor speed will no longer depend on load. We start our study by examining the details of synchronous machine construction.

7.1 Machine Construction

Recall that it is not our objective to provide comprehensive material on how to design EM machines. However, a basic understanding of how machines are physically constructed is necessary to understand the principles of operation and control, and to provide insight into the derivation of mathematical models. As with all EM machines, the device has two major parts: the stator (the part that does not rotate) and the rotor (the part that does rotate). Occasionally, the stator is called the "armature," which is defined as the part of the machine in which AC voltage is induced due to the motion of a DC magnetic field. However, since the term armature is also sometimes used to define a rotating structure, the term will be avoided in this discussion.

[1] We defer analysis of asynchronous operation to Chapter 8.

7.1.1 Stator Design

For our purpose, the stator of a polyphase synchronous machine is essentially identical to that of a polyphase induction machine. It is recommended that the reader reread Section 4.1.1. In summary, the stator is cylindrical (annular in cross-section, as shown in Figure 4.1b) and composed of laminated ferromagnetic material with equally spaced slots positioned around the inner "hole" parallel to the machine axis. Insulated coils will be placed in the slots, and interconnected to form windings. The windings are then connected in a balanced three-phase delta or wye configuration. For a detailed description of the stator winding layout, refer to Section 4.2.

7.1.2 Rotor Design

Examine the structures of Figure 7.1, which are basically cylindrical in shape and composed of either laminated or solid ferromagnetic material. The arrangement shown in Figure 7.1a is the salient pole rotor design. The term "salient" means "projecting"; observe that the pole "project," or "stick out" of the rotor structure, so that they may be easily counted. The rotor field is produced either with current-carrying windings or from permanent magnets. For the winding case, each pole is surrounded by an insulated winding, which contains a DC current (I_F the "field current") and permits direct control of the strength of the

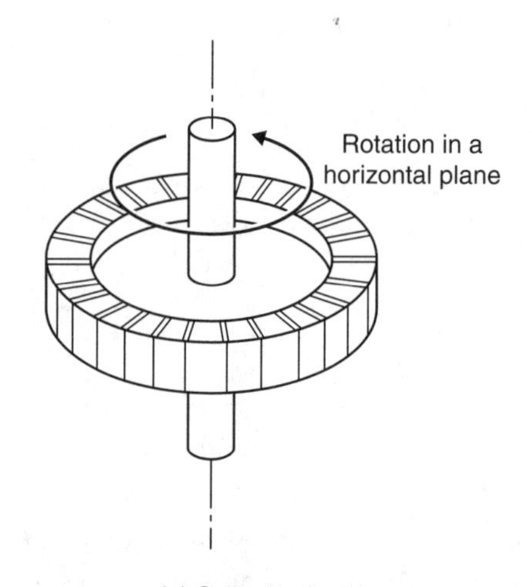

(a) Salient pole rotor
Hydro turbine driven; 26 poles; ω_{sm} = 276.9 rpm @ 60 Hz

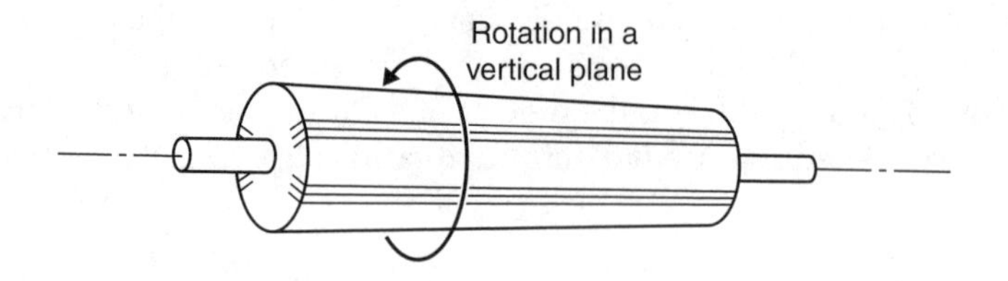

(b) Nonsalient pole rotor
Steam turbine driven; 4 poles; ω_{sm} = 1800 rpm @ 60 Hz

FIGURE 7.1. Synchronous machine rotor design.

corresponding magnetic field. There is always an even number of poles of symmetrical design, and consecutive poles are of alternating north and south polarity.

The arrangement shown in Figure 7.1b is called the nonsalient pole rotor design. The poles do not project from the rotor, and are not obvious from a casual visual inspection. Note that this design results in a uniform air-gap. Again, the rotor field is produced either with current-carrying windings or from permanent magnets. For the wound rotor design, the field windings are imbedded in slots that penetrate the rotor structure. Salient rotors are normally used in most synchronous machines designed to serve as synchronous motors and slow-speed generators. Nonsalient rotors are normally used in synchronous machines designed to serve as high-speed generators (typically two- or four-pole designs).

Imbedded in the pole faces are conducting bars that, together with the connecting end rings, form cage-like structures, called damper (or amortisseur[2]) windings. The primary function of the damper windings is to damp out mechanical oscillations about synchronous speed, a phenomenon called "hunting." For motors, the damper windings are frequently over-designed to permit starting as a cage rotor induction motor. It is important to understand that when the machine is operating in a normal balanced three-phase constant speed mode, the damper windings are inactive, containing zero current and voltage. However, under unbalanced or asynchronous speed conditions, the damper windings have a dramatic impact on machine performance.

The rotor field windings are excited from a controlled DC source, referred to as "the excitation system." Excitation systems are of two basic types: rotating (DC generators) for older applications and solid state for newer applications. In both cases, their function is to provide an adjustable DC source of sufficient capacity to provide requisite values of field current. Adjustable field current is desirable to facilitate reactive power control in the stator. If field control is not important, the rotor poles may be permanent magnets.

The rotor is placed in the stator "hole," supported by bearings, free to rotate, and positioned such that the air gap is symmetrical. The shaft is rigidly connected to the rotor.

7.2 Evolution of the Machine Model from the Induction Machine

Because the synchronous machine is so similar to the induction machine, one would suspect that the circuit models are quite similar, and this is basically true. Recall the stator positive sequence equivalent circuit for a three-phase induction machine, as developed in Chapter 4, which is presented in Figure 7.2a. In particular, recall that the rotor resistance is divided by the slip(s).

Now consider the situation as the rotor speed approaches synchronous speed, the speed of the rotating stator magnetic field. From the rotor's perspective, the stator field is moving slower and slower as the rotor catches up to the stator field, inducing a progressively weaker motional voltage into the rotor phases. As the induced rotor voltages become weaker, so do the rotor currents. Finally, when the rotor reaches synchronous speed, the rotor is now stationary with respect to the stator field. Hence, all motion-induced voltages and current in the rotor drop to zero, as do their frequencies. That is to say, if there is any rotor current, its frequency should be zero (which corresponds to DC).

Note that the model elegantly predicts all of this automatically. As the rotor speed approaches synchronous speed, the slip (s) approaches zero, and R_2'/s increases, becoming

[2] The term "amortisseur" is derived from the French term for deadening.

an open circuit at $s = 0$, producing the circuit of Figure 7.2b. The current I'_A (which is the rotor current reflected into the stator circuit) is clearly forced to zero by the open circuit. Recall that the rotor frequency is sf_s, which likewise approaches zero.

The rotor quantities are now completely inconsequential. All currents and voltages are zero, at zero frequency. There is no rotating rotor field. The developed torque is zero and the energy conversion process has broken down. Consider the following question:

If we wish to re-establish the energy conversion process, what rotor quantities are necessary?

Well, first of all, the rotor frequency should be zero. That is, since the stator field is stationary relative to the rotor, so the rotor field should be stationary too, since the rotor and stator fields must be synchronized to produce nonzero average torque. *But zero frequency AC is simply DC.* That is, the rotor winding voltages and currents must be DC. Secondly, to convert substantial energy, we need a powerful rotor field (roughly of the same strength as the stator field). Hence, we need strong rotor currents. To produce strong DC rotor current, we must apply significant DC voltage to the rotor windings. Without concerning ourselves with the details, suppose we replaced the induction machine rotor with a redesigned structure that permitted the application of a DC voltage that causes strong rotor currents to flow through windings, creating an N_P pole DC rotor magnetic field (The structures described in Section 7.1.2 are appropriate). Further, suppose we did it in such a manner that the rotating stator field was unaffected.[3]

How would such a modification be accommodated in the circuit model? Well, we have re-established the current I_A, and hence I'_A, which is now independent of stator quantities. The only way to establish I'_A, independently of stator circuitry, is to use a current source, as shown in Figure 7.2c. Because current sources are awkward to manipulate, most analysts prefer to convert into the equivalent voltage source form shown in Figure 7.2d:

$$\bar{E}_f = jX_m \bar{I}'_A \tag{7.2}$$

$$X_d = X_1 + X_m \tag{7.3}$$

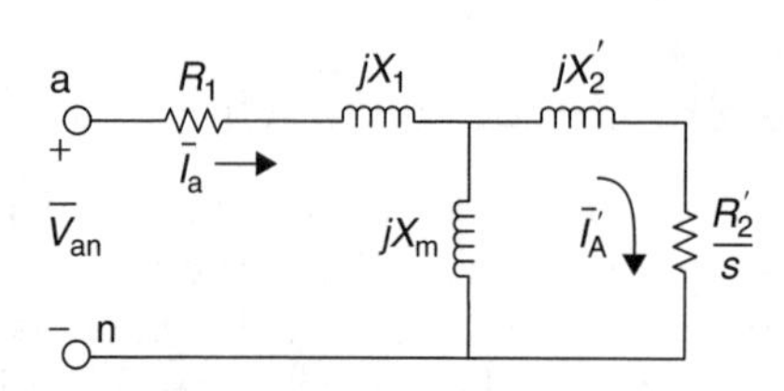

(a) Three-phase induction machine per-phase positive sequence equivalent circuit

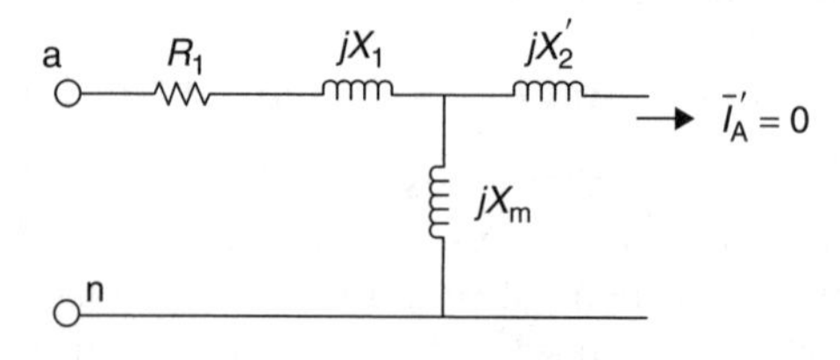

(b) Circuit of (a) at zero slip

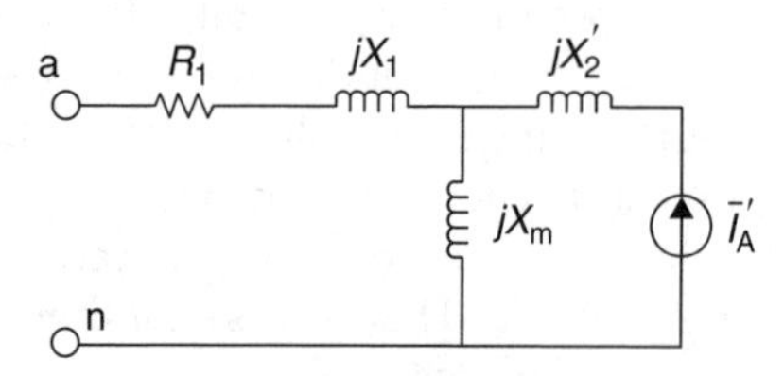

(c) Circuit of (b) with rotor replaced with synchronous machine rotor (rotor DC field current forced by excitation system)

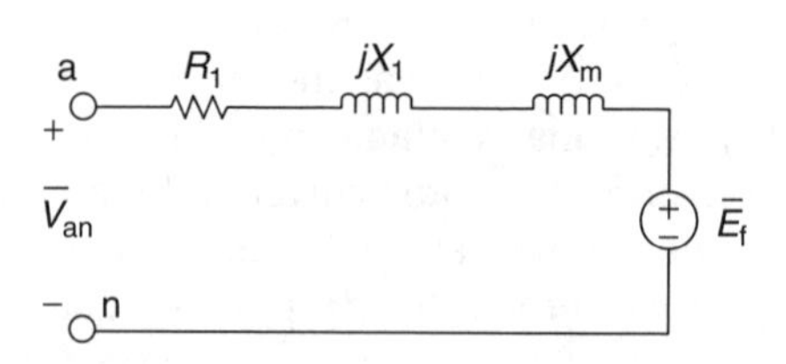

(d) Circuit of (c) with rotor current source transformed to voltage source

FIGURE 7.2. Evolution of induction to synchronous machine stator positive sequence equivalent circuit.

[3] What would be required is that effective path reluctance would be the same for both rotor designs.

It is important to recall that the source E_f is AC at stator frequency in the circuit of Figure 7.2d, even though it is caused by, and dependent on, the rotor current, which is DC. Before we demonstrate how the circuit of Figure 7.2d can be used to predict machine performance, we should examine the relation between the DC rotor field current, which we chose to symbolize as I_F, and E_f.

7.3 Interaction of the Rotor and Stator Circuits: The Magnetization Characteristic

Consider measuring the stator terminal voltage with the stator at no load ($I_a = 0$) as the (rotor) field current is varied from zero to its rated value, with the machine running at synchronous speed. If the results of such a test are plotted, a graph such as that shown in Figure 7.3 is produced.[4] This graph is routinely supplied by the machine manufacturer, and is called by various names, including the magnetization characteristic, the saturation curve, the open-circuit characteristic, and the no-load characteristic.[5] We shall use the term "magnetization characteristic," as discussed in Section 1.4, and is abbreviated as "MagC." Experience shows that the MagC can be represented adequately with the following empirical equation:

$$I_F = \frac{E_f}{K_{ag}} + A_x e^{B_x E_f} \tag{7.4}$$

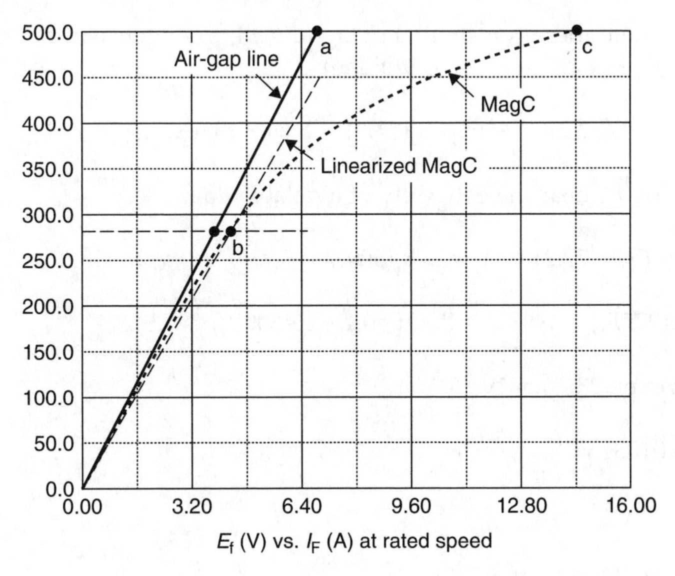

FIGURE 7.3. The magnetization curve (MagC).

[4] Actually, if the field current is varied over its full range in both positive and negative directions, a narrow hysteresis loop will be measured. What is used here is the single-valued expression. See Section 1.4 for a detailed explanation.

[5] The line (phase-to-phase) voltage is commonly plotted by manufacturers. However, since the phase-to-neutral voltage is more germane to our model, we shall use the latter. The conversion is simple: $V_{LINE} = \sqrt{3} V_{PHASE}$.

where

K_{ag} = slope of the air-gap line (unsaturated MagC) (V/A)
A_x = ΔI_F added to account for saturation (A)
B_x = exponential constant (1/V)

Equation 7.1 is the same form as Equation 1.10, except in different variables (E_f, I_F instead of B, H). The linearized MagC passes through the same point at rated voltage as the actual MagC (see Figure 7.3). Given the MagC, determination of K_{ag}, A_x, and B_x is straightforward, as was demonstrated in Example 1.5. Example 7.1 will clarify the situation.

Example 7.1

Figure 7.3 is the MagC for a 480 V three-phase 100 kVA 60 Hz four-pole synchronous machine. (a) Determine K_{ag}, A_x, and B_x as required in Equation 7.4. (b) Determine the linearized MagC.

SOLUTION

(a) Draw a linear extension of the MagC (line o-a, or the "air-gap line"):

$$E_f = K_{ag} I_F$$

Hence

$$K_{ag} = 500/6.93 = 72.2\,\text{V/A}$$

Select a point b where there is a small, but significant, divergence of the characteristic from the air-gap line (usually at rated voltage):

Point b: $4.22 = (277/72.2) + A_x \exp(B_x(277)) = 3.84 + A_x \exp(277B_x)$

Select a second point c near the extremity of available data.

Point c: $14.44 = (500/72.2) + A_x \exp(B_x(500)) = 6.93 + A_x \exp(500B_x)$

Point b: $A_x \exp(277B_x) = 0.38$ Point c: $A_x \exp(500B_x) = 7.51$

Dividing c by b: $\exp(223B_x) = 19.76$

$$B_x = 0.01334\,\text{V}^{-1}$$

Hence

$$A_x \exp(500(0.01334)) = 7.51 \quad \text{and} \quad A_x = 0.00952$$

The expression is valid only in the first quadrant.

(b) At rated voltage, $E_f = 277.1$ V. From the MagC, $I_F = 4.22$ A. The linearized MagC will be represented by

$$E_f = K_{ag} I_F$$

Hence

$$K_{ag} = 277.1/4.22 = 65.66\ \text{V/A or the linearized MagC is}\ E_f = 65.66\ I_F.$$

7.4 The Nonsalient Pole Synchronous Machine: Generator Operation

We are far enough along to summarize the nonsalient synchronous machine model when operating in a balanced three-phase mode at constant speed. Since there are significant differences between generator and motor operation, we chose to present these as two different scenarios, starting with generator operation. Refer to Figure 7.3, which summarizes the generator situation. The purpose of an equivalent circuit is to provide a mathematical model which may be used to assess machine performance. We will consider what we can determine from the model and then apply it to a specific application. The development follows that presented for the three-phase induction machine (see Section 4.5). The following terminology is specific to operation in the generator mode, and the coordinate system used is that appropriate to generator operation.

To consider this last point, consider the situations illustrated in Figure 7.4. As in all systems, the positive sense of all quantities must be clearly defined. The clearest and most efficient way to deal with this is to use a reference diagram. In machine models, there are two basic operating modes: generator and motor. If one were primarily interested in generation operation, it would make sense to define positive power flow to be from machine to system, as shown in Figure 7.4a, defining a so-called generator coordinate system. If V_{an} and I_a are defined as shown, this correlates to positive current flow from machine to system. If the machine was operating as a generator at leading power factor, the proper phasor diagram is shown in Figure 7.4c, in generator coordinates.

However, since the assigned positive sense of all variables is arbitrary, we could elect to define the variables as shown in Figure 7.4b, in effect replacing "I_a" with "$-I_a$." We shall call this choice a "motor coordinate system." Again, if the machine was operating as a generator at leading power factor, the proper phasor diagram is shown in Figure 7.4d, in motor coordinates. In particular, note that the currents are 180° out of phase in Figure 7.4c and Figure 7.4d. All other possibilities appear in Figure 7.4e to 7.4j. It is not a question of which coordinate system is correct; actually, they all are. It is clear, however, that choosing a generator coordinate system for generator operation, and a motor coordinate system for motor operation, is more "natural," and this is the normal practice, to which we shall conform. Finally, there are more variables involved than just V_{an} and I_a; Figure 7.4 and Figure 7.6 provide comprehensive generator and motor coordinate assignments, respectively.

Continue with a consideration of generator operation in generator coordinates (Figure 7.5). In the field circuit,

$$I_F = \frac{V_F}{R_F} \tag{7.5}$$

Equation 7.4 may be solved for E_f. In the stator,

$$\begin{aligned} \bar{E}_f &= R_1\bar{I}_a + j(X_1 + X_m)\bar{I}_a + \bar{V}_{an} \\ \bar{E}_f &= R_1\bar{I}_a + jX_d\bar{I}_a + \bar{V}_{an} \\ \bar{V}_a &= R_1\bar{I}_a + \bar{V}_{an} \end{aligned} \tag{7.6}$$

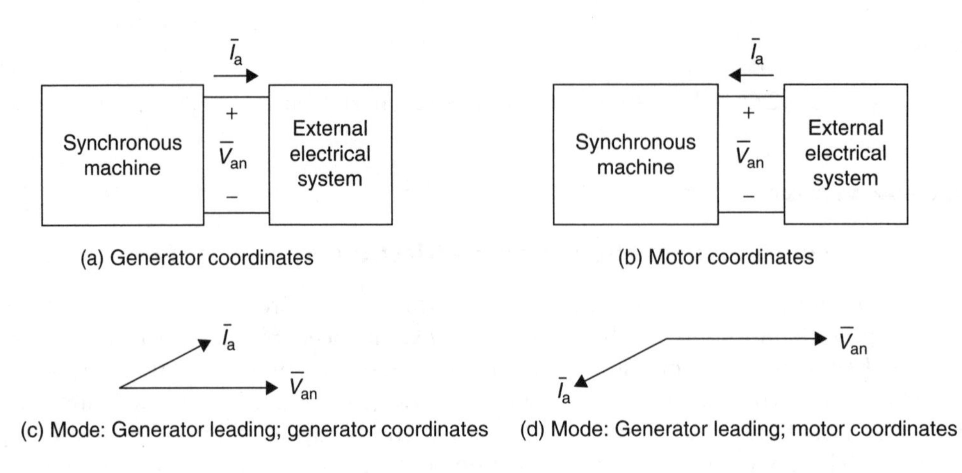

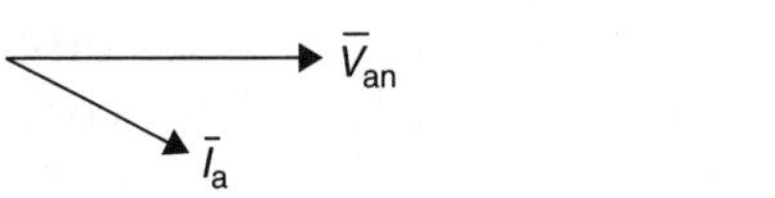

FIGURE 7.4. Clarification of generator–motor coordinate systems.

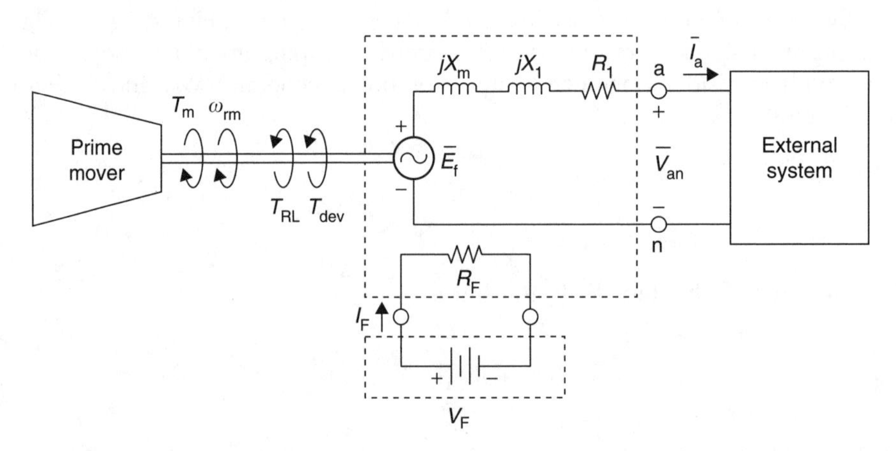

FIGURE 7.5. Three-phase synchronous machine model: generator coordinates.

Consider the powers,

$$\begin{aligned}
P_{\text{out}} &= 3V_{\text{an}}I_{\text{a}}\cos\theta_{\text{a}} \\
\text{Stator winding loss} &= \text{SWL} = 3I_{\text{a}}^2R_1 \\
\text{Rotor winding loss} &= \text{RWL} = \text{Field loss} = \text{FL} = I_{\text{F}}^2R_{\text{F}} \\
\text{Rotational loss} &= P_{\text{RL}} = T_{\text{RL}}\omega_{\text{rm}} = K_{\text{RL}}(\omega_{\text{rm}})^2 \\
\text{Sum of losses} &= \Sigma L = \text{SWL} + \text{RWL} + P_{\text{RL}} \\
P_{\text{dev}} &= 3E_{\text{f}}I_{\text{a}}\cos(\delta - \theta_{\text{a}}) \\
P_{\text{m}} &= P_{\text{dev}} + P_{\text{RL}} \\
P_{\text{in}} &= P_{\text{out}} + \Sigma L
\end{aligned} \tag{7.7}$$

Consider the efficiency,

$$\text{Efficiency} = \eta = \frac{P_{\text{out}}}{P_{\text{in}}} \tag{7.8}$$

Consider the torques,

$$\begin{aligned}
T_{\text{m}} &= \frac{P_{\text{m}}}{\omega_{\text{rm}}} \\
\text{Rotational loss torque} &= T_{\text{RL}} = K_{\text{RL}}\omega_{\text{rm}} \\
T_{\text{dev}} &= T_{\text{m}} - T_{\text{RL}}
\end{aligned} \tag{7.9}$$

A comprehensive analysis (Example 7.2) of the model will prove useful.

Example 7.2
Consider the synchronous machine defined in the following table.

Three-Phase Synchronous Machine Data		
	Ratings	
*V*Line = 480 V;	*I*Line = 120.3 A;	$S_{3\text{ph}}$ = 100.0 kVA
Stator frequency = 60 Hz;	Rotor type: NONSALIENT	
Synchronous speed = 1800.0 rpm = 188.5 rad/s		
No. of poles = 4;	DC excitation voltage = 125.0 V	
	Equivalent Circuit Values (R, X in Ω)	
R_a = 0.0461;	K_{RL} = 0.0844 N m s/rad;	P_{RL} = 2998.8 W
X_1 = 0.2304;	R_{F} = 7.8125	
X_d = 2.534;	X_q = 2.534	
Note: If machine is NONSALIENT, $X_d = X_q$.		
MagC at 1800 rpm:	$I_{\text{F}} = E_{\text{f}}/K_{\text{ag}} + A_x\exp(B_x * E_{\text{f}})$.	
K_{ag} = 72.17 Ω;	A_x = 0.009519 A;	B_x = 0.0133411 1/V;
Note: A_x = 0 indicates that the MagC has been linearized.		

For operation as a generator, at rated stator output, pf = 0.866 lagging, find (a) synchronous speed in rad/s and rev/min; (b) all machine currents and voltages (draw the stator phasor diagram), powers, and torques; (c) the efficiency; and (d) repeat (a) through (c) for pf = 0.866 leading.

SOLUTION

(a) Synchronous speed in rad/s and rev/min

$$\omega \;= \omega_{se} = 2\pi f = 2\pi(60) = 377.0\,\text{rad/s, or }3600\,\text{rev/min}$$

$$\omega_{sm} = 2\omega/N_P = 2(377)/4 = 188.5\,\text{rad/s, or }1800\,\text{rev/min, ccw}$$

(b) Stator line voltage, 480 V; frequency, 60 Hz; operating mode: generator

$$\text{Power factor} = \text{pf} = \cos\theta = 0.8660 \text{ lagging}$$

$$\theta = \cos^{-1}(0.866) = 30°;\ \bar{I}_a \text{ lags } \bar{V}_{an}$$

$$I_a = \frac{S_{3ph}}{\sqrt{3}V_L} = \frac{100}{\sqrt{3}(0.48)} = 120.3\text{ A}$$

$$\bar{I}_a = 120.3\angle{-30°}\text{ A}$$

$$\bar{E}_f = R_1\bar{I}_a + j(X_1 + X_m)\bar{I}_a + \bar{V}_{an}$$

$$= (0.0461 + j2.534)(120.3\angle{-30°}) + 277.1$$

$$\bar{E}_f = 506.9\angle 31°\text{ V}$$

$$\bar{V}_a = R_1\bar{I}_a + \bar{V}_{an} = (0.0461)(120.3\angle{-30°}) + 277.1$$

$$= 282.0\angle{-0.6°}\text{ V}$$

Solving Equation 7.4 for the field current,

$$I_F = 15.25\text{ A}$$

$$V_F = I_F R_F = (15.25)(7.8125) = 119.2\text{ V}$$

Solving for the powers,

$$\bar{S}_{3ph} = 100\angle 30° = 86.6 + j50$$

$$P_{out} = 86.6\text{ kW};\quad Q_{out} = 50\text{ kvar}$$

$$\text{Stator winding loss} = \text{SWL} = 3I_a^2R_1 = 3(120.3)^2(0.042) = 2.001\text{ kW}$$

$$\text{Rotor winding loss} = \text{Field winding loss} = \text{FWL} = I_F^2R_F = (15.25)^2(7.8125) = 1.817\text{ kW}$$

$$\text{Rotational loss} = P_{RL} = K_{RL}\omega_{rm} = (0.06299)(188.5) = 2.999\text{ kW}$$

$$\text{Total machine loss} = P_{LOSS} = P_{RL} + \text{SWL} + \text{RWL} = 2.001 + 1.817 + 2.999 = 6.817\text{ kW}$$

$$\text{Developed power} = P_{dev} = 3E_fI_a\cos(30° + 31°) = 88.60\text{ kW}$$

$$\text{Mechanical power} = P_m = P_{dev} + P_{RL} = 88.60 + 2.999 = 91.60\text{ kW (122.8 hp)}$$

$$\text{Power input} = P_m + \text{FWL} = 93.42\text{ kW}$$

Solving for the machine torques:

$$\text{Developed torque} = T_{dev} = P_{dev}/\omega_{rm}$$

$$= 88.60/188.5 = 470.0\text{ N m}$$

$$\text{Rotational loss torque} = T_{RL} = P_{RL}/\omega_{rm} = 2.999/188.5 = 15.9\text{ N m}$$

$$\text{Mechanical input torque} = T_m = T_{dev} + T_{RL} = 485.9\text{ N m}$$

The relevant phasor diagram is shown in Figure 7.6a.

(c) Efficiency $= \eta = P_{out}/P_{in} = 86.60/93.42 = 92.70\%$

(d) The analysis is to be repeated for pf = 0.866 leading.

Stator line voltage, 480 V; frequency, 60 Hz; Operating mode: Generator

$$\text{Power factor} = \text{pf} = \cos\theta = 0.8660 \text{ leading}$$

$$\theta = \cos^{-1}(0.866) = 30°;\ I_a \text{ leads } V_{an}$$

$$I_a = \frac{S_{3ph}}{\sqrt{3}V_L} = \frac{100}{\sqrt{3}(0.48)} = 120.3\text{ A}$$

$$\bar{I}_a = 120.3\angle{+30°}\text{ A}$$

$$\bar{E}_f = R_1\bar{I}_a + j(X_1 + X_m)\bar{I}_a + \bar{V}_{an}$$
$$= (0.0461 + j2.534)(120.3\angle{+30°}) + 277.1$$
$$\bar{E}_f = 296.5\angle 64.1°\text{ V}$$
$$\bar{V}_a = R_1\bar{I}_a + \bar{V}_{an} = 282.0\angle{+0.6°}\text{ V}$$

Solving Equation 7.4 for the field current,

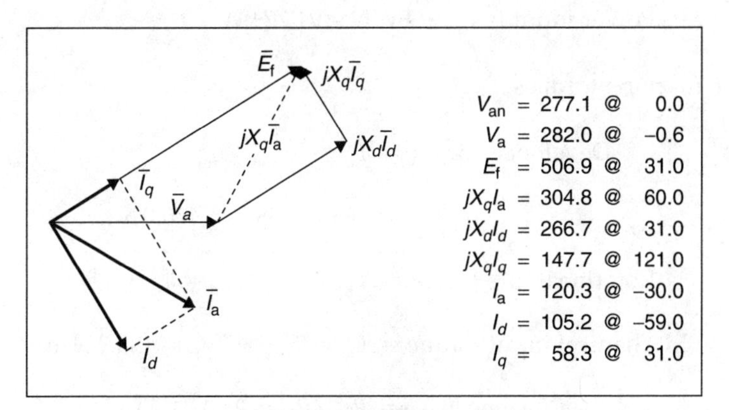

(a) Generator lagging mode

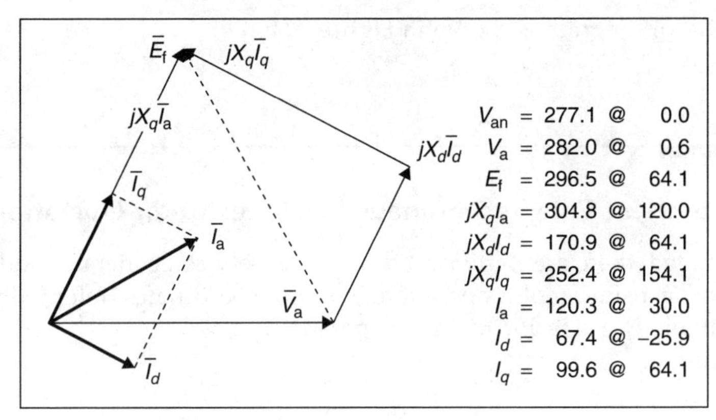

(b) Generator leading mode

FIGURE 7.6. Phasor diagrams.

$$I_F = 4.61\,\text{A}$$
$$V_F = I_F R_F = (4.61)(7.8125) = 36\,\text{V}$$

Solving for the powers,

$$\bar{S}_{3ph} = 100\angle{-30°} = 86.6 - j50$$
$$P_{out} = 86.6\,\text{kW};\ Q_{out} = -50\,\text{kvar}$$

$$\text{Stator winding loss} = \text{SWL} = 3I_a^2 R_1 = 2.001\,\text{kW}$$

$$\text{Field winding loss} = \text{FWL} = I_F^2 R_F = 0.166\,\text{kW}$$

$$\text{Rotational loss} = P_{RL} = K_{RL}\omega_{rm} = 2.999\,\text{kW}$$

$$\text{Total machine loss} = P_{RL} + \text{SWL} + \text{RWL} = 5.165\,\text{kW}$$

$$\text{Developed power} = P_{dev} = 3E_f I_a \cos(-30° + 64.1°) = 88.60\,\text{kW}$$

$$\text{Mechanical power} = P_m = P_{dev} + P_{RL} = 91.60\,\text{kW (122.8 hp)}$$

$$\text{Power input} = P_m + \text{FWL} = 91.77\,\text{kW}$$

Solving for the machine torques,

$$\text{Developed torque} = T_{dev} = P_{dev}/\omega_{rm}$$
$$= 88.60/188.5 = 470.0\,\text{N m}$$

$$\text{Rotational loss torque} = T_{RL} = P_{RL}/\omega_{rm} = 15.9\,\text{N m}$$

$$\text{Mechanical input torque} = T_m = T_{dev} + T_{RL} = 485.9\,\text{N m}$$

$$\text{Efficiency} = \eta = P_{out}/P_{in} = 94.37\%$$

The relevant phasor diagram is shown in Figure 7.6b.

7.5 The Nonsalient Pole Synchronous Machine: Motor Operation

As previously noted, there are significant differences between generator and motor operation. We will now consider motor operation, in motor coordinates (refer to Figure 7.7). The following terminology is specific to motor operation.

In the field circuit,

$$I_F = \frac{V_F}{R_F} \tag{7.10}$$

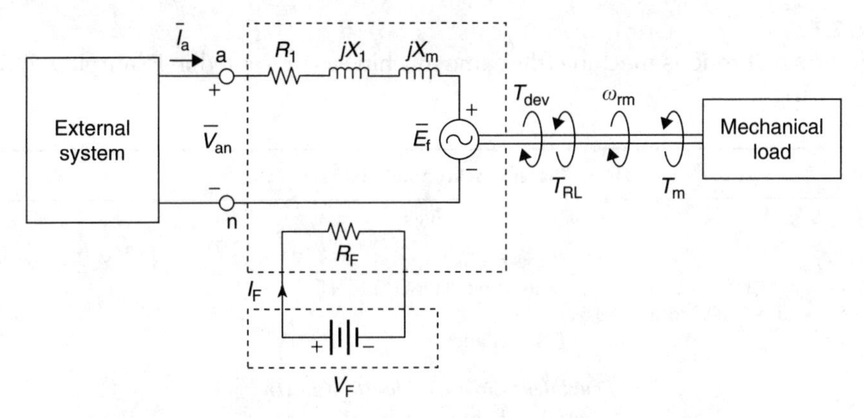

FIGURE 7.7. Three-phase synchronous machine model: motor coordinates.

Equation 7.4 may be solved for E_f. In the stator,

$$\begin{aligned}
\bar{V}_{an} &= R_1\bar{I}_a + j(X_1 + X_m)\bar{I}_a + \bar{E}_f \\
\bar{V}_{an} &= R_1\bar{I}_a + jX_d\bar{I}_a + \bar{E}_f \\
\bar{V}_{an} &= R_1\bar{I}_a + \bar{V}_a
\end{aligned} \tag{7.11}$$

Consider the powers:

$$\begin{aligned}
P_{stator} &= 3V_{an}I_a \cos\theta_a \\
\text{Stator winding loss} &= \text{SWL} = 3I_a^2R_1 \\
\text{Rotor winding loss} &= \text{RWL} = \text{Field Loss} = \text{FL} = I_F^2R_F \\
\text{Rotational loss} &= P_{RL} = T_{RL}\omega_{rm} = K_{RL}(\omega_{rm})^2 \\
\text{Sum of losses} &= \Sigma L = \text{SWL} + \text{RWL} + P_{RL} \\
P_{dev} &= 3E_fI_a\cos(\delta - \theta_a) \\
P_m &= P_{out} = P_{dev} - P_{RL} \\
P_{in} &= P_{out} + \Sigma L
\end{aligned} \tag{7.12}$$

Consider the efficiency,

$$\text{Efficiency} = \eta = \frac{P_{out}}{P_{in}} \tag{7.13}$$

Consider the torques,

$$\begin{aligned}
&T_m = \frac{P_m}{\omega_{rm}} = T_{out} \\
&\text{Rotational loss torque} = T_{RL} = K_{RL}\omega_{rm} \\
&T_{dev} = T_m + T_{RL}
\end{aligned} \tag{7.14}$$

A comprehensive analysis (Example 7.3) of the model will prove useful.

Example 7.3

Consider the synchronous machine (the same machine as presented in Example 7.2) defined in the following table.

Three-Phase Synchronous Machine Data		
	Ratings	
VLine = 480 V;	ILine = 120.3 A;	S_{3ph} = 100.0 kVA
Stator frequency = 60 Hz;	Rotor type: NONSALIENT	
Synchronous speed = 1800.0 rpm = 188.5 rad/s		
No. of poles = 4;	DC excitation voltage = 125.0 V	
	Equivalent Circuit Values (R, X in Ω)	
$R_a = 0.0461$;	K_{RL} = 0.0844 N m s/rad;	P_{RL} = 2998.8 W
$X_l = 0.2304$;	$R_F = 7.8125$	
$X_d = 2.534$;	$X_q = 2.534$	
Note: If machine is NONSALIENT, $X_d = X_q$.		
MagC at 1800 rpm: $I_F = E_f/K_{ag} + A_x \exp(B_x * E_f)$.		
K_{ag} = 72.17 Ω;	A_x = 0.009519 A;	B_x = 0.0133411 1/V;
Note: $A_x = 0$ indicates that the MagC has been linearized.		

For operation as a motor, at rated stator input, pf = 0.866 leading, find (a) synchronous speed in rad/s and rev/min; (b) all machine currents and voltages (draw the stator phasor diagram), powers, and torques; (c) the efficiency; and (d) repeat (a) through (c) for pf = 0.866 lagging.

SOLUTION

(a) Synchronous speed in rad/s and rev/min.

$$\omega = \omega_{se} = 2\pi f = 2\pi(60) = 377.0\ \text{rad/s, or } 3600\ \text{rev/min}$$

$$\omega_{sm} = 2\omega/N_P = 2(377)/4 = 188.5\ \text{rad/s, or } 1800\ \text{rev/min, ccw}$$

(b) Stator line voltage, 480 V; frequency, 60 Hz; Operating mode: Motor

$$\text{Power factor} = \text{pf} = \cos\theta = 0.8660\ \text{leading}$$

$$\theta = \cos^{-1}(0.866) = 30°;\ I_a \text{ leads } V_{an}$$

$$\begin{aligned}\bar{E}_f &= \bar{V}_{an} - R_1\bar{I}_a - j(X_1 + X_m)\bar{I}_a \\ &= 277.1 - (0.0461 + j2.534)(120.3\angle 30°)\end{aligned}$$

$$\begin{aligned}\bar{E}_f &= 501.6\angle{-32.1°}\ \text{V} \\ \bar{V}_a &= \bar{V}_{an} - R_1\bar{I}_a = 277.1 - (0.0461)(120.3\angle 30°) \\ &= 272.3\angle{-0.6°}\ \text{V}\end{aligned}$$

Solving Equation 7.4 for the field current,

$$\begin{aligned}I_F &= 14.62\ \text{A} \\ V_F &= I_F R_F = (14.62)(7.8125) = 114.2\ \text{V}\end{aligned}$$

Solving for the powers,

$$\begin{aligned}\bar{S}_{3ph} &= 100\angle 30° = 86.6 - j50 \\ P_{3ph} &= 86.6\ \text{kW}; \quad Q_{3ph} = -50\ \text{kvar}\end{aligned}$$

Stator winding loss = SWL = $3I_a^2R_1$ = $3(120.3)^2(0.042)$ = 2.001 kW

Rotor winding loss = Field winding loss = FWL = $I_F^2R_F$ = $(14.62)^2(7.8125)$ = 1.670 kW

Rotational loss = $P_{RL} = K_{RL}\omega_{rm}$ = (0.06299)(188.5) = 2.999 kW

Total machine loss = $P_{LOSS} = P_{RL}$ + SWL + RWL = 2.001 + 1.67 + 2.999 = 6.669 kW

Developed power = $P_{dev} = 3E_fI_a\cos(30 + 32.1°)$ = 84.60 kW

Mechanical power = $P_m = P_{out} = P_{dev} - P_{RL}$ = 84.60 − 2.999 = 81.60 kW (109.4 hp)

Power input = $P_{out} + \Sigma L$ = 88.27 kW

Solving for the machine torques,

$$\text{Developed torque} = T_{dev} = P_{dev}/\omega_{rm} = 84.60/188.5 = 448.8\,\text{N m}$$

$$\text{Rotational loss torque} = T_{RL} = P_{RL}/\omega_{rm} = 2.999/188.5 = 15.9\,\text{N m}$$

$$\text{Mechanical input torque} = T_m = T_{dev} - T_{RL} = 432.9\,\text{N m}$$

The relevant phasor diagram is shown in Figure 7.8a.

(c) Efficiency = $\eta = P_{out}/P_{in}$ = 81.60/88.27 = 92.44%

(d) The analysis is to be repeated for pf = 0.866 lagging.

Stator line voltage, 480 V; frequency, 60 Hz; Operating mode: Motor

$$\text{Power factor} = \text{pf} = \cos\theta = 0.8660 \text{ lagging}$$

$$\theta = \cos^{-1}(0.866) = 30°;\ I_a \text{ lags } V_{an}$$

$$I_a = \frac{S_{3ph}}{\sqrt{3}V_L} = \frac{100}{\sqrt{3}(0.48)} = 120.3\text{ A}$$

$$\bar{I}_a = 120.3\angle{-30°}\text{ A}$$

$$\bar{E}_f = \bar{V}_{an} - R_1\bar{I}_a - j(X_1 + X_m)\bar{I}_a$$

$$= 277.1 - (0.0461 + j2.534)(120.3\angle{-30°})$$

$$\bar{E}_f = 287.4\angle{-65.3°}\text{V}$$

$$\bar{V}_a = \bar{V}_{an} - R_1\bar{I}_a = 277.1 - (0.0461)(120.3\angle{-30°})$$

$$= 272.3\angle 0.6°\text{ V}$$

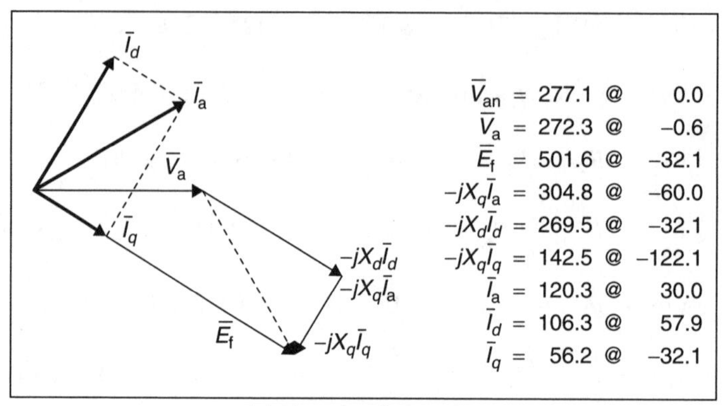

(a) Motor leading mode

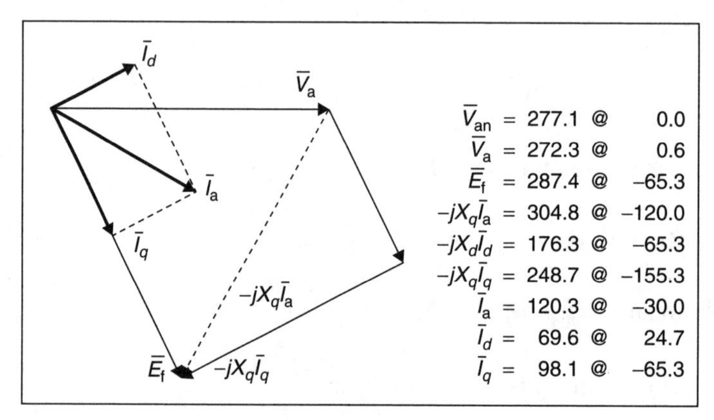

(b) Motor lagging mode

FIGURE 7.8. Phasor diagrams.

Solving Equation 7.4 for the field current,

$$I_F = 4.42 \text{ A}$$

$$V_F = I_F R_F = (4.42)(7.8125) = 34.6 \text{ V}$$

Solving for the powers,

$$\bar{S}_{3ph} = 100\angle 30° = 86.6 + j50$$

$$P_{3ph} = 86.6 \text{ kW}; \; Q_{3ph} = 50 \text{ kvar}$$

$$\text{Stator winding loss} = \text{SWL} = 3I_a^2 R_1 = 3(120.3)^2(0.042) = 2.001 \text{ kW}$$

$$\text{Rotor winding loss} = \text{Field winding loss} = \text{FWL} = I_F^2 R_F = (4.42)^2(7.8125) = 0.153 \text{ kW}$$

$$\text{Rotational loss} = P_{RL} = K_{RL}\omega_{rm} = (0.06299)(188.5) = 2.999 \text{ kW}$$

$$\text{Total machine loss} = P_{LOSS} = P_{RL} + \text{SWL} + \text{RWL} = 2.001 + 0.153 + 2.999 = 5.152 \text{ kW}$$

$$\text{Developed power} = P_{dev} = 3E_f I_a \cos(30 - 65.3°)$$

$$\text{Developed power} = 3E_f I_a \cos(30° - 65.3°) = 84.60\,\text{kW}$$

$$\text{Mechanical power} = P_m = P_{out} = P_{dev} - P_{RL} = 84.60 - 2.999 = 81.60\,\text{kW (109.4 hp)}$$

$$\text{Power input} = P_{out} + \Sigma L = 86.75\,\text{kW}$$

Solving for the machine torques,

$$\text{Developed torque} = T_{dev} = P_{dev}/\omega_{rm} = 84.60/188.5 = 448.8\,\text{N m}$$

$$\text{Rotational loss torque} = T_{RL} = P_{RL}/\omega_{rm} = 2.999/188.5 = 15.9\,\text{N m}$$

$$\text{Mechanical input torque} = T_m = T_{dev} - T_{RL} = 432.9\,\text{N m}$$

The relevant phasor diagram is shown in Figure 7.8b.

$$\text{Efficiency} = \eta = P_{out}/P_{in} = 81.60/86.75 = 94.06\%$$

7.6 The Salient Pole Synchronous Machine

The nonsalient machine is reasonably straightforward to analyze, at least when operating in a balanced three-phase mode at constant speed. The fact that an equivalent circuit exists is a major advantage, since much of the analysis can be formulated as an AC circuits problem. Our next topic can be posed as a question:

What changes in the stator circuit model must be made to account for rotor saliency?

To resolve this issue we return to the derivation of the model, which was implemented considering the stator and rotor rotating fields. Recall from Sections 7.3 and 7.4 that it makes a difference as to whether we "think generator" or "think motor" and likewise what coordinate system we select. For our purposes, assume we are operating in the generator mode, and are using generator coordinates.

Consider the situation illustrated in Figure 7.9a. We now introduce two new concepts: the direct, or *d*-axis, which is the magnetic axis of the rotor field winding; and the quadrature, or *q*-axis, which lags the *d*-axis by 90 electrical degrees.[6] The rotor produces a vector magnetic field,[7] defined as $\hat{F}$, which rotates at ω_{rm}, along with the rotor. Assuming that balanced stator AC currents flow, a second rotating magnetic field $\hat{A}$, will be produced. The two fields combine to produce a resultant field $\hat{R}$, as shown in Figure 7.9a. If the machine is operating in a balanced three-phase mode at constant speed, the stator field and the rotor both turn at synchronous speed and are stationary relative to each other. Note that in Figure 7.9a, we "just happened" to position the rotor *d*-axis, and hence the rotor magnetic field, as it was passing through the plane of the Phase a winding.[8] It follows that the voltage induced into stator Phase a by the rotor field (i.e., the voltage $\bar{E}_f$) would be at a positive maximum at this very instant. Using this condition to define our time origin, $e_f(t)$ would be a positive cosine function, and the phasor $\bar{E}_f$ would be in the phase reference position, as shown in Figure 7.9b.

[6] The definitions of the *d* and *q*-axes are arbitrary, and hence not unique in the literature. In fact, at some point in the historical evolution of machine theory, the matter was hotly debated, and strongly held opinions were espoused by advocates of one system over another.

[7] More precisely, a sinusoidally distributed mmf, represented by the vector $\hat{F}$.

[8] What a wonderful coincidence!

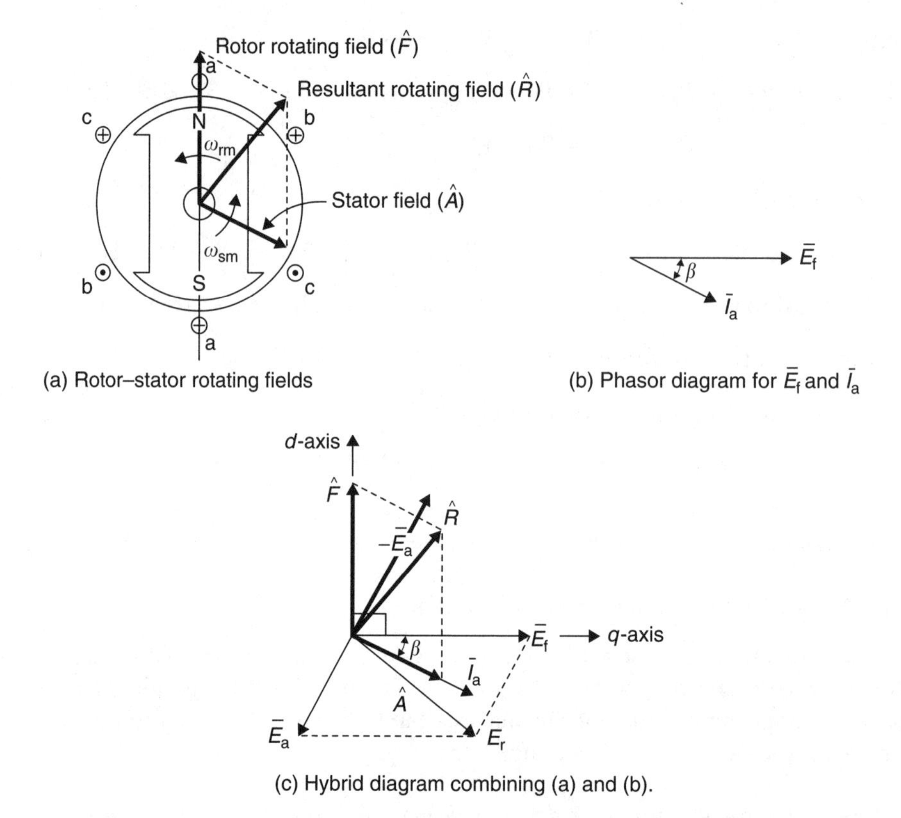

FIGURE 7.9. Space–time relations between magnetic fields and associated phasors.

Suppose that the stator is terminated in balanced three-phase impedances, such that balanced three-phase currents flow and that the Phase-a current $\bar{I}_a$ is in phase with $\bar{E}_F$. Where the rotating stator field is located at $t = 0$? *On the Phase-a magnetic axis!* Now suppose that the Phase-a current $\bar{I}_a$ lags $\bar{E}_f$ by 30°, as shown in Figure 7.9b. The rotating stator field would be 30° "behind" (i.e., CW from) the Phase-a magnetic axis at $t = 0$, as shown in Figure 7.9a.[9] These observations lead us to the following general conclusion:

The phase position of the phasor Phase-a current directly correlates to the spatial location of the stator rotating magnetic field.

These insights can be compactly and elegantly illustrated by combining the phasor diagram of Figure 7.9b and the vector diagram of Figure 7.9a into one hybrid diagram as shown in Figure 7.9c.[10] The key to the proper construction of Figure 7.9c is to realize that E_f, must fall on the q-axis. Indeed, we realize that

$$\hat{R} = \hat{F} + \hat{A}$$
$$\bar{E}_r = \bar{E}_f + \bar{E}_a$$

[9] The reader is invited to refresh his/her understanding of the properties of the rotating magnetic field by rereading Section 4.3.

[10] The author regretfully admits that this elegant diagram is certainly not original with him; unfortunately he was not able to determine to whom it should be attributed. We all should be grateful for such an insightful window into synchronous machine performance.

where the subscripts r, f, and a indicate which voltage is caused by which field. We define "β" to be the spatial angle of "$\hat{A}$" relative to the q-axis, and equivalently, the phase angle of $\bar{I}_a$ relative to $\bar{E}_f$. If $\bar{I}_a$ is in phase with $\bar{E}_f$ ($\beta = 0$), the rotating stator mmf ($\hat{A}$) is positioned on the q-axis; likewise, if $\bar{I}_a$ leads or lags $\bar{E}_f$ by 90° ($\beta = \pm 90°$), the rotating stator mmf is positioned on the d-axis.

We now return to considering the voltage (E_a), produced by the rotating stator field (A). Since A is in direct proportion to I_a, and E_a to A, it follows that E_a is in direct proportion to I_a. Therefore, let $E_a = XI_a$, where X is a constant of proportionality, and must be in ohms, since I_a and E_a are in amperes and volts, respectively. Furthermore, $\bar{E}_a = -jX\bar{I}_a$, since $\bar{E}_a$ must lag $\hat{A}$, and hence $\bar{I}_a$, by 90°. Consequently, $-\bar{E}_a = +jXI_a$. Results are summarized in Figure 7.10a and Figure 7.10b.

For the nonsalient rotor, the air-gap was uniform around the entire circumference of the stator; hence the magnetic field produced by the stator mmf was independent of its position relative to the magnetic axis (i.e., the d-axis) of the rotor. However, this is not true of the salient machine. If in fact the rotating stator field were aligned with the d-axis, the subsequent magnetic flux would be at maximum strength, since the air gap is minimum in this direction. Likewise, if the rotating stator field were aligned with the q-axis, the subsequent magnetic flux would be at minimum strength, since the air-gap is maximum in this direction. Hence, X must be a function of "β," minimizing when $\beta = 0$, and maximizing when $\beta = \pm 90°$, as shown in Figure 7.10c. Define the maximum and minimum values of X to be $\hat{X}_d$ and $\hat{X}_q$, respectively. To compute an accurate value of $-E_a$ we break $\hat{A}$, and equivalently $\bar{I}_a$, into d and q components. Therefore,

$$(-\bar{E}_A) = j\hat{X}_d\bar{I}_d + j\hat{X}_q\bar{I}_q$$

$$\bar{I}_a = \bar{I}_d + \bar{I}_q$$

$$X(\beta) = \tfrac{1}{2}[\hat{X}_d + \hat{X}_q + (\hat{X}_q - \hat{X}_d)\cos 2\beta]$$

It is desirable to replace the boxes in the circuit of Figure 7.10b with traditional circuit elements. Box "F" is easy. Since the voltage E_f is not dependent on the current through it (I_a), the proper circuit element is a voltage source. However, the box "A" is a bit strange. Since the voltage $-\bar{E}_a = +jX(\beta)\bar{I}_a$, it is an inductor, but one whose value depends on β, the phase position of $\bar{I}_a$ relative to $\bar{E}_f$. Finally, we have yet to consider leakage flux and winding

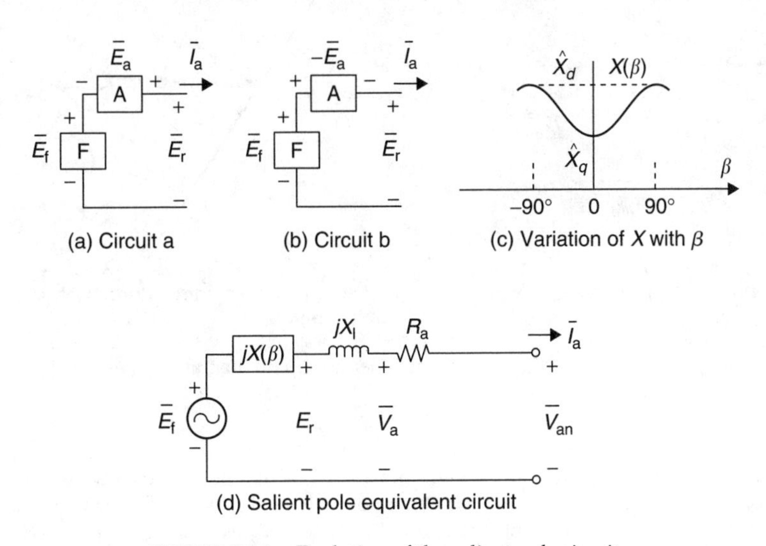

FIGURE 7.10. Evolution of the salient pole circuit.

resistance. These two effects are included by adding appropriate series components. The finished circuit appears in Figure 7.10d. Finally, applying KVL to the circuit,

$$\begin{aligned}\bar{E}_f &= j\hat{X}_d\bar{I}_d + j\hat{X}_q\bar{I}_q + jX_1\bar{I}_a + R_a\bar{I}_a + \bar{V}_{an} \\ &= j\hat{X}_d\bar{I}_d + j\hat{X}_q\bar{I}_q + jX_1(\bar{I}_d + \bar{I}_q) + R_a\bar{I}_a + \bar{V}_{an} \\ &= j(\hat{X}_d + X_1)\bar{I}_d + j(\hat{X}_q + X_1)\bar{I}_q + R_a\bar{I}_a + \bar{V}_{an} \\ &= jX_d\bar{I}_d + jX_q\bar{I}_q + R_a I_a + \bar{V}_{an} \\ &= jX_d\bar{I}_d + jX_q\bar{I}_q + \bar{V}_a\end{aligned} \tag{7.15}$$

where

$$\begin{aligned}X_d &= \hat{X}_d + X_1 \\ X_q &= \hat{X}_q + X_1 \\ \bar{V}_a &= R_a\bar{I}_a + \bar{V}_{an}\end{aligned}$$

Note that a new voltage (V_a) has been defined in Figure 7.10d for convenience.[11] A phasor diagram representing Equation 7.15 is shown in Figure 7.11a. Shifting the phase reference to V_a produces Figure 7.11b. Now consider the projection of V_a onto the d and q axes in Figure 7.11b:

$$\begin{aligned}V_d &= V_a \sin\delta \\ V_q &= V_a \cos\delta\end{aligned}$$

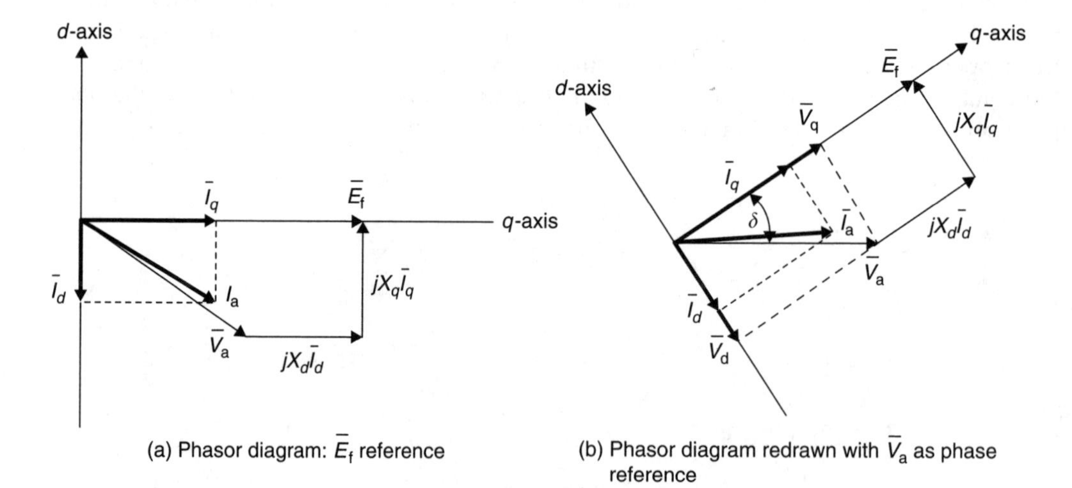

FIGURE 7.11. Salient generator phase diagram.

[11] Since the voltage R_aI_a is normally less than 2% of the other voltages, it is not easily shown on a phasor diagram. Likewise, the corresponding power, while small, is not normally negligible, particularly in efficiency calculations. Inclusion of these terms would dramatically complicate the analysis. Casting the equations in terms of $\bar{V}_a$ neatly sidesteps the problem.

Now consider the average power P_a:

$$P_a = I_a \cos\theta_a = V_d I_d + V_q I_q = (V_a \sin\delta) I_d + (V_a \cos\delta) I_q$$
$$= (V_a \sin\delta)\left(\frac{E_f - V_a \cos\delta}{X_d}\right) + (V_a \cos\delta)\left(\frac{V_a \sin\delta}{X_q}\right) I_q$$

$$P_a = \frac{E_f V_a}{X_d}\sin\delta + \frac{V_a^2 (X_d - X_q)}{2X_d X_q}\sin(2\delta) \tag{7.16a}$$

A similar approach can be used to derive the reactive power Q_a:

$$Q_a = \frac{E_f V_a}{X_d}\cos\delta + \frac{V_a^2 (X_d - X_q)}{2X_d X_q}\cos 2\delta - \frac{V_a^2 (X_d + X_q)}{2X_d X_q} \tag{7.16b}$$

But P_a is also the power that flows out of the "E_f" source. Hence:

$$P_{dev} = 3\frac{E_f V_a}{X_d}\sin\delta + \frac{3V_a^2 (X_d - X_q)}{2X_d X_q}\sin(2\delta)$$

Note that the key to employment of Equations 7.16 is to determine the location of the d and q axes, or alternatively the angle δ. Consider:

$$\bar{E}_f = jX_d \bar{I}_d + jX_q \bar{I}_q + \bar{V}_a = jX_d \bar{I}_d + jX_q(\bar{I}_a - \bar{I}_q) + \bar{V}_a$$
$$\bar{E}_f - j(X_d - X_q)\bar{I}_d = jX_q \bar{I}_a + \bar{V}_a$$

But $\bar{E}_f$ and $-j(X_d - X_q)\bar{I}_d$ are both on the q-axis! Hence, $jX_q\bar{I}_a + \bar{V}_a$ must locate the q-axis! The remainder of the machine equations are essentially the same as for the nonsalient machine.

Also note if $X_q = X_d$,

$$P_{dev} = 3\frac{E_F V_a \sin\delta}{X_d} \tag{7.17a}$$

$$Q_a = \frac{3E_F V_a \cos\delta}{X_d} - \frac{3V_a^2}{X_d} \tag{7.17b}$$

$$\bar{E}_f = jX_d \bar{I}_a + \bar{V}_a \tag{7.17c}$$

and the salient machine morphs into the nonsalient machine. We will demonstrate application of these concepts through a comprehensive example.

Example 7.4

Consider the synchronous machine defined in the following table:

Three-Phase Synchronous Machine Data		
Ratings		
VLine = 480 V;	ILine = 120.3 A;	S_{3ph} = 100.0 kVA
Stator frequency = 60 Hz;	Rotor type: SALIENT	
Synchronous speed = 1800.0 rpm = 188.5 rad/s		
No. of poles = 4;	DC excitation voltage = 125.0 V	
Equivalent Circuit Values (R, X in Ω)		
R_a = 0.0;	K_{RL} = 0.0844 N m s/rad;	P_{RL} = 2998.8 W
X_1 = 0.2304;		R_F = 7.8125
X_d = 2.534;		X_q = 1.647
Note: If machine is NONSALIENT, $X_d = X_q$.		
MagC at 1800 rpm: $I_F = E_f/K_{ag} + A_x \exp(B_x * E_f)$.		
K_{ag} = 72.17 Ω;	A_x = 0.009519 A;	B_x = 0.0133411 1/V;
Note: A_x = 0 indicates that the MagC has been linearized.		

For operation as a generator, at rated stator output, pf = 0.866 lagging, find (a) δ; (b) all machine currents and voltages (draw the stator phasor diagram), powers, and torques; (c) the efficiency; (d) repeat (a) through (c) for pf = 0.866 leading; (e) for operation as a motor, at rated stator output, draw the stator phasor diagram for pf = 0.866 lagging and leading; and (f) for E_F = 273 V, consider the P–δ plots.

SOLUTION

(a) Observe that $\bar{V}_a = \bar{V}_{an} = 277.1\angle 0°$. Locating the d and q axes:

$$jX_q\bar{I}_a + \bar{V}_a = j(1.647)(120.3\angle -30°) + 277.1\angle 0° = 413.5\angle 24.5°$$

Hence, the q-axis is located at 24.5° and the d-axis at 24.5° + 90° = 114.5°

(b) The d and q currents can easily be calculated:

$$I_d = I_a \sin(30 + 24.5°) = 97.94\text{ A}; \qquad \bar{I}_d = 97.94\angle -65.5°\text{ A}$$
$$I_q = I_a \cos(30 + 24.5°) = 69.81\text{ A}; \qquad \bar{I}_q = 69.81\angle +24.5°\text{ A}$$

$$\bar{E}_f = jX_d\bar{I}_d + jX_q\bar{I}_q + \bar{V}_a = 500.4\angle 24.5°\text{ V}$$

Solving Equation 7.4 for the field current,

$$I_F = 14.48\text{ A}; \qquad V_F = I_F R_F = (14.48)(7.8125) = 113.1\text{ V}$$

Solving for the powers,

$$\bar{S}_{3\,ph} = 100\angle 30° = 86.6 + j50$$
$$\bar{P}_{out} = 86.6\text{ kW}; \qquad Q_{out} = 50\text{ kvar}$$

$$\text{Stator winding loss} = \text{SWL} = 3\, I_a^2\, R_1 = 0$$

$$\text{Rotor winding loss} = \text{Field winding loss} = \text{FWL} = I_F^2\, R_F = (14.48)^2\,(7.8125) = 1.638\text{ kW}$$

$$\text{Rotational loss} = P_{\text{RL}} = K_{\text{RL}}\omega_{\text{rm}} = (0.06299)(188.5) = 2.999\,\text{kW}$$

$$\text{Total machine loss} = P_{\text{LOSS}} = P_{\text{RL}} + \text{SWL} + \text{RWL} = 1.638 + 2.999 = 4.637\,\text{kW}$$

$$\text{Developed power} = P_{\text{dev}} = 86.60\,\text{kW}$$

$$\text{Mechanical power} = P_{\text{m}} = P_{\text{dev}} + P_{\text{RL}} = 89.60\,\text{kW (120.1 hp)}$$

$$\text{Power input} = P_{\text{m}} + \text{FWL} = 91.24\,\text{kW}$$

Solving for the machine torques,

$$\text{Developed torque} = T_{\text{dev}} = P_{\text{dev}}/\omega_{\text{rm}} = 86.60/188.5 = 459.4\,\text{N m}$$

$$\text{Rotational loss torque} = T_{\text{RL}} = P_{\text{RL}}/\omega_{\text{rm}} = 2.999/188.5 = 15.9\,\text{N m}$$

$$\text{Mechanical input torque} = T_{\text{m}} = T_{\text{dev}} + T_{\text{RL}} = 475.3\,\text{N m}$$

The relevant phasor diagram is shown in Figure 7.12a.

(c) Efficiency $= \eta = P_{\text{out}}/P_{\text{in}} = 86.60/91.24 = 94.92\%$

(d) The analysis is repeated for pf = 0.866 leading. Results are Locating the d and q axes:

$$jX_q\bar{I}_{\text{a}} + \bar{V}_{\text{a}} = j(1.647)(120.3\angle{+30°}) + 277.1\angle 0° = 247.3\angle 43.9°$$

Hence, the q-axis is located at 43.9° and the d-axis at 43.9 + 90° = 133.9°
The d, q currents can easily be calculated:

$$\bar{I}_d = 28.97\angle{-46.1°}\text{ A}; \qquad \bar{I}_q = 116.7\angle{+43.9°}\text{ A}$$
$$\bar{E}_{\text{f}} = jX_d\bar{I}_d + jX_q\bar{I}_q + \bar{V}_{\text{a}} = 273.0\angle 43.9°\text{ V}$$

Solving Equation 7.4 for the field current,

$$I_{\text{F}} = 4.15\text{ A}; \qquad V_{\text{F}} = I_{\text{F}}R_{\text{F}} = 32.4\text{ V}$$

Solving for the powers,

$$\bar{S}_{\text{3ph}} = 100\angle 30° = 86.6 + j50$$
$$\bar{P}_{\text{out}} = 86.6\text{ kW}; \qquad Q_{\text{out}} = 50\text{ kvar}$$
$$\text{SWL} = 3I_{\text{a}}^2R_1 = 0 \qquad \text{FWL} = I_{\text{F}}^2R_{\text{F}} = 0.134\text{ kW}$$

$$\text{Rotational loss} = 2.999\,\text{kW}; \quad P_{\text{LOSS}} = 3.133\,\text{kW}$$

$$\text{Developed power} = P_{\text{dev}} = 86.60\,\text{kW}$$

$$\text{Mechanical power} = P_{\text{m}} = P_{\text{dev}} + P_{\text{RL}} = 89.60\,\text{kW (120.1 hp)}$$

$$\text{Power input} = 89.73\,\text{kW}$$

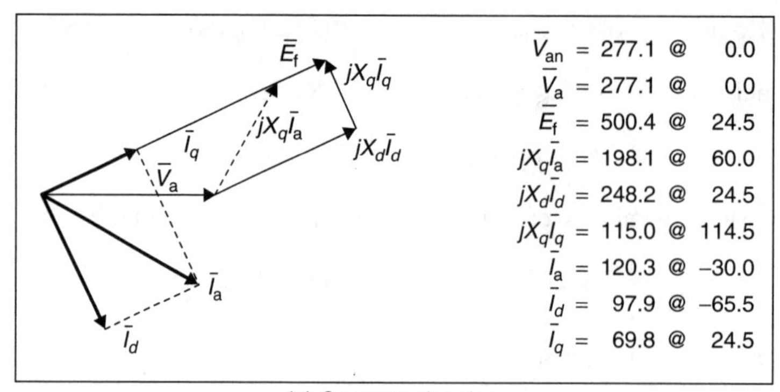

(a) Generator lagging

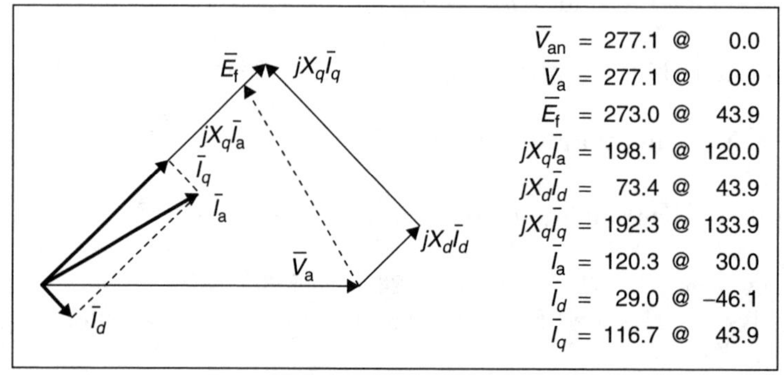

(b) Generator leading

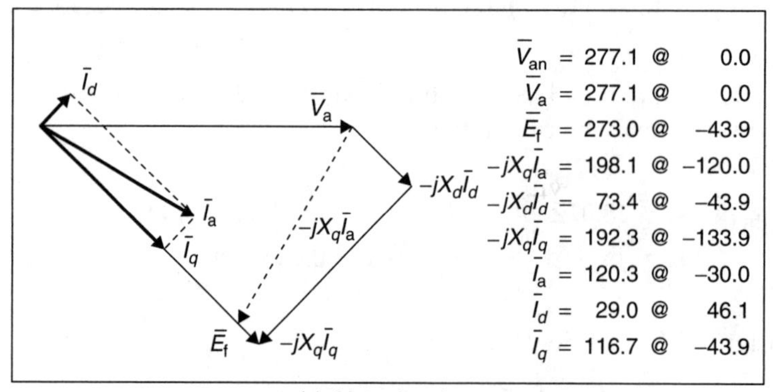

(c) Motor lagging

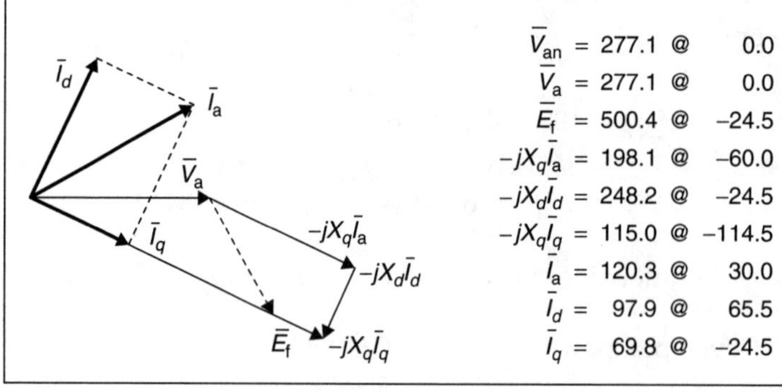

(d) Motor leading

FIGURE 7.12. Phasor diagrams for Example 7.4.

$$T_{\text{dev}} = 459.4\,\text{N m} \quad T_{\text{RL}} = 15.9\,\text{N m} \quad T_{\text{m}} = T_{\text{dev}} + T_{\text{RL}} = 475.3\,\text{N m}$$

The relevant phasor diagram is shown in Figure 7.12b.

$$\text{Efficiency} = \eta = P_{\text{out}}/P_{\text{in}} = 96.51\%.$$

(e) For motor operation, use motor coordinates, as presented in Figure 7.7. The phasor diagrams are presented in Figure 7.12c and Figure 7.12d.
(f) Consider the P–δ relations

$$E_{\text{f}} = 273.0\,\text{V}; \qquad V_{\text{a}} = 277.1\,\text{V}$$

$$P_1 = (3E_{\text{f}}V_{\text{a}}/X_d)\sin\delta = 89.56\sin\delta$$

$$P_2 = 3(X_d - X_q)(V_{\text{a}})^2/(2X_d X_q) = 24.4753\sin 2\delta$$

$$P_{\text{dev}} = P_1 + P_2;\ P\text{'s are in \% of } S \text{ rated (100.0 kVA)}$$

δ	P_1	P_2	P_{dev}	δ	P_1	P_2	P_{dev}
0	0.00	0.00	0.00	90	89.56	0.00	89.56
10	15.55	8.37	23.92	100	88.20	−8.37	79.83
20	30.63	15.73	46.36	110	84.16	−15.73	68.42
30	44.78	21.20	65.97	120	77.56	−21.20	56.36
40	57.57	24.10	81.67	130	68.61	−24.10	44.50
50	68.61	24.10	92.71	140	57.57	−24.10	33.46
60	77.56	21.20	98.76	150	44.78	−21.20	23.58
70	84.16	15.73	99.89	160	30.63	−15.73	14.90
80	88.20	8.37	96.57	170	15.55	−8.37	7.18
90	89.56	−0.00	89.56	180	−0.00	0.00	0.00

Plots of P_1, P_2, and P_{dev} are provided in Figure 7.13.

The P–δ plots of Example 7.4f can be used to make some general points. The plots can be used for torque as well as power, since power and torque are proportional for synchronous operation. Note that saliency adds the $\sin 2\delta$ term to the $\sin\delta$ term, providing somewhat more synchronous power (torque) transfer capability. The maximum power transfer point is sometimes called the steady-state stability limit. For the machine of Example 7.4, this occurs at point 1 ($P = 100\%$; $\delta_1 = 67°$). For a comparable nonsalient machine, we would use plot P1, for which the limit occurs at point 2 ($P = 90\%$; $\delta_1 = 90°$). Note that the salient machine has some power transfer capability, even if the rotor field is zero (see plot P2, steady-state stability limit: $P = 24\%$; $\delta_1 = 45°$).

For the salient machine of Example 7.4, motor operation, with $E_{\text{F}} = 273\,\text{V}$, consider a total load of $P'_{\text{m}} = 60\%$. Note that there are apparently two possible δ values at this load level (points 3 and 4, $\delta = 27$ and 117°). Both locations are called equilibrium values, since the machine is in torque (or power) balance at both points. Now, consider that we are at point 3, and do to some transient disturbance, the system moves slightly to the right (e.g., to point 3′, $\delta = 30°$). Clearly, the developed torque increases, upsetting the balanced condition, and accelerating the rotor. Since we are in the motor mode, this closes (decreases) the angle δ, and the system is driven back to point 3. However, consider that we are point 4, and the system moves to nearby point 4′. Now the developed torque decreases, decelerating the rotor. This opens (increases) the angle δ, and the system is driven away from point 4. Hence, the machine cannot operate at point 4, and it represents an unstable condition. Therefore, points 3 and 4 are called points of stable and unstable equilibrium, respectively.

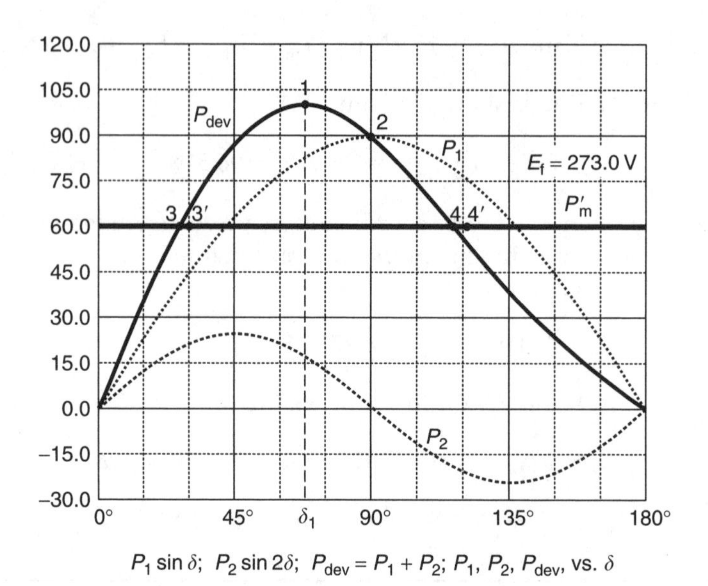

$P_1 \sin\delta$; $P_2 \sin 2\delta$; $P_{dev} = P_1 + P_2$; P_1, P_2, P_{dev}, vs. δ

FIGURE 7.13. Power transfer as rotor angle (δ) varies.

7.7 Synchronous Machine Constants from Tests

As we noted for the induction machine, mathematical models have little practical value unless it is possible to obtain accurate values for the required element constants. It is relatively straightforward to compute such constants from data acquired from certain fairly simple tests. The required constants are the element values (R_1, X_d, X_q, R_F, X_2'), the constant K_{RL} and the MagC. Figure 7.14 summarizes the laboratory setups for the required tests.

7.7.1 The DC Test: R_a

By applying a DC source between terminals (a-b, b-c, c-a, the neutral is usually not available), and measuring the DC voltage and current, the DC stator resistance can be determined. Considering a wye connection, the resistance per phase is half of that measured. Multiplication by two correction factors, for temperature and for AC skin effect, produces a value for R_1:

$$R_{ab} = \frac{V_{ab}}{I_{ab}}; \quad R_{bc} = \frac{V_{bc}}{I_{bc}}; \quad R_{ab} = \frac{V_{ca}}{I_{ca}}; \quad R_{dc} = \frac{R_{ab} + R_{bc} + R_{ca}}{6};$$

$$R_1 = K_{TEMP} K_{ac} R_{dc}$$

$K_{TEMP} K_{ac}$ can vary from about 1.2 to 2.5, depending on conductor temperature and geometry. For rough work, K_{TEMP} K_{ac} = 1.5 is a reasonable guess.

7.7.2 The Open-Circuit Test: The Magnetization Characteristic and Field Resistance

See Figure 7.14a for the laboratory setup. Consider measuring the stator terminal voltage with the stator at no load ($I_a = 0$) as the (rotor) field current is varied from zero to its rated value, with the machine running at synchronous speed. Recall that a plot of these measurements is referred to by various terms, including the magnetization characteristic, the

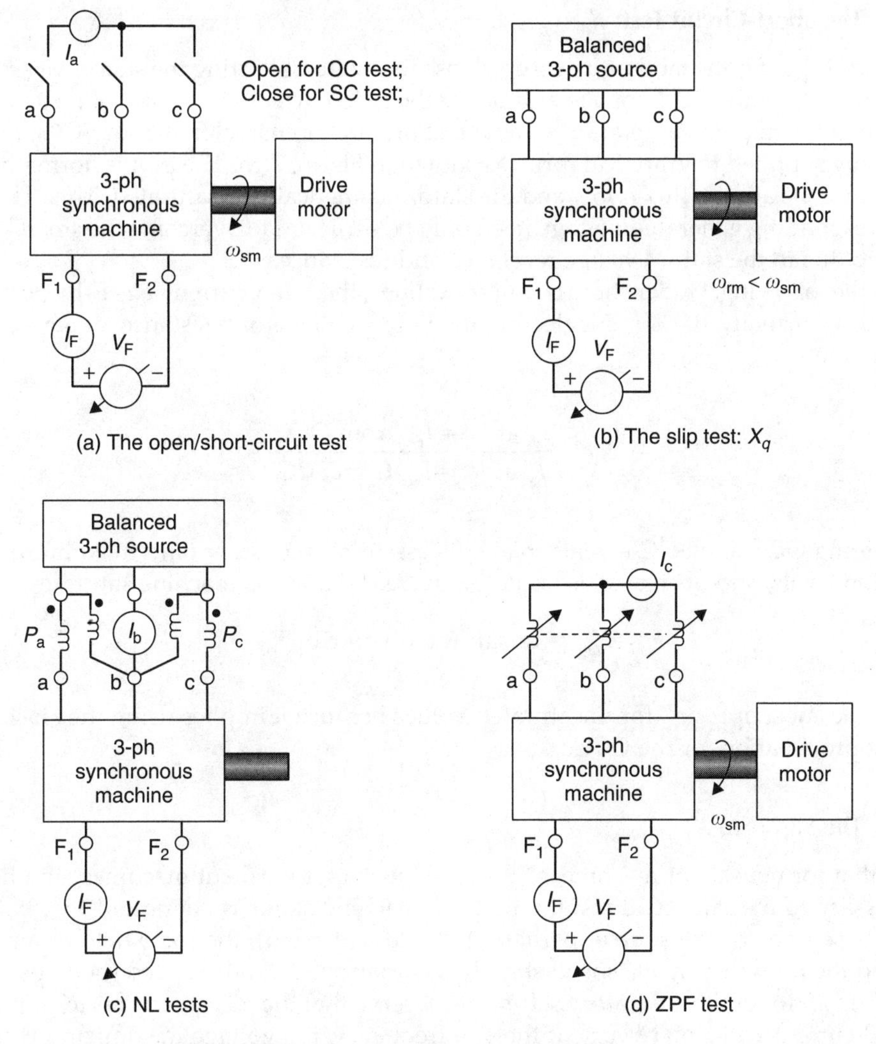

FIGURE 7.14. Three-phase synchronous machine testing.

saturation curve, the open-circuit characteristic, and the no-load characteristic.[12] We shall use the term "magnetization characteristic", and abbreviate as "MagC" (or "OCC"). If the DC field voltage is also measured, the ratio of field voltage to field current is the field resistance (R_F). It is advisable to measure R_F at three points (low, medium, and high), and average the results. Recall that the MagC can be represented adequately with the following empirical equation, as discussed in Section 7.3:

$$I_F = \frac{E_f}{K_{ag}} + A_x e^{B_x E_f} \tag{7.4}$$

The procedure for computing K_{ag}, A_x, and B_x was illustrated in Example 1.4. Also see Example 7.5.

[12] The line (phase-to-phase) voltage is commonly plotted by manufacturers. However, since the phase-to-neutral voltage is more germane to our model, we shall use the latter. The conversion is simple: $V_{LINE} = \sqrt{3}V_{PHASE}$.

7.7.3 The Short-Circuit Test: X_d

See Figure 7.14a for the laboratory setup. Close the switch, shorting the stator. Measure the stator currents as the field voltage and hence the field current (I_F). A plot of I_a *vs.* I_F under short-circuit conditions is referred to as the short-circuit characteristic (or SCC), and is a commonly supplied in graphical form, as shown in Figure 7.15. The plot is normally quite linear, since the air-gap flux is low and the stator magnetically unsaturated. This is because the corresponding generated voltage need only be sufficient to overcome the small ZI voltage drop due to the stator leakage reactance and resistance.

Note the following. Under short-circuit conditions, the stator current lags E_f by nearly 90°. Hence, I_a is virtually 100% I_d. Neglecting the very small stator resistance, at a given level of excitation:

$$X_d = \frac{E_f \text{ at } I_F = I_{Fo} \text{ from OCC}}{I_a \text{ at } I_F = I_{Fo} \text{ from SCC}}$$

Considering OCC and SCC reveals that X_d is essentially constant (and maximum) at low excitation levels, and decreases as excitation increases and the machine saturates.

$$X_{dag} = \text{unsaturated value of } X_d$$

We use the subscript "ag" for unsaturated values because it implies using the air-gap line as an approximation for the OCC.

7.7.4 The Slip Test: X_q

Recall that for nonsalient machines, $X_q = X_d$. However, for salient machines, another test is necessary to measure X_q. Presume that the machine stator is connected to a balanced three-phase low-voltage source (perhaps 10 to 20% of rated), the rotor field on open circuit, and the rotor driven at a speed slightly less than synchronous speed (perhaps 99.9%). See Figure 7.14b for the laboratory setup. We observe that the measured rms terminal voltage and current will slowly vary at the slip frequency, the voltage maximizing when the current minimizes, and vice versa. What is happening is that the air-gap flux is slowly drifting at slip speed from the d to the q axes. Hence the machine impedance is slowly varying from 100% X_d to 100% X_q (neglecting stator resistance). It follows that

$$X_{dag} = \frac{V_{an} \text{ (maximum)}}{I_a \text{ (minimum)}}; \qquad X_{qag} = \frac{V_{an} \text{ (minimum)}}{I_a \text{ (maximum)}}$$

We define

$$K_{qd} = \frac{X_{qag}}{X_{dag}}$$

which in effect relates X_q to X_d (for unsaturated conditions). Frequently, the ratio is assumed to be independent of saturation, providing

$$X_q = K_{qd} X_d$$

7.7.5 The No-Load Test: Rotational Losses

See Figure 7.14c for the laboratory setup. Apply a balanced AC rated voltage source at the stator terminals (a-b, b-c, c-a, the neutral is usually not available), synchronize the machine to the source (run as a motor) and measure the voltage, current, and power. Since the output is zero, the input power consists of only the winding losses plus the rotational loss.

$$P_{NL} = P_a + P_c$$

$$P_{RL} = P_{NL} - (3I_{NL}^2)R_1$$

and

$$K_{RL} = P_{RL}/(\omega_{rm})^2 \approx P_{RL}/(\omega_{sm})^2$$

As was discussed earlier, the rotational losses are a combination of magnetic losses, mechanical losses, and stray loss. In our machine model, there is no incentive to separate the rotational losses into its constituent parts. However, if this is desirable, we can proceed as follows. Under no-load test conditions, the stray losses are negligible. If the excitation is varied, the rotational losses will vary as well. If we extrapolate the rotational loss to zero, what remains should be the mechanical losses, and will be constant with excitation. At nonzero excitation levels, the difference between the rotational losses and the mechanical losses will be the magnetic losses. These magnetic losses can be further subdivided into hysteresis and eddy current and hysteresis components, understanding that these vary with the first and second power of the field strength, respectively.

7.7.6 The Zero-Power Factor Lagging Test: X_l

Terminate the synchronous machine to be tested with a balanced three-phase adjustable inductor, as shown in Figure 7.14d.[13] Starting from a short circuit, increase the test machine field current ($I_F = I_{Fo}$) until the stator current is rated ($I_a = I_d$, since the current lags the voltage by 90°). Record the field current and terminal voltage. Increase the terminating impedance by a small amount, and increase I_F until the stator current returns to rated. Repeat, plotting the resultant data to form the curve shown in Figure 7.15 (Curve ZPF).

The data may be interpreted as follows. At zero terminal voltage, I_{Fo} must correspond to the mmf needed to supply the voltage needed to balance the X_dI_d voltage drop (neglecting stator resistance, as usual). Recall that

$$X_d = X_l + \hat{X}_d$$
$$I_{Fo} = I_{Fl} + \hat{I}_{Fo}$$

where

I_{Fl} is proportional to the mmf necessary to overcome X_lI_d drop
$\hat{I}_{Fo}$ is proportional to the mmf necessary to overcome $\hat{X}_dI_d$ drop

[13] A similar (identical, if possible) synchronous machine with adjustable excitation may be used as an adjustable inductive load instead, and may be more practical for large machines.

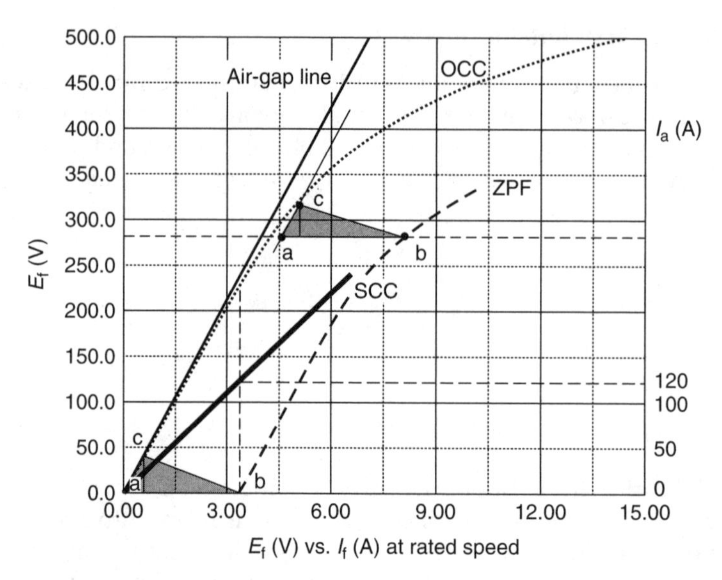

FIGURE 7.15. Synchronous machine characteristic curves.

Note that the altitude of the triangle abc (the Potier triangle) is in fact $X_d I_d$. Also note that if we slide point c upwards along the OCC keeping the abc triangle rigid, (1) point b should trace out the ZPF curve, and (2) side ac is parallel to the air-gap line. This insight provides us with a method (Potier's method) for determining the leakage reactance graphically from the machine characteristics:

1. At zero voltage, read I_{Fo} = base ab on the Potier triangle.
2. Draw a horizontal line on the machine characteristics at rated voltage.
3. On (2), lay off "ab," locating point "b" on the ZPF curve.
4. Construct a line with the slope of the air-gap line passing through point "a."
5. Line (4) should intersect the OCC at point "c."
6. The altitude of triangle abc is $X_d I_d$:

$$X_l = \frac{(X_l I_{\text{Rated}})}{I_{\text{Rated}}}$$

The reactance thus determined is sometimes called the "Potier reactance," and serves as Potier's estimation of the leakage reactance. As is our custom, we will apply these ideas in an example.

Example 7.5

Consider a 480 V 10 kVA 60 Hz four-pole salient synchronous machine, which is similar to that of Example 7.4. The field is rated at 125 V DC. Test data are provided below.

Determine the machine constants from the machine data and characteristics.

SOLUTION

(a) The DC Test (measurements made at 120.3 A (rated current)):

$$V_{ab} = 7.39 \text{ V};\ V_{bc} = 7.40 \text{ V};\ V_{ca} = 7.39 \text{ V}$$

$$R_{ab} = \frac{V_{ab}}{I_{ab}} = \frac{7.39}{120.3} = 0.06143\ \Omega$$

$$R_{bc} = \frac{7.40}{120.3} = 0.06151\,\Omega; \qquad R_{ca} = \frac{7.39}{120.3} = 0.06143\,\Omega$$

$$R_{dc} = \frac{R_{ab} + R_{bc} + R_{ca}}{2 \cdot 3} = 0.03073\,\Omega$$

$$R_1 = K_{TEMP} K_{ac} R_{dc} = 1.5(0.03073) = 0.0461\,\Omega$$

(b) The Open circuit test: MagC and field resistance

OOC			ZPF (I_F at E_f)
I_F	E_f	V_F	I_F
0	0	0	3.3
1	70.4	7.81	4.3
2	139.4	15.6	5.3
3	205.8	23.4	6.5
4.22	277.1	33.0	7.7
6	354.9	46.9	9.6
8	411.3	62.5	
10	448.6	78.1	
12	475.5	93.8	
14	496.1	109.4	

The MagC (OCC) and ZPF curve are plotted in Figure 7.15.
Finding R_F:

$$\text{at } I_F = 1\text{ A:} \quad R_F = \frac{V_F}{I_F} = \frac{7.81}{1} = 7.81; \qquad \text{at } I_F = 8\text{ A:} \quad R_F = \frac{62.5}{8} = 7.813;$$

$$\text{at } I_F = 14\text{ A:} \quad R_F = \frac{109.4}{14} = 7.814; \qquad \text{Averaging:} \quad R_F = 7.812\,\Omega$$

The procedure of Example 1.4 is used to evaluate MagC:

$$K_{ag} = 72.17\,\Omega; \qquad B_x = 0.01334\text{ V}^{-1}; \qquad A_x = 0.009519\text{ A}$$

The potier triangle is constructed in Figure 7.15. Reading its altitude:

$$X_l = \frac{41}{120} = 0.34\,\Omega$$

(c) The Short-Circuit Test: X_d
SC measurements (SCC is linear):

I_F	I_a
0	0
6.18 A	120 A

Plotting I_a vs. I_F produces the SCC (see Figure 7.15). Finding X_d:

$$X_d = \frac{E_f\text{ (from OCC at } I_F = 3.26\text{ A)}}{I_a\text{ (from SCC at } I_F = 3.26\text{ A)}} = \frac{221.7}{120.3} = 1.843\,\Omega$$

$$X_{dag} = \frac{E_f\text{ (from air-gap line at } I_F = 6.18\text{ A)}}{I_a\text{ (from SCC at } I_F = 6.18\text{ A)}} = \frac{235.3}{120.3} = 1.956\,\Omega$$

$$k_{dsat} = \frac{X_d}{X_{dag}} = \frac{1.843}{1.956} = 0.9424$$

(d) The Slip Test: X_q

The following waveforms were measured in the slip test at low excitation level:

$$X_{dag} = \frac{V_{an}\text{ (maximum)}}{I_a\text{ (minimum)}} = \frac{100}{51} = 1.96$$

$$X_{qag} = \frac{V_{an}\text{ (maximum)}}{I_a\text{ (minimum)}} = \frac{81}{64} = 1.27$$

$$K_{qd} = \frac{X_{qag}}{X_{dag}} = \frac{1.27}{1.96} = 0.65$$

Hence $X_q = K_{qd}X_d = 0.65(1.843) = 1.198\,\Omega$

(e) The No-Load Test: Rotational losses

Running the machine as an unloaded motor at unity pf:

$$V_{ab} = 480\text{ V},\quad I_a = 3.6\text{ A},\quad P_a = 1501,\quad P_c = 1501$$

$$P_{NL} = P_a + P_c = 3002$$

$$P_{RL} = P_{NL} - (3I_{NL}^2)R_1 = 3002 - 3 = 2999\text{ W}$$

$$K_{RL} = P_{RL}/(\omega_{sm})^2 = 0.0844$$

Results are summarized in the following table.

Three-Phase Synchronous Machine Data		
	Ratings	
VLine = 480 V;	ILine = 0.0 A;	S_{3ph} = 100.0 kVA
Frequency = 60 Hz;	Poles = 4;	Sync speed = 1800.0 rpm
Rotor type: SALIENT;	DC field voltage = 125.0 V	
	Equivalent Circuit Values (R, X in Ω)	
$R_a = 0.0461$;	$T_{RL} = K_{RL} * \omega_{rm}$;	K_{RL} = 0.0844 N m s/rad
$X_d = 1.843$;	$X_q = 1.198$	
$X_l = 0.3456$;	$R_F = 7.8125$	
MagC at 1800 rpm:	$I_F = E_f/K_{ag} + A_x\exp(B_x * E_f)$.	
$K_{ag} = 72.17\,\Omega$	A_x = 0.009519 A;	B_x = 0.01334111/V;

Thus, the necessary machine constants for balanced three-phase steady-state constant speed operation can be determined.[14]

[14] These are by no means all the constants required for general machine analysis, a discussion of which is continued in Chapter 8.

Consideration of magnetic saturation complicates machine modeling. If saturation is to be considered, we start by separating the leakage reactance from X_d, according to

$$\hat{X}_{dag} = X_{dag} - X_l$$
$$\hat{X}_{qag} = X_{qag} - X_l$$

We argue that the leakage reactance does not saturate since a large part of the related magnetic flux paths are in air. Now

$$\begin{aligned}
\hat{X}_d &= k_{dsat}\,\hat{X}_{dag} \\
\hat{X}_q &= k_{qsat}\,\hat{X}_{qag} \\
X_d &= \hat{X}_d + X_l \\
X_q &= \hat{X}_q + X_l \\
k_{dsat} &= d\text{-axis saturation factor} \leq 1 \text{ ("1" means unsaturated)} \\
k_{qsat} &= q\text{-axis saturation factor} \leq 1
\end{aligned}$$

The issue then is to accurately compute the d and q saturation factors. What is required is to accurately determine the air-gap flux at a given operating condition. The voltage E_r is that which is most nearly representative of the true air-gap flux. Hence, we compute:

$$\bar{E}_r = (R_a + jX_l)\,\bar{I}_a + \bar{V}_{an} \quad \text{(generator coordinates)}$$

Locate the d–q axes. Then compute

$$\begin{aligned}
E_d &= E_r \cos\beta = \text{projection of } \bar{E}_r \text{ onto the } d\text{-axis} \\
E_d &= E_r \sin\beta = \text{projection of } \bar{E}_r \text{ onto the } q\text{-axis}
\end{aligned}$$

Recall that there is a 90° phase shift between mmf and the voltage created by that mmf. That is, it provides a q-axis voltage proportional to air-gap flux for an mmf acting on the d-axis. Also, note that the traditional OCC is the *d-axis OCC*. Then

$$k_{dsat} = \frac{E_q = \text{voltage on the } d\text{-axis OCC } (I_F = I_{F1})}{E_q \text{ on the air-gap line at } I_F = I_{F1}}$$

$$k_{qsat} = \frac{E_d = \text{voltage on the } q\text{-axis OCC } (I_F = I_{F2})}{E_d \text{ on the air-gap line at } I_F = I_{F2}}$$

To properly compute saturation along the q-axis, we need a *q-axis OCC*, which is difficult to obtain and not routinely available. Denied access to the q-axis OCC, analysts normally (1) ignore q-axis saturation altogether, or (2) use the d-axis OCC as an approximation for the q-axis OCC. As a coarser approximation, one can simply calculate

$$k_{sat} = \frac{E_r = \text{voltage on the } d\text{-axis OCC } (I_F = I_{F1})}{E_r \text{ on the air-gap line at } I_F = I_{F1}}$$

$$X_d = k_{sat}\hat{X}_{dag} + X_l$$

and either ignore q-axis saturation altogether, or use the same "k_{sat}" to saturate X_q. There is no universally "best" method; one would select the simplest method to implement that meets the accuracy requirements of the application.

7.7.7 Vee Curves

Consider a nonsalient pole lossless synchronous machine operating as a motor at about 50% rated load. The simplified circuit model is shown in Figure 7.16a. The power transferred from system to machine is

$$P_{\text{dev}} = 3\frac{E_F V_a \sin\delta}{X_d} = P_m$$

Application of KVL produces

$$\bar{V}_{\text{an}} = jX_d\bar{I}_a + \bar{E}_f, \quad \text{leading to} \quad \bar{I}_a = \frac{\bar{V}_{\text{an}} - \bar{E}_f}{jX_d}$$

Now consider three levels of excitation: (1) low, (2) medium, and (3) high. The corresponding phasor diagrams are shown in Figure 7.16b. Observe that the corresponding current levels are (1) high, (2) low, and (3) high. A plot of I_a vs. I_F is provided in Figure 7.16c. If the situation is related to pf, we note that conditions (1), (2), or (3) correlate to lagging, unity, and leading pf conditions. In particular, note that the current minimizes at the unity pf point (2).

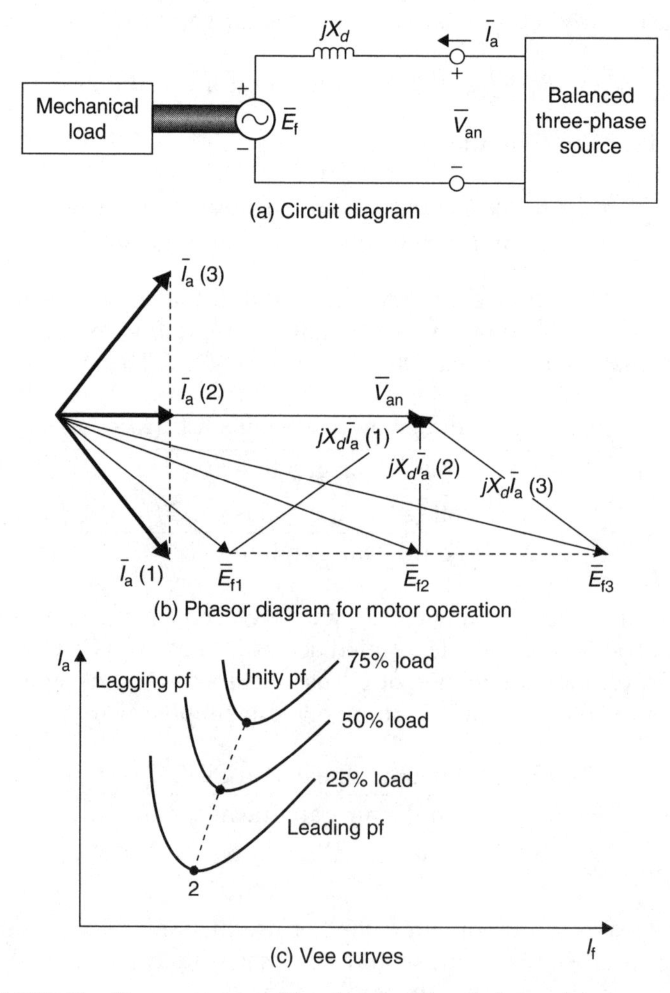

FIGURE 7.16. Vee curves: controlling the operating pf of a synchronous motor.

Change the load level and repeat the analysis. The corresponding plots of I_a vs. I_F are collectively called "Vee" curves, and provide clear insight into the following principle:

> At a given load condition, the operating power factor (and hence the flow of reactive power) may be exclusively controlled by controlling the excitation. Low excitation results lagging pf (Q flow into the motor), and high excitation results leading pf (Q flow out of the motor).

7.8 The Synchronous Generator Operating in a Utility Environment

Arguably, the most important application of the synchronous machine is as a generator in an electric utility environment. Upwards of 97% of the electrical energy supplied by utilities is supplied by synchronous machines.[15] Whether the energy source is hydro, fossil fuel, nuclear, wind, geothermal, or solar-thermal, the synchronous machine is the device used to convert mechanical power in the form of torque applied to a rotating shaft into the electrical form (Figure 7.17). There are several factors that make this application of special interest to our study.

- The stator is terminated into the "power grid," the external power system consisting of all the components (transformers, line, generators, loads, etc.) that make up the system, and which may be modeled as a "Thevenin Equivalent" circuit, whose voltage is independent in magnitude and frequency of our machine.
- The machine is operated as the central component of a double control loop, one which controls the excitation system in order to control the flow of reactive power (Q) into the grid, and a second which acts to control real power flow from machine to grid.

Although a comprehensive study of the operation of this system is beyond the scope of this book, there are some basic points we can consider.

7.8.1 Prime Movers

The generator of course does not create energy; it is an energy *conversion* device. The energy enters the generator in the form of a torque applied by an external device to its rotating shaft. This external device is generically called a prime mover, and normally is a turbine of some type. Prime movers were discussed in some detail in Section 3.9. There are three common situations:

Hydraulic turbines

- Low head plants ($H < 50\,\text{m}$)
- Medium head plants ($50 < H < 250\,\text{m}$)
- High head plants ($H > 250\,\text{m}$)

Steam turbines

- Fossil fuel technologies
  - Coal-fired boilers
  - Oil-fired boilers
  - Natural-gas-fired boilers

[15] An important exception is solar-electric or photovoltaic energy production, which converts solar radiation directly into electricity. Although this technology shows great promise and is continuing to develop, at this time of writing, the total energy production is rather small.

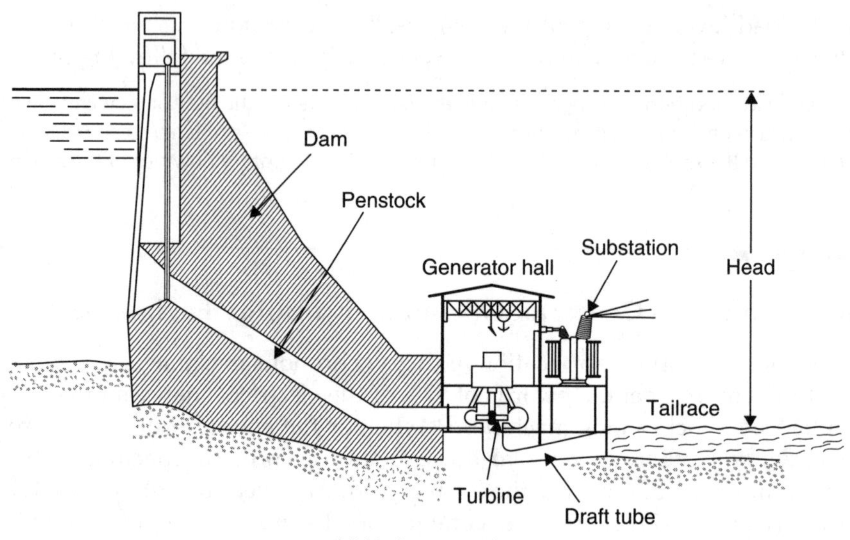

(a) Hydrogeneration

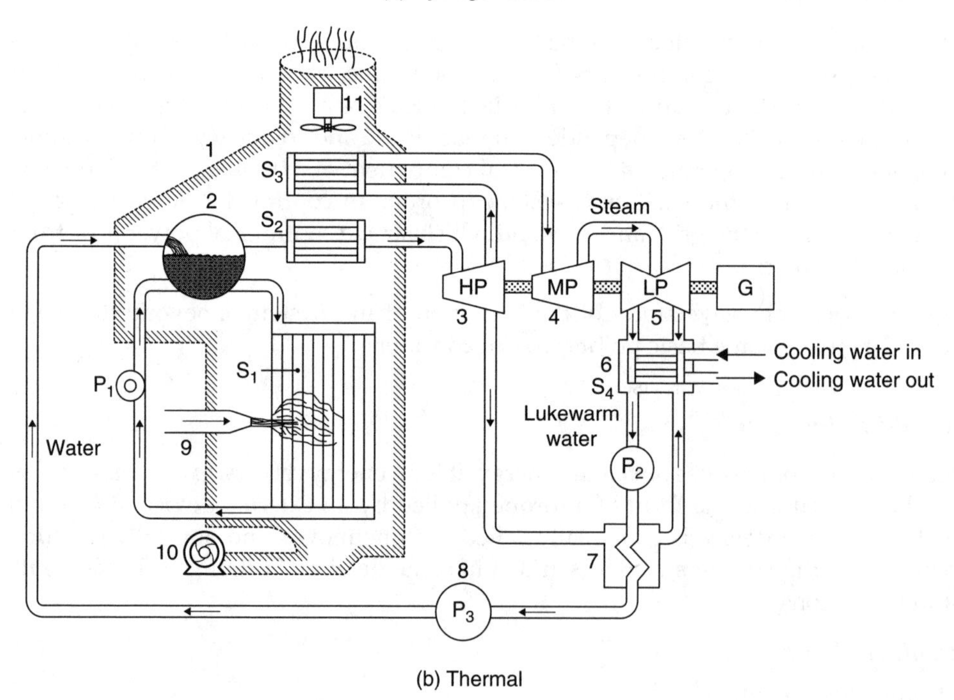

(b) Thermal

FIGURE 7.17. Electric power generation.

- Nuclear fission
- Geothermal steam generation
- Solar-thermal steam generation

Gas turbines

7.8.2 Excitation Systems

We have noted that a controllable DC source is required to supply the machine field current. Consider a 100 MVA machine. If 1% of this power is required for the field supply, this

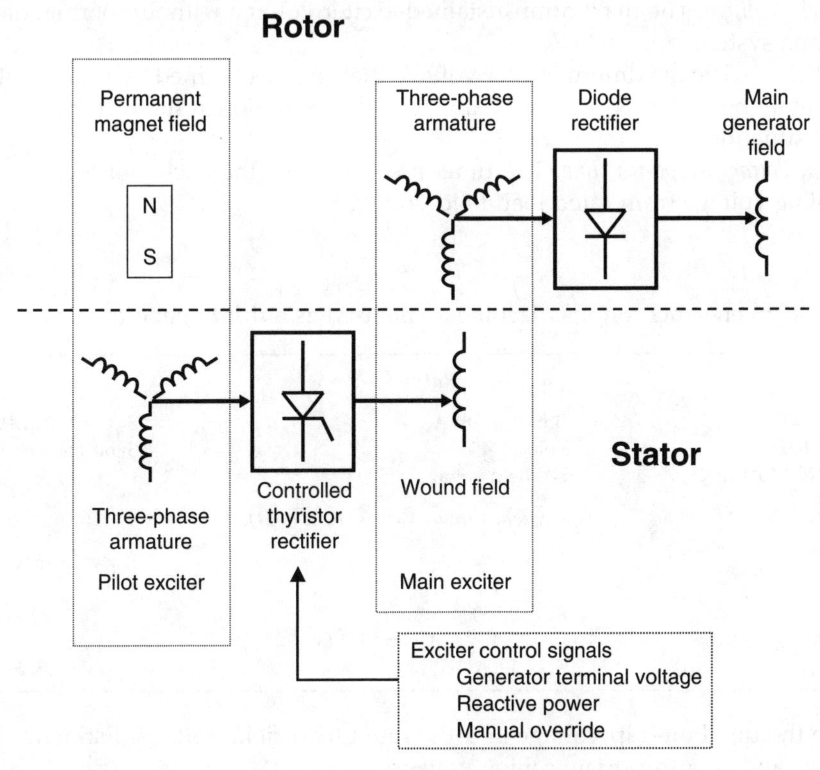

FIGURE 7.18. A brushless excitation system.

amounts to 1 MW. At 500 V, this amounts to 2000 A. Thus, what is required is a variable 500 V 2000 A DC source. Likewise, the response time is an important issue when one considers the dynamic response of a generating unit. Realizing that large generators may be rated upwards of 1000 MVA, we can appreciate that this subsystem of the generator, called an "excitation system," is quite important, and presents a major engineering design challenge in its own right.

There are several designs that solve this problem. One approach, a so-called brushless excitation system, is represented in Figure 7.18. The pilot exciter is a small permanent magnet rotor three-phase synchronous generator that supplies an unregulated low power balanced three-phase output voltage, which is converted into controlled DC using a three-phase thyristor (or IGBT) bridge rectifier. Note that the controlled rectifier is on the stator. This DC output drives the field of the exciter, a medium power three-phase synchronous generator, whose three-phase windings are on the rotor. The exciter output feeds a three-phase diode bridge rectifier, which supplies the generator field winding. Since the rectifier is on the rotor, its output can be directly connected to the main generator field, without the need for brushes.

Field control is exercised with the controlled thyristor rectifier. For a 100 MVA generator, the exciter might be rated at 1 MVA, and the pilot exciter at 10 kVA. There are several exciter ratings of interest:

Rated air-gap voltage: The exciter voltage that produces rated generator voltage on the generators air-gap line.

Rated load field voltage: The exciter voltage that produces rated generator stator terminal conditions.

Rated Field Voltage: The maximum sustained exciter voltage without thermal damage to the excitation system.

Ceiling Voltage: The maximum exciter voltage that can be obtained by the excitation system that cannot be sustained without damage to the excitation system. To be used only in a transient situation.

Excitation voltage response time: The time in seconds for the excitation system to reach 95% of ceiling voltage from rated load field voltage.

Example 7.6

Consider the given synchronous machine, to be used as a utility generator.

	Ratings	
VLine = 13.80 kV;	ILine = 4184 A;	S_{3ph} = 100.0 MVA
Frequency = 60 Hz;	Poles = 4;	Sync speed = 1800.0 rpm
Rotor type: NONSALIENT;	DC field voltage = 500 V	
	Equivalent Circuit Values (R, X in Ω)	
$R_a = 0.0095$;	$T_{RL} = K_{RL} * \omega_{rm}$;	$K_{RL} = 28.15\,\text{N m s/rad}$
$X_d = 2.481$;	$X_q = 2.481$	
$X_l = 0.2481$;	$R_F = 0.5000$	
MagC at 1800 rpm;	$I_F = E_f/K_{ag} + A_x \exp(B_x * E_f)$.	
K_{ag} = at 33.2 Ω;	$A_x = 1.8\,\text{A}$;	$B_x = 0.00032511/\text{V}$;

Determine the rated air-gap voltage and the rated load field voltage. Estimate the rated field voltage and an appropriate ceiling voltage.

SOLUTION

At rated voltage ($E_F = 7.967\,\text{kV}$), $I_F = 240\,\text{A}$ from the ag line, so that

$$V_F = R_F I_F = 120\,\text{V} \quad \text{(rated air-gap voltage)}$$

At rated stator conditions ($V_{an} = 7.967\,\text{kV}$; $I_a = 4184$; pf = 0.8660 lagging)

$$E_F = 15.95\,\text{kV. From the MagC, } I_F = 803\,\text{A (neglecting } R_a\text{)}$$

$$V_F = 401\,\text{V} \quad \text{(rated load field voltage)}$$

The exciter is normally rated about 25% above rated load field voltage.

$$V_F = 500\,\text{V} \quad \text{(rated field voltage)}$$

The ceiling voltage is normally about 25% above rated field voltage.

$$\text{Ceiling voltage} = 500 \times 1.25 = 625\,\text{V}$$

7.8.3 Capability Curves

A fundamental constraint on power grid operation is that the total power absorbed must equal the total power generated. In equation form, this appears as

$$P_{GEN} = P_{LOAD} + P_{LOSS}$$

$$Q_{GEN} = Q_{LOAD} + Q_{LOSS}$$

P_{GEN}, Q_{GEN} = Total generated real, reactive power injected into the grid

P_{LOAD}, Q_{LOAD} = Total load real, reactive power extracted from the grid

P_{LOSS}, Q_{LOSS} = Total grid real, reactive power losses

Since P_{GEN} can only come from generators, there is naturally a concern as to how much power we can get from each generator. There is a similar concern with the availability of Q_{GEN}. Hence, generators may be viewed simply as "power pumps," i.e., sources of complex power that inject P and Q into the grid, and the limits on the production of P and Q are of interest.

7.8.3.1 Stator Thermal Limits

Recall that stator winding losses (SWL) are I^2R that occur in the stator conductors, causing a temperature rise. At some point, high temperature will degrade the stator winding insulation, and in extreme cases, will melt the stator conductors. Hence, there must be some upper limit on the stator current. Assuming operation at rated voltage, this translates into a upper limit on stator power:

$$S_{RATED} = \sqrt{3} V_{Lrated} I_{Lrated}$$

7.8.3.2 Rotor Thermal Limits

Likewise, rotor winding losses (RWL = FL) are I^2R that occur in the rotor conductors, causing a temperature rise, potentially damaging the rotor winding insulation. Hence, there must be some upper limit on the rotor current I_F. This translates into a maximum voltage on the MagC. Neglecting stator resistance, the limits on stator power are

$$P = \frac{3E_{f\,max} V_a}{X_d} \sin(\delta) + \frac{3V_a^2 (X_d - X_q)}{2X_d X_q} \sin(2\delta)$$

$$Q = \frac{3E_{f\,max} V_a}{X_d} \cos\delta + \frac{3V_a^2 (X_d - X_q)}{2X_d X_q} \cos(2\delta) - \frac{3V_a^2 (X_d + X_q)}{2X_d X_q}$$

7.8.3.3 End-Region Heating

Under high excitation conditions, the iron at the ends of air gap is saturated so that the leakage field at the end-turn region (the stator winding turns immediately outside of the rotor–stator air gap) is relatively small. Hence, the corresponding eddy current losses are within normal ranges. However, under low excitation, the end-region iron is not saturated, permitting much larger leakage fields. Compounding the problem is that the lagging pf stator current produces an mmf which reinforces the rotor field mmf to produce even larger leakage flux. This large leakage flux produces large localized core loss in the end-region iron. The larger the stator current, and the more it lags the voltage (i.e., the greater the negative Q dispatched), the more pronounced this effect becomes. Thus, this can limit generator operation in the fourth quadrant (i.e., under leading pf conditions). End-region heating can be avoided by limiting Q, according to

$$Q \geq Q_{MIN} \quad \text{where } Q_{MIN} \text{ is negative.}$$

7.8.3.4 *Steady-State Stability Limits*

We noted that as the angle δ increases, we reach the maximum power transfer capability of the generator. Likewise, the $\partial P/\partial\delta$ becomes smaller, reaching zero at the maximum power point. This maximum point is traditionally called the steady-state stability limit. As this point is approached, the generator is increasingly vulnerable to loss of synchronism, which would be catastrophic. As was the case for end-region heating, this effect is most troublesome for generator operation in the fourth quadrant (i.e., under leading pf conditions). Hence, prudence would suggest that the generator should be operated so as to avoid large values of δ, or

$$\delta \leq \delta_{MAX}$$

Neglecting stator resistance, the corresponding limits on stator power are

$$P = \frac{3E_f V_a}{X_d}\sin(\delta_{max}) + \frac{3V_a^2(X_d - X_q)}{2X_d X_q}\sin(2\delta_{max})$$

$$Q = \frac{3E_f V_a}{X_d}\cos(\delta_{max}) + \frac{3V_a^2(X_d - X_q)}{2X_d X_q}\cos(2\delta_{max}) - \frac{3V_a^2(X_d - X_q)}{2X_d X_q}$$

7.8.3.5 *Boiler Thermal Limits*

If a thermal power plant is base-loaded, it is possible to regulate the system such that the temperatures of all thermal components are within normal ranges, and stabilize the associated thermodynamics. However, as we reduce load, this becomes increasingly difficult, and at low power levels, we may lose temperature control. This can result in damage to thermal components, such as boiler slagging. Hence, we may wish to constrain the system to operate at or above some lower power level, so that

$$P \geq P_{MIN}$$

These limits on the production of P and Q are collectively referred to as "capability curves," and are typically provided in graphical form. An example is useful.

Example 7.7

Consider the synchronous machine of Example 7.6. The machine operates as an electric utility generator. Plot the unit's capability curves, if we choose

$$\delta_{MAX} = 60°;\ P_{MIN} = 10\%;\ Q_{MIN} = -60\%;$$

SOLUTION

$E_{fmax} = 15935.3$ V; $V_{an} = 7967.4$ V
δ (max) $= 60.0°$; Q_{3ph} (min) $= -60.0\%$
$X_d = 2.4809\ \Omega$; $X_q = 2.4809\ \Omega$

$S_1 = 3 * E_f * V_a/X_d = 153.53\%$ (of S rated)
$S_2 = 3 * (X_d - X_q) * (V_a)^2/(2X_d * X_q) = 0.00\%$
$Q_o = 3 * (X_d + X_q) * (V_a)^2/(2X_d * X_q) = 76.76\%$
$P_{3ph} = S_1 \sin\delta + S_2 \sin 2\delta$
$Q_{3ph} = S_1 \cos\delta + S_2 \cos 2\delta - Q_o$

δ	P_{3ph}	Q_{3ph}	Limit
0.0	0.00	76.77	Boiler limited
5.2	13.80	76.15	Rotor heating
10.3	27.49	74.29	Rotor heating
15.5	40.95	71.20	Rotor heating
20.6	54.08	66.93	Rotor heating
25.8	66.78	61.48	Rotor heating
30.9	78.94	54.92	Rotor heating
36.1	90.45	42.65	Stator heating
41.3	100.00	0.00	Stator heating
—	100.00	0.00	Stator heating
—	90.45	−24.54	Stability limited
—	78.94	−31.19	Stability limited
—	66.78	−38.21	Stability limited
—	54.08	−45.54	Stability limited
—	40.95	−53.12	Stability limited
—	27.49	−60.00	End-turn heating
—	13.80	−60.00	End-turn heating
—	0.00	−60.00	End-turn heating

The calculated power factor at stator–rotor crossover = 0.8697 lagging.

The Capability Curves are plotted in Figure 7.19.

There is a curious issue raised in conjunction with capability curves, namely the concept of "rated power factor." It relates to the crossover point between stator and rotor limits (see point "x" in Figure 7.16). The P,Q coordinates of this point correspond to an operating power factor, called "rated power factor." In Example 7.7, the rated power factor was 0.8697, lagging.

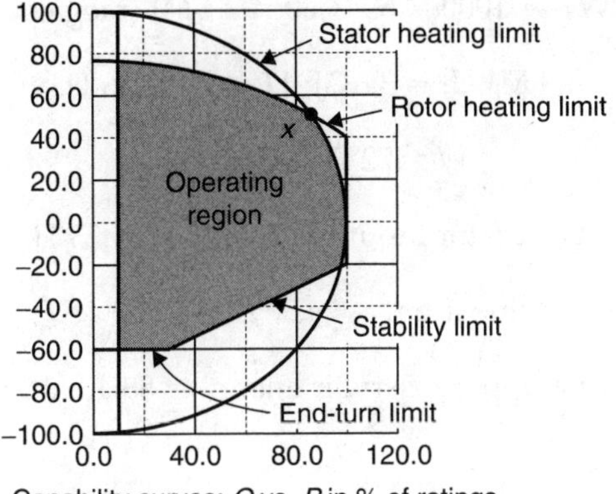

FIGURE 7.19. Capability curves.

7.8.4 Pumped Storage

A second constraint on power grid operation is that the system must be in power balance instant by instant in real time. That is, electrical energy cannot be stored; it must be supplied at the same time that it is used.[16] It is highly desirable to use the system generation capabilities to generate "surplus" energy in times of off-peak load, store it, and arrange to inject this energy back into the grid later during peak load conditions.

An ideal illustration of this concept is the pumped storage application. Here, water is pumped up to an "upper reservoir" during a period of low system load, and use the water as a hydro energy source during peak load conditions. A suitable site requires a mountain and a river or lake not too far from a large load center. Three examples include the Tennessee Valley Authority Raccoon Mountain facility near Chattanooga, Union Electric's Taum Sauk facility in Missouri, and the Glenheim-Bilboa facility, operated by the New York Power Authority in New York. The application is germane to our interests since it serves as an excellent example of a synchronous machine operating in both motor and generator modes.

Example 7.8

A decision has been made to construct a pumped storage facility that can supply 1000 MW for 4 h per day for 5 days (Monday through Friday) weekly. The location is 90 miles north of a major metropolitan area. An average head of 300 m is available. (a) Size the area (in acres) of the upper reservoir (average depth 10 m), given the following data:

Machine efficiency = 97%
Reservoir losses (due to evaporation, leakage, etc.) = 6%
Transport losses (moving water from upper to lower reservoir, or vice versa) = 4%
Turbine efficiency = 89%

(b) Given that we shall use four units, each rated for an output of 250 MW, both in motor and generator mode, develop a schedule for pumping based on 1, 2, 3, or 4 units pumping.

SOLUTION

(a) *For generating mode operation*: Grid energy delivered = $W = 5 \times 1000\,\text{MW} \times 4 = 20\,\text{GW h}$

$$1\,\text{kW h} = 1000 \times 60 \times 60 = 3.6\,\text{MJ (megajoule)}$$

$$1\,\text{MW h} = 3.6\,\text{GJ};\ 1\,\text{GW h} = 3.6\,\text{TJ}$$

$$W = 72\,\text{TJ (terajoules)}$$

$$\text{Generator input} = 72/0.97 = 74.23\,\text{TJ}$$

$$\text{Turbine input} = 74.23/0.89 = 83.40\,\text{TJ}$$

$$\text{Corresponding upper reservoir energy} = 83.40/0.96 = 86.88\,\text{TJ}$$

[16] Strictly speaking, this is not really true. Electrical energy can indeed be stored in the inductive and capacitive elements of the system. Ongoing research in superconducting inductors and "supercapacitors" are expanding our capabilities in these technologies. Still, for the most part, system generation must roughly match system load in real time.

$$\text{Upper reservoir energy, allowing for reservoir losses} = 92.43\,\text{TJ}$$

$$\text{Mass of water} \times g \times \text{head} = Mgh = 93.46\,\text{TJ}$$

$$M = 92.43/(9.802 \times 300) = 31.43 \times 10^9\,\text{kg}$$

$$\rho = \text{density of water} = 1000\,\text{kg/m}^3$$

$$V = \text{volume} = M/\rho = 31.43 \times 10^6\,\text{m}^3$$

$$A = \text{area} = V/d = 3.143 \times 10^6\,\text{m}^2$$
$$3.281\,\text{ft} = 1\,\text{m};\ 10.76\,\text{ft}^2 = 1\,\text{m}^2$$

$$A = 33.82 \times 10^6\,\text{ft}^2$$

There are 43560 ft^2 in an acre.
Therefore $A = 776.4$ acres.

(b) *Pumping (motor) mode operation*: Four units, each rated at 250 MW, will be used. To pump 92.43 TJ into the upper reservoir:

$$\text{Turbine output} = 92.43/0.96 = 96.28\,\text{TJ}$$

$$\text{Motor output} = 96.28/0.89 = 108.2\,\text{TJ} = 30.06\,\text{GWh}$$

$$\text{Motor input} = 108.2/0.97 = 111.5\,\text{TJ} = 30.98\,\text{GWh}$$

Units ON	Pumping Time Required (h)
1	120.2
2	60.11
3	40.07
4	30.06

7.9 Permanent Magnetic Synchronous Machines

For permanent magnet machines, there are no field windings, and hence no excitation system, no field current, and no field losses. Hence, we have no control over the operating power factor. Applications are for the most part motors driven from solid-state drives. Except for field related models, the machine models are the same, except that ϕ, and therefore E_f, is constant.

$$E_f = K_a \phi \omega_{rm} = K_F \omega_{rm}$$

Recall that in Chapters 5 and 6, we investigated induction motor performance when driven from variable voltage, variable frequency sources (solid-state motor drives). It is possible to replace the induction motor with a permanent magnet synchronous motor. An example will demonstrate.

Example 7.9

Consider a synchronous machine as defined in the following table.

Three-Phase Synchronous Machine Data		
	Ratings	
*V*Line = 480 V;	*I*Line = 120.3 A;	S_{3ph} = 100 kVA
Frequency = 60 Hz;	Poles = 4;	Sync speed = 1800 rpm
Rotor type: Permanent magnet		
	Equivalent Circuit Values (R, X in Ω)	
R_a = 0.0000;	$T_{RL} = K_{RL} * \omega_{rm}$;	K_{RL} = 0.0844 N m s/rad
X_d = 2.534;	X_q = 1.647	
X_l = 0.2304	E_f at 1800 rpm = 400 V	

The machine operates as a motor driven from a variable voltage, variable frequency drive, supplying a load which requires 40 kW at 1080 rpm. Determine all currents and the input power and power factor.

SOLUTION

$k = 1080/1800 = 0.6$

Therefore $f = (0.6)60 = 36\,\text{Hz}$ and $V_a = 0.6(277.1) = 166.3\,\text{V}$

Likewise $E_F = 0.6(400) = 240\,\text{V}$

$X_d = 0.6(2.534) = 1.5204\,\Omega; \quad X_q = 0.9882\,\Omega$

Now

$$3E_f V_a/X_d = 78.73\,\text{kW}; \qquad 3(V_a)^2(X_d - X_q)/2X_dX_q = 14.69\,\text{kW}$$

and $78.73 \sin\delta + 14.69 \sin 2\delta = 40 + (0.6)^2(2.999) = 41.08\,\text{kW}$

$\delta = 22.8°$

$I_d = (E_F - V_a \cos\delta)/X_d = 57.06$

$I_q = V_a \sin\delta/X_q = 65.32$

$$I_a = I_d + I_q = 57.06/112.8° + 65.32/22.8°$$
$$= 86.73/18.3°\,\text{A}$$

$\text{pf} = \cos 18.3° = 0.9495$ lead

Likewise $\bar{S}_{IN} = 41.08\,\text{kW} - j13.58\,\text{kvar}$

It was implied in Example 7.9 that our model is valid at any frequency and voltage, just as long as the voltage is balanced three-phase and sinusoidal, and the speed was synchronous. This is correct, since the rotor damper windings experienced no induced voltage under those conditions. However, if the voltage is nonsinusoidal and unbalanced, the model must be modified to properly account for the effect of damper windings. We defer that consideration until Chapter 8.

7.10 The Polyphase Synchronous-Reluctance Machine

Recall the equation

$$P_{dev} = 3\frac{E_f V_a \sin\delta}{X_d} + 3V_a^2 \sin 2\delta\left(\frac{X_d - X_q}{2X_d X_q}\right)$$

Now suppose that $E_f = 0$:

$$P_{dev} = 3V_a^2 \sin 2\delta\left(\frac{X_d - X_q}{2X_d X_q}\right)$$

$$T_{dev} = \frac{3V_a^2 \sin 2\delta(X_d - X_q)}{2X_d X_q \omega_{rm}}$$

This means that the machine will convert power, and develop electromagnetic torque, *even if there is no rotor field at all.* Such a device is called a "reluctance machine" because it relies on the variation in air-gap reluctance in order to produce torque. Obviously it must be a salient pole. Again, an example should prove instructive.

Example 7.10
Consider a synchronous machine, identical to that of Example 7.7, but with no field. The machine operates as a motor driven from a variable voltage, variable frequency AC drive.

(a) Can the motor supply a load which requires 40 kW at 1080 rpm?
(b) Can the motor supply a load which requires 10 kW at 1080 rpm. If so, determine all currents and the input power and power factor.

SOLUTION

From Example 7.9,

$f = 36$ Hz and $V_a = 166.3$ V

$X_d = 1.5204\ \Omega$; $X_q = 0.9882\ \Omega$

$3(V_a)^2(X_d - X_q)/2X_d X_q = 14.69$ kW

(a) The maximum power developed at 1080 rpm must be 14.69 kW. Hence, the motor cannot drive a load of 40 kW at that speed.
(b) For a 10 kW load, operation is possible.

$$P_{dev} = 14.69 \sin 2\delta = 10 + (0.6)^2(2.999) = 11.08\ \text{kW}$$
$$\delta = 24.5°$$

$$T_{dev} = P_{dev}/\omega_{rm} = 11080/113.1 = 97.96\ \text{N m}$$

$$I_d = V_a \cos\delta / X_d = 99.51\ \text{A}$$

$$I_q = V_a \sin\delta / X_q = 69.73$$

$$I_a = I_d + I_q = 99.51\underline{/-114.5°} + 69.73/-24.5° = 121.5\underline{/-79.5°}\ \text{A}$$

$$\text{pf} = \cos 18.3° = 0.1828\,\text{lag}$$

$$\text{Likewise } \bar{S}_{\text{IN}} = 11.08\,\text{kW} + j59.60\,\text{kvar}$$

7.11 The Brushless DC Motor

To this point, we have confined our discussion to AC machines, deferring the discussion of DC machines to Chapter 9. This is because AC machines (1) are far more common, and (2) have simpler structures. We will learn that a DC motor has rotor and stator magnetic fields (F and A) inherently positioned 90° apart, and that a DC motor is particularly amenable to precision speed and position control. It is possible to operate a synchronous motor in such a way that it can be controlled in a similar manner, which allows for control options that mimic traditional DC motor performance (see Figure 7.20a for details).

Starting from an ordinary balanced three-phase constant voltage, constant frequency AC source, the power is converted into a balanced three-phase variable voltage and frequency (constant V/f) AC source.[17] This output is applied to a permanent magnet three-phase synchronous motor, which drives a mechanical load. The AC drive may be controlled in such a way that $\hat{F}$ and $\hat{A}$ are positioned 90° apart, and that desired shaft speed and position conditions are achieved.

Consider the condition that $\hat{F}$ and $\hat{A}$ are 90° apart. Recall that the phase position of I_a determines the spatial position of $\hat{A}$, and that E_f lags $\hat{F}$ by 90°. Hence, $\bar{I}_a$ must be in phase with $\bar{E}_f$, and on the q-axis. Therefore, $\bar{I}_a = \bar{I}_q$ ($\bar{I}_d$ must be 0). These relations are summarized in Figure 7.20b.

The drive is controlled in such a way that the frequency meets the speed requirements and the voltage magnitude and phase achieves the desired torque and field orientation requirements. In particular, note that at a given speed and load torque, E_f and I_a are in phase. Now suppose that the load torque is increased, and we wish to operate at the same speed. Since the speed does not change, neither does E_f, and the increase in torque must be met with a proportional increase in I_a. Hence, at a given speed, developed torque and armature current are directly proportional, just as they are in a traditional DC machine. An example should prove instructive.

Example 7.11

Consider a brushless DC motor system, as shown in Figure 7.20, using the synchronous machine of Example 7.9, driving a pump whose torque speed characteristic is

$$T_m = 0.02765(\omega_{rm})^2$$

It is desired to operate the pump at 1080 rpm. Determine the machine current, voltage, input power, and power factor.

SOLUTION

$$k = 1080/1800 = 0.6; \quad \text{Therefore } f = 36\,\text{Hz and } \omega_{rm} = 113.1\,\text{rad/s}$$

so that $T_m = 0.02765(\omega_{rm})^2 = 353.7\,\text{N m}$, and

[17] Refer to Section 5.8 to review the AC motor drive.

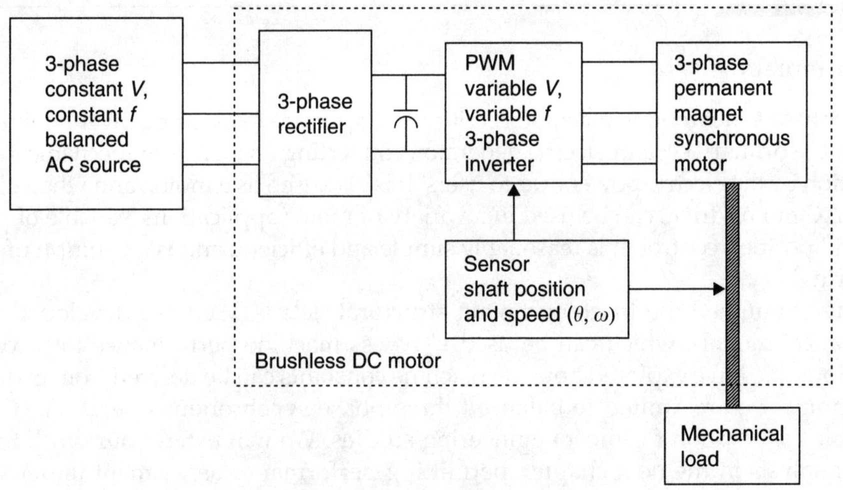

(a) Brushless DC motor system

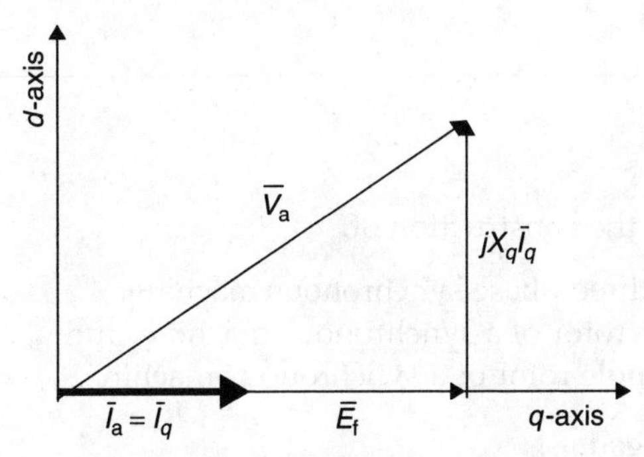

(b) Phasor diagram for brushless DC motor operation

FIGURE 7.20. The brushless DC motor.

$$P_m = T_m \omega_{rm} = 40\,\text{kW}$$

$$P_{DEV} = P_m + P_{RL} = 41.08\,\text{kW}$$

Hence

$$I_a = I_q = P_{DEV}/3E_F = 41080/(3 \times 240) = 57.06\,\text{A}$$

$$V_a = \sqrt{(E_F)^2 + (X_q I_q)^2} = 246.5\,V$$

$$\tan\delta = I_q X_q / E_F = 0.2349;\ \delta = 13.2°$$

Converting V_a into the phase reference

$$\bar{I}_a = \bar{I}_q = 57.06\underline{/13.2°}\,\text{A}$$

$$\bar{V}_a = 246.5\underline{/0}\,\text{V}$$

$$\bar{E}_f = 240\underline{/-13.2°}\,\text{V}$$

$$\text{pf} = 0.9737 \text{ lagging}$$

$$S_{IN} = 41.08\,\text{kW} + j9.65\,\text{kvar}$$

7.12 Summary

The polyphase synchronous machine is critically important to electric power engineering. It serves as the primary system electric generator, converting energy from mechanical form for distribution by the electric power grid to users. It is also used as a motor, and when combined with an AC motor drive, can be used in a variety of motor applications, capable of precision speed and position control. It is reasonably simple and efficient, and is a triumph of modern engineering.

We have examined the machine's main structural details. We have developed accurate mathematical models, which can be used to assess machine performance for a variety of applications. We have explored how the machine constants can be derived from certain tests. At this point, we are limited to balanced three-phase synchronous operation, which is a reasonable assumption for most engineering studies. We will extend our work to a more general analysis in the next chapter, permitting performance assessment under virtually any operating conditions.

Problems

7.1 Describe briefly the construction of:

(a) the stator of a three-phase synchronous machine,
(b) the salient pole rotor of a synchronous machine, and
(c) the nonsalient pole rotor of a synchronous machine.

Diagrams are encouraged.

7.2 Construct a table of synchronous speeds (in rad/s and rpm) for 60 Hz synchronous machines, from 2 to 30 poles.

7.3 Repeat Problem 7.2 for 50 Hz synchronous machines.

7.4 Determine rated stator current for a 460 V 75 kVA three-phase six-pole 60 Hz synchronous machine.

7.5 A 460 V 75 kVA three-phase six-pole 60 Hz synchronous machine is driven at 1200 rpm, and the open circuit phase-to-neutral voltage is measured as field current is varied:

I_F (A)	V_{an} (V)
0.022	0.0
0.612	53.1
1.213	106.2
1.830	159.4
2.476	212.5
3.168	265.6
3.939	318.7
4.844	371.8
5.971	424.9
7.471	478.1
9.600	531.2

Determine A_x, B_x, and K_{ag}.

7.6 Consider the synchronous machine defined in Example 7.2. For operation as a generator, at rated stator output, pf = 0.80 lagging, find (a) synchronous speed in rad/s and rev/min; (b) all machine currents and voltages (draw the stator phasor diagram), powers, and torques; and (c) the efficiency. Work the problem twice, first by hand, and then using EMAP. Certify that the hand solution checks with EMAP. If the solution does not check, tabulate the discrepancies in the solutions.

7.7 Repeat Problem 7.6 for unity pf.

7.8 Repeat Problem 7.6 for pf = 0.8 leading.

7.9 Repeat Problem 7.6 for pf = 0.5 lagging.

7.10 Repeat Problem 7.6 for pf = 0.5 leading.

7.11 Repeat Problem 7.6 for an instructor-defined pf.

7.12 Consider the synchronous machine defined in Example 7.3. For operation as a motor, at rated stator input, pf = 0.85 leading, find (a) synchronous speed in rad/s and rev/min; (b) all machine currents and voltages (draw the stator phasor diagram), powers, and torques; and (c) the efficiency. Work the problem twice, first by hand, and then using EMAP. Certify that the hand solution checks with EMAP. If the solution does not check, tabulate the discrepancies in the solutions.

7.13 Repeat Problem 7.12 for unity pf.

7.14 Repeat Problem 7.12 for pf = 0.85 leading.

7.15 Repeat Problem 7.12 for pf = 0.5 lagging.

7.16 Repeat Problem 7.12 for pf = 0.5 leading.

7.17 Repeat Problem 7.6 for an instructor-defined pf.

7.18 Consider the synchronous machine defined in Example 7.4. For operation as a generator, at rated stator output, pf = 0.85 lagging, find (a) synchronous speed in rad/s and rev/min; (b) all machine currents and voltages (draw the stator phasor diagram), powers, and torques; and (c) the efficiency. Check with EMAP.

7.19 Repeat Problem 7.18 for unity pf.

7.20 Repeat Problem 7.18 for pf = 0.85 leading.

7.21 Consider the synchronous machine defined in Example 7.4. For operation as a motor, at rated stator input, pf = 0.85 lagging, find (a) synchronous speed in rad/s and rev/min; (b) all machine currents and voltages (draw the stator phasor diagram), powers, and torques; and (c) the efficiency. Check with EMAP.

7.22 Repeat Problem 7.21 for unity pf.

7.23 Repeat Problem 7.21 for pf = 0.85 leading.

7.24 Consider the synchronous machine defined in Example 7.4. Make P–δ plots, as shown in Figure 7.13, for:

(a) $E_f = 0$ V,
(b) $E_f = 300$ V, and
(c) $E_f = 500$ V.

7.25 Repeat Problem 7.24, but modify the machine data such that the machine is non-salient ($X_d = X_q = 2.534\,\Omega$).

7.26 Consider the volumetric flow rate of water required for a 150 MW hydroelectric generation facility, ignoring all losses, for a head of:

(a) 10 m, (b) 100 m, and (c) 1000 m.

7.27 Consider a coal-fired 400 MW thermal electric generation facility, using coal whose heating value is 32 MJ/kg. If the overall plant efficiency is 35%, find:

(a) The heat rate.
(b) The daily fuel consumption in metric tons.
(c) Suppose the retail electric energy rate is $0.10 per kW h, and the fuel is 30% of the rate ($0.03 per kW h), determine the cost of coal per metric ton.

7.28 Consider the situation in Example 7.6. Using given machine data, at rated stator load, pf = 0.866 lagging, confirm that E_f is in fact 15.95 kV, and next that $I_F = 803$ A and, finally, that $V_F = 401$ V.

7.29 Consider the following test results for a 2400 V 750 kVA 60 Hz four-pole salient synchronous machine. The field is rated at 250 V DC.

Ratings

*V*Line = 2400 V;	*I*Line = 180.4 A;	S_{3ph} = 750 kVA
Frequency = 60 Hz;	Poles = 4;	Sync speed = 1800 rpm
Rotor type: SALIENT;	DC field voltage = 250.0 V	

Equivalent Circuit Values (R, X in Ω)

$R_a = 0.0768$;	$T_{RL} = K_{RL} * \omega_{rm}$;	$K_{RL} = 0.4222$ N m s/rad
$X_d = 7.680$;	$X_q = 4.992$	
$X_l = 1.1520$;	$R_F = 8.3333$	
MagC at 1800 rpm:	$I_F = E_f/K_{ag} + A_x \exp(B_x * E_f)$.	
$K_{ag} = 192.46\,\Omega$;	$A_x = 0.017849$ A;	$B_x = 0.0026682$ 1/V;

Note: $A_x = 0$ indicates that the MagC has been linearized.

DC Test (measurements made at rated current):

$V_{ab} = V_{bc} = V_{ca} = 18.47$ V;

$K_{TEMP}K_{ac} = 1.5$

Open-circuit and ZPF tests:

I_F (A)	E_f (V)	I_F (ZPF)
0.02	0	0
0.69	128	
1.37	256	
2.05	385	
2.73	513	
3.43	641	
4.13	769	
4.86	897	
5.60	1025	
6.38	1154	
7.21	1282	
7.92	**1386**	**66**
9.53	1593	
10.18	1666	
11.47	1795	
13.01	1922.6	
14.90	2050.8	
17.30	2179.0	
20.40	2307.2	
24.50	2435.3	
30.00	2563.5	

Short-circuit test:

I_F	I_a
0	0
7.2 A	167 A

Slip test (measured at low excitation level, provides waveforms shown in Example 7.5d):

V_{an} (maximum) = 500 V; V_{an} (minimum) = 403 V
I_a (maximum) = 100 A; I_a (minimum) = 81 A

No-load test (running the machine as an unloaded motor at unity pf):

$$V_{ab} = 2400\,\text{V} \quad I_a = 12\,\text{A} \quad P_a = 7\,\text{kW} \quad P_c = 7\,\text{kW}$$

(a) Plot the machine characteristics, as shown in Figure 7.14.
(b) Determine the machine constants.

7.30 A nonsalient six-pole 480 V 60 Hz 150 hp lossless synchronous motor has $X_d = 1.5\,\Omega$. Plot the "Vee curves" by computing the I_F, I_a coordinates at five pf's: 0.6 lagging; 0.8 lagging; unity; 0.8 leading; 0.6 leading. Select six load levels: 0, 20, 40, 60, 80, and 100% rated. Assume magnetic linearity.

7.31 Consider the following generator:

Three-Phase Synchronous Machine Data		
	Ratings	
Vline = 13.80 kV;	S_{3ph} = 120 MVA	
Frequency = 60 Hz;	Poles = 4;	Sync speed = 1800 rpm
Rotor type: NONSALIENT;	DC field voltage = 500.0 V	
	Equivalent Circuit Values (R, X in Ω)	
$R_a = 0.0$;	$T_{RL} = K_{RL} * \omega_{rm}$;	$K_{RL} = 0$ N m s/rad
$X_d = 2.3$;	$X_q = 2.3$	
$X_l = 0.2304$;	$R_F = 0.6000$	
MagC at 1800 rpm;	$I_F = E_f/K_{ag} + A_x \exp(B_x * E_f)$.	
$K_{ag} = 90\,\Omega$;	$A_x = 0.4$ A;	$B_x = 0.0005$ 1/V;

(a) Plot the unit's capability curves to scale, for
$\delta_{MAX} = 60°;\ \ P_{MIN} = 0\%;\ \ Q_{MIN} = -50\%$
(b) Determine the rated power factor.

7.32 Consider a permanent magnet synchronous machine as defined in Example 7.9. The machine operates as a motor driven from a $k = V/f$; variable voltage, variable frequency, constant V/f drive, supplying a load which requires 30 kW at 900 rpm. Determine all currents, the input power, and power factor.

7.33 Consider the reluctance motor described in Example 7.10. The machine operates as a motor driven from a variable voltage, variable frequency, constant V/f drive, and runs at 900 rpm.

(a) What is the maximum torque it can develop at this speed?
(b) At the torque found in (a), determine all currents, the input power, and power factor.

7.34 Consider the brushless DC motor and load discussed in Example 7.11. Determine all currents, the input power, and power factor when driving the load at 900 rpm, assuming that the rotor and stator fields are perpendicular.

8

The Polyphase Synchronous Machine: The General Coupled Circuit Model

In Chapter 7, we discussed the synchronous machine operating in a variety of environments and configurations, but always in a balanced three-phase mode with sinusoidal excitation, and at constant speed. The machine model presented was quite satisfactory for the applications discussed, and should by all means be used in those situations. But what if one, or more, of these restrictions are violated? For example, suppose we wish to predict machine performance if one of its phases is suddenly opened, or shorted? Suppose we are interested in the dynamic performance of the machine, if it were suddenly decoupled from its load (operating as a motor), or its prime mover (operating as a generator)? How about performance on nonsinusoidal excitation, as would be the case when operated from a PWM drive?

The primary flaw in the model developed so far is that it does not adequately account for the effect of the rotor damper windings, nor does it adequately account for mutual flux coupling between stator and rotor windings. The coupled-circuit model overcomes these problems. However, bear in mind that no mathematical model *exactly* predicts performance of any physical system, a concept worth special attention:

> ***Any mathematical model of any physical device is always based on a set of simplifying assumptions. Furthermore, one analyzes the model, never the physical device.***

The most important assumptions fundamental to the coupled circuit model are that

- all windings are sinusoidally distributed,
- magnetic linearity is assumed throughout the machine.

8.1 The General Coupled Circuit Model of the Synchronous Machine

We start with Figure 8.1a, which serves to identify all windings and reference quantities. It is important to note that *MOTOR* sign conventions are used on both the stator and rotor (i.e., positive power flows *into* the machine). Observe that the machine has six windings: *a,b,c*, which are located on the stator; and *F,D,Q*, which are located on the rotor. All windings are inductively coupled to all other windings, indicated by the dots in Figure 8.1b.

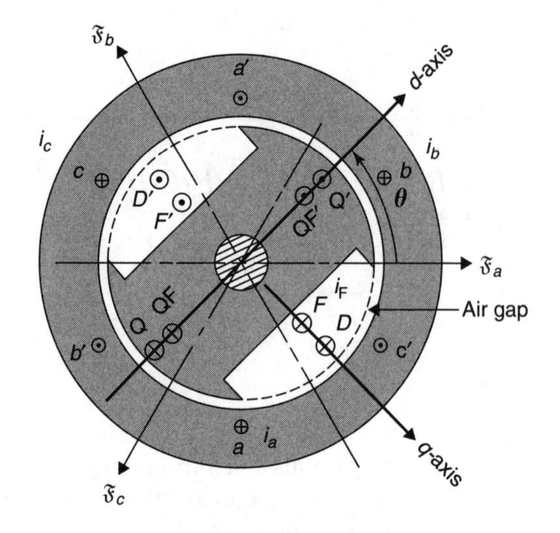

(a) Winding layout

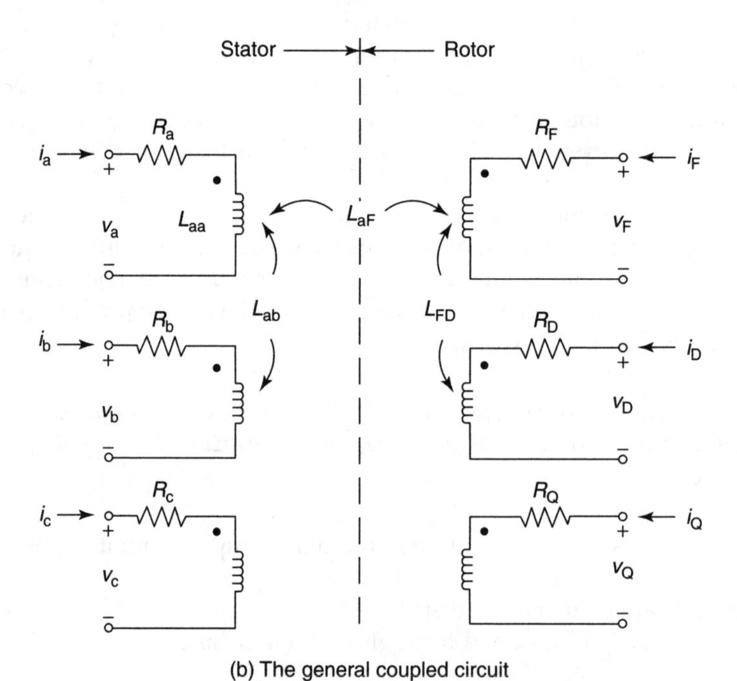

(b) The general coupled circuit

FIGURE 8.1. The general six-winding synchronous machine.

Note that lowercase and uppercase letters are used for stator and rotor quantities, respectively. We write

$$
\begin{aligned}
v_a &= R_a i_a + p\lambda_a \\
v_b &= R_b i_b + p\lambda_b \\
v_c &= R_c i_c + p\lambda_c \\
v_F &= R_F i_F + p\lambda_F \\
v_D &= R_D i_D + p\lambda_D \\
v_Q &= R_Q i_Q + p\lambda_Q
\end{aligned}
$$

where $p = \frac{d}{dt}(\cdots)$

so that $p\lambda = \frac{d}{dt}(\lambda)$

There is a clear incentive to use matrix notation. Thus, define

$$\hat{v}_{abcFDQ} = [v_a \; v_b \; v_c \; v_F \; v_D \; v_Q]^T \quad (6\times 1) \text{ voltage vector}$$
$$\hat{i}_{abcFDQ} = [i_a \; i_b \; i_c \; i_F \; i_D \; i_Q]^T \quad (6\times 1) \text{ current vector}$$
$$\hat{\lambda}_{abcFDQ} = [\lambda_a \; \lambda_b \; \lambda_c \; \lambda_F \; \lambda_D \; \lambda_Q]^T \quad (6\times 1) \text{ flux linkage vector}$$
$$[R_{abcFDQ}] = \text{diag}[R_a \; R_b \; R_c \; R_F \; R_D \; R_Q] \quad (6\times 6) \text{ resistance matrix}$$

Hence, we write

$$\hat{v}_{abcFDQ} = [R_{abcFDQ}]\hat{i}_{abcFDQ} + p\hat{\lambda}_{abcFDQ} \tag{8.1}$$

Now

$$\begin{aligned}
\lambda_a &= \lambda_{aa} + \lambda_{ab} + \lambda_{ac} + \lambda_{aF} + \lambda_{aD} + \lambda_{aQ} \\
\lambda_b &= \lambda_{ba} + \lambda_{bb} + \lambda_{bc} + \lambda_{bF} + \lambda_{bD} + \lambda_{bQ} \\
\lambda_c &= \lambda_{ca} + \lambda_{cb} + \lambda_{cc} + \lambda_{cF} + \lambda_{cD} + \lambda_{cQ} \\
\lambda_F &= \lambda_{Fa} + \lambda_{Fb} + \lambda_{Fc} + \lambda_{FF} + \lambda_{FD} + \lambda_{FQ} \\
\lambda_D &= \lambda_{Da} + \lambda_{Db} + \lambda_{Dc} + \lambda_{DF} + \lambda_{DD} + \lambda_{DQ} \\
\lambda_Q &= \lambda_{Qa} + \lambda_{Qb} + \lambda_{Qc} + \lambda_{QF} + \lambda_{QD} + \lambda_{QQ}
\end{aligned}$$

Also, observe that

$$\lambda_{ij} = L_{ij} i_j \quad \text{for } i, j = a, b, c, F, D, \text{ and } Q$$

It follows that

$$\begin{aligned}
\lambda_a &= L_{aa}i_a + L_{ab}i_b + L_{ac}i_c + L_{aF}i_F + L_{aD}i_D + L_{aQ}i_Q \\
\lambda_b &= L_{ba}i_a + L_{bb}i_b + L_{bc}i_c + L_{bF}i_F + L_{bD}i_D + L_{bQ}i_Q \\
\lambda_c &= L_{ca}i_a + L_{cb}i_b + L_{cc}i_c + L_{cF}i_F + L_{cD}i_D + L_{cQ}i_Q \\
\lambda_F &= L_{Fa}i_a + L_{Fb}i_b + L_{Fc}i_c + L_{FF}i_F + L_{FD}i_D + L_{FQ}i_Q \\
\lambda_D &= L_{Da}i_a + L_{Db}i_b + L_{Dc}i_c + L_{DF}i_F + L_{DD}i_D + L_{DQ}i_Q \\
\lambda_Q &= L_{Qa}i_a + L_{Qb}i_b + L_{Qc}i_c + L_{QF}i_F + L_{QD}i_D + L_{QQ}i_Q
\end{aligned}$$

or in matrix notation

$$\hat{\lambda}_{abcFDQ} = [L_{abcFDQ}]\hat{i}_{abcFDQ} \tag{8.2}$$

Thus it will take 6 + 36 = 42 element values (6 resistances and 36 inductances to formulate the machine coupled circuit model). The resistances are no problem. Conceptually, they are straightforward and relatively simple to directly measure. Also, since the machine is three-phase symmetric by design, the resistance of the three stator phases are equal:

$$R_a = R_b = R_c = R_s = \text{stator phase winding resistance}$$

However, the inductances are more complicated. Recall the definition of inductance

$$L = \frac{\lambda}{i} = \frac{N\phi}{i} = \frac{N\mathfrak{F}}{\mathfrak{R}i} = \frac{N^2}{\mathfrak{R}} = N^2\mathfrak{P}$$

Hence the magnetic path reluctance ($\mathfrak{R}$) and permeance ($\mathfrak{P} = 1/\mathfrak{R}$) are variable, and so must also the inductance. Consider Figure 8.1a. Note that as the rotor moves, the flux path reluctance available to flux produced by Phase-a current varies. Observe that for $\theta = 0$, the d-axis is aligned with the phase-a axis, maximizing $\mathfrak{P}$ (and L_{aa}), and for $\theta = \pi/2$, the q-axis is aligned with the phase-a axis, minimizing $\mathfrak{P}$ (and L_{aa}). Hence L_{aa} is $L_{aa}(\theta)$, and varies sinusoidally, as shown in Figure 8.2a.

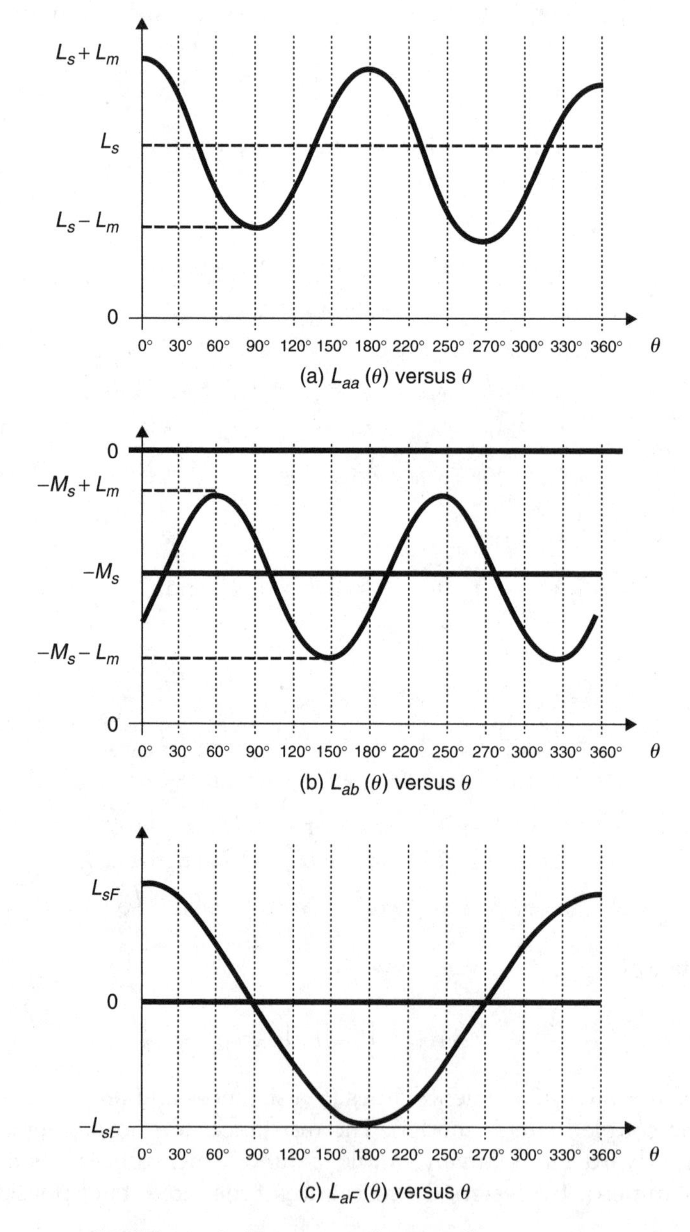

FIGURE 8.2. Inductance as a function of rotor position.

Hence

$$L_{aa} = L_s + L_m \cos(2\theta)$$

Also, as the rotor moves, the flux path reluctance available to flux mutual to phase-a and phase-b varies, minimizing at $\theta = 60°$, which places the greatest air gap on the a and b axes. Also, note that the flux linking phase-a due to phase-b current is negative, since the windings are more than 90° apart. The variation again is sinusoidal (see Figure 8.2b). Hence

$$L_{ab} = -M_s - L_m \cos 2(\theta + 30°)$$

The machine has complete three-phase symmetry on the stator. The angle of 120° is involved so frequently that a symbol for this angle is justified. Defined as

$$\alpha = 120° = 2\pi/3 \text{ rad}$$ [1]

The remaining stator self- and mutual inductances can be derived in a similar manner. Results are presented in matrix form

$$[L_{SS}] = \begin{bmatrix} L_s + L_m \cos(2\theta) & -M_s - L_m \cos 2(\theta + 30°) & -M_s - L_m \cos 2(\theta - 30°) \\ -M_s - L_m \cos 2(\theta + 30°) & L_s + L_m \cos 2(\theta - \alpha) & -M_s - L_m \cos 2(\theta - 90°) \\ -M_s - L_m \cos 2(\theta - 30°) & -M_s - L_m \cos 2(\theta - 90°) & L_s + L_m \cos 2(\theta + \alpha) \end{bmatrix}$$

Now consider the situation from the rotor's perspective. The d and q-axis reluctances are constant (independent of rotor position) for all rotor windings! Also, since the d and q-axis windings are at right angles, corresponding mutual flux is 0. Since the D and F windings share the same magnetic axis, there is strong mutual coupling. Again, results are presented in matrix form

$$[L_{RR}] = \begin{bmatrix} L_F & L_{FD} & 0 \\ L_{FD} & L_D & 0 \\ 0 & 0 & L_Q \end{bmatrix}$$

Finally, consider mutual coupling between stator and rotor windings. As the rotor moves, the flux path reluctance available to flux produced by phase-a current linking winding F varies. Observe that for $\theta = 0$, the d-axis (which is also the F-axis) is aligned with the Phase-a axis, maximizing $\mathfrak{P}$ (and L_{aF}) for $\theta = \pi/2$, the F-axis is orthogonal with the Phase-a axis, driving $\mathfrak{P}$ (and L_{aa}) to 0; and finally for $\theta = \pi$, the *negative* d-axis is aligned with the Phase-a axis, maximizing $\mathfrak{P}$ (and driving L_{aF} to a *negative* maximum). The variation with θ is sinusoidal, as shown in Figure 8.2c. Hence

$$L_{aF} = L_{SF} \cos(\theta)$$

[1] A purist would insist on the consistent use of radians throughout, which indeed is rigorously correct: E.g., "ωt + 30°" creates anomalies such as "$\pi/6$ rad + 30°". However, the use of degrees is frequently much clearer when considering the physical machine. We will use both forms, relying on the reader's sophistication to avoid obvious errors. If there is any doubt, always use radians. Blessed are the rigorous for they shall inhibit the Earth.

Presenting results in matrix form

$$[L_{SR}] = \begin{bmatrix} L_{SF}\cos(\theta) & L_{SD}\cos(\theta) & L_{SQ}\sin(\theta) \\ L_{SF}\cos(\theta-\alpha) & L_{SD}\cos(\theta-\alpha) & L_{SQ}\sin(\theta-\alpha) \\ L_{SF}\cos(\theta+\alpha) & L_{SD}\cos(\theta+\alpha) & L_{SG}\sin(\theta+\alpha) \end{bmatrix}$$

The system is reciprocal

$$[L_{SR}] = \begin{bmatrix} L_{SF}\cos(\theta) & L_{SF}\cos(\theta-\alpha) & L_{SF}\cos(\theta+\alpha) \\ L_{SF}\cos(\theta) & L_{SD}\cos(\theta-\alpha) & L_{SD}\cos(\theta+\alpha) \\ L_{SF}\sin(\theta) & L_{SQ}\cos(\theta-\alpha) & L_{SQ}\cos(\theta+\alpha) \end{bmatrix} = [L_{RS}]^T$$

We collect this information into one large matrix

$$[L_{abcFDQ}] = \begin{Bmatrix} [L_{SS}] & [L_{SR}] \\ [L_{RS}] & [L_{RR}] \end{Bmatrix} \tag{8.3}$$

We now have accumulated all of the information that is necessary to describe general synchronous machine performance. Exploiting symmetry we need

Resistances (4): R_s, R_F, R_D, R_Q
Inductances (9): $L_s, L_m, M_s, L_F, L_D, L_Q, L_{FD}, L_{SD}, L_{SQ}$

Unfortunately, the machine model is in a form which is somewhat cumbersome to use. The situation is analogous to that encountered when attempting to analyze three-phase circuits. We learned that by transforming the problem from phase (abc) to sequence (012) variables using the symmetrical component transformation, many problems are greatly simplified.[2] The simplifications are most dramatic when the system is operating under "normal" (balanced three-phase steady state) conditions.

There is a transformation similar to symmetrical components, which will materially simplify the machine model. Experience shows that the transformed model is much more amenable to analysis, particularly hand analysis.

8.2 The 0*dq* Transformation[3]

Consider projecting stator winding mmf's onto the *d* and *q* axes in Figure 8.1a:

$$\mathfrak{F}_d = \cos(\theta)N_S i_a + \cos(\theta-\alpha)N_S i_b + \cos(\theta+\alpha)N_S i_c$$
$$\mathfrak{F}_q = \sin(\theta)N_S i_a + \sin(\theta-\alpha)N_S i_b + \sin(\theta+\alpha)N_S i_c$$

[2] See Appendix B for an explanation of the symmetrical component transformation.

[3] Most workers give credit to R.H. Park for this tremendous contribution to machine theory, as it was first presented in his historic 1929 paper. In fact, many refer to it as Park's transformation. Park's original work was presented in per-unit form, and used a different *k* value from the one we shall select. We shall refer to the version presented here as the 0*dq* transformation.

Suppose these *d,q* mmf's are caused by ficticious *d,q* currents flowing in ficticious *d,q* windings, each with kN_S turns. Furthermore, let us add an arbitrary third equation[4] to the set creating a "0" winding, also with kN_S turns. We produce

$$\mathfrak{F}_0 = kNi_0 = \frac{N_S}{\sqrt{2}} i_a + \frac{N_S}{\sqrt{2}} i_b + \frac{N_S}{\sqrt{2}} i_c$$

$$\begin{aligned} \mathfrak{F}_d = kNi_d &= \cos(\theta)N_S i_a + \cos(\theta-\alpha)N_S i_b + \cos(\theta+\alpha)N_S i_c \\ \mathfrak{F}_q = kNi_q &= \sin(\theta)N_S i_a + \sin(\theta-\alpha)N_S i_b + \sin(\theta+\alpha)N_S i_c \end{aligned}$$

Canceling N_S, we have

$$\begin{aligned} ki_0 &= \frac{1}{\sqrt{2}} i_a + \frac{1}{\sqrt{2}} i_b + \frac{1}{\sqrt{2}} i_c \\ ki_d &= \cos(\theta)i_a + \cos(\theta-\alpha)i_b + \cos(\theta+\alpha)i_c \\ ki_q &= \sin(\theta)i_a + \sin(\theta-\alpha)i_b + \sin(\theta+\alpha)i_c \end{aligned}$$

Converting into matrix form

$$\begin{bmatrix} i_0 \\ i_d \\ i_q \end{bmatrix} = \frac{1}{k} \begin{bmatrix} 1/\sqrt{2} & 1/\sqrt{2} & 1/\sqrt{2} \\ \cos(\theta) & \cos(\theta-\alpha) & \cos(\theta+\alpha) \\ \sin(\theta) & \sin(\theta-\alpha) & \sin(\theta+\alpha) \end{bmatrix} \begin{bmatrix} i_a \\ i_b \\ i_c \end{bmatrix}$$

or

$$\tilde{i}_{0dq} = [P]\,\tilde{i}_{abc}$$

Consider

$$[P][P]^T = \frac{1}{k^2} \begin{bmatrix} (3/2) & 0 & 0 \\ 0 & (3/2) & 0 \\ 0 & 0 & (3/2) \end{bmatrix}$$

Now if we define "*k*" as follows:

$$\frac{1}{k^2} = \frac{2}{3} \quad \text{or} \quad k = \sqrt{\frac{3}{2}}$$

then, $[P][P]^T = \begin{bmatrix} 1 & 0 & 0 \\ 0 & 1 & 0 \\ 0 & 0 & 1 \end{bmatrix} = [I]$ = identity (or unit) matrix, and $[P]^T = [P]^{-1}$ = inverse of $[P]$.

[4] It may confound the reader that this is even possible. How can we simply pull an equation "out of thin air," with arbitrary coefficients, and add it to the set? And why this particular choice? Nonetheless, as long as it is linearly independent of the other two equations, there is no *mathematical* restriction against this, if the goal is simply to create three independent equations in two sets of three variables. As one might suspect, certain "arbitrary" choices are better than others. We cannot foresee the consequences of this choice until we develop the work further. Experience shows that this is a good choice. As the old hymn admonishes, "Trust and Obey."

That is, the transpose of $[P]$ is equal to the inverse of $[P]$, if $k = \sqrt{3/2}$. A matrix with this particular feature is said to be "orthogonal". Now consider the stator winding instantaneous power

$$\begin{aligned} p_S(t) &= p_a(t) + p_b(t) + p_c(t) = v_a(t)\,i_a(t) + v_b(t)\,i_b(t) + v_c(t)\,i_c(t) \\ &= \{\hat{v}_{abc}\}^T\,\hat{i}_{abc} \end{aligned}$$

Suppose phase (abc) voltage were transformed to $0dq$ voltage, according to some unknown transformation[5]:

$$\hat{v}_{abc} = [\aleph]\,\hat{v}_{0dq}$$

Substituting into the power expression

$$p_S(t) = \{\hat{v}_{abc}\}^T\hat{i}_{abc} = \{[\aleph]\hat{v}_{0dq}\}^T[P]^{-1}\hat{i}_{0dq} = \{\hat{v}_{0dq}\}^T[\aleph]^T[P]^{-1}\hat{i}_{0dq}$$

If $[\aleph]^T[P]^{-1} = [I]$, then $\{\hat{v}_{abc}\}^T\hat{i}_{abc} = \{\hat{v}_{0dq}\}^T\hat{i}_{0dq}$, and the transformation is said to be "power-invariant"; that is the stator power computed in phase (abc) variables is equal to the power computed in $0dq$ variables. The results are quite elegant and lead to a beautiful machine model.[6]

To summarize our work then, we have selected "k" so that the transformation is orthogonal; voltage, current, and flux linkage are all transformed by the same $[P]$; and the transformation is power invariant. The results are

$$\tilde{i}_{0dq} = [P]\tilde{i}_{abc}; \quad \hat{v}_{0dq} = [P]\hat{v}_{abc}; \quad \tilde{\lambda}_{0dq} = [P]\tilde{\lambda}_{abc}; \quad [P]^{-1} = [P]^T$$

$$[P] = \sqrt{\frac{2}{3}}\begin{bmatrix} 1/\sqrt{2} & 1/\sqrt{2} & 1/\sqrt{2} \\ \cos(\theta) & \cos(\theta-\alpha) & \cos(\theta+\alpha) \\ \sin(\theta) & \sin(\theta-\alpha) & \sin(\theta+\alpha) \end{bmatrix} \tag{8.4}$$

[5] We are not sure what is the "best" transformation for voltage. A given choice will have certain consequences which may, or may not, be desirable. For the moment, we treat it as unknown.

[6] Many authorities, particularly machine designers, use an alternate (traditional) version of $[P]$:

$$[P] = \frac{2}{3}\begin{bmatrix} 1/2 & 1/2 & 1/2 \\ \cos(\theta) & \cos(\theta-\alpha) & \cos(\theta+\alpha) \\ \sin(\theta) & \sin(\theta-\alpha) & \sin(\theta+\alpha) \end{bmatrix}$$

$$[P]^{-1} = \begin{bmatrix} 1 & \cos(\theta) & \sin(\theta) \\ 1 & \cos(\theta-\alpha) & \sin(\theta-\alpha) \\ 1 & \cos(\theta+\alpha) & \sin(\theta+\alpha) \end{bmatrix}$$

This traditional version of the $0dq$ transformation has merit, and it can be argued that it is "closer" to the physical machine, particularly if one is concerned with details of coil design, turns ratios, and mmf and flux density distributions. However, it is neither orthogonal nor power-invariant. For example, the corresponding stator power becomes

$$p_S(t) = 3v_0i_0 + \tfrac{3}{2}v_di_d + \tfrac{3}{2}v_qi_q$$

Later in this chapter, we consider scaling the problem into pu. When the traditional transformation is scaled into pu by traditional methods, the corresponding machine equations have two similar but different forms: those which are valid in SI units, and those valid in pu, creating a quite complicated and potentially confusing situation. Because of this, and because our emphasis is application as opposed to design, we choose to use the power invariant form.

The $0dq$ transformation only involves the *stator* variables. To include the three rotor equations, we define a larger 6×6 transformation matrix

$$[T] = \begin{bmatrix} [P] & [0] \\ [0] & [I] \end{bmatrix}, \qquad \text{where } [0] = \begin{bmatrix} 0 & 0 & 0 \\ 0 & 0 & 0 \\ 0 & 0 & 0 \end{bmatrix}$$

Extending to the full machines set

$$\begin{aligned} \hat{i}_{0dqFDQ} &= [T]\,\hat{i}_{abcFDQ} \\ \hat{v}_{0dqFDQ} &= [T]\,\hat{v}_{abcFDQ} \\ \hat{\lambda}_{0dqFDQ} &= [T]\,\hat{\lambda}_{abcFDQ} \\ [T]^{-1} &= [T]^{T} \end{aligned}$$

We are now ready to develop the $0dqFDQ$ machine model. Analyzing the machine as presented in Figure 8.1

$$\hat{v}_{abcFDQ} = [R_{abcFDQ}]\,\hat{i}_{abcFDQ} + p\hat{\lambda}_{abcFDQ}$$

or

$$[T]^{T}\hat{v}_{0dqFDQ} = [R_{abcFDQ}][T]^{T}\hat{i}_{0dqFDQ} + p\{[L_{abcFDQ}]\hat{i}_{abcFDQ}\}$$

Premultiplying by $[T]$

$$\begin{aligned} \hat{v}_{0dqFDQ} &= [T][R_{abcFDQ}][T]^{T}\,\hat{i}_{0dqFDQ} + [T]\{p\{[L_{abcFDQ}][T]^{T}\,\hat{i}_{0dqFDQ}\}\} \\ &= [T][R_{abcFDQ}][T]^{T}\,\hat{i}_{0dqFDQ} + \{[T][L_{abcFDQ}][T]^{T}\}p\{\hat{i}_{0dqFDQ}\} + [T][p\{[L_{abcFDQ}][T]^{T}\}]\{\hat{i}_{0dqFDQ}\} \end{aligned}$$

which is quite complicated. However, it happens that[7]

$$[T][R_{abcFDQ}][T]^{T} = [R_{abcFDQ}] = [R_{0dqFDQ}]$$

Also, define

$$[L_{0dqFDQ}] = [T][L_{abcFDQ}][T]^{T}$$

simplifying the expression to

$$\tilde{v}_{0dqFDQ} = [R_{0dqFDQ}]\,\hat{i}_{0dqFDQ} + [L_{0dqFDQ}]\,p\{\hat{i}_{0dqFDQ}\} + [T][p\{[L_{abcFDQ}][T]^{T}\}]\{\hat{i}_{0dqFDQ}\} \tag{8.5}$$

[7] See Problem 8.1.

A major task is to determine $[L_{0dqFDQ}]$[8]. The results are

$$[L_{0dqFDQ}] = \begin{bmatrix} L_{00} & 0 & 0 & 0 & 0 & 0 \\ 0 & L_{dd} & 0 & L_{dF} & L_{dD} & 0 \\ 0 & 0 & L_{qq} & 0 & 0 & L_{qQ} \\ 0 & L_{Fd} & 0 & L_{FF} & L_{FD} & 0 \\ 0 & L_{Dd} & 0 & L_{DF} & L_{DD} & 0 \\ 0 & 0 & L_{Qq} & 0 & 0 & L_{QQ} \end{bmatrix} \tag{8.6}$$

where

$$\begin{aligned} L_{00} &= L_0 = L_s - 2L_m \\ L_{dd} &= L_d = L_s + M_s + \frac{3}{2}L_m \\ L_{qq} &= L_q = L_s + M_s - \frac{3}{2}L_m \\ L_{dF} &= L_{Fd} = kL_{SF} = \sqrt{3/2}\,L_{SF} \\ L_{dD} &= L_{Dd} = kL_{SD} = \sqrt{3/2}\,L_{SD} \\ L_{qQ} &= L_{Qq} = kL_{SQ} = \sqrt{3/2}\,L_{SQ} \end{aligned}$$

Compare $[L_{abcFDQ}]$ with $[L_{0dqFDQ}]$. The simplification is dramatic! $[L_{abcFDQ}]$ has only four zero entries, with 27 that were functions of θ (i.e., rotor position)! $[L_{0dqFDQ}]$ has 22 zero entries, and none of the remaining 14 is a function of θ!

We now focus on simplifying the last term in Equation 8.5.[9] Realize that by the chain rule

$$\frac{\mathrm{d}}{\mathrm{d}t}\{f(\theta)\} = \frac{\partial}{\partial\theta}\{f(\theta)\}\frac{\mathrm{d}\theta}{\mathrm{d}t} = \omega\frac{\partial}{\partial\theta}\{f(\theta)\}$$

The results are

$$[T][p\{[L_{abcFDQ}][T]^T\}] = \omega[G_{0dqFDQ}] = \begin{bmatrix} 0 & 0 & 0 & 0 & 0 & 0 \\ 0 & 0 & +\omega L_q & 0 & 0 & +\omega L_{qQ} \\ 0 & -\omega L_d & 0 & -\omega L_{dF} & -\omega L_{dD} & 0 \\ 0 & 0 & 0 & 0 & 0 & 0 \\ 0 & 0 & 0 & 0 & 0 & 0 \\ 0 & 0 & 0 & 0 & 0 & 0 \end{bmatrix} \tag{8.7}$$

Again, the simplification is dramatic! $\omega[G_{0dqFDQ}]$ has 31 zero entries, and none of the remaining 5 terms is a function of θ!

---

[8] See Problem 8.2.

[9] See Problem 8.3.

The final version of our $0dqFDQ$ machine model is

$$\hat{v}_{0dqFDQ} = [R_{0dqFDQ}]\hat{i}_{0dqFDQ} + [L_{0dqFDQ}]p\{\hat{i}_{0dqFDQ}\} + \omega[G_{0dqFDQ}]\{\hat{i}_{0dqFDQ}\} \tag{8.8a}$$

$$\hat{\lambda}_{0dqFDQ} = [L_{0dqFDQ}]\hat{i}_{0dqFDQ} \tag{8.8b}$$

Writing the equations in nonmatrix form

$$\begin{aligned} v_0 &= R_s i_0 + p\lambda_0 \\ v_d &= R_s i_d + p\lambda_d + \omega\lambda_q \\ v_q &= R_s i_q + p\lambda_q - \omega\lambda_d \\ v_F &= R_F i_F + p\lambda_F \\ v_D &= R_D i_D + p\lambda_D \\ v_Q &= R_Q i_Q + p\lambda_Q \end{aligned} \tag{8.8c}$$

where

$$\begin{aligned} \lambda_0 &= L_0 i_0 \\ \lambda_d &= L_d i_d + L_{dF} i_F + L_{dD} i_D \\ \lambda_q &= L_q i_q + L_{qQ} i_Q \\ \lambda_F &= L_F i_F + L_{dF} i_d + L_{FD} i_D \\ \lambda_D &= L_D i_D + L_{Dd} i_d + L_{FD} i_F \\ \lambda_Q &= L_Q i_Q + L_{Qq} i_q \end{aligned} \tag{8.8d}$$

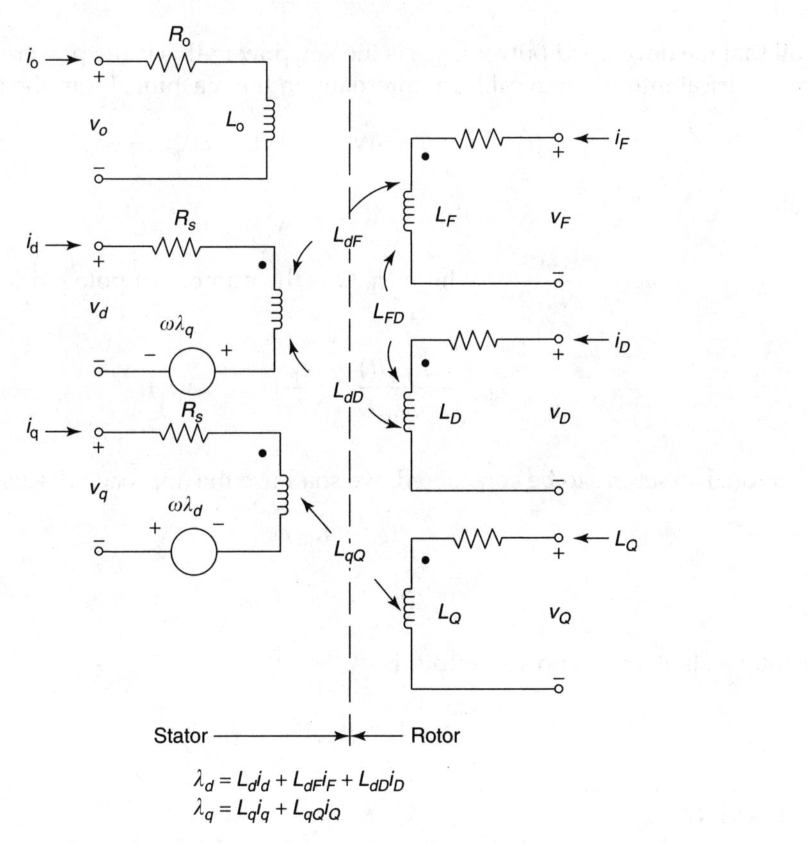

FIGURE 8.3. $0dqFDQ$ Equivalent circuit motor convention.

We also recall that θ is measured in *electrical* radians; therefore ω is in *electrical* rad/s. If we want the rotor speed as we have previously used the term

$$\omega_{rm} = \frac{\omega}{N_P/2} \quad \text{where } N_P/2 \text{ is the number of pole-pairs.} \tag{8.9}$$

These equations inspire the equivalent circuit shown in Figure 8.3. Note also that the extension of the model to accommodate additional circuits on the rotor *D,Q* axes is clear. We will exercise that option in Section 8.4 when we convert to generator sign convention.

8.3 Powers and Torques in the 0*dqFDQ* Model

Recall that the 0*dqFDQ* model was derived using motor convention. Hence, the total instantaneous power input to the model is

$$p_{IN}(t) = \{\hat{v}_{abcFDQ}\}^T \hat{i}_{abcFDQ} = \{\hat{v}_{0dqFDQ}\}^T \hat{i}_{0dqFDQ} \tag{8.10a}$$

It is straightforward to show the machine winding losses occur in the i^2R terms.[10] Thus

$$\text{SWL} = R_s i_0^2 + R_s i_d^2 + R_s i_a^2 \tag{8.10b}$$

$$\text{RWL} = R_F i_F^2 + R_D i_D^2 + R_Q i_Q^2 \tag{8.10c}$$

Recall that the developed power (p_{dev}) is the key power; this is the power that is converted from electrical into mechanical form internally in the machine. From the model, it is

$$p_{dev}(t) = p_{IN}(t) - \text{SWL} - \text{RWL} = \omega\lambda_q i_d - \omega\lambda_d i_q \tag{8.10d}$$

Recall

$$\omega_{rm} = \frac{\omega}{N_P/2} \quad \text{where } N_P/2 \text{ is the number of pole pairs.}$$

$$T_{dev} = \frac{p_{dev}(t)}{\omega_{rm}} = \frac{N_P}{2}(\lambda_q i_d - \lambda_d i_q) \tag{8.11a}$$

If rotational losses are to be considered, we shall use the approach discussed earlier.

$$T_{RL} = K_{RL}\omega_{rm} \tag{8.11b}$$

$$p_{RL}(t) = T_{RL}\omega_{rm} \tag{8.10e}$$

The total instantaneous power output is

$$p_{OUT}(t) = p_{dev}(t) - p_{RL}(t) \tag{8.10f}$$

[10] See Problem 8.4.

Up to this point, there are no restrictions on winding or shaft terminations. However, for "normal" machines, the D,Q windings are always shorted, and the F winding typically terminated in the exciter (frequently modeled as an ideal DC voltage source).

8.4 The 0*dqFDQ* Model Using Generator Sign Conventions

We recall Figure 8.1a, which serves to identify all windings and reference quantities, utilizing *MOTOR* sign conventions. All equations developed through Section 8.3 were based on this choice of sign convention. It is noted that these relations are sufficient to express machine performance in any and all actual operating modes. If the machine actually operates in the motor mode, this is a logical and "natural" way to reference an engineering analysis, since normal values of the most important quantities are positive. However, if we are investigating a generator application, the key quantities of most interest to engineers become negative. Matters become more awkward, more abstract, and more difficult to communicate. For example, we could get into a situation where we determine that the "input" is " −100 kW." Is it still correct to refer to such a value as "input"? Technically yes, but in practice, this is a confusing way to express machine performance. Hence, there is motivation to consider what constitutes a more natural reference frame in which to investigate generator operation.

Conversion of coordinate systems is straightforward, but not trivial. On the stator windings, we realize we have two options if we wish to reverse the sense of positive power flow: we could reverse the assigned positive direction of either voltage or current (but not both). The consequences of reversing current are more involved: magnetic axes and associated flux linkages are affected (unless we can tolerate strange constructions like $\lambda = -Li$). We chose the simpler alternative: *reverse the positive sense of the stator phase voltages* v_a, v_b, and v_c. Note that we do *NOT* change assigned positive directions of any rotor quantities. Since we desire that the relations

$$\hat{i}_{0dq} = [P]\,\hat{i}_{abc}; \quad \hat{v}_{0dq} = [P]\,\hat{v}_{abc}; \quad \hat{\lambda}_{0dq} = [P]\,\hat{\lambda}_{abc}$$

$$[P]^{-1} = [P]^T$$

$$[P] = \sqrt{\frac{2}{3}}\begin{bmatrix} 1/\sqrt{2} & 1/\sqrt{2} & 1/\sqrt{2} \\ \cos(\theta) & \cos(\theta-\alpha) & \cos(\theta+\alpha) \\ \sin(\theta) & \sin(\theta-\alpha) & \sin(\theta+\alpha) \end{bmatrix}$$

are to remain valid, the positive sense of v_0, v_d, v_q has also been reversed. We also noted in Section 8.2 that the extension of the model to accommodate additional rotor circuits appeared to be straightforward. We will exercise that option now. Note that there are two rotor circuits on the d-axis, and are mutually coupled to each other, and the "d" stator winding. We add a seventh winding to the machine; specifically rotor winding "U," on the rotor Q axis (which also aligns with the q-axis).

The "new" circuit model appears in Figure 8.4. It appears that the currents have been reversed, but actually they have not. In two steps, starting in Figure 8.2, first reverse the voltages, and then draw the stator circuits "upside down." This flips the dots to the bottom. We have also altered the rotor, stator winding locations so that the normal flow of power is from right to left, as is customary in circuit diagrams. We have also added a seventh winding to the machine; specifically rotor winding U.

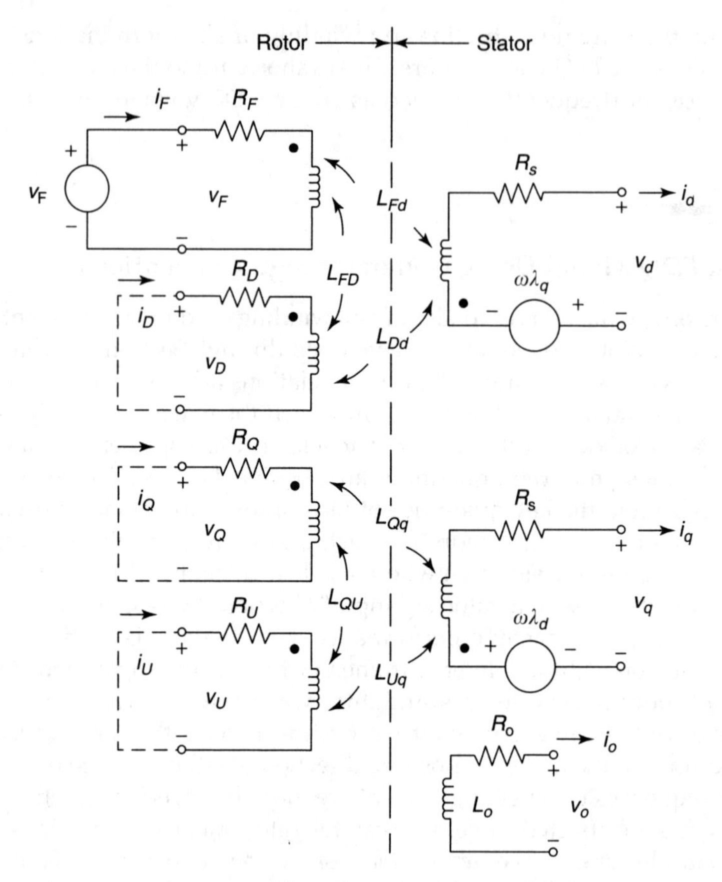

FIGURE 8.4. 0*dqFDQU* synchronous machine model: generator sign conventions.

The corresponding machine equations are

$$\begin{aligned}
-v_0 &= R_s i_0 + p\lambda_0 \\
-v_d &= R_s i_d + p\lambda_d + \omega\lambda_q \\
-v_q &= R_s i_q + p\lambda_q - \omega\lambda_d \\
v_F &= R_F i_F + p\lambda_F \\
v_D &= R_D i_D + p\lambda_D \\
v_Q &= R_Q i_Q + p\lambda_Q \\
v_U &= R_U i_U + p\lambda_U
\end{aligned} \tag{8.12a}$$

where

$$\begin{aligned}
\lambda_0 &= L_0 i_0 \\
\lambda_d &= L_d i_d + L_{dF} i_F + L_{dD} i_D \\
\lambda_q &= L_q i_q + L_{qQ} i_Q + L_{UQ} i_U \\
\lambda_F &= L_F i_F + L_{dF} i_d + L_{FD} i_D \\
\lambda_D &= L_D i_D + L_{Dd} i_d + L_{FD} i_F \\
\lambda_Q &= L_Q i_Q + L_{Qq} i_q + L_{QU} i_U \\
\lambda_U &= L_U i_U + L_{Uq} i_q + L_{QU} i_Q
\end{aligned} \tag{8.12b}$$

Redefining the developed power (p_{dev}) to be the power converted from mechanical into electrical form (i.e., the negative of the motor convention power):

$$p_{dev}(t) = \omega\lambda_d i_q - \omega\lambda_q i_d \tag{8.13a}$$

The machine losses are the same. However, the sense of "output" and "input" has changed:

$$p_{OUT}(t) = v_0 i_0 + v_d i_d + v_q i_q \tag{8.13b}$$

8.5 Balanced Three-Phase Constant Speed Generator Performance

We now realize that the situations studied in Chapter 7 are simply special cases of general machine performance. To demonstrate the point, consider a synchronous machine operating in a balanced three-phase generator mode at constant speed. We use generator sign convention. Assume the voltages are

$$\begin{aligned} v_a &= \frac{V_L\sqrt{2}}{\sqrt{3}}\cos(\omega t + \beta) \\ v_b &= \frac{V_L\sqrt{2}}{\sqrt{3}}\cos(\omega t + \beta - \alpha) \\ v_c &= \frac{V_L\sqrt{2}}{\sqrt{3}}\cos(\omega t + \beta + \alpha) \end{aligned} \tag{8.14}$$

where V_L is rms phase-to-phase (line) voltage.

Consider the case where $\theta = (1 - s)\omega$. Note that if $s = 0$, the machine is synchronized to the source. Computing the $0dq$ voltages,

$$\hat{v}_{0dq} = [P]\hat{v}_{abc} = \sqrt{\frac{2}{3}}\begin{bmatrix} 1/\sqrt{2} & 1/\sqrt{2} & 1/\sqrt{2} \\ \cos(\theta) & \cos(\theta-\alpha) & \cos(\theta+\alpha) \\ \sin(\theta) & \sin(\theta-\alpha) & \sin(\theta+\alpha) \end{bmatrix}\begin{bmatrix} \frac{V_L\sqrt{2}}{\sqrt{3}}\cos(\omega t+\beta) \\ \frac{V_L\sqrt{2}}{\sqrt{3}}\cos(\omega t+\beta-\alpha) \\ \frac{V_L\sqrt{2}}{\sqrt{3}}\cos(\omega t+\beta+\alpha) \end{bmatrix}$$

$$\begin{aligned} v_0 &= 0 \\ v_d &= \frac{2V_L}{3}[\cos((1-s)\omega t)\cos(\omega t+\beta) + \cos((1-s)\omega t-\alpha)\cos(\omega t+\beta-\alpha) \\ &\quad + \cos((1-s)\omega t+\alpha)\cos(\omega t+\beta+\alpha)] = V_L[\cos(s\omega t+\beta)] \\ v_q &= \frac{2V_L}{3}[\sin((1-s)\omega t)\cos(\omega t+\beta) + \sin((1-s)\omega t-\alpha)\cos(\omega t+\beta-\alpha) \\ &\quad + \sin((1-s)\omega t+\alpha)\cos(\omega t+\beta+\alpha)] = -V_L[\sin(s\omega t+\beta)] \end{aligned} \tag{8.15}$$

Now consider running at synchronous speed (i.e., $s = 0$). Note that the d and q voltages are constant (DC): Likewise, the field is terminated in a DC voltage source, and the three damper windings all terminate in short circuits.

$$v_F = V_F; \quad v_D = v_Q = v_U = v_0 = 0$$
$$v_d = V_L \cos\beta; \quad v_q = -V_L \sin\beta$$

Since all machine windings are either shorted or have DC voltage applied, all currents are either DC or 0. The only possibility for current flow in windings D, Q, or U is due to mutual induced voltage. But since all currents are DC, this possibility is eliminated, resulting in[11]

$$i_D = i_Q = i_U = i_0 = 0$$
$$i_d = J_d; \quad i_q = J_q; \quad i_F = I_F = \frac{V_F}{R_F}$$

Now

$$\lambda_0 = 0$$
$$\lambda_d = L_d J_d + L_{dF} I_F$$
$$\lambda_q = L_q J_q$$
$$\lambda_F = L_F I_F + L_{dF} J_d + L_{FD} J_D$$
$$\lambda_D = L_{Dd} J_d + L_{FD} I_F$$
$$\lambda_Q = L_{Qq} J_q$$
$$\lambda_U = L_{Uq} J_q$$

and

$$V_F = R_F I_F$$
$$-V_L \cos\beta = R_s J_d + \omega\lambda_q = R_s J_d + \omega(L_q J_q)$$
$$+V_L \sin\beta = R_s J_q - \omega\lambda_d = R_s J_q - \omega(L_d J_d + L_{dF} I_F)$$

Multiply through by j

$$-jV_L \cos\beta = R_s jJ_d + j\omega L_q J_q$$
$$+V_L \sin\beta = R_s J_q + j\omega L_d (jJ_d) - \omega(L_{dF} I_F)$$

Adding the two equations together and rearranging terms

$$\omega L_{dF} I_F = R_s J_q + R_s jJ_d + j\omega L_q J_q + j\omega L_d (jJ_d) - V_L \sin\beta + jV_L \cos\beta$$

---

[11] The author apologizes for what seems to be a bizarre choice of notation (J for current, and later U for voltage). The problem is that if we choose $i_d = I_d$, etc., this leads to a conflict of notation with Chapter 7. It is extremely undesirable to have the same symbol have different meanings in different parts of a book, particularly when the concepts are closely related. We choose the lesser of two evils.

Define

$$
\begin{aligned}
\bar{J}_q &= J_q + j0; \quad \bar{J}_d = 0 + jJ_d \\
\bar{J}_a &= \bar{J}_q + \bar{J}_d = J_q + jJ_d \\
\bar{U}_f &= \omega L_{dF} I_F + j0; \quad \bar{U}_{an} = V_L \angle(\beta - 90°) \\
\bar{U}_a &= R_s \bar{I}_a + \bar{U}_{an} \\
X_d &= \omega L_d; \quad X_q = \omega L_q
\end{aligned}
$$

Substituting

$$
\begin{aligned}
\bar{U}_f &= R_s \bar{I}_a + \bar{U}_{an} + jX_q \bar{J}_q + jX_d \bar{J}_d \\
&= R\bar{U}_a + jX_q \bar{J}_q + jX_d \bar{J}_d
\end{aligned} \tag{8.16}
$$

The corresponding phasor diagrams are shown in Figure 8.5. Now Suppose

$$
\begin{aligned}
\bar{J}_q &= \sqrt{3}\,\bar{I}_q; \quad & \bar{J}_d &= \sqrt{3}\,\bar{I}_d; \quad & \bar{J}_a &= \sqrt{3}\,\bar{I}_a \\
\bar{U}_f &= \sqrt{3}\,\bar{E}_f; \quad & \bar{U}_{an} &= \sqrt{3}\,\bar{V}_{an}; \quad & \bar{U}_a &= \sqrt{3}\,\bar{V}_a
\end{aligned}
$$

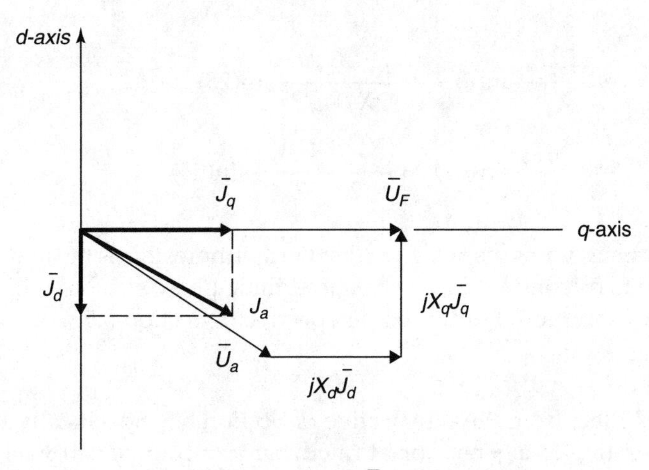

(a) Phasor diagram: $\bar{U}_F$ reference

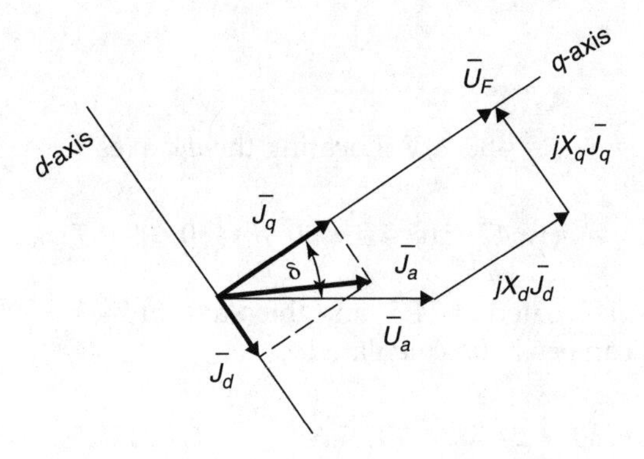

(b) (a) Redrawn with $\bar{U}_a$ as phase reference

FIGURE 8.5. Salient generator phasor diagram.

Substituting

$$\begin{aligned}\sqrt{3}\,\bar{E}_f &= R_s\sqrt{3}\,\bar{I}_a + \sqrt{3}\,\bar{V}_{an} + jX_q\sqrt{3}\,\bar{I}_q + jX_d\sqrt{3}\,\bar{I}_d \\ \bar{E}_f &= R_s\bar{I}_a + \bar{V}_{an} + jX_q\bar{I}_q + jX_d\bar{I}_d \\ &= \bar{V}_a + jX_q\bar{I}_q + jX_d\bar{I}_d\end{aligned}$$

The results are identical with those presented in Chapter 7 (see Equation 7.15; Figure 7.11)! We have in effect derived the Chapter 7 model from the more generalized model of Chapter 8! To study the matter further, consider the power developed:

$$\begin{aligned}p_{\text{dev}}(t) &= \omega\lambda_d i_q - \omega\lambda_q i_d = \omega(L_d J_d + L_{dF} I_F)J_q - \omega(L_q J_q)J_d \\ &= U_f J_q + X_d J_d J_q - X_q J_q J_d = U_f J_q + (X_d - X_q)J_d J_q\end{aligned}$$

But from Figure 8.5

$$J_d = \frac{U_f - U_a\cos\delta}{X_d} \quad \text{and} \quad J_q = \frac{U_a \sin\delta}{X_q}$$

$$\begin{aligned}p_{\text{dev}}(t) &= U_f\left[\frac{U_a\sin\delta}{X_q}\right] + (X_d - X_q)\left[\frac{U_f - U_a\cos\delta}{X_d}\right]\left[\frac{U_a\sin\delta}{X_q}\right] \\ &= \frac{U_f U_a}{X_d}\sin(\delta) + \frac{(X_d - X_q)U_a^2}{2X_d X_q}\sin(2\delta) \\ &= 3\frac{E_f V_a}{X_d}\sin(\delta) + 3\frac{(X_d - X_q)V_a^2}{2X_d X_q}\sin(2\delta)\end{aligned} \tag{8.17}$$

which completely agrees with Chapter 7 results! Furthermore, this is the *instantaneous* power, which just happens to be constant for the balanced three-phase synchronous case!

We shall rework Example 7.4, from the perspective of Section 8.5.

Example 8.1

Rework Example 7.4abc, from the perspective of Section 8.5. Specifically for the machine of Example 7.4 operating as a generator, at rated stator output, $pf = 0.866$ lagging, find (a) delta; (b) all machine currents and voltages (draw the stator phasor diagram), powers, and torques; (c) the efficiency.

SOLUTION

(a) Observe that $\bar{U}_a = \bar{U}_{an} = 480\,\angle 0°$. Locating the d,q axes

$$jX_q\bar{J}_a + \bar{U}_a = j(1.647)(208.4\angle -30°) + 480\angle 0° = 716.2\angle 24.5°$$

Hence, the q-axis is located at 24.5° and the d-axis at 24.5° + 90° = 114.5°

(b) The d,q currents can easily be calculated:

$$\begin{aligned}J_d &= J_a\sin(30° + 24.5°) = 138.5\text{ A}; \qquad \bar{J}_d = 138.5\angle{-65.5°}\text{ A} \\ J_q &= J_a\cos(30° + 24.5°) = 120.9\text{ A}; \qquad \bar{J}_q = 120.9\angle{+24.5°}\text{ A} \\ \bar{U}_F &= jX_d\bar{J}_d + jX_q\bar{J}_q + \bar{U}_a = 866.7\angle 24.5°\text{ V}\end{aligned}$$

The MagC is provided in terms of the E_F of Chapter 7. Solving 7.4 for the field current

$$I_F = 14.48\ \text{A}; \qquad V_F = I_F R_F = (14.48)(7.8125) = 113.1\ \text{V}$$

Solving for the powers

$$\bar{S}_{3\text{ph}} = \bar{V}_a \bar{I}_a^* = 100\angle 30° = 86.6 + j50$$
$$P_{\text{out}} = 86.6\ \text{kW}; \qquad Q_{\text{out}} = 50\ \text{kvar}$$

Stator winding loss = SWL = $I_a^2 R_1 = 0$
Field winding loss = FWL = $I_F^2 R_F = 1.638\ \text{kW}$
Rotational loss = $P_{\text{RL}} = K_{\text{RL}}\omega_{\text{rm}} = 2.999\ \text{kW}$
Total machine loss = $P_{\text{LOSS}} = P_{\text{RL}}$ + SWL + RWL = 4.637 kW
Developed power = $P_{\text{dev}} = 86.60\ \text{kW}$
Mechanical power = $P_{\text{m}} = P_{\text{dev}} + P_{\text{RL}} = 89.60\ \text{kW}$ (120.1 hp)
Power input = P_{m} + FWL = 91.24 kW

Solving for the machine torques
Developed torque = $T_{\text{dev}} = P_{\text{dev}}/\omega_{\text{rm}} = 86.60/188.5 = 459.4\ \text{N m}$
Rotational loss torque = $T_{\text{RL}} = P_{\text{RL}}/\omega_{\text{rm}} = 2.999/188.5 = 15.9\ \text{N m}$
Mechanical input torque = $T_{\text{m}} = T_{\text{dev}} + T_{\text{RL}} = 475.3\ \text{N m}$
The relevant phasor diagram is shown in Figure 7.13a.

(c) Efficiency = $\eta = P_{\text{out}}/P_{\text{in}} = 86.60/91.24 = 94.92\%$

The approach in Chapter 8 is clearly more sophisticated. Not only is the total power captured in Equation 8.17, this is the *instantaneous power*, and is valid for any operating condition (balanced, unbalanced, sinusoidal, nonsinusoidal, etc.). When the situation merits this level of sophistication, we shall use the *0dqFDQU* model. However, for "ordinary" balanced three-phase constant speed analysis, we return to the methods of Chapter 7.

Consider the balanced three-phase, but with the machine running at a speed other than synchronous speed, i.e., $s \neq 0$

$$\omega_{\text{rm}} = (1-s)\omega_{\text{rm}}; \qquad \theta = (1-s)\omega t$$

The *0dq* voltages are

$$v_0 = 0$$
$$v_d = V_L \cos(s\omega t + \beta)$$
$$v_q = -V_L \sin(s\omega t + \beta)$$

Six of the seven windings are now active. Three windings (*d,q,F*) will contain DC and AC voltages and currents at frequency sf; three (*D,Q,U*) will contain AC voltages and currents at frequency sf. It is practical to solve such problems only with computer assistance.

8.6 Per-Unit Scaling as Applied to Synchronous Machines

It is customary for workers in synchronous machine analysis and simulation to use scaled data, scaled according to the so-called per-unit (pu) system. Historically, this scaling simplified to some extent rather complex calculations, which were of necessity made by hand using tables and the slide rule. Furthermore, machine constants expressed in pu fall in a fairly narrow range, independent of the machine's ratings. The use of the computer has largely eliminated the former advantage, but the latter remains as a justification for the method. Also, many computer programs perform their analysis in pu. In fact, manufacturer's data are frequently provided in pu form. Actually, pu scaling has a broader sphere of application, extending to the whole of power system analysis.

In fairness, we must point out that there are also disadvantages to pu scaling. We are one step further removed from the physical machine, rendering the analysis to be even more abstract. We are deprived of a time-honored engineering check; namely that of checking units, since scaling strips the units from all variables. In some scaling systems, the actual form of the machine model is altered by the act of scaling. In view of these difficulties, one might simply decide to abandon the whole idea, but realistically this is not an option. Too many colleagues, including those who supply machine data, continue to use the system for us to avoid it. Our purpose here is to understand its application to synchronous machines.[12]

The basic concept is simple enough

$$\text{per-unit value} = \frac{\text{actual value}}{\text{base value}}$$

where

actual value: the actual value of voltage, current, power, resistance, inductance, impedance, flux linkage, torque, speed, frequency, and time, in SI units. Actual values may be time-domain, frequency-domain, real or complex variables.

base value: the base value of voltage, current, power, resistance, inductance, impedance, flux linkage, torque, speed, frequency, and time, in SI units. Base values are real constants.

per-unit value: the scaled value of voltage, current, power, resistance, inductance, impedance, flux linkage, torque, speed, frequency, and time. Carries no units (dimensionless). "Per-unit" is abbreviated as "pu." Pu values may be time-domain, frequency-domain, real or complex variables.

The key to proper pu scaling is to understand base selection. The bases are divided into three sets: the stator bases, which apply to only stator windings ($a,b,c,0,d,q$); rotor bases, which apply to only rotor windings (F,D,Q,U); and the "universal" bases, which apply to both stator and rotor. First, define

S_{RATED} = total (3-phase) stator apparent power rating, in VA

V_{RATED} = nominal rms phase-to-phase (line) stator voltage, in V

$I_{\text{RATED}} = \dfrac{S_{\text{RATED}}}{\sqrt{3}V_{\text{RATED}}}$ = nominal rms phase (line) stator current, in A

f_{RATED} = normal stator operating frequency, in Hz

$\omega_{\text{RATED}} = 2\pi f_{\text{RATED}}$ = rated electrical radian frequency in electrical rad/s

$\Omega_{\text{RATED}} = \dfrac{\omega_{\text{RATED}}}{N_P/2}$ = rated (synchronous) speed, in mechanical rad/s

[12] The reader is cautioned that pu scaling practices are not universally the same. What is presented here is used by many workers and manufacturers, and differs only in minor details from that which a large majority of experts use. When interpreting pu data, make sure that you know all base values.

Now define *Universal Bases* (these bases apply in all machine windings):

$$S_{BASE} = \frac{S_{RATED}}{3}$$

$$\Omega_{BASE} = \Omega_{RATED}$$

$$\text{Torque}_{BASE} = \frac{S_{BASE}}{\Omega_{BASE}}$$

$$f_{BASE} = f_{RATED}, \quad \omega_{BASE} = \omega_{RATED}$$

$$t_{BASE} = \frac{1}{\omega_{BASE}}$$

One additional base is required for each winding to specify all additional bases. Since the three stator windings are identical, one stator voltage base is sufficient.

Stator Bases[13]

$$V_{BASE} = \frac{V_{RATED}}{\sqrt{3}}$$

All remaining stator bases can now be computed:

$$I_{BASE} = \frac{S_{BASE}}{V_{BASE}}, \quad Z_{BASE} = \frac{V_{BASE}}{I_{BASE}} = \frac{V_{BASE}^2}{S_{BASE}}$$

$$L_{BASE} = \frac{Z_{BASE}}{\omega_{BASE}}, \quad \lambda_{BASE} = L_{BASE}\, I_{BASE}$$

Rotor F Winding Current Base

The *F* winding base current is selected on the basis of constant mutual flux linkage. That is, the rotor base current is selected so as to produce the same air gap flux linkage as that caused by the stator base current. From the manufacturer's data, a constant called L_{ad} is available. The air gap flux linkage is

$$\lambda_{ag} = L_{ad} I_{BASE}$$

From the air gap line on the MagC, we can determine I_F at $E_f = V_{BASE}$. Call this current I_{AG}. The corresponding air gap flux is

$$\lambda_{ag} = \frac{V_{BASE}\sqrt{3}}{\omega_{BASE}} = \frac{V_{RATED}}{\omega_{BASE}}$$

[13] This is essentially the widely used "Rankin per-unit scaling system," proposed by A.W. Rankin in 1945, with one major difference. Rankin uses maximum (or peak), rather than rms, values for the voltage and current bases. Hence, in the Rankin system,

$$V_{BASE} = \frac{V_{RATED}\sqrt{2}}{\sqrt{3}}; \quad I_{BASE} = \frac{I_{RATED}\sqrt{2}}{\sqrt{3}}; \quad \text{so that} \quad S_{BASE} = 2V_{BASE}I_{BASE}$$

Since both V and I differ by the same factor ($\sqrt{2}$), Z_{BASE} (and L_{BASE}) is the same in both systems. If our primary concern is conversion of impedance (and inductance), this difference, while significant, has no impact on R, X, Z, and L values.

The *F* winding base current required for equal air gap flux linkage is therefore

$$I_{FBASE} = \frac{(L_{ad}\ I_{BASE})}{\left(\dfrac{V_{BASE}\sqrt{3}}{\omega_{BASE}}\right)} I_{AG} = \frac{(L_{ad}\omega_{BASE}\ I_{BASE})}{V_{RATED}} I_{AG}$$

All Remaining Rotor Windings

If the remaining rotor windings were electrically accessible, we could directly measure winding currents and voltages, and continue the procedure established for winding F (i.e., equal mutual flux linkage). Unfortunately, windings *DQU* are not normally electrically accessible, and information relevant to their constants must be obtained indirectly from measurements made in the stator windings or winding F. Therefore, it is impossible to obtain "true" base values for windings *DQU*. We can, however use the F base values for ALL rotor windings, and determine constants for equivalent *DQU* windings, equivalent to winding F in the sense that the turn ratios between all equivalent rotor windings is 1:1. This idea will be developed in more detail in the next section.

A final word with regard to notation. Base values will be identified with the subscript "BASE". However, no attempt will be made to discriminate between stator and rotor base values with special notation. The appropriate location will be clear from context. Likewise, no special notation will be used to discriminate between "actual" and "pu" values. In all cases, units shall be provided, using pu to indicate per-unit values. An example will be useful.

Example 8.2

Consider the generator below. For reciprocal pu

Machine Ratings
3ph apparent power rating: 100 MVA; frequency: 60 Hz
rms line-to-line voltage: 13.8 kV; I_{ag}: 250.0 A
$X_d = 1.60$ pu; $X_l = 0.19$ pu

For reciprocal pu scaling, determine

(a) the universal bases,
(b) the stator bases,
(c) the rotor bases.

SOLUTION

(a) Universal bases:

$$S_{BASE} = \frac{100}{3} = 33.33\,\text{MVA}; \qquad \Omega_{BASE} = 188.5\,\text{rad/s} = 1800\,\text{rpm}$$

$$\text{Torque}_{BASE} = \frac{S_{BASE}}{\Omega_{BASE}} = 176.8\,\text{kN m}$$

$$f_{BASE} = 60\,\text{Hz}; \qquad \omega_{BASE} = 377\,\text{rad/s}$$

$$t_{BASE} = \frac{1}{377} = 2.653\,\text{ms}$$

(b) Stator bases:

$$V_{BASE} = \frac{13.8}{\sqrt{3}} = 7.967\,\text{kV}$$

$$I_{BASE} = \frac{S_{BASE}}{V_{BASE}} = \frac{33.33}{7.967} = 4.184\,\text{kA}; \quad Z_{BASE} = \frac{V_{BASE}}{I_{BASE}} = 1.904\,\Omega$$

$$L_{BASE} = \frac{Z_{BASE}}{\omega_{BASE}} = 5.052\,\text{mH}; \quad \lambda_{BASE} = L_{BASE}\,I_{BASE} = 21.13\,\text{Wb}$$

(c) Rotor Bases (Referenced to Winding F):
From the manufacturer's data, L_{ad} can be computed:

$$X_{ad} = X_d - X_l = 1.60 - 0.19 = 1.41\,\text{pu}$$

Therefore

$$L_{ad} = 1.41\,\text{pu} = 7.123\,\text{mH}$$

$$I_{BASE} = \frac{(L_{ad}\omega_{BASE}\,I_{RATED})}{V_{RATED}} I_{AG} = \frac{(7.123)(0.377)(4.184)}{13.8}(250) = 203.5\,\text{A}$$

$$V_{BASE} = \frac{S_{BASE}}{I_{BASE}} = \frac{33.33}{0.2035} = 163.8\,\text{kV} \quad Z_{BASE} = \frac{V_{BASE}}{I_{BASE}} = 804.8\,\Omega$$

$$L_{BASE} = \frac{Z_{BASE}}{\omega_{BASE}} = 2135\,\text{mH} \quad \lambda_{BASE} = L_{BASE}\,I_{BASE} = 434.5\,\text{Wb}$$

8.7 The Tee Equivalent Circuits

Although the coupled circuit forms of Figure 8.3 and Figure 8.4 are well defined and accepted, another equivalent form (called the "Tee" form) has certain advantages. Consider the coupled equations

$$v_1 = L_{11}\frac{di_1}{dt} + L_{12}\frac{di_2}{dt}i_2 = (L_{11} - L_{12})\frac{di_1}{dt} + L_{12}\frac{d(i_1 + i_2)}{dt}$$
$$v_2 = L_{21}\frac{di_1}{dt} + L_{22}\frac{di_2}{dt} = L_{12}\frac{d(i_1 + i_2)}{dt} + (L_{22} - L_{12})\frac{di_2}{dt}$$

The first form suggests the circuit of Figure 8.5a, while the second conforms to the circuit of Figure 8.6b, the former referred to as the "coupled" circuit form, and the latter the "Tee" circuit form. From ports 1 and 2, they are equivalent. Two new inductances are formed (the leakage values),

$$L_{1\ell} = L_{11} - L_{12}; \quad L_{2\ell} = L_{22} - L_{12}$$

Now consider extension to three windings, as shown in Figure 8.6c:

$$v_1 = L_{11}\frac{di_1}{dt} + L_{12}\frac{di_2}{dt}i_2 + L_{13}\frac{di_3}{dt}$$
$$v_2 = L_{21}\frac{di_1}{dt} + L_{22}\frac{di_2}{dt} + L_{23}\frac{di_3}{dt}$$
$$v_3 = L_{31}\frac{di_1}{dt} + L_{32}\frac{di_2}{dt} + \hat{L}_{33}\frac{di_3}{dt}$$

Note that matters do not work out so neatly now. An extension to circuit of Figure 8.6d is possible only if all L_{ij}, $i \neq j$, are equal. However, all L_{ij}, $i \neq j$, are not equal in general for actual machines. Let us consider using pu scaling to force this condition.

Recall that there is a common power and frequency bases for all windings. One degree of freedom remains for each winding. We may select either V_{BASE}, I_{BASE}, Z_{BASE}, or L_{BASE} and all remaining bases are specified. Select V_{B1} (V_{BASE} in winding 1). Dividing v_1 by V_{B1},

$$\frac{v_1}{V_{B1}} = \frac{L_{11}(di_1/dt) + L_{12}(di_2/dt) + L_{13}(di_3/dt)}{V_{B1}}$$

Temporarily, use a special notation to indicate values which have been scaled into pu.[14] The term on the left-hand side is

$$\frac{v_1}{V_{B1}} = \hat{v}_1 \ (v_1 \text{ in pu})$$

Now

$$V_{B1} = Z_{B1}I_{B1} = L_{B1}\omega_B I_{B1} = L_{B1}\frac{I_{B1}}{t_B}$$

The first term on the right-hand side is

$$\frac{L_{11}(di_1/dt)}{L_{B1}(I_{B1}/t_B)} = \hat{L}_{11}\frac{d\hat{i}_1}{d\hat{t}}$$

which scales neatly into pu. However, the second and third terms are

$$\frac{L_{12}(di_2/dt)}{L_{B1}(I_{B1}/t_B)} = \frac{L_{12}(di_2/d\hat{t})}{L_{B1}I_{B1}} = \frac{L_{12}(di_2/d\hat{t})}{L_{B1}I_{B1}(I_{B2}/I_{B2})} = \frac{L_{12}(d\hat{i}_2/d\hat{t})}{L_{B1}(I_{B1}/I_{B2})}$$

$$\frac{L_{13}(di_3/dt)}{L_{B1}(I_{B1}/t_B)} = \frac{L_{13}(di_3/d\hat{t})}{L_{B1}I_{B1}} = \frac{L_{13}(di_3/d\hat{t})}{L_{B1}I_{B1}(I_{B3}/I_{B3})} = \frac{L_{13}(d\hat{i}_3/d\hat{t})}{L_{B1}(I_{B1}/I_{B3})}$$

To force the same mutual inductance

$$\hat{L}_M = \frac{L_{12}}{L_{B1}(I_{B1}/I_{B2})} = \frac{L_{13}}{L_{B1}(I_{B1}/I_{B3})} \quad \text{or} \quad L_{12}I_{B2} = L_{13}I_{B3}$$

The same argument starting with the v_2 and v_3 equations, produces

$$L_{21}I_{B1} = L_{23}I_{B3}$$
$$L_{32}I_{B2} = L_{31}I_{B1}$$

[14] This practice will be discontinued outside of this section. Since the same machine equations are valid in both SI units, and pu, there seems little point in two sets of variable symbols, particularly when appropriate units will always be indicated for all variables.

If the current bases are selected according to these constraints

$$\hat{L}_m = \hat{L}_{21} = \hat{L}_{23} = \hat{L}_{13}$$

and the equation set simplifies to $\hat{L}_{1\ell}$

$$\hat{v}_1 = \hat{L}_{1\ell}\frac{d\hat{i}_1}{d\hat{t}} + \hat{L}_m\frac{d\hat{i}_m}{d\hat{t}}$$
$$\hat{v}_2 = \hat{L}_{2\ell}\frac{d\hat{i}_2}{d\hat{t}} + \hat{L}_m\frac{d\hat{i}_m}{d\hat{t}}$$
$$\hat{v}_3 = \hat{L}_{3\ell}\frac{d\hat{i}_3}{d\hat{t}} + \hat{L}_m\frac{d\hat{i}_m}{d\hat{t}}$$

where

$$\hat{i}_m = \hat{i}_1 + \hat{i}_2 + \hat{i}_3$$
$$\hat{L}_{1\ell} = \hat{L}_{11} - \hat{L}_m; \quad \hat{L}_{2\ell} = \hat{L}_{22} - \hat{L}_m; \quad \hat{L}_{3\ell} = \hat{L}_{33} - \hat{L}_m$$

The corresponding equivalent Tee circuit appears in Figure 8.6b. We shall see how Tee circuits can be used to advantage for the d, q machine equivalent circuits in the next section.[15]

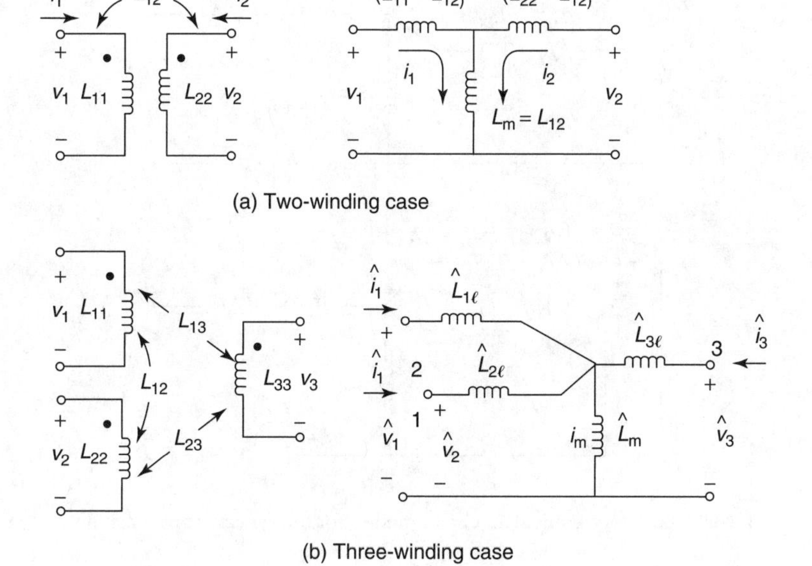

FIGURE 8.6. Equivalent Tee circuits.

[15] Unfortunately, for four or more coupled windings, this condition cannot be met.

8.8 *0dqFDQU* Constants Derived from Manufacturer's Data

As we have previously noted, mathematical models are of little use without accessibility to the constants required to apply the model to a specific machine. Such constants are determined either by direct computation from machine design parameters, by direct measurement, or indirectly from standard machine tests. Of course, the machine owner in theory could directly perform the necessary tests him/herself. However, usually in practice, this approach is either too costly, or the owner lacks the necessary test facilities, or expertise. The typical procedure for obtaining necessary data is to request it from the machine manufacturer. Unfortunately, the necessary data to directly implement the *0dqFDQU* model are not routinely available.[16] However, information can be obtained from certain standard tests, and certain machine constants are available. From this, the requisite constants can be determined, or at least estimated. We focus in this section on how to obtain *0dqFDQU* constants from "manufacturer's data," that is, which is routinely available from machine manufacturers. It is instructive to view the machine model in terms of its Tee equivalent circuits, as shown in Figure 8.7.

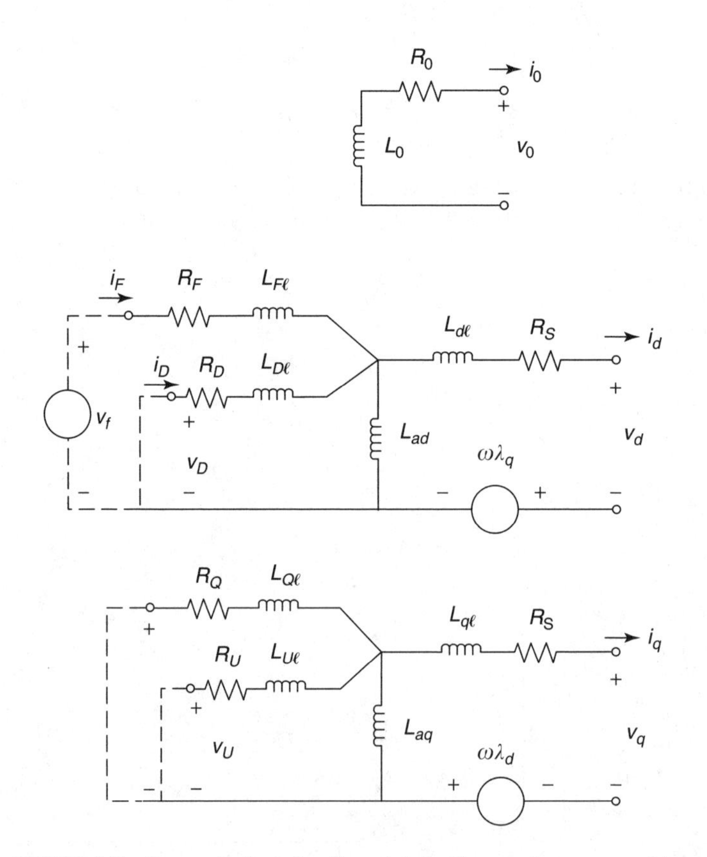

FIGURE 8.7. Tee-equivalent circuit models for the synchronous machine.

[16] This has always been somewhat of a curiosity. Most electrical components, such as capacitors, transistors, integrated circuits, etc. have data sheets that provide relevant comprehensive technical information. EM machines, which are much more complicated and costly, rarely have easily accessible and comprehensive data which is needed for engineering analysis. Of course, much data are available. Unfortunately, much is not.

Data required for the direct implementation of the *0dqFDQU* model are not routinely available for several reasons:

- *Historical practice*: for many years, machine data have been provided in a particular format, primarily because it is easily obtained from standard test data. As with any established procedure, there is an inherent reluctance to change.
- *Practicality*: as was noted, it may be difficult, and therefore costly, to determine complete data.
- *Demand*: until recently, there has been little demand for complete machine data, since engineers did not have the perceived need for, or capability to perform, general analysis.
- *Existing databases for older machines*: even if complete data were provided for all new applications, analyses performed on existing installations would create a need for data.

Therefore, there is a need to examine how *0dqFDQU* data may be obtained from manufacturer's data. The manufacturer's data set may be divided into three subsets: ratings, curves, and constants.

Ratings:

S_{RATED} = total (3-phase) stator apparent power rating, in VA
V_{RATED} = nominal rms phase-to-phase (line) stator voltage, in V
f_{RATED} = normal stator operating frequency, in Hz
$\omega_{RATED} = 2\pi f_{RATED}$ = rated electrical frequency, in electrical rad/s

$$\Omega_{RATED} = \frac{\omega_{RATED}}{N_P/2} = \text{rated (synchronous) speed, in mechanical rad/s}$$

Curves:

The magnetization curve (MagC)
Air gap (ag) line (linear extension of the unsaturated MagC)
The short circuit characteristic (SCC)
Zero power factor lagging curve (ZPF)

Constants:

R_a = stator winding phase (Y) resistance, in pu
X_l = stator winding phase (Y) leakage reactance, in pu
X_d = direct axis synchronous reactance, in pu
X'_d = direct axis transient reactance, in pu
X''_d = direct axis subtransient reactance, in pu
X_q = quadrature axis synchronous reactance, in pu
X'_q = quadrature axis transient reactance, in pu
X''_q = quadrature axis subtransient reactance, in pu
τ_a = armature (stator) winding time constant, in s
τ'_{d0} = direct axis open-circuit transient time constant, in s
τ''_{d0} = direct axis open-circuit subtransient time constant, in s
τ'_{q0} = quadrature axis open-circuit transient time constant, in s
τ''_{q0} = quadrature axis open-circuit subtransient time constant, in s
τ'_d = direct axis short-circuit transient time constant, in s
τ''_d = direct axis short-circuit subtransient time constant, in s
τ'_q = quadrature axis short-circuit transient time constant, in s
τ''_q = quadrature axis short-circuit subtransient time constant, in s

The problem is to determine the $0dqFDQU$ data.

Resistances (6): $R_s, R_0, R_F, R_D, R_Q, R_U$
Inductances (15): $L_d, L_{dF}, L_{dD}, L_q, L_{qQ}, L_{qU}, L_0, L_F, L_D, L_Q, L_U, L_{FD}, L_{SD}, L_{SQ}, L_{QU}$

Step 1. Recognize that the given data is in pu and:

$$L = X \text{ (in pu and for } \omega = \omega_{BASE});$$
$$R_s = R_0 = R_a$$
$$L_d = X_a; \quad L_q = X_q$$
$$X_{ad} = L_{ad} = X_d - X_\ell;$$
$$L_{dF} = L_{ad} = L_{dD} = L_{DF}$$
$$X_{aq} = L_{aq} = X_q - X_\ell;$$
$$L_{qU} = L_{qd} = L_{qQ} = L_{UQ}$$

Step 2. The d-axis circuit inductance, looking in from the d-port, neglecting the resistance, including the F-winding, but excluding the D-winding, is approximately

$$X'_d = L'_d = L_\ell + \frac{1}{\frac{1}{L_{ad}} + \frac{1}{L_{F\ell}}} \quad \text{or} \quad L_{F\ell} = \frac{1}{\left(\frac{1}{L'_d - L_\ell}\right) - \frac{1}{L_{ad}}}$$

then

$$L_{FF} = L_{ad} + L_{F\ell}$$

Step 3. The d-axis circuit inductance, looking in from the d-port, neglecting the resistance, and including the F- and D-windings, is approximately

$$X''_d = L''_d = L_\ell + \frac{1}{\frac{1}{L_{ad}} + \frac{1}{L_{F\ell}} + \frac{1}{L_{D\ell}}} \quad \text{or} \quad L_{D\ell} = \frac{1}{\left(\frac{1}{L''_d - L_\ell}\right) - \frac{1}{L_{ad}} - \frac{1}{L_{F\ell}}}$$

then

$$L_{DD} = L_{ad} + L_{D\ell}$$

Step 4. The q-axis circuit inductance, looking in from the q-port, neglecting the resistance, including the U-winding, but excluding the Q-winding, is approximately

$$X'_q = L'_q = L_\ell + \frac{1}{\frac{1}{L_{aq}} + \frac{1}{L_{U\ell}}} \quad \text{or} \quad L_{U\ell} = \frac{1}{\left(\frac{1}{L'_q - L_\ell}\right) - \frac{1}{L_{aq}}}$$

then

$$L_{UU} = L_{aq} + L_{U\ell}$$

Step 5. The q-axis circuit inductance, looking in from the q-port, neglecting the resistance, and including the U- and Q-windings, is approximately

$$X''_q = L''_q = L_\ell + \frac{1}{\frac{1}{L_{aq}} + \frac{1}{L_{U\ell}} + \frac{1}{L_{Q\ell}}} \quad \text{or} \quad L_{D\ell} = \frac{1}{\left(\frac{1}{L''_q - L_\ell}\right) - \frac{1}{L_{aq}} - \frac{1}{L_{U\ell}}}$$

then

$$L_{QQ} = L_{aq} + L_{Q\ell}$$

All inductances are now determined. To find the resistances, we can use either the (a) open-circuit time constants, or (b) the short-circuit time constants. First, recall that the time constants are in seconds, and we wish to use pu throughout. To convert:

$$\tau_{\text{PER-UNIT}} = \frac{\tau_{\text{SECONDS}}}{t_{\text{BASE}}} = \omega_{\text{BASE}}\tau_{\text{SECONDS}}$$

Step 6a. Looking in from the terminals of R_F, neglecting all other resistances and the *D*-winding, and placing the *d*-winding on open circuit, L_{EQ} is approximately

$$L_{EQ} = L_{ad} + L_{F\ell}$$

then

$$R_F = \frac{L_{EQ}}{\tau'_{d0}}$$

Step 6b. Looking in from the terminals of R_F, neglecting all other resistances and the *D*-winding, and placing the *d*-winding on short circuit, L_{EQ} is approximately

$$L_{EQ} = L_{F\ell} + \frac{1}{\dfrac{1}{L_{ad}} + \dfrac{1}{L_\ell}}$$

then

$$R_F = \frac{L_{EQ}}{\tau'_d}$$

Step 7a. Looking in from the terminals of R_D, neglecting all other resistances, and placing the *d*-winding on open circuit, L_{EQ} is approximately

$$L_{EQ} = L_{D\ell} + \frac{1}{\dfrac{1}{L_{ad}} + \dfrac{1}{L_{F\ell}}}$$

then

$$R_D = \frac{L_{EQ}}{\tau''_{d0}}$$

Step 7b. Looking in from the terminals of R_D, neglecting all other resistances, and placing the *d*-winding on short circuit, L_{EQ} is approximately

$$L_{EQ} = L_{D\ell} + \frac{1}{\dfrac{1}{L_{ad}} + \dfrac{1}{L_{F\ell}} + \dfrac{1}{L_\ell}}$$

then

$$R_D = \frac{L_{EQ}}{\tau''_d}$$

Step 8a. Looking in from the terminals of R_U, neglecting all other resistances and the Q-winding, and placing the q-winding on open circuit, L_{EQ} is approximately

$$L_{EQ} = L_{aq} + L_{U\ell}$$

then

$$R_U = \frac{L_{EQ}}{\tau'_{q0}}$$

Step 8b. Looking in from the terminals of R_U, neglecting all other resistances and the Q-winding, and placing the q-winding on short circuit, L_{EQ} is approximately

$$L_{EQ} = L_{U\ell} + \frac{1}{\dfrac{1}{L_{aq}} + \dfrac{1}{L_\ell}}$$

then

$$R_U = \frac{L_{EQ}}{\tau'_q}$$

Step 9a. Looking in from the terminals of R_Q, neglecting all other resistances, and placing the d-winding on open circuit, L_{EQ} is approximately

$$L_{EQ} = L_{Q\ell} + \frac{1}{\dfrac{1}{L_{aq}} + \dfrac{1}{L_{U\ell}}}$$

then

$$R_Q = \frac{L_{EQ}}{\tau''_{q0}}$$

Step 9b. Looking in from the terminals of R_Q, neglecting all other resistances, and placing the q-winding on short circuit, L_{EQ} is approximately

$$L_{EQ} = L_{Q\ell} + \frac{1}{\dfrac{1}{L_{aq}} + \dfrac{1}{L_{U\ell}} + \dfrac{1}{L_\ell}}$$

then

$$R_Q = \frac{L_{EQ}}{\tau''_q}$$

All constants are now determined. An example would be useful.

Example 8.3
Manufacturer's data for the machine of Example 8.2 follows:

Machine Ratings

3ph apparent power rating: 100 MVA; Frequency: 60 Hz
rms line-to-line voltage: 13.8 kV; I_{ag}: 250.0 A

Base Values

$S_{BASE} = S_b = 33.333$ MVA; Time = 2.653 ms; $w_b = 377$ rad/s

Winding	V_{BASE} (kV)	I_{BASE} (kA)	Z_{BASE} (Ω)	L_{BASE} (mH)	λ_{BASE} (Wb)
Stator (*abc0dq*)	7.967	4.184	1.9044	5.052	21.134
Rotor (*FDQU*)	163.787	0.20352	804.8	2134.8	434.5

Manufacturer's Generator Data (in pu or s)

X_d = 1.60000	X'_d = 0.40000	X''_d = 0.20000
X_q = 1.50000	X'_q = 0.45000	X''_q = 0.20000
X_1 = 0.19000	r_a = 0.00200	T_a = 0.26525
X_0 = 0.20000	X_2 = 0.14850	H = 5.00000
T'_d = 2.50000	T''_d = 0.01500	T'_q = 0.15000
T'_{d0} = 10.00000	T''_{d0} = 0.03000	T'_{q0} = 0.50000
T''_q = 0.01775	T''_{q0} = 0.04000	

(a) Convert manufacturer's data into SI units.
(b) Determine *0dqFDQU* constants from manufacturer's data.

SOLUTION

(a) Base values were determined in Example 8.2 and are repeated above. Conversions are straightforward. The results are

Manufacturer's Generator Data (in Ω or s)

X_d = 3.04704	X'_d = 0.76176	X''_d = 0.38088
X_q = 2.85660	X'_q = 0.85698	X''_q = 0.38088
X_1 = 0.36184	r_a = 0.00381	T_a = 0.26525
X_0 = 0.38088	X_2 = 0.28280	H = 5.00000
T'_{d0} = 10.00000	T''_{d0} = 0.03000	T'_{q0} = 0.50000
T'_d = 2.50000	T''_d = 0.01500	T'_q = 0.15000
T''_{q0} = 0.04000	T''_q = 0.01775	

(b) Using the methods of Section 8.8

Step 1. Recognize that

$$R_s = R_0 = R = 0.0381\ \text{pu}$$
$$L_d = X_d = 1.60\ \text{pu}; \quad L_q = X_q = 1.50\ \text{pu}$$
$$X_{ad} = X_d - X_f = 1.41\ \text{pu} = L_{dF} = X_{ad} = L_{ad} = L_{dD} = L_{DF}$$
$$X_{aq} = X_q - X_f = 1.31\ \text{pu} = L_{qU} = X_{aq} = L_{aq} = L_{qQ} = L_{QU}$$

Step 2.

$$X'_d = L'_d$$

$$L_{F\ell} = \frac{1}{\left(\dfrac{1}{L'_d - L_\ell}\right) - \dfrac{1}{L_{ad}}} = \frac{1}{\left(\dfrac{1}{0.40 - 0.19}\right) - \dfrac{1}{1.41}} = 0.2468\,\text{pu}$$

then $L_{FF} = L_{ad} + L_{F1} = 1.657\,\text{pu}$

Step 3.

$$X''_d = L''_d$$

$$L_{D\ell} = \frac{1}{\left(\dfrac{1}{L''_d - L_\ell}\right) - \dfrac{1}{L_{ad}} - \dfrac{1}{L_{F\ell}}} = \frac{1}{\left(\dfrac{1}{0.20 - 0.19}\right) - \dfrac{1}{1.41} - \dfrac{1}{0.2468}} = 0.0105\,\text{pu}$$

then $L_{DD} = L_{ad} + L_{D1} = 0.0105 + 1.41 = 1.4205\,\text{pu}$

Step 4.

$$X'_q = L'_q$$

$$L_{U\ell} = \frac{1}{\left(\dfrac{1}{L'_q - L_\ell}\right) - \dfrac{1}{L_{aq}}} = \frac{1}{\left(\dfrac{1}{0.45 - 0.19}\right) - \dfrac{1}{1.31}} = 0.3244\,\text{pu}$$

then $L_{UU} = L_{aq} + L_{U\ell} = 1.634\,\text{pu}$

Step 5.

$$X''_q = L''_q$$

$$L_{Q\ell} = \frac{1}{\left(\dfrac{1}{L''_q - L_\ell}\right) - \dfrac{1}{L_{aq}} - \dfrac{1}{L_{U\ell}}} = \frac{1}{\left(\dfrac{1}{0.20 - 0.19}\right) - \dfrac{1}{1.31} - \dfrac{1}{0.3244}} = 0.0101\,\text{pu}$$

then $L_{QQ} = L_{aq} + L_{Q\ell} = 1.31 + 0.0109 = 1.3201\,\text{pu}$

All inductances are now determined. We compute resistances from the time constants, which must be converted from seconds into pu.

$$\tau_{\text{PER-UNIT}} = \omega_{\text{BASE}}\tau_{\text{SECONDS}}$$

Step 6a. We choose the open-circuit time constants

$$L_{EQ} = L_{ad} + L_{F1} = 1.60\,\text{pu}; \quad \tau'_{d0} = (377)(10) = 3770\,\text{pu};$$

$$R_F = \frac{L_{EQ}}{\tau'_{d0}} = \frac{1.60}{3770} = 0.00044\,\text{pu}$$

Step 7a.

$$L_{EQ} = L_{D\ell} + \frac{1}{\dfrac{1}{L_{ad}} + \dfrac{1}{L_{F\ell}}} = 0.0105 + \frac{1}{\dfrac{1}{1.41} + \dfrac{1}{0.2468}} = 0.2205\,\text{pu}$$

then

$$R_D = \frac{L_{EQ}}{\tau''_{d0}} = \frac{0.2205}{0.03(377)} = 0.0195\,\text{pu}$$

Step 8a.

$$L_{EQ} = L_{aq} + L_{U\ell} = 1.31 + 0.3244 = 1.634\,\text{pu}$$
$$R_U = \frac{L_{EQ}}{\tau'_{q0}} = \frac{1.634}{0.5(377)} = 0.0087\,\text{pu}$$

Step 9a.

$$L_{EQ} = L_{Q\ell} + \frac{1}{\dfrac{1}{L_{aq}} + \dfrac{1}{L_{U\ell}}} = 0.0101 + \frac{1}{\dfrac{1}{1.31} + \dfrac{1}{0.3244}} = 0.2701\,\text{pu}$$

then

$$R_Q = \frac{L_{EQ}}{\tau''_{q0}} = \frac{0.2701}{0.04(377)} = 0.01791\,\text{pu}$$

We shall re-compute resistances from the short-circuit time constants

Step 6b.

$$L_{EQ} = L_{F\ell} + \frac{1}{\dfrac{1}{L_{ad}} + \dfrac{1}{L_{\ell}}} = 0.2468 + \frac{1}{\dfrac{1}{1.41} + \dfrac{1}{0.19}} = 0.4142\,\text{pu}$$
$$R_F = \frac{L_{EQ}}{\tau'_d} = \frac{0.4142}{2.5(377)} = 0.00044\,\text{pu}$$

Step 7b.

$$L_{EQ} = L_{D\ell} + \frac{1}{\dfrac{1}{L_{ad}} + \dfrac{1}{L_{F\ell}} + \dfrac{1}{L_{\ell}}} = 0.0105 + \frac{1}{\dfrac{1}{1.41} + \dfrac{1}{0.2468} + \dfrac{1}{0.19}} = 0.1103\,\text{pu}$$

then

$$R_D = \frac{L_{EQ}}{\tau''_d} = \frac{0.1103}{0.015(377)} = 0.0195\,\text{pu}$$

Step 8b.

$$L_{EQ} = L_{U\ell} + \frac{1}{\dfrac{1}{L_{aq}} + \dfrac{1}{L_\ell}} = 0.3244 + \frac{1}{\dfrac{1}{1.31} + \dfrac{1}{0.19}} = 0.4903\,\text{pu}$$

then

$$R_U = \frac{L_{EQ}}{\tau'_q} = \frac{0.4903}{0.15(377)} = 0.00867\,\text{pu}$$

Step 9b.

$$L_{EQ} = L_{Q\ell} + \frac{1}{\dfrac{1}{L_{aq}} + \dfrac{1}{L_{U\ell}} + \dfrac{1}{L_\ell}} = 0.0101 + \frac{1}{\dfrac{1}{1.31} + \dfrac{1}{0.3244} + \dfrac{1}{0.19}} = 0.1199\,\text{pu}$$

then

$$R_Q = \frac{L_{EQ}}{\tau''_q} = \frac{0.1199}{0.01775(377)} = 0.01791\,\text{pu}$$

A summary of results follows:

Analytical Generator Data (in pu)

L_d = 1.60000	l_d = 0.19000	L_{ad} = 1.41000
L_q = 1.50000	l_q = 0.19000	L_{aq} = 1.31000
L_F = 1.65675	l_F = 0.24675	r_F = 0.00044
L_D = 1.42050	l_D = 0.01050	r_D = 0.01950
L_P = 1.63438	l_U = 0.32438	r_U = 0.00867
L_Q = 1.32008	l_Q = 0.01008	r_Q = 0.01791
L_0 = 0.20000		$r_d = r_q = r_0$ = 0.00200

Conversions to SI units are straightforward. The results are

Analytical Generator Data (in mH or Ω)

L_d = 8.08252	l_d = 0.95980	L_{ad} = 7.12272
L_q = 7.57737	l_q = 0.95980	L_{aq} = 6.61757
L_F = 3536.78	l_F = 526.75	r_F = 0.35368
L_D = 3032.44	l_D = 22.42	r_D = 15.69054
L_P = 3489.02	l_P = 692.48	r_P = 6.97805
L_Q = 2818.06	l_Q = 21.51	r_Q = 14.41
L_0 = 1.01032		$r_d = r_q = r_0$ = 0.00381

8.9 0*dqFDQU* Model Performance

The complexity of the 0*dqFDQU* model prohibits hand analysis of all but the simplest cases. Beyond this, computer assistance is required. As usual, we study the problem through an example.

Example 8.4
For the machine of Example 8.2

(a) The machine considered operates as a generator running at synchronous speed at rated load. *pf* = 0.866 lagging, driven by an ideal prime mover (speed cannot change) and an ideal exciter (ideal DC voltage source). The *DQU* windings are shorted. Determine

$$\hat{v}_{0dqFDQU}; \quad \hat{i}_{0dqFDQU}; \quad \hat{\lambda}_{0dqFDQU}; \quad \hat{v}_{abc}; \quad \hat{i}_{abc}$$

(b) From (a), compute P_{dev}.
(c) Now at $t = 0$, a short circuit is suddenly applied to the machine stator terminals. Summarize the equations that must be solved to determine the currents.

SOLUTION
We will work in pu throughout.

(a) At rated conditions, *pf* = 0.866 lagging

$$v_a = 1.414\cos(\omega t); \quad v_b = 1.414\cos(\omega t - 120°); \quad v_c = 1.414\cos(\omega t + 120°)$$
$$i_a = 1.414\cos(\omega t - 30°); \quad i_b = 1.414\cos(\omega t - 150°); \quad i_c = 1.414\cos(\omega t + 90°)$$

Using the $0dq$ transformation

$$v_0 = 0; \quad v_d = -1.0362; \quad v_q = 1.3951$$
$$i_0 = 0; \quad i_d = -1.5884; \quad i_q = 0.6908$$
$$v_D = v_Q = v_U = 0; \quad i_D = i_Q = i_U = 0$$

Now all "$p\lambda$" terms are 0, since all "λ" terms are DC.

$$-v_d = 1.0362 = R_s i_d + 0 + \omega\lambda_q = (0.002)(-1.5884) + \lambda_q; \quad \lambda_q = 1.0395$$
$$-v_q = -1.3951 = R_s i_q + 0 - \omega\lambda_d = (0.002)(0.6908) - \lambda_d; \quad \lambda_d = 1.3946$$

Solving for the field current

$$\lambda_d = 1.3946 = L_d i_d + L_{dF} i_F + L_{dD} i_D = 1.6(-1.5584) + 1.41\, i_F + 0; \quad i_F = 2.792$$

Solving for the remaining λ's

$$\lambda_F = L_F i_F + L_{dF} i_d + L_{FD} i_D = 1.657(2.792) + 1.41(-1.0362) + 0 = 3.165$$
$$\lambda_D = L_D i_D + L_{Dd} i_d + L_{FD} i_F = 0 + (1.6)(-1.5884) + 1.41(2.792) = 1.395$$
$$\lambda_Q = L_Q i_Q + L_{Qq} i_q + L_{QU} i_U = 0 + 1.31(0.6908) + 0 = 0.9049$$
$$\lambda_U = L_U i_U + L_{Uq} i_q + L_{QU} i_Q = 0 + 1.31(0.6908) + 0 = 0.9049$$
$$v_F = R_F i_F = 0.0004395(2.792) = 0.001227$$

The entries in

$$\hat{v}_{0dqFDQU}; \quad \hat{i}_{0dqFDQU}; \quad \hat{\lambda}_{0dqFDQU}; \quad \hat{v}_{abc}; \quad \hat{i}_{abc}$$

are now determined.

(b) From (a), compute P_{dev}.

$$P_{\text{dev}}(t) = \omega\lambda_d i_q - \omega\lambda_q i_d = 1.0(1.3946)(0.6908) - 1.0(1.0395)(-1.5884)$$
$$= 2.604\ \text{pu}$$

Note that $\text{SWL} = i_d^2 R_s + i_q^2 R_s = 0.0060$

$$P_{\text{out}}(t) = P_{\text{dev}}(t) - \text{SWL} \ = \ i_d^2 R_s + i_q^2 R_s = 2.604 - 0.0060 = 2.598\ \text{pu}$$

Observe that $P_{\text{out}} = pfS_{\text{out}} = 3.0(0.8661) = 2.598\ \text{pu}$

(c) Now at $t = 0$, a short circuit is suddenly applied to the machine stator terminals. It follows that:

$$v_0 = v_d = v_q = v_D = v_Q = v_U = 0; \quad v_F = 0.001227$$

The machine equations are:

$$\begin{aligned}
-v_0 &= 0 = R_s i_0 + p\lambda_0 \\
-v_d &= 0 = R_s i_d + p\lambda_d + \omega\lambda_q \\
-v_q &= 0 = R_s i_q + p\lambda_q - \omega\lambda_d \\
v_F &= 0.01227 = R_F i_F + p\lambda_F \\
v_D &= 0 = R_D i_D + p\lambda_D \\
v_Q &= 0 = R_Q i_Q + p\lambda_Q \\
v_U &= 0 = R_U i_U + p\lambda_U
\end{aligned}$$

where $\hat{i}_{0dqFDQU} = [L_{0dqFDQU}]^{-1}\hat{\lambda}_{0dqFDQU}$

and $\hat{i}_{abc} = [P]^T \hat{i}_{0dq}$

The necessary initial conditions for the λ's are the values determined in (a). To solve the set, MATLAB® is recommended.

8.10 Summary

It was noted that the synchronous machine analysis presented in Chapter 7 was essentially restricted to balanced three-phase steady-state constant speed operation. In Chapter 8, we have extended the work to general synchronous machine modeling, placing no restrictions on speed or terminal voltage. The model was based on a coupled circuits approach, and required two fundamental assumptions:

- all windings are sinusoidally distributed, and
- magnetic linearity is assumed throughout the machine.

We discovered that implementation of a mathematical transformation (i.e., the $0dq$ transformation) drastically simplified the analysis. We also discussed the problem of scaling the analysis into pu. We investigated the problem of procurement of traditional constants from machine manufacturers, and conversion into the constants required by the coupled circuits model. Finally, we reconciled the model with the work presented in Chapter 7.

Problems

8.1 Create a drawing of a four winding (rotor windings $A'A'$; B–B'; stator windings a–a'; b–b') salient machine, in the manner of Figure 8.1a. Develop the coupled circuit machine equations, producing a development that parallels that of Section 8.1.

8.2 Show that

$$[T][R_{abcFDQ}][T]^T \quad \text{reduces to} \quad [R_{abcFDQ}]$$

8.3 Show that

$$[L_{0dqFDQ}] = [T][L_{abcFDQ}][T]^T$$

simplifies to

$$[L_{0dqFDQ}] = \begin{bmatrix} L_{00} & 0 & 0 & 0 & 0 & 0 \\ 0 & L_{dd} & 0 & L_{dF} & L_{dD} & 0 \\ 0 & 0 & L_{qq} & 0 & 0 & L_{qQ} \\ 0 & L_{Fd} & 0 & L_{FF} & L_{FD} & 0 \\ 0 & L_{Dd} & 0 & L_{DF} & L_{DD} & 0 \\ 0 & 0 & L_{Qq} & 0 & 0 & L_{QQ} \end{bmatrix}$$

where

$$L_{00} = L_0 = L_s - 2L_m$$
$$L_{dd} = L_d = L_s + M_s + \frac{3}{2}L_m$$
$$L_{qq} = L_q = L_s + M_s - \frac{3}{2}L_m$$
$$L_{dF} = L_{Fd} = kL_{SF} = \sqrt{3/2}\,L_{SF}$$
$$L_{dD} = L_{Dd} = kL_{SD} = \sqrt{3/2}\,L_{SD}$$
$$L_{qQ} = L_{Qq} = kL_{SQ} = \sqrt{3/2}\,L_{SQ}$$

8.4 Show that

$$[T][p\{[L_{abcFDQ}][T]^T\}] = \omega[G_{0dqFDQ}] = \begin{bmatrix} 0 & 0 & 0 & 0 & 0 & 0 \\ 0 & 0 & +\omega L_q & 0 & 0 & +\omega L_{qQ} \\ 0 & -\omega L_d & 0 & -\omega L_{dF} & -\omega L_{dD} & 0 \\ 0 & 0 & 0 & 0 & 0 & 0 \\ 0 & 0 & 0 & 0 & 0 & 0 \\ 0 & 0 & 0 & 0 & 0 & 0 \end{bmatrix}$$

8.5 Rework Example 7.4(a) to (c), from the perspective of Section 8.5. Specifically for the machine of Example 7.4 operating as a generator, at rated stator output, $pf = 0.866$ lagging, find (a) delta; (b) all machine currents and voltages (draw the stator phasor diagram), powers, and torques; (c) the efficiency.

8.6 Repeat Problem 8.5 for unity *pf*.

8.7 Repeat Problem 8.5 for $pf = 0.866$ leading.

8.8. For the machine of Example 7.4 operating as a motor, at rated stator input, $pf = 0.866$ leading, find (a) delta; (b) all machine currents and voltages (draw the stator phasor diagram), powers, and torques; (c) the efficiency.

8.9 Repeat Problem 8.8 for unity *pf*.

8.10 Repeat Problem 8.8 for $pf = 0.866$ lagging.

8.11 Given the following Manufacturer's data:

Machine Ratings			
3ph apparent power rating:	120.000 MVA;	Frequency:	60.0 Hz
rms line-to-line voltage:	13.800 kV;	$I_{f_{ag}}$:	275.0 A

Magnetization characteristic: $I_f = E/K_{ag} + A_x \exp(B_{xE})$
E = Stator LN rms voltage, kV; I_f = DC field current, A
$K_{ag} = 28.975$; $A_x = 0.00000$; $B_x = 0.00054406$

Determine the *reciprocal* pu base values.
Answers

Per-unit System: Reciprocal

$S_{BASE} = S_b = 40.000$ MVA; Time = 2.653 ms; $w_b = 377.0$ rad/s

Winding	V_b (kV)	I_b (kA)	Z_b (Ω)	L_b (mH)	λ (Wb)
Stator (*abc0dq*)	7.967	5.020	1.5870	4.210	21.134
Rotor (*FDQP*)	182.561	0.21910	833.2	2210.2	484.3

8.12 For the machine of Problem 8.11, determine the *Rankin* pu base values.
Answers

Per-unit System: Rankin

$S_{BASE} = S_b = 120.000$ MVA; Time = 2.653 ms; $w_b = 377.0$ rad/s

Winding	V_b (kV)	I_b (kA)	Z_b(Ω)	L_b (mH)	λ (Wb)
Stator (*abc0dq*)	11.268	7.100	1.5870	4.210	29.888
Rotor (*FDQP*)	316.206	0.37950	833.2	2210.2	684.8

8.13 For the machine of Problem 8.11, the following manufacturer's data are given:

Manufacturer's Generator Data (in reciprocal pu or s)

X_d = 1.55000	X'_d = 0.35000	X''_d = 0.18000
X_q = 1.40000	X'_q = 0.42000	X''_q = 0.18000
X_1 = 0.17000	r_a = 0.00300	T_a = 0.15915
X_0 = 0.08000	X_2 = 0.14850	H = 7.00000
T'_d = 2.03226	T''_d = 0.01543	T'_q = 0.15000
T'_{d0} = 9.00000	T''_{d0} = 0.03000	T'_{q0} = 0.50000
T''_q = 0.01711	T''_{q0} = 0.04000	
R_e = 0.00000	X_e = 0.00000	

Convert the data into SI units (Ω, s).

8.14 For the machine of Problem 8.13, determine the machine's analytical data

(a) in (reciprocal) pu.
(b) in SI units (Ω, s).

Answers

Analytical Generator Data (in pu)

L_d = 1.55000	l_d = 0.17000	L_{ad} = 1.38000
L_q = 1.40000	l_q = 0.17000	L_{aq} = 1.23000
L_F = 1.58700	l_F = 0.20700	r_F = 0.00047
L_D = 1.39059	l_D = 0.01059	r_D = 0.01685
L_P = 1.54378	l_P = 0.31378	r_P = 0.00819
L_Q = 1.24008	l_Q = 0.01008	r_Q = 0.01725
L_0 = 0.08000		$r_d = r_q = r_0$ = 0.00300

8.15 For the machine of Problem 8.13, determine the machine's analytical data in SI units (Ω, s).
Answers

Analytical Generator Data (in mH or Ω)

L_d = 6.52495	l_d = 0.71564	L_{ad} = 5.80931
L_q = 5.89351	l_q = 0.71564	L_{aq} = 5.17787
L_F = 3507.55	l_F = 457.51	r_F = 0.38973
L_D = 3073.44	l_D = 23.40	r_D = 14.04111
L_P = 3412.01	l_P = 693.50	r_P = 6.82403
L_Q = 2740.80	l_Q = 22.28	r_Q = 14.37
L_0 = 0.33677		$r_d = r_q = r_0$ = 0.00476

8.16 The machine of Problem 8.13 is driven by an ideal prime mover (speed is synchronous and cannot change) and an ideal exciter (ideal DC voltage source). The *DQU* windings are shorted. The machine operates as a generator at rated load, $pf = 0.866$ lagging. Determine

$\hat{v}_{0dqFDQU}$; $\hat{i}_{0dqFDQU}$; $\hat{\lambda}_{0dqFDQU}$; $\hat{v}_{abc}$; $\hat{i}_{abc}$

8.17 From the results of Problem 8.16, compute P_{dev} and T_{dev}

(a) in per-unit
(b) in SI units

8.18 Using Chapter 7 methods, for the situation described in Problem 8.16, compute P_{dev} and T_{dev}

(a) in per-unit
(b) in SI units

Compare answers with Problem 8.17.

8.19 Consider the situation in Problem 8.16. At $t = 0$, a short circuit is suddenly applied to the machine stator terminals, such that $(\hat{v})_{abc} = \hat{0}$

(a) What is $\hat{v}_{0dq}$?
(b) What is $\hat{v}_{0dqFDQU}$?
(c) For $t > 0$, compute and plot $\hat{i}_{0dqFDQU}$
(d) For $t > 0$, compute and plot $\hat{i}_{abc}$

8.20 Continuing Problem 8.19, confirm that the final value of $i_a(t)$ (I_a) converges to the same value that obtained using Chapter 7 analysis methods.

9

The DC Machine

We now consider the last of the three basic types of rotating electromagnetic (EM) machines, the so-called "DC machine." Reflecting on the nature of rotating machines, a coil rotating in a magnetic field, or a field rotating in a coil, inherently produces an AC voltage. One would then suspect that the production of a constant, or "DC" voltage, would be somewhat more complicated, which in fact it is. It is of historical interest that DC machines were the first invented, mainly because the only available electrical source was the battery. Prior to about 1960, DC machines were widely used throughout industry, and served virtually all applications that required wide range speed control. As power electronic technology matured, it became practical and cost-effective to implement full range speed control of AC motors, and DC motors declined in importance, a trend that continues till today. DC machines are also somewhat more expensive, and require more maintenance than their AC equivalents. Nonetheless, there remain many applications for which a DC machine is ideally suited, and their continued investigation is justified in any comprehensive study of EM machines.

9.1 Machine Construction

Again, it is not our intent to provide comprehensive material on how to design EM machines. However, a basic understanding of how machines are physically constructed is necessary to understand the principles of operation and control, and to provide insight into the derivation of mathematical models. As with all EM machines, the device has two major parts: the stator (the part that does not rotate) and the rotor (the part that does rotate). The rotor is called the "armature," which is defined as the part of the machine in which AC voltage is induced due to motion through a DC magnetic field.[1] A photograph of a typical DC machine appears in Figure 9.1.

9.1.1 Stator Design

The stator of a DC machine consists of a frame, or yoke, cylindrical in shape, with a large "hole" in the middle, in which the rotor is placed. The machine is designed to minimize the clearance between the rotor and stator, a region called the "air gap." The stator is constructed of ferromagnetic material, on which are attached an even number of protruding structures called "poles," which protrude into the hole, and consist of either permanent magnets or

[1] Observe that this is opposite to the terminology used for AC machines.

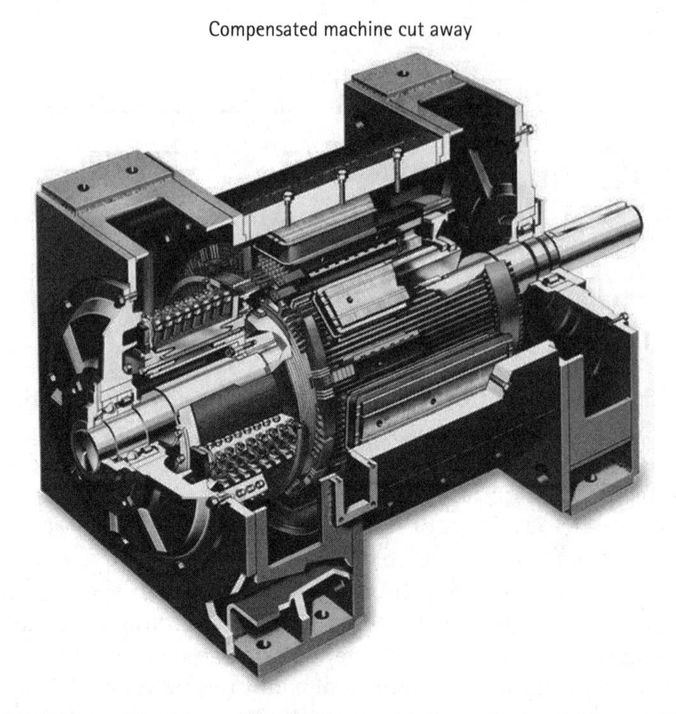

FIGURE 9.1. The DC machine. Courtesy of Reliance-Rockwell Corporation.

ferromagnetic structures around which are placed one or more insulated coils per pole. The coils are connected in series to make up a winding, in such a way that the poles alternate in magnetic polarity. These poles, and associated windings, are collectively referred to as the "field;" hence the machine is said to have either a "permanent magnet field" or a "wound field." For the wound field machine, two basic field winding designs are used:

- *The series field winding*, characterized by few turns, large conductors, high current, and low resistance.
- *The shunt field winding*, characterized by many turns, small conductors, low current, and high resistance.

The stator also includes other structures, such as the bearing races, the bearings (which support the rotor), the brush rigging, interpoles, and the brushes, about which more will be said later. Hence, the following terminology is common:

- *DC series machine.* A DC machine, which has exclusively a series field winding.
- *DC shunt machine.* A DC machine, which has exclusively a shunt field winding.
- *DC permanent magnet machine.* A DC machine, which has exclusively a permanent magnet field.
- *DC compound machine.* A DC machine, which has series and shunt field windings.
- *DC machine.* A DC machine of unknown field design, but frequently using a permanent magnet field.

It is always important to know the nature of the field design in a given application. Also, understand that *DC motors* and *DC generators* are merely *DC machines* used for a specific application that require normal operation in that mode.

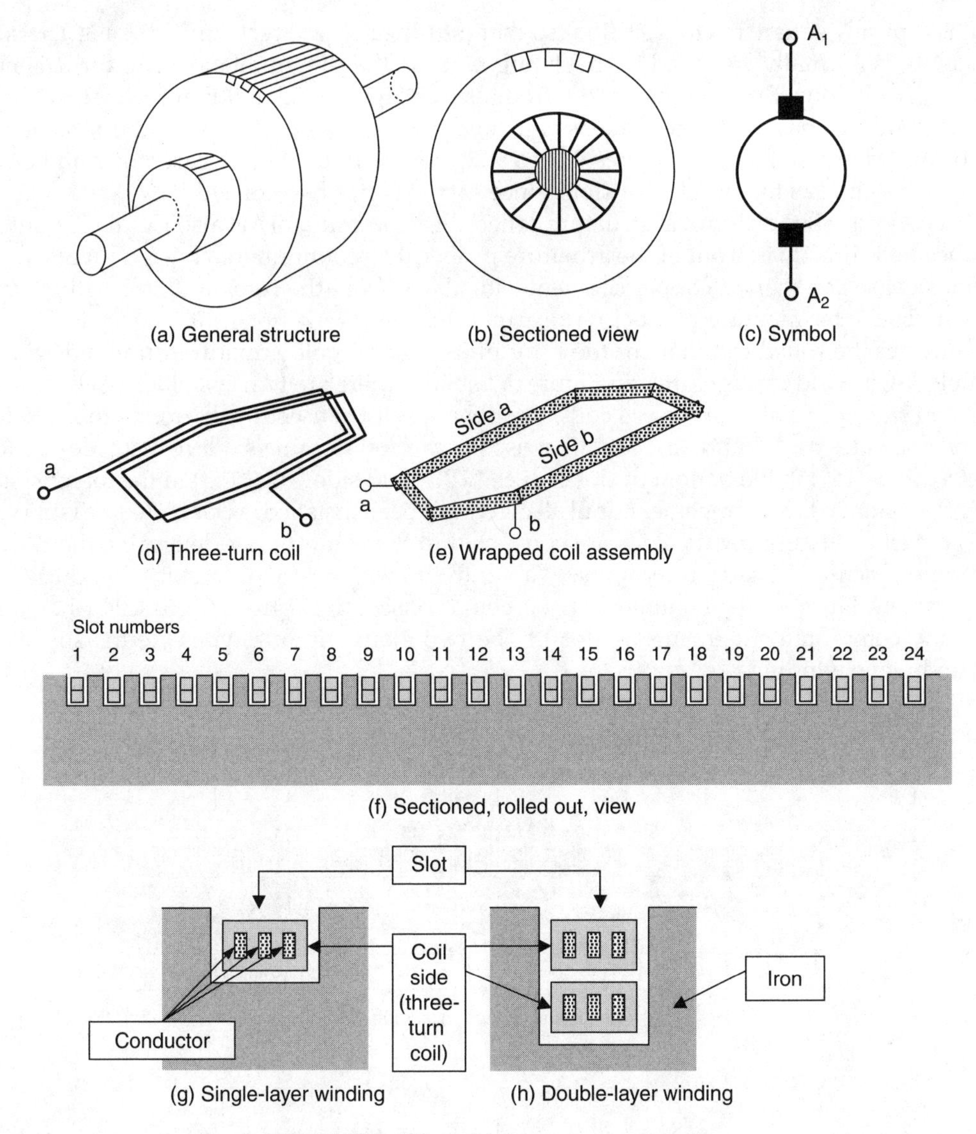

FIGURE 9.2. DC machine armature construction.

9.1.2 Rotor (Armature) Design

The bulk of the rotor,[2] usually called the armature, is a laminated ferromagnetic cylindrical structure, rigidly coupled to a central shaft, with slots, parallel to the rotor axis, and equally spaced around the surface (see Figures 9.2a to 9.2c). The armature is placed in the stator "hole," supported by bearings, free to rotate, and positioned such that the air gap is uniform. The brushes are stationary and mounted on brush riggings, which are attached to the stator.

Insulated hexagonal structures called "coils" are placed in the slots, and are interconnected to form the armature winding. Coils have four basic parts: the coil "sides" (2) that are parallel and pressed into the armature slots, and the end connections (2) that protrude out of the slots at the slot ends. Coils have a few (one to ten) turns of insulated conductor

[2] Technically, the rotor includes everything that rotates, including the cylindrical slotted part, the commutator, the shaft, and the armature winding.

that is typically square in cross section (so that it fits neatly into rectangular slots). The coil assembly is normally wrapped in insulating tape, so that the number of coil turns is not externally obvious. Each coil has two terminals (see Figure 9.2d and Figure 9.2e).

There are two basic arrangements: single layer (one coil side per slot) and double layer (two coil sides per slot) (see Figures 9.2f to 9.2h). In the double-layer case, the number of coils is the same as the number of slots, since each coil has two coil sides: $N_{coil} = N_{slot}$.

A second smaller cylindrical structure called the "commutator" is also rigidly mounted on the shaft directly in front of the armature proper. The commutator consists many small cylindrical sector-shaped copper segments, insulated from the rotor, and each other, and serving as the termination points for the armature coils.

Consider the rolled out view of the rotor of a 24 slot 4-pole armature, supplied with a double-layer winding, as shown in Figure 9.3a. Since there are two coil sides per slot, and two coil sides per coil, there are 24 coils. Number the slot consecutively from 1 to 24. Refer to the coil sides as "a" and "b," and likewise for the coil terminals. Place coil side "a" for Coil #1 (i.e., "1a") in the bottom of slot #1 (i.e., "1B"), and side "b" ("1b") in the top of slot 7 ("7T"). Continue this arrangement until all coils have been installed, as tabulated in Table 9.1.

The particulars of how the coils are connected to the commutator determine the winding type. Each coil has two terminals, "a" and "b" which lead into sides "a" and "b", respectively. Number the commutator segments consecutively from C1 to C24 (there are as many commutator segments as slots).[3] The two main winding types are the so-called lap and wave windings.

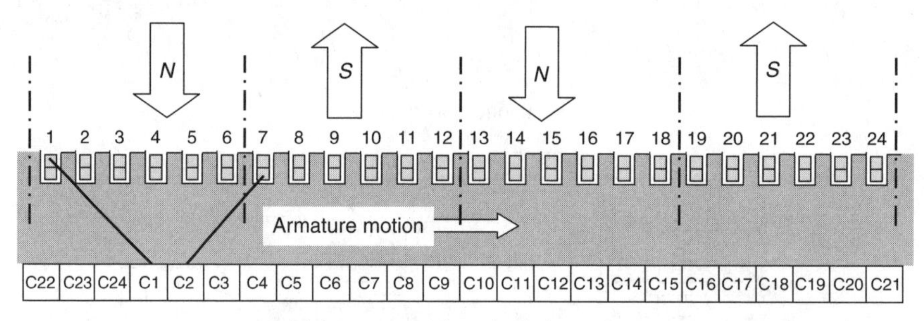

(a) Armature sectioned rolled out, view

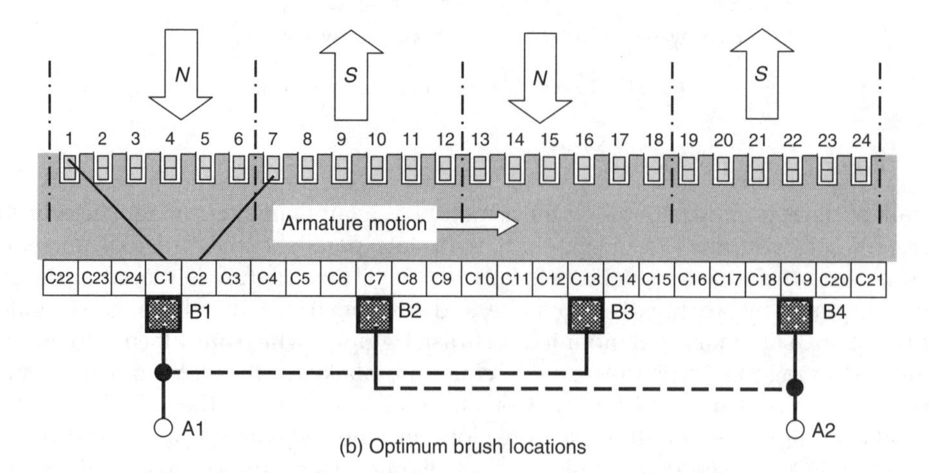

(b) Optimum brush locations

FIGURE 9.3. Brush, coil and commutator connections; lap winding.

[3] At least for simplex windings. There are several other types, including duplex, triplex, progressive, retrogressivs, frogleg, etc. Our intent is not to provide a comprehensive discussion of the topic.

9.1.2.1 The Lap Winding

Connect terminals a and b for Coil #1, to segments C1 and C2, respectively. Connect Coil #2, a to C2, and b to C3. Continue until all coils have terminated, as shown in Table 9.2.

9.1.2.2 The Wave Winding

Connect Coil #1, terminals a and b to segments C1 and C13; connect Coil #2, terminals a and b to C2 and C14, respectively. Continue this algorithm until all coils have terminated, as tabulated in Table 9.3.

9.2 Generation of DC Voltage

To understand how such a exotic structure can be used for constructive purposes, consider a 4-pole stator combined with a 24-slot lap wound armature, shown in "rolled-out" form in Figure 9.3. Likewise, consider that each coil has three turns. Consider the armature to be

TABLE 9.1
Coil Side Slot Locations

Coil Side	a	b	Coil Side	a	b	Coil Side	a	b	Coil Side	a	b
#1	1B	7T	#2	2B	8T	#3	3B	9T	#4	4B	10T
#5	5B	11T	#6	6B	12T	#7	7B	13T	#8	8B	14T
#9	9B	15T	#10	10B	16T	#11	11B	17T	#12	12B	18T
#13	13B	19T	#14	14B	20T	#15	15B	21T	#16	16B	22T
#17	17B	23T	#18	18B	24T	#19	19B	1T	#20	20B	2T
#21	21B	3T	#22	22B	4T	#23	23B	5T	#24	24B	6T

TABLE 9.2
Coil–Commutator Connections: The Lap Winding

Coil Term	a	b	Coil Term	a	b	Coil Term	a	b	Coil Term	a	b
#1	C1	C2	#2	C2	C3	#3	C3	C4	#4	C5	C6
#5	C5	C6	#6	C6	C7	#7	C7	C8	#8	C9	C10
#9	C9	C10	#10	C10	C11	#11	C11	C12	#12	C12	C13
#13	C13	C14	#14	C14	C15	#15	C15	C16	#16	C16	C17
#17	C17	C18	#18	C18	C19	#19	C19	C20	#20	C20	C21
#21	C21	C22	#22	C22	C23	#23	C23	C24	#24	C24	C1

TABLE 9.3
Coil–Commutator Connections: The Wave Winding

Coil Term	a	b	Coil Term	a	b	Coil Term	a	b	Coil Term	a	b
#1	C1	C13	#2	C2	C14	#3	C3	C15	#4	C5	C16
#5	C5	C17	#6	C6	C18	#7	C7	C19	#8	C9	C20
#9	C9	C21	#10	C10	C22	#11	C11	C23	#12	C12	C24
#13	C13	C1	#14	C14	C2	#15	C15	C3	#16	C16	C4
#17	C17	C5	#18	C18	C6	#19	C7	C20	#20	C20	C8
#21	C21	C9	#22	C22	C10	#23	C23	C11	#24	C24	C12

rotating, which means the rolled out structure is moving linearly (to the right, for example). The four poles alternate in north, south polarity, and produce magnetic fields oriented as shown. The fields are distributed throughout the four regions, indicated, alternating North, South in magnetic polarity.

Focus on the left-most region, including slots 1–6. To simplify the situation, assume the field strength is uniform, with north, south polarity as indicated (flux into the armature from slots 1 to 6; flux out of the armature from slots 7 to 13; etc.). Also assume a voltage drop (+ to −) "E" is induced in each conductor as it moves through the field. Starting at C1, wish to trace out a conducting path to C7, summing induced voltage as we go. From the front, starting at C1, we enter side "a" of turn #1 in Coil #1 and in the background move to the foreground on side "b" of turn #1, accumulating a total voltage of E + E = 2E. Since there are 3 turns, we retrace the path, accumulating a total voltage of 3 × 2E = 6E, arriving at C2. From C2, we proceed to Coil #2, (slots 2 and 8), with Coils #3, 4, 5, and 6 following, terminating on C7. Since all conductors are in series, the total voltage accumulated is:

$$V_{C1\text{–}C7} = 6E + 6E + 6E + 6E + 6E + 6E = +36E,$$ where "+" indicates a drop from C1 to C7.

Continuing through coils 7 through 13, we compute

$$V_{C7\text{–}C13} = -36E,$$ where "−" indicates a rise from C7 to C13.

Likewise:

$$V_{C13\text{–}C19} = +36E$$

$$V_{C19\text{–}C1} = -36E$$

Apparently, we have designed the world's most useless device, developing some 144 equal and opposite voltages which cancel! On the other hand, it is fortuitous that the voltages do sum to zero, since we have a closed path of copper conductors. Even a small voltage imbalance would certainly cause a high circulating current within the winding.

However, notice that we do have substantial voltage (36E) between certain commutator segments (e.g., C1 and C7). Suppose we tap this voltage. But how? The main problem is that the rotor, including the commutator, is in motion. How do we make a stationary connection to a moving structure? The problem is solved with a copper impregnated carbon block, called a "brush," which makes sliding contact with the commutator. We locate the brushes B1 and B2 at the point that taps off the maximum available voltage (C1–C7) as illustrated in Figure 9.3b. Likewise locate the brushes B3 and B4 at C13–C19. We realize that the voltage from C1 to C13 9 (and C7 to C19) is zero; hence B1 and B3 can be interconnected, as can B2 and B4. These connections constitute an access port into the armature, which we label as A1–A2.

But this analysis was performed only at one instant of time! OK, so let's consider a later time, when the rotor has turned through some arbitrary multiple of 15°, 120°, or eight commutator segments, for example. C17 is now under brush B1, with symmetrical shifts of all other segments. Repeat the analysis and we get the same results (i.e. the same total voltages)! For other angles, brushes will straddle adjacent commentator segments, shorting out those coil voltages, dropping the terminal voltage somewhat. However, suppose we redesign the field such that it is not uniform, but rather strongest in the middle, tapering to near-zero at the N–S boundaries (called the "magnetic neutrals"). It follows that the voltage is the shorted coils would be small (perhaps negligible). The plan then is to locate the brushes at shorted coils are located at the so-called "magnetic neutral", the location where the flux crossing the air-gap is essentially zero.[4]

Since the terminal voltage is essentially the same for any rotor position, we characterize the machine as "DC". It is interesting, and important, to note that any coil voltage is AC (but usually not even close to sinusoidal). Each coil is switched out of a path, and into a reverse path, at the very instant that its generated voltage reverses in polarity! The commutator therefore is functionally a rectifier, mechanically switching coils at polarity reversing points in much the same way that diodes electrically switch to provide a DC output voltage.

The lab winding has inherently divided itself into four symmetrical (or "parallel") paths, (p_p), parallel because they are indeed elctrically in parallel. Thus the current flowing out of either armature terminal (A1 or A2) divides evenly among the four parallel paths: any given coil current is ¼ of the total armature current.

Note that if the brushes are positioned at the wrong location, disaster occurs. The brushes would then short coils which have **nonzero** generated voltage, which would result in extremely large internal circulating current. Worse yet, such currents would be rapidly switched on and off as commutator segments slide under the brushes, causing severe arcing at the commutator.

Ultimately, the brushes, commutator, and associated conduction problems prove to be limiting factors for DC machine design and operation. The associated losses, wear, and associated maintenance prove to be significant disadvantages.[5]

A corresponding detailed analysis of the wave winding is reserved in Problem 9.3. In the wave, we find that the four paths are now asymmetrical, with one path consisting of one zero-voltage coil between positive brushes; one path consisting of one zero-voltage coil between negative brushes; and the remaining coils forming two symmetrical paths between the armature positive and negative terminals.

We are now in a position to quantify the generator voltage produced in the armature. Consider a one-turn coil rotating in a uniform and static magnetic field, as shown in Figure 9.4a.

As the coil rotates from position 1 ("closed") to position 2 ("open"), the change in magnetic flux linkage, goes from zero to "ϕ": In equation form:

$$\Delta\lambda = 0 + \phi = \phi$$

This happens in time Δt, the time it takes the coil to rotate through 1/4 revolution. Hence the average, or DC, voltage induced in that 1/4 revolution is

$$E_{\text{coil}} = \text{average (DC) coil voltage} = \text{average of } e_{\text{coil}}(t) = \frac{\Delta\lambda}{\Delta t} = \frac{4\phi}{T}$$

where T is the time required for one revolution. But this is for a two-pole field. We recognize that for a N_P pole field, T more generally becomes the period of $e_{\text{coil}}(t)$, so that

$$E_{\text{coil}} = 4f\phi = \frac{4\omega\phi}{2\pi} = \frac{4N_P\omega_{\text{rm}}\phi}{4\pi} = \frac{N_P\omega_{\text{rm}}\phi}{\pi}$$

---

[4] This is true only at no load. When the machine is loaded, the armature current produces an mmf in quadrature with the main field, shifting the magnetic neutral, a phenomenon called "armature reaction." Since the brush locations are fixed, if uncompensated, this causes severe arcing at the commutator. The problem is corrected by installing small poles (called "interpoles") between the main poles, and supplied with armature current polarized to cancel the quadrature field at the brush locations. Since all practical DC machines are designed to compensate for armature reaction, it may be ignored for modeling purposes.

[5] Nonetheless, one cannot help but admire the ingenious and creative work that the early pioneers accomplished in DC machine design. There were many contributors, including Davenport and T.A. Edison.

DC Machine Data

Ratings

Voltage (V_a) = 250.0 V; horsepower = 50.0 hp
Current (I_a) = 160.7 A; no. of poles = 4
Field current = 5.74 A; rated speed = 2000 rpm = 209.4 rad/s
Base speed = 1150 rpm; J motor = 0.569 kg-m^2

Equivalent Circuit Values

$R_a = 0.0622\ \Omega$; $L_a = 0.3$ mH; $R_f = 43.57\ \Omega$
Rotational loss torque = $T_{RL} = K_{RL} * \omega_{rm} = 0.01871 * \omega_{rm}$ in N m
ω_{rm} = rotor speed in rad/s
MagC at 2000 rpm: $I_{feq} = E_a/K_{ag} + A_x \exp(B_x * E_a)$.
$K_{ag} = 139.89\ \Omega$; $A_x = 0.0000580173$ A; $B_x = 0.025798$
Note: $A_x = 0$ indicates that the OCC has been linearized.

(a) DC machine data

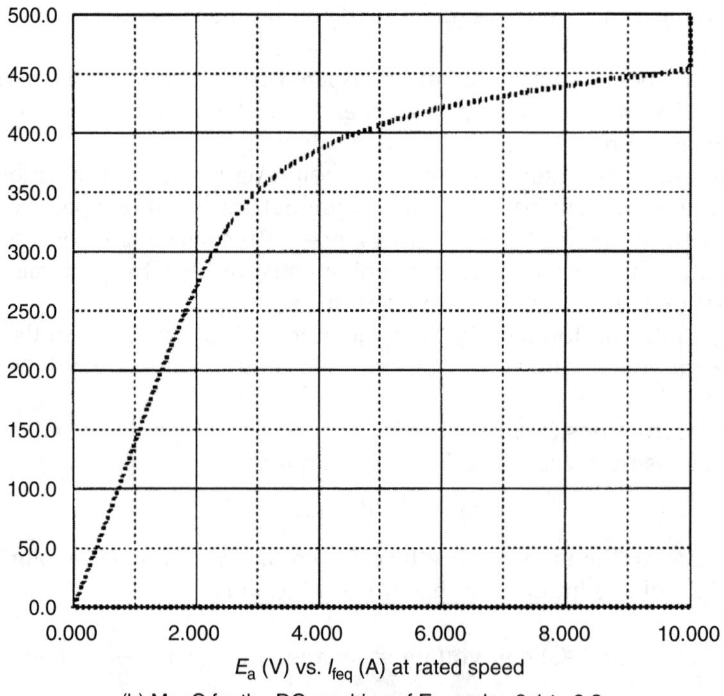

(b) MagC for the DC machine of Examples 9.1 to 9.3

FIGURE 9.4. Data for the DC machine of Examples 9.1 to 9.3.

We can elect to assign this voltage equally to each coil side (or conductor), so that

$$E_{cond} = \frac{E_{coil}}{2} = \frac{N_P \omega_{rm} \phi}{2\pi}$$

To determine the armature voltage we need to consider the total number of *series* armature conductors. We will do this in two steps: first, determine the total number of armature conductors (N_{total}) and then divide the number of parallel paths.

Finally, then:

$$E_a = \left(\frac{N_{total}}{P_P}\right) E_{cond} = \left(\frac{N_{total} N_P}{2\pi P_P}\right) \phi\omega_{rm} = K_a \phi\omega_{rm} \tag{9.1}$$

where

E_a = DC (average) generated internal armature voltage (in V)
ϕ = flux per pole (in W)
ω_{rm} = armature (rotor) mechanical speed (in mech rad/s)

$K_a = \dfrac{N_{total} N_P}{2\pi P_P}$ = armature constant

N_P = number of poles (even integer)
P_P = number of parallel paths
Lap windings: $P_P = N_P$
Wave windings: $P_P = 2$
N_{total} = total number of armature conductors
= number of conductors/slot × number of slots

An example would be useful.

Example 9.1
Consider a 24-slot double-layer lap wound armature, wound to be used in a four-pole 250 V 1800 rpm DC machine. Each coil has three turns.

(a) How many coils does the armature have?
(b) What is the total number of armature conductors?
(c) Find K_a.
(d) Find the flux per pole if the armature-generated voltage has to be 250 V at 1800 rpm.
(e) Repeat (c) and (d) if the armature is wave-wound.

(a) Since the armature is double layer, there are two coil sides per slot. Since each coil has two sides, there are 24 coils.
(b) There are 2 × 3 = 6 conductors per slot. Hence,

$$N_{total} = 6 \times 24 = 144 \text{ conductors}$$

(c) $K_a = \dfrac{N_{total} N_P}{2\pi P_P} = \dfrac{144(4)}{2\pi(4)} = 22.92$

(d) $\omega_{rm} = 1800 \text{ rpm} = 188.5 \text{ rad/s}$

$$E_a = K_a \phi\, \omega_{rm} = 22.92\phi\ (188.5) = 250$$

$$\phi = 57.87 \text{ mWb}$$

(e) For the wave-wound case,

$$K_a = \frac{N_{total} N_P}{2\pi P_P} = \frac{144(4)}{2\pi(2)} = 45.84$$

$$E_a = K_a\phi\omega_{rm} = 45.84\phi\,(188.5) = 250$$
$$\phi = 28.94\,\text{mWb}$$

Now consider how the magnetic field is produced. On the stator are an even number of projecting structures called "poles" from which the magnetic flux flows across the air gap into the armature structure. On these poles are windings which provide the requisite mmf. See Figure 9.5.

The two windings differ in design

Series field winding: Few turns (N_S), large conductor, high current rating.

Shunt field winding: Many turns (N_F), small conductor, low current rating.

Consider the mmf per pole applied to the system:

$$\mathfrak{F}_{EQ} = N_S I_S + N_F I_F$$

If the series field winding is missing, the machine is called a "shunt machine"; if the shunt field winding is missing, the machine is called a "series machine"; if both field windings are present, the machine is called a "compound machine". To deal with this, we define a concept called the "equivalent field current". (I_{FEQ}) which is the current that must flow in the shunt winding to produce the same mmf (and hence the same flux) as that in an actual machine at a given operating point. The subscripts "F, S" refer to the shunt and series windings, respectively. Hence:

$$\mathfrak{F}_{EQ} = N_S I_S + N_F I_F = N_F I_{FEQ}$$
$$I_{FEQ} = I_F + \frac{N_S I_S}{N_F}, \quad \text{compound machine}$$
$$I_{FEQ} = I_F, \quad \text{shunt machine}$$
$$I_{FEQ} = \frac{N_S I_S}{N_F}, \quad \text{series machine}$$
$$K_a\phi = \text{constant}, \quad \text{permanent magnet field machine}$$

To analyze the machine it is necessary to determine the relationship between I_{FEQ} and ϕ.

This can be determined by a simple test. Consider a DC compound machine. Leave the series field on open circuit. Arrange to drive the machine at a constant known speed.[6] Measure and record the armature terminal voltage at no load ($I_a = 0$) as the shunt field current ($I_F = I_{FEQ}$) is varied from zero to its rated value. Graphing the results of such a test (called the "no load test") produces a plot as shown in Figure 9.5.[7] This graph is routinely supplied the machine manufacturer, and is called by various names, including the magnetization characteristic, the saturation curve, the open circuit characteristic, and the no load characteristic. We shall use the term "Magnetization Characteristic", as was discussed

[6] In principle, any speed may be used as long as it is defined. In practice, the MagC is supplied at either base speed or rated speed. Base speed is the speed at which the machine would run at no load as a motor at maximum field current. Rated speed, a higher value, is the maximum speed recommended for normal operation.

[7] Actually, if the field current is varied over its full range in both positive and negative directions, a narrow hysteresis loop will be measured. What is used here is the single-valued expression. See Section 1.4 for a detailed explanation.

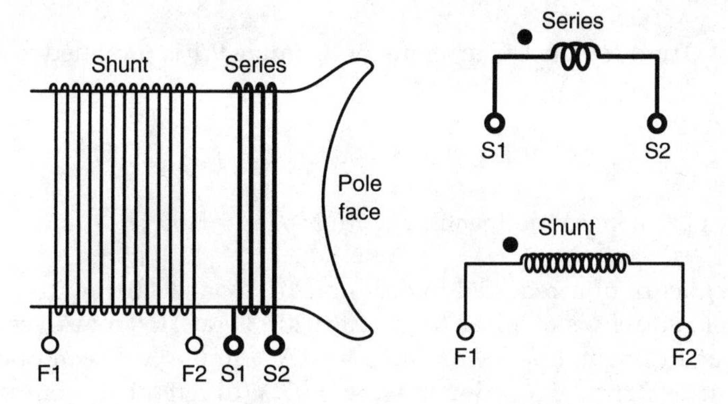

FIGURE 9.5. The stator pole structure.

in Chapters 1 and 7, and is abbreviated as "MagC". Experience shows that the MagC can be represented adequately with the following empirical equation:

$$I_{\mathrm{FEQ}} = \frac{E_{\mathrm{a}}}{K_{\mathrm{ag}}} + A_x \mathrm{e}^{B_x E_{\mathrm{a}}} \tag{9.2}$$

where

K_{ag} = slope of the air gap line (unsaturated MagC), volts/ampere.
A_x = ΔI_{FEQ} added to account for saturation, ampere.
B_x = exponential constant, 1/volts.

Equation 9.2 is the same form as equation 1.10, except in different variables (E_{a}, I_{FEQ} instead of B, H).

If the machine is a shunt DC machine, of course inherently $I_{\mathrm{FEQ}} = I_{\mathrm{F}}$; if a series machine, consider that $N_{\mathrm{S}}/N_{\mathrm{F}} = 1$ and $I_{\mathrm{FEQ}} = I_{\mathrm{S}}$.

Let us consider the meaning of the MagC in more depth. We measured and plotted E_{a} versus I_{FEQ}. But what fundamentally varied as we varied I_{FEQ}? Recall

$$E_{\mathrm{a}} = K_{\mathrm{a}}\phi\omega_{\mathrm{rm}}$$

The speed did't change, nor did K_{a} (obviously)! The fundamental dependent variable was the magnetic field ϕ. And exactly what caused ϕ to vary? Field current? Actually it was the mmf ($\mathfrak{F}$) applied to the magnetic circuit! At its most basic level, think of the MagC as providing the necessary information regarding the machine's magnetic properties; that is, think of it as a plot of "ϕ versus $\mathfrak{F}$."

There is one other important stator pole structure. We could use permanent magnets, with no field windings. For this case I_{FEQ} is meaningless and

$$K_{\mathrm{a}}\phi = \text{constant} \qquad \text{permanent magnet field machine}$$

We just need to measure the open circuit armature voltage at a known speed. Then:

$$K_a\phi = \frac{E_a}{\omega_{rm}}$$

Example 9.2 should clarify the situation.

Example 9.2

The MagC at 2000 rpm for the DC machine of Example 9.1 is provided in Figure 9.4b, for which

$$K_{ag} = 139.9\,\Omega;\ A_x = 0.00005802\ \text{A};\ B_x = 0.0258$$

The machine is a DC compound machine, with $N_S/N_F = 0.05$.

(a) What field current is required to produce 201.3 V at 2000 rpm? $I_S = 0$.
(b) What field current is required to produce 201.3 V at 1150 rpm? $I_S = 0$.
(c) Find E_a at 2000 rpm, if $I_F = 4$ A and $I_S = 50$ A (cumulatively compounded).
(d) Find E_a at 2000 rpm, if $I_F = 4$ A and $I_S = -50$ A (differentially compounded).
(e) Find E_a at 2000 rpm, if $I_F = 0$ A and $I_S = 100$ A (series operation).

(a) Look up $E_a = 201$ V at 2000 rpm on the MagC. $I_F = I_{FEQ} = 1.4$ A
(b) 200 V at 1150 rpm is equivalent to (2000/1150)200 = 350 V at 2000 rpm. Look up $E_a = 350$ V on the MagC. $I_F = I_{FEQ} = 3.0$ A.
(c) $I_{FEQ} = I_F + N_S I_S/N_F = 4 + 0.05(50) = 6.50$ A. From the MagC, $E_a = 430$ V.
(d) $I_{FEQ} = I_F + N_S I_S/N_F = 4 - 0.05(50) = 1.50$ A. From the MagC, $E_a = 205$ V.
(e) $I_{FEQ} = I_F + N_S I_S/N_F = 0 + 0.05(100) = 5.0$ A. From the MagC, $E_a = 405$ V.

9.3 The DC Machine Model: Generator Operation

As with the synchronous machine, the DC machine is best analyzed in a coordinate system, compatible with its expected mode of operation. We will consider the device in generator coordinates first. Since compound machines are becoming increasingly rare, and since extension to this case is straightforward, we will confine our study to the DC shunt machine. Historically, it was common to provide an external resistor in series with the field (R_X is the field rheostat) for voltage control. In the interests of generality, we will leave it in the circuit for now.

Consider the DC machine model in generator coordinates, as in Figure 9.6.

$$e_a = R_a i_a + L_a \frac{di_a}{dt} + v_a$$

$$v_f = (R_F + R_X) i_F + L_f \frac{di_F}{dt}$$

For steady-state DC operation, the inductors become short circuits, producing

$$E_a = R_a I_a + V_a$$

$$V_f = (R_F + R_X) I_F$$

Consider the powers:

$$P_a = V_a I_a$$

$$\text{Armature winding loss (AWL)} = I_a^2 R_a$$

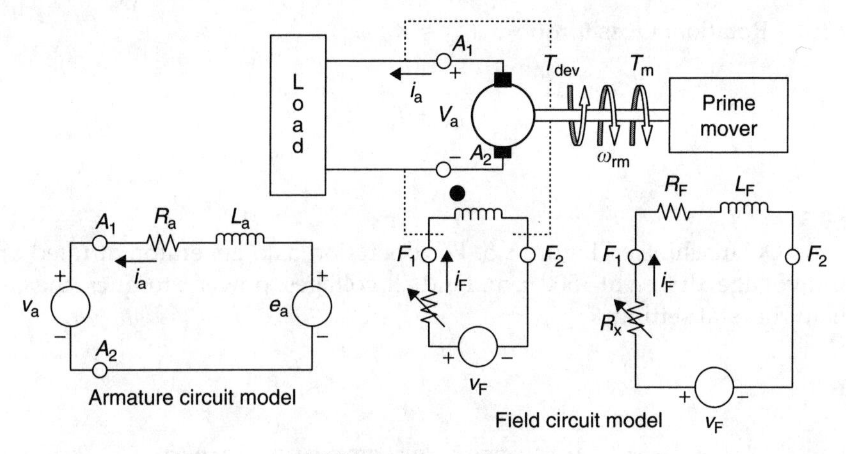

(a) Generator coordinates

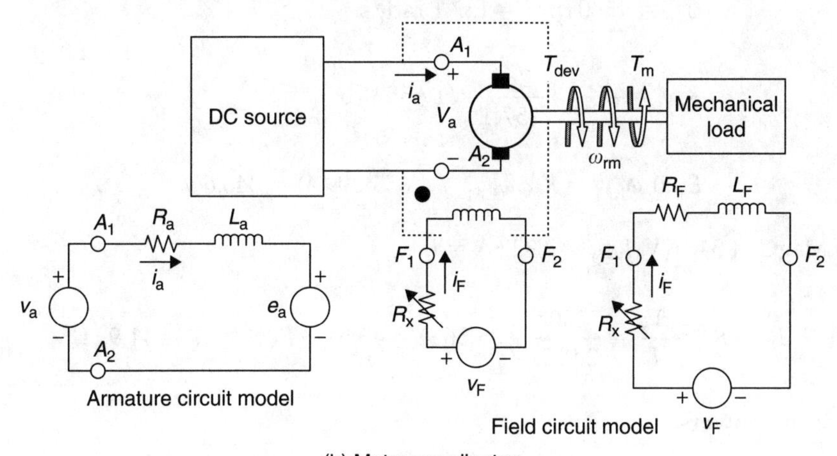

(b) Motor coordinates

FIGURE 9.6. DC machine model.

$$\text{Field winding Loss (FWL)} = I_F^2 R_F$$

$$\text{Rotational loss} = P_{RL} = T_{RL}\omega_{rm} = K_{RL}(\omega_{rm})^2$$

$$\text{Sum of losses} = \sum L = \text{AWL} + \text{FWL} + P_{RL}$$

$$P_{dev} = E_a I_a$$

$$P_m = P_{out} = P_{dev} + P_{RL}$$

$$P_{in} = P_{out} + \sum L$$

$$\text{Efficiency} = \eta = \frac{P_{out}}{P_{in}}$$

Consider the torques:

$$T_{dev} = \frac{P_{dev}}{\omega_{rm}} = \frac{E_a I_a}{\omega_{rm}} = \frac{K_a \phi \omega_{rm} I_a}{\omega_{rm}} = K_a \phi I_a \tag{9.3}$$

$$\text{Rotational loss torque} = T_{RL} = K_{RL}\omega_{rm}$$

$$T_m = T_{dev} + T_{RL} = \frac{P_m}{\omega_{rm}}$$

Example 9.3

Consider the DC machine of Figure 9.5. For operation as a generator, at rated armature current and voltage, driven at 1500 rpm. Find all voltages, powers, torques, the efficiency, and the field rheostat setting.

SOLUTION

$$E_a = V_a + R_a I_a = 250 + (0.0622)(160.7) = 260.0\,\text{V}$$

$$\omega_{rm} = 1500\,\text{rpm} = 157.1\,\text{rad/s}$$

$$K_a\phi = \frac{E_a}{\omega_{rm}} = \frac{260}{157.1} = 1.655\,\text{Wb}$$

$$E_a \text{ at } \omega_{RAT} = K_a\phi\omega_{RAT} = 1.655(209.4) = 346.6\,\text{V}$$

From the MagC at 346.6 V, $I_{FEQ} = 2.922\,\text{A} = I_F$

$$R_F + R_X = \frac{V_a}{I_F} = \frac{250}{2.922} = 85.56\,\Omega = 43.57 + R_X, \quad R_X = 41.99\,\Omega$$

Solving for the powers:

$$\text{AWL} = I_a^2 R_a = 1606\,\text{W}$$

$$\text{FWL} = I_F^2 R_F = 372\,\text{W}$$

$$P_{RL} = K_{RL}\omega_{rm} = 462\,\text{W}$$

$$\Sigma L = P_{loss} = P_{RL} + \text{AWL} + \text{FWL} = 2440\,\text{W}$$

$$\text{Developed power} = P_{dev} = E_a I_a = 41.78\,\text{kW}$$

$$\text{Output power} = P_{out} = P_{dev} - \text{SWL} = 40.18\,\text{kW}$$

$$P_m = P_{dev} + P_{RL} = 42.24\,\text{kW};$$

$$\text{Input} = P_{out} + \Sigma L = 42.62\,\text{kW}$$

$$\text{Efficiency} = \eta = P_{out}/P_{in} = 40.18/42.62 = 94.27\%$$

Solving for the machine torques:

$$\text{Developed torque} = T_{dev} = P_{dev}/\omega_{rm} = 266.0\,\text{N m}$$

$$\text{Rotational loss torque} = T_{RL} = P_{RL}/\omega_{rm} = 2.94\,\text{N m}$$

$$\text{Mechanical input torque} = T_m = T_{dev} + T_{RL} = 268.9\,\text{N m}$$

9.4 The DC Machine Model: Motor Operation

Motor operation is the most common and important application for DC machines. We will now consider the device in motor coordinates. Historically, it was common to provide an external resistor in series with the field (R_X is the field rheostat) for speed control purposes. With solid-state DC motor drive control, this resistor is unnecessary, but in the interests of generality, we will leave it in the circuit for now. We will remove it later by the simple expediency of setting it to zero. Consider Figure 9.7b.

$$v_a = R_a i_a + L_a \frac{di_a}{dt} + e_a$$

$$v_f = (R_F + R_X) i_F + L_f \frac{di_F}{dt}$$

For steady-state DC operation, the inductors become short circuits, producing

$$V_a = R_a I_a + E_a$$

$$V_f = (R_F + R_x) I_F$$

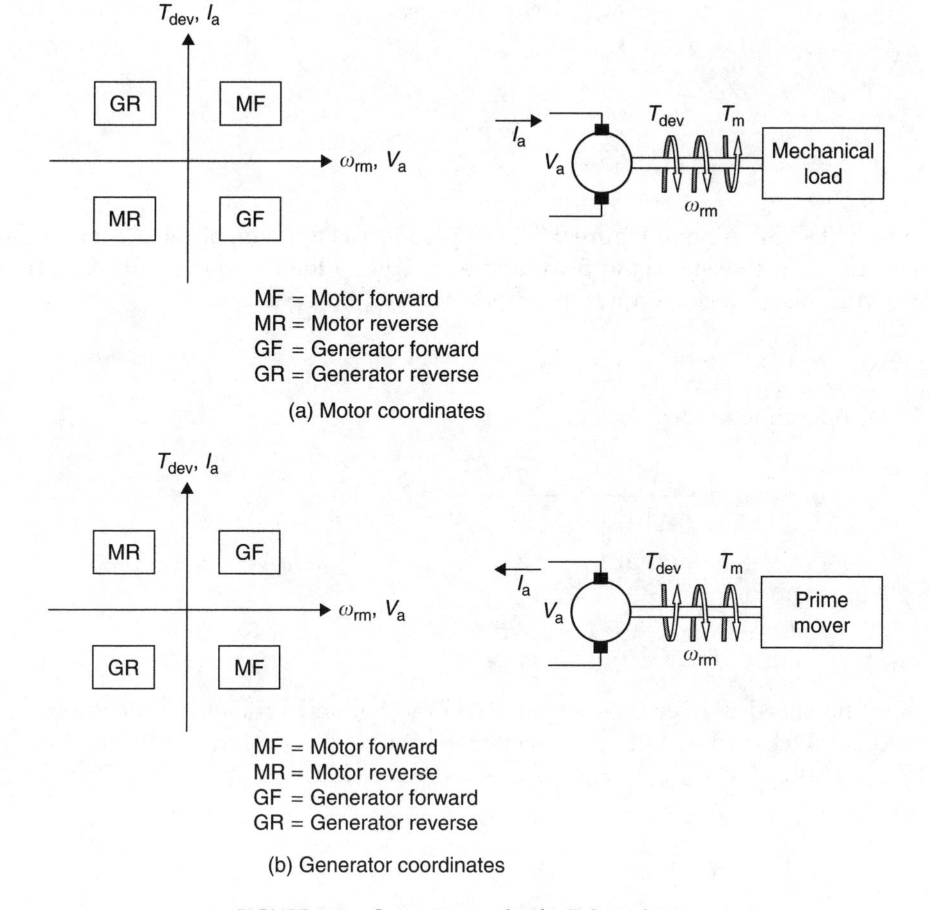

FIGURE 9.7. Operating modes for DC machines.

Consider the powers:

$$P_a = V_a I_a$$

$$\text{AWL} = I_a^2 R_a$$

$$\text{FWL} = I_F^2 R_F$$

$$P_{RL} = T_{RL}\omega_{rm} = K_{RL}(\omega_{rm})^2$$

$$\Sigma L = \text{AWL} + \text{FWL} + P_{RL}$$

$$P_{dev} = E_a I_a$$

$$P_m = P_{out} = P_{dev} - P_{RL}$$

$$P_{in} = P_{out} + \Sigma L$$

$$efficiency = \eta = \frac{P_{out}}{P_{in}}$$

Consider the torques:

$$T_{dev} = \frac{P_{dev}}{\omega_{rm}} = \frac{E_a I_a}{\omega_{rm}} = \frac{K_a \phi\, \omega_{rm} I_a}{\omega_{rm}} = K_a \phi\, I_a$$

$$T_{RL} = K_{RL}\omega_{rm}$$

$$T_m = T_{dev} - T_{RL} = \frac{P_m}{\omega_{rm}}$$

Example 9.4

Consider the DC machine of Figure 9.5. For operation as a motor, at rated armature voltage, with the field rheostat set at 50 Ω, and supplying a load torque of 200 N m, find all voltages, currents, powers, torques, the efficiency, and the speed.

SOLUTION

$V_a = 250$ V; operating mode: motor

$$I_{FEQ} = I_F = \frac{V_a}{R_F + R_X} = \frac{250}{43.57 + 50} = 2.672 \text{ A}$$

From the MagC at $I_{FEQ} = 2.672$ A; E_a at $\omega_{RAT} = 331.5$ V

$$K_a\phi = \frac{E_a \text{ at } \omega_{RAT}}{\omega_{RAT}} = 1.583 \text{ Wb}$$

We know the speed is over base speed (120.4 rad/s), and probably under-rated speed (209.4 rad/s). Pick some value between the two (e.g., 150 rad/s). Therefore $T_{RL} = 0.02(150) = 3$ N m. Hence $T_{dev} = T_m + T_{RL} = 200 + 3 = 203$ N m:

$$I_a = \frac{T_{dev}}{K_a\phi} = \frac{203}{1.583} = 128.2 \text{ A}$$

$$E_a = V_a - R_a I_a = 250 - (0.0622)(128.2) = 242 \text{ V}$$

Refining our speed estimate,

$$\omega_{rm} = \frac{E_a}{K_a\phi} = 152.9\,\text{rad/s} = 1460\,\text{rpm}$$

at 152.9 rad/s, and $T_{RL} = 0.0187(152.9) = 2.86\,\text{N m}$. Hence $T_{dev} = T_m + T_{RL} = 200 + 2.86 = 202.86\,\text{N m}$:

$$I_a = \frac{T_{dev}}{K_a\phi} = \frac{202.86}{1.583} = 128.1\,\text{A}$$

Solving for the powers,

$$\text{AWL} = I_a^2 R_a = 1021\,\text{W}$$

$$\text{FWL} = I_F^2 R_F = 311\,\text{W}$$

$$P_{RL} = K_{RL}\omega_{rm} = 437\,\text{W}$$

$$\Sigma L = P_{loss} = P_{RL} + \text{AWL} + \text{FWL} = 1769\,\text{W}$$

$$\text{Developed power} = P_{dev} E_a I_a = 31.01\,\text{kW}$$

$$\text{Output power} = P_{out} = P_m = P_{dev} - P_{RL} = 30.57\,\text{kW (40.98 hp)}$$

$$\text{Input} = P_{out} + \Sigma L = 32.34\,\text{kW}$$

$$\text{Efficiency} = \eta = P_{out}/P_{in} = 94.53\%$$

Solving for the machine torques

$$T_{dev} = P_{dev}/\omega_{rm} = 202.9\,\text{N m}$$

$$T_{RL} = P_{RL}/\omega_{rm} = 2.86\,\text{N m}$$

$$T_m = T_{dev} - T_{RL} = 200.0\,\text{N m}$$

9.5 Speed Control of DC Motors

Consider the DC machine operating as a motor, using a motor coordinate system. In the steady state, neglecting armature resistance,

$$\omega_{rm} = \frac{V_a}{K_a\phi} \tag{9.1}$$

Equation (9.1) reveals the options for DC motor speed control. Notice that (1) speed is inversely proportion to ϕ, and (2) directly proportional to V_a. Exploitation of the first option is called "field control" and the second, "armature control." Let us consider both.

9.5.1 Field Control

In the early days of electric motor application, a major advantage of the DC motor was its amenability to speed control, particularly using field control. We will demonstrate through an example.

Example 9.5

Consider the DC machine of Figure 9.4, driving a constant torque load of 200 N m. Determine the field rheostat setting to make the system run at (a) 1500 rpm, (b) 2000 rpm, (c) 1000 rpm. Comment on results.

SOLUTION

This is a difficult problem to solve in closed form. We have sufficient equations. However, since R_X and I_a are unknowns, the MagC equation will be one of the equations of constraint, and the set is nonlinear. We resort to a "trial and error" approach:

1. Select R_X.
2. Use the procedure described in Example 9.4 to find the speed.
3. Based on the R_X, ω_{rm} combination of (1) and (2), select a "better" value for R_X.
4. Repeat steps 1 to 3 until we converge to the required speed values.

Executing the above procedure, results are as follows:[8]

R_X (Ω)	ω_{rm} (rpm)	I_F (A)	I_a (A)	T_{load} (N m)
0	1168	5.74	102	200
50	1448	2.67	128	200
54.8	1500	2.54	132	200
100.5	2000	1.74	179	200

Note that

- Increasing R_X decreased I_F (and hence $K_a\phi$), thereby increasing the speed.
- As speed increased, so did the armature current, overloading the machine at 2000 rpm.
- We could not reach the requested speed of 1000 rpm. The most we could reduce R_X was to zero, resulting in a minimum speed of 1168 rpm.

The results of Example 9.5 have general implications.

- Speed control can clearly be achieved by controlling the field current. One simple and cheap method is to insert a variable resistor in the field circuit.
- The relationship is inverse. High-field current causes low speed and vice versa.
- There is a lower limit to field control. Setting R_X to 0 maximizes I_F, resulting in minimum speed (the so-called base speed). Even if we increased the field voltage, which would further increase I_F, at some point, we will exceed the thermal rating of the field.
- There appears to be no upper limit on the speed. However, as the speed increases, so does the armature current. If we limit armature current to its rated value, this will indirectly establish an upper limit on speed. This limit is load-dependent.
- Continuing with the last point, there is a definite possibility of speed instability ("runaway"). What if the field circuit were opened, causing I_F to drop to zero? $K_a\phi$ would drop to near zero, causing the speed to become extremely high. Of course, the armature current would also have to be extremely large, both to satisfy the load and to accelerate the system. Overcurrent protection should trip the motor offline,

[8] Lest the reader think that the author is a computational junkie, the analysis was done with the assistance of the computer program EMAP.

should this occur. Nonetheless, the concern is legitimate, as this can and does happen and proper attention should be given to the issue for all DC motor applications.

9.5.2 Armature Control

Field control appears to have a fatal flaw. We cannot slow down below base speed! We now consider our second option: specifically, to vary the armature voltage. As usual, we resort to an example.

Example 9.6

Continuing the situation defined in Example 9.5, set the field rheostat to zero, and consider the armature voltage to be variable, driving a constant torque load of 200 N m.

(a) Determine V_a to make the system run at: 1500; 1000; 500; 0; −500 rpm
(b) Comment on results.

SOLUTION

Recall the DC machine of Example 9.2, driving a constant torque load of 200 N m. Since we are mostly operating in the motor mode, we shall use motor coordinates throughout. The field is separately excited from a 250 V source.

$$I_{FEQ} = I_F = \frac{V_F}{R_F} = \frac{250}{43.57} = 5.738\text{ A}$$

From the MagC at $I_{FEQ} = 5.758$ A,

$$K_a\phi = \frac{E_a \text{ at } \omega_{RAT}}{\omega_{RAT}} = \frac{417.4}{209.4} = 1.993\text{ Wb}$$

Now at 1000 rpm, $T_{RL} = 0.02(104.7) = 2.1$ N m. Hence $T_{dev} = T_m + T_{RL} = 200 + 2.1 = 202.1$ N m:

$$I_a = \frac{T_{dev}}{K_a\phi} = \frac{202.1}{1.993} = 101.4\text{ A}$$

$$V_a = E_a + R_a I_a = 1.993(104.7) + (0.0622)(101.4) = 215.0\text{ V}$$

Repeating the above procedure for the requested speeds,[9] results are as follows:

ω_{rm} (rpm)	Mode	I_a (A)	V_a (V)	T_{load} (N m)
1000	MF	101	215.0	200
500	MF	101	110.6	200
0	MF	100	6.25	200
1500	MF	102	319.4	200
−500	GR	100	−99.14	200

Note that

- Increasing V_a increases the speed.
- The armature current is virtually constant, which makes sense for a constant torque load, since $K_a\phi$ is constant. Rotational loss is the only reason that this condition is not exact.

[9] Again, EMAP to the rescue.

- We can run above the base speed. However, the penalty is that we must operate above the rated voltage, which normally is not desirable.
- The last value (−500 rpm) is particularly interesting. Recall that negative speed means reverse or backward rotation. Since we interpret constant torque to mean unidirectional and independent of speed, the mode of operation changes from "motor forward (MF)" to "generator reverse (GR)". To operate at this point, negative V_a was applied. The important point is that armature control can accommodate backward rotation.

The results of Example 9.6 have general implications:

- Speed control can clearly be achieved by armature voltage control.
- The relationship is direct: increasing armature voltage increases the speed.
- There is an upper limit to armature control. Operating above the base speed requires operating above the rated voltage, which is not desirable.
- There is no lower limit on the speed. The armature current will normally drop with speed, or at most remain constant, depending on the load.
- Operation in all four quardrants (MF, MR, GF, and GR) can be accommodated with armature control.

Since armature control is clearly superior to field control, why bother with the latter? Because as noted, it is the easiest and most cost-effective method to operate above the base speed. Furthermore, armature control requires a variable reversible controlled DC source of capacity comparable to the power rating DC machines. Such a source is expensive, equaling or exceeding the DC machine cost. If field control can meet the application requirements, it is usually much cheaper. Still, most applications are ideally suited to armature control, which is used extensively.

9.5.3 Four Quadrant Operation

Example 9.6 raised a point of general interest, namely that the DC machine might be required to operate in any one of, or all, four quadrants. To review the issue, the reader is referred to Section 3.6.

Now examine Figure 9.7. "Quadrants" were defined in terms of torque and speed. But recall that

$$T_{dev} = K_a \phi I_a \quad \text{and} \quad E_a = K_a \phi \, \omega_{rm}$$

Furthermore

$$T_m \cong T_{dev} \quad \text{and} \quad V_a \cong E_a$$

Therefore, for armature control (constant field), torque and speed are essentially proportional to armature current and voltage, respectively. Hence four-quadrant operation will require that the armature be supplied with a bipolar voltage and bidirectional current controllable DC source. The interpretation of positive and negative voltage, current, torque, and speed depends of course on the reference coordinates.

9.5.4 Reversing DC Motors

To drive a load in reverse, we must reverse the developed torque. Apparently we have two options: either reverse the field or armature current (but not both, of course). The field may be reversed by reversing the polarity of field voltage, which in turn reverses the field current, and hence $K_a\phi$. This is a practical option, and is sometimes used if only two-quadrant operation (MF and MR) is required.

When only armature control is utilized, we must reverse the polarity of the armature voltage. To operate in the "MR" mode, we must provide for reverse current flow as well.

9.6 DC Machine Constants from Tests

Mathematical models have little practical value unless it is possible to obtain accurate values for the required element constants. Again, it is relatively straightforward to compute such constants from data acquired using certain fairly simple tests. The required constants are the element values (R_a, L_a, X_q, R_F, L_F), the constant K_{RL}, and the MagC.

9.6.1 The DC Test: R_a

By applying an external DC source between armature terminals (A_1–A_2) with the machine at standstill, and measuring the terminal DC voltage and current, the DC stator resistance can be determined ($R_a = V_{A1-A2}/I_a$). The measurement should be made at the temperature and current level that best corresponds to normal operation:[10]

9.6.2 The Transient Test: L_a

If DC armature voltage is suddenly applied at stand still, and the armature current transient is measured (by oscilloscope), the time required for the current to stabilize (T_0) can be determined. Then

$$L_a = \text{armature inductance} = \frac{T_0}{5} R_a$$

The current should not exceed rated.

9.6.3 The Open-Circuit Test: The Magnetization Characteristic and Field Resistance

This test is essentially the same as was described for the synchronous machine. Measure the open-circuit armature terminal voltage with the machine at no load as the field current is varied from zero to its rated value, with the machine running at rated speed. Recall that a plot of these measurements is referred to by various terms, including the magnetization characteristic (MagC), the saturation curve, the OCC, and the no-load characteristic. If the DC field voltage is also measured, the ratio of field voltage to field current is the field resistance (R_F). Recall that the MagC can be represented adequately with the following empirical equation, as discussed in Chapter 1.

$$I_F = \frac{E_f}{K_{ag}} + A_x e^{B_x E_f}$$

[10] The situation is actually more complicated than one might assume. Not only does the value of R_a vary with temperature, the current must flow through the pigtail connections, the interpole windings, the commutator, through the brushes into the armature windings, and back. In particular, the voltage drop across the brushes is not linear with current, but is roughly constant over a wide range. Hence, lumping all these effects into one linear resistor is an inherent defect of the model. Fortunately the error is small for most applications.

The procedure for computing K_{ag}, A_x, and B_x was illustrated in Example 1.4; see also Example 7.5. If rated DC field voltage is suddenly applied, and the field current transient is measured (by oscilloscope), the time required for the current to stabilize (T_0) can be determined. Then

$$L_F = \text{field inductance} = \frac{T_0}{5} R_F$$

9.6.4 The No-Load Test: Rotational Losses

Appling rated voltage at the armature terminals, run the machine as a motor. Adjust the field until the machine runs at rated speed, and measure the armature voltage and current. Since the output is zero, the input power consists of only the armature winding loss plus the rotational loss:

$$P_{NL} = V_a I_a, \qquad P_{RL} = P_{NL} - I_a^2 R_a, \qquad K_{RL} = \frac{P_{RL}}{\omega_{rm}^2}$$

As was discussed earlier, the rotational losses are a combination of magnetic losses, mechanical losses, and stray loss. In our machine model, there is no incentive to separate the rotational losses into its constituent parts.

9.7 DC Motor Drives: Half-Wave Converters

Armature control requires a controllable DC source. For most applications, the available power source is normally a single- or three-phase AC constant voltage constant frequency source. Recall the development of rectifiers, presented in Sections 5.7 to 5.9. We can apply these circuits directly, interpreting the load as the armature of the DC machine. To study the problem, we shall use an example DC motor:

DC Machine Data

Ratings

Voltage (V_a) = 125.0 V; horsepower = 1.0 hp
Current (I_a) = 7.8 A; no. of Poles = 4
Permanent magnet field: $K_a\phi = 0.6631$ Wb;
Rated, base speed = 1800, 1200 rpm
J motor = 0.018 kg-m^2.

Equivalent Circuit Values

$R_a = 0\,\Omega$; $L_a = 20.0$ mH
Rotational loss torque = $T_{RL} = 0$
ω_{rm} = rotor speed in rad/s

Note that armature resistance is zero, as is the rotational loss. The machine has a permanent magnet field, with $K_a\phi$ = constant = 0.6631 Wb. Assume that the system inertia is large enough that the system speed cannot change within an AC cycle. Therefore

$$e_a(t) = E_a \quad \text{(a constant)}$$

$$K_{RL} = P_{RL}/(\omega_{rm})^2 \approx P_{RL}/(\omega_{sm})^2$$

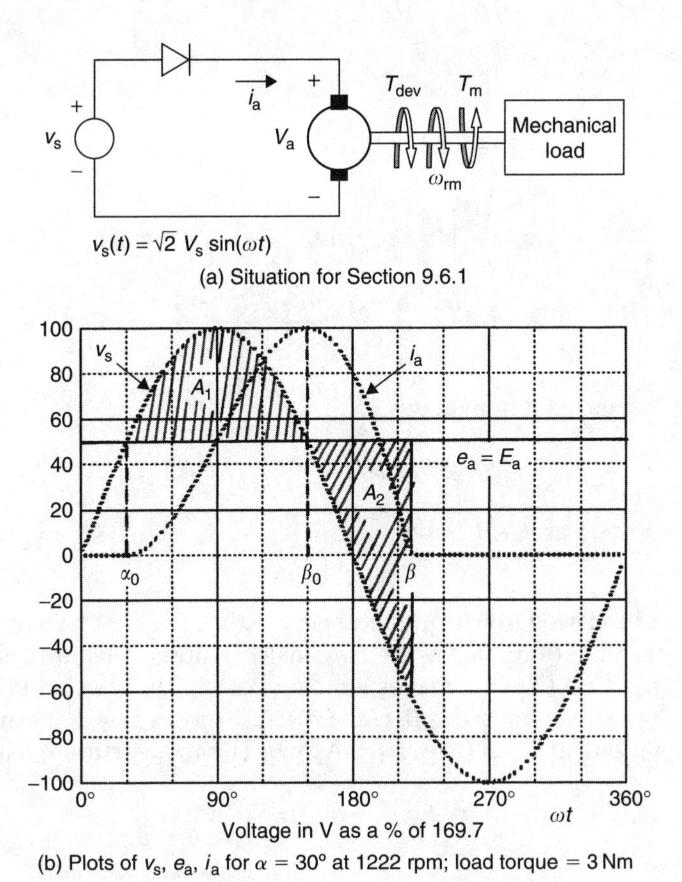

FIGURE 9.8. The half-wave rectifier applied as a DC motor drive.

As was discussed earlier, the rotational losses are a combination of magnetic losses, mechanical losses, and stray losses. In our machine model, there is no incentive to separate the rotational losses into its constituent parts.

9.7.1 The Half-Wave Diode DC Motor Drive

Examine the system of Figure 9.8. The AC source is 120 V rms 60 Hz. Suppose the system runs at 1222 rpm (128.0 rad/s). Then

$$E_a = K_a \phi \omega_{rm} = 0.6631(128.0) = 84.85\ \text{V}$$

For $v_s < E_a$ ($\omega t < \alpha_0$), the diode is reverse-biased, blocking current flow. As $\omega t > \alpha_0$, the diode turns ON, and positive current increases until $\omega t = \beta_0$. Armature inductance prevents step changes in current. The current drops for $\beta_0 < \omega t < \beta$, with $L_a di_a/dt$ providing just enough positive voltage to keep the diode ON. At $\omega t = \beta$, the current reaches zero, at which point the diode turns OFF. Solving for the relevant values,

$$E_a = 120\sqrt{2}\sin(\alpha_0) = 84.85\ \text{V}, \quad \alpha_0 = \sin^{-1}\left(\frac{84.85}{120\sqrt{2}}\right) = 30°$$

Likewise

$$E_a = 120\sqrt{2}\sin(\beta_0) = 84.85\ \text{V}, \quad \beta_0 = 150°$$

The current i_a is

$$i_a = \frac{1}{L_a}\int_{t_\alpha}^{t}(v_s - E_a)\,dt, \qquad t_0 < t < t_\beta$$

$$i_a = \frac{1}{\omega L_a}\int_{\alpha}^{\theta}(v_s - E_a)\,d\theta, \qquad \alpha_0 < \theta < \beta$$

$$i_a = \frac{V_{max}(\cos\alpha - \cos\theta) + E_a(\alpha - \theta)}{X_a}$$

The current reaches its maximum at $\theta = 150°$:

$$i_a \text{ at } \theta = \beta_0 = 150° = 15.4\ \text{A}$$

The current falls to zero at $\theta = \beta$. Solving for β,

$$\beta = 218.4°$$

The situation may be viewed from a graphical perspective. The current is proportional to the area between $v_s(t)$ and E_a (indicated as the cross-hatched areas in Figure 9.8b, considered as positive when $v_s(t) > E_a$). Observe that as θ moves past α_0, the area (and thus the current) increases, reaching a maximum at $\theta = \beta_0$ (Area A_1). As θ moves past β_0, the net area (current) decreases, reaching zero at $\theta = \beta$ (Area $A_1 - A_2 = 0$).The *average* current may be computed:

$$I_a = \frac{1}{2\pi}\int_{\alpha}^{\beta} i_a(\theta)\,d\theta, \qquad \alpha_0 < \theta < \beta$$

$$I_a = 4.53\ \text{A}$$

Since there is no rotational loss, and the speed is constant,

$$T_m = T_{dev} = 3\ \text{N m}$$

The *average* developed torque is

$$T_{dev} = K_a\phi I_a = (0.6631)(4.53) = 3.005 \cong 3\ \text{N m}$$

We now make some general observations. Define the conduction angle $\gamma = \beta - \alpha$. Suppose the load increases. Since the torques are no longer balanced, the system slows down, which proportionally drops E_a. But this decreases α, increases area A_1, and increases the current. To balance area A_1, area A_2 also increases which increases β, γ, the average current, and the average torque. If the load torque drops, the system speeds up, reducing area A_1, the current, and torque. Thus the system is inherently stable, and automatically seeks out the equilibrium running speed. However, the system has one major flaw: we have no control over speed. The speed is exclusively determined by the load, running faster at light loads and slower for heavier loads. To control the speed, we need to control α.

9.7.2 The Half-Wave Thyristor DC Motor Drive

Examine the system of Figure 9.9, which is identical to Figure 9.8, except that the diode has been replaced with a thyristor (silicon controlled rectifier [SCR]). The AC source is 120 Vrms 60 Hz. Assume that the motor supplies a constant torque load (3 N m), comparable to the load of Section 9.6.1. Remember conduction cannot commence unless the SCR is forward-biased when the gate is pulsed at $\theta = \alpha$. Assume that the system is running at 1222 rpm, and we elect

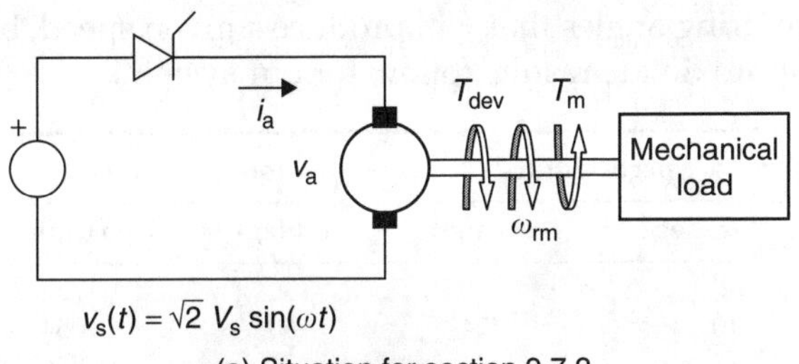

(a) Situation for section 9.7.2

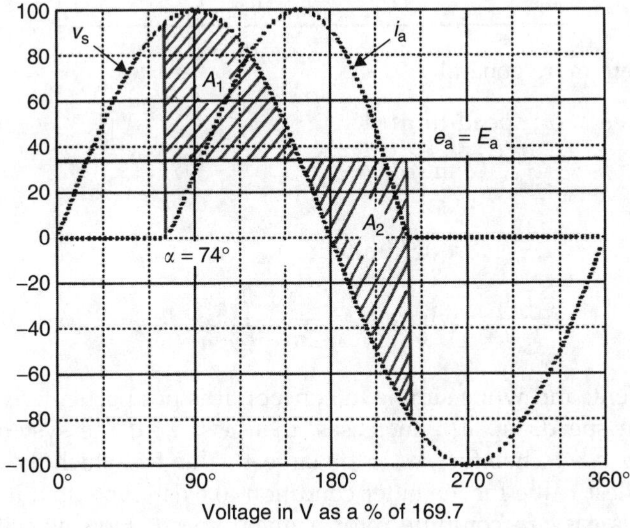

(b) Plots of v_s, e_a, i_a for α = 74° at 836 rpm: torque limited

FIGURE 9.9. Half-wave thyristor DC motor drive.

to fire the SCR at $\alpha = 30°$, turning on the SCR just in time to allow just enough current to supply the load requirements. At this point, the situation is identical to the diode case.

Now fire the SCR at $\alpha = 20°$. At 1222 rpm, $E_a > v_s(t)$, and the SCR fails to turn ON. Hence no current flows, $T_{dev} = 0$, and the system must slow down. The system slows down until it reaches

$$E_a = 120\sqrt{2}\sin(20°) = 58.04\ \text{V},$$

$$\omega_{rm} = \frac{E_a}{K_a\phi} = \left(\frac{58.04}{0.6631}\right) = 87.53\ \text{rad/s} = 836\ \text{rpm}$$

As speed drops below 836 rpm, $E_a < v_s(t)$ and the SCR turns ON. In the first cycle where this is the case, more than enough current to support the load will flow. But as the system accelerates above 836 rpm, again the SCR fails to turn ON. Hence the system stabilizes at an average speed of 836 rpm, producing too much torque if it falls under 836 rpm, and too little if it exceeds 836 rpm. The argument is valid for any angle in the range $0 < \alpha < 30°$.

Again, suppose the system runs at 1222 rpm, and we now fire the SCR at 74°. The SCR is strongly forward-biased, and we easily initiate current flow. However, since we are later into the cycle, the source voltage begins to drop sooner, and the current has less time to build up. Hence the average developed torque is reduced, and the system slows down. This permits more current to flow, producing more torque, and the system seeks out an equilibrium point at 836 rpm. The later we fire the SCR, the slower the system runs. The argument is valid for any α in the range $0 < \alpha < 30°$.

Thus there are two firing angles that will produce a given speed, but for different physical reasons. Some approximate results follow for our system:

Speed-Limited		Torque-Limited	
α (deg)	ω (rpm)	α (deg)	ω (rpm)
0	0	30	1222
10	424	60	1054
15	632	74	836
20	836	90	478
30	1222	106	0

To state the problem more generally:

$0 \le \alpha \le \alpha_0$: speed-limited,

$$\omega_{\mathrm{rm}} = \frac{V_s\sqrt{2}\sin(\alpha)}{K_a\phi}$$

$\alpha_0 \le \alpha \le \alpha_{\max}$: torque-limited:

$$I_a = \frac{T_m}{K_a\phi} \quad \text{where } I_a = \frac{1}{2\pi X_a}\int_{\alpha}^{\beta}(v_s - R_a i_a - K_a\phi\omega_{\mathrm{rm}})\,\mathrm{d}\theta$$

α_0 is load-dependent, and while simple in concept, it is not particularly simple to calculate. If the system speeds up if α increases, then $\alpha < \alpha_0$; if the system slows down if α increases, then $\alpha > \alpha_0$. Therefore, α_0 is the largest value for which the former condition exists, or the smallest value for the latter condition to exist. Note that if we know we are speed-limited, it is easy to compute α for a given speed. However, if we are torque-limited, the problem is much more difficult, particularly if resistance is included, since the *average* torque (and hence current) is involved, which is a function of the speed. Again, in Figure 9.9b, note that the areas A_1 and A_2 are equal.

9.7.3 The Half-Wave IGBT DC Motor Drive

Examine the system of Figure 9.10, which is identical to Figure 9.8, except that the diode has been replaced with an IGBT, and a freewheeling diode has been added in parallel with the armature. Conduction cannot commence unless the IGBT is forward biased, and it is ON (at $\theta = \alpha$). The source is disconnected when the IGBT is turned OFF (at $\theta = \beta$). If $i_a > 0$, then the negative voltage caused by the armature circuit inductance will forward bias the freewheeling diode, proving a path for the armature current to decay to zero. Analyzing the situation, and neglecting armature circuit resistance,

$0 < \theta < \alpha$:

$$i_S = i_a = 0, \qquad i_{\mathrm{FWD}} = 0$$

$\alpha < \theta < \beta$:

$$i_S = i_a = \frac{1}{X_a}\int_{\alpha}^{\theta}(v_s - E_a)\,\mathrm{d}\theta, \qquad i_{\mathrm{FWD}} = 0$$

$$\text{at } \theta = \beta; \quad i_S = i_a = I_\beta$$

$\beta < \theta < \beta_{\max}$:

$$i_a = i_{\mathrm{FWD}} = I_\beta - \frac{E_a}{X_a}\theta, \qquad i_S = 0$$

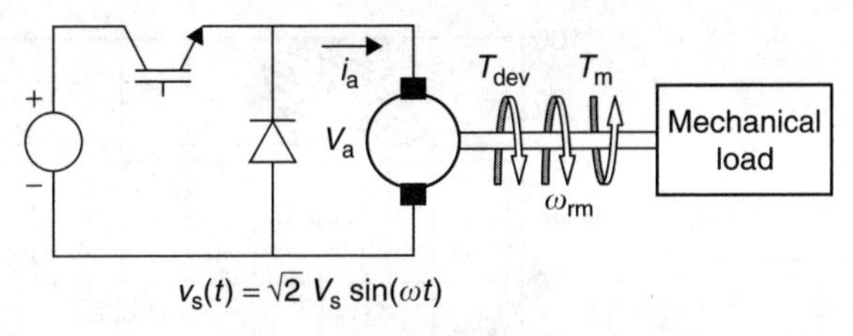

(a) Situation for Section 9.7.3

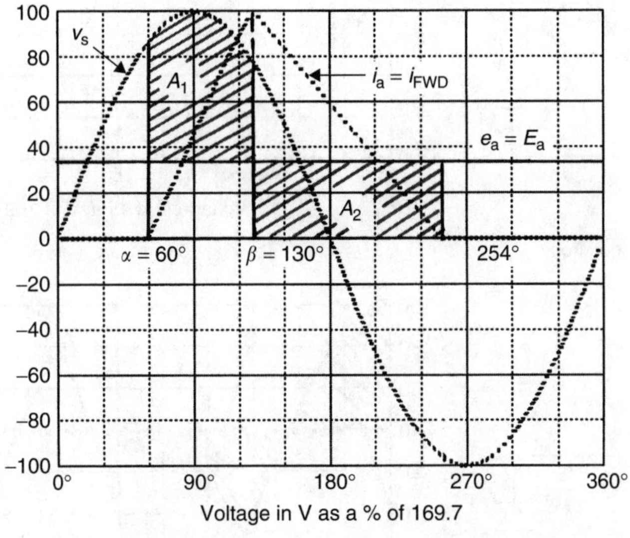

(b) Plots of v_s, e_a, i_a for $\alpha = 60°$; $\beta = 130°$ at 836 rpm

FIGURE 9.10. Half-wave IGBT DC motor drive.

$\beta_{max} < \theta < 2\pi$:

$$i_S = i_a = 0, \qquad i_{FWD} = 0$$

Applying the analysis to our example system for $\alpha = 60°$ and $\beta = 130°$, the system speed becomes 836 rpm. Plots of the relevant voltages and currents are presented in Figure 9.10b. Speed control may be achieved by varying α and β.

9.8 DC Motor Drives: Full-Wave Converters

Examination of the half-wave circuits discussed to this point leads to the observation that they only appear to be working "half the time;" the system appears to "coast" through the second half of the AC cycle. As we go to higher power levels, it is reasonable to consider full-wave converters. We shall use the same DC machine as presented in Section 9.7 for demonstration purposes.

9.8.1 The Full-Wave Diode DC Motor Drive

Consider the situation in Figure 9.11a. The load is a constant torque of 3 N m. The situation is a simple extension of the half-wave case. The only difference is that we receive two equal pulses of current per cycle.

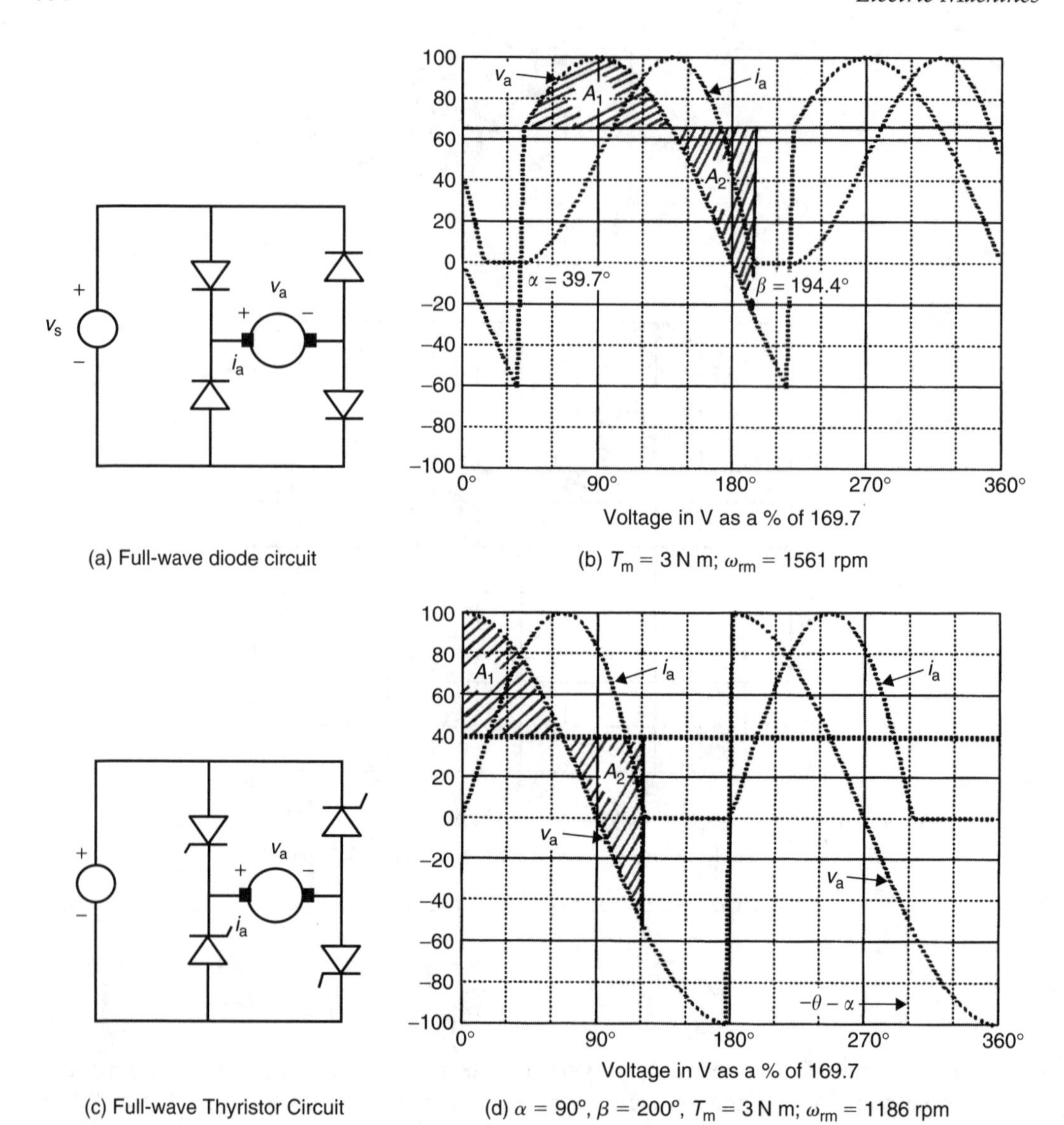

FIGURE 9.11. Full-wave DC motor drives circuit.

For a given load (T_m must be a known function of speed), the following equations must be solved simultaneously to describe steady-state system performance:

$$E_a = K_a\phi\omega_{rm} = V_s\sqrt{2}\sin(\alpha)$$

$$L_a\frac{di_a}{dt} = v_s - R_a i_a - K_a\phi\omega_{rm}, \quad \alpha < \theta < \beta, \quad \theta = \omega t$$

$$i_a = 0, \quad \beta < \theta < \alpha + \pi, \quad i_a(\theta + \pi) = i_a(\theta)$$

$$I_a = \frac{1}{\pi X_a}\int_{\alpha}^{\beta}(v_s - R_a i_a - K_a\phi\omega_{rm})\,d\theta$$

$$T_{dev} = K_a\phi I_a = T_m + T_{RL}$$

For our demonstration system, at $T_m = 3\,\text{N m}$

$$I_a = \frac{T_m}{K_a\phi} = \frac{3}{0.6631} = 4.53$$

where

$$I_a = \frac{1}{\pi X_a}\int_\alpha^\beta (v_s - K_a\phi\omega_{rm})\,d\theta = \frac{1}{7.54\pi}\int_\alpha^\beta (169.7\sin\theta - 0.6631\omega_{rm})\,d\theta$$

which reduces to

$$4.53 = 7.164(\cos\alpha - \cos\beta) - 0.028\,\omega_{rm}(\beta - \alpha)$$

also

$$E_a = 0.6631\omega_{rm} = 169.7\sin\alpha$$

producing

$$\alpha = 39.7°, \qquad \beta = 194.4°$$

Details are presented in Figure 9.11b. Note that we have no control over the speed.

9.8.2 The Full-Wave Thyristor DC Motor Drive

Now replace the diodes with thyristors, as shown in Figure 9.11c, with no other changes. Fire SCRs 1,4 at $\alpha = 39.7°$, and SCRs 2,3 at $\alpha = 180° + 39.7° = 219.7°$. The situation is the same as with the diodes. The system runs at 1561 rpm, as before, supplying a load torque of 3 N m.

Now fire SCRs 1,4 at $\alpha = 90°$, and SCRs 2,3 at $\alpha = 180° + 90° = 270°$. For area A_1 to equal area A_2, the base line (E_a) must be lower, since we are getting "a later start," i.e., the SCRs are fired later. Hence the system runs slower. The situation is shown in Figure 9.11d. Note that we have shifted the plot 90° to the left for convenience. The general principle is clear; the longer we delay firing the SCRs the slower the system runs.

Details are presented in Figure 9.11b.

The applicable steady-state equations are almost the same as for the diode case.

$$E_a = K_a\phi\omega_{rm}; \quad \alpha \text{ is known, but } \beta \text{ is not.}$$

$$L_a\frac{di_a}{dt} = v_s - R_a i_a - K_a\phi\omega_{rm}, \qquad \alpha < \theta < \beta, \quad \theta = \omega t$$

$$i_a = 0, \quad \beta < \theta < \alpha + \pi, \qquad i_a(\theta + \pi) = i_a(\theta)$$

$$I_a = \frac{1}{\pi X_a}\int_\alpha^\beta (v_s - R_a i_a - K_a\phi\,\theta_{rm})\,d\theta$$

$$T_{dev} = K_a\phi I_a = T_m + T_{RL}$$

9.8.3 The Six-Step Three-Phase Thyristor DC Motor Drive

Up to now we have only considered single-phase sources. As we move to higher power levels, three-phase sources make more sense. Also, the only cases considered were operational in the so-called discontinuous mode, that is the armature current was flowing for only part of the cycle. We continue to use the demonstration machine of Section 9.6.

Consider the circuit of Figure 9.12a. For convenience we elect to use a "θ" reference according to

$$v_{ab} = V_{\max}\cos(\omega t - \pi/6) = V_{\max}\cos(\theta - 30°); \qquad v_{ba} = -v_{ab}$$

$$v_{bc} = V_{\max}\cos(\theta - 150°); \qquad v_{cb} = -v_{bc}$$

$$v_{ca} = V_{\max}\cos(\theta + 90°); \qquad v_{ac} = -v_{ca}$$

Now fire the SCRs according to the following table:

At α (deg)	Fire SCR	SCRs ON
30	1	1–4
90	6	1–6
150	3	3–6
210 = −150	2	3–2
270 = −90	5	5–2
330 = −30	4	5–4

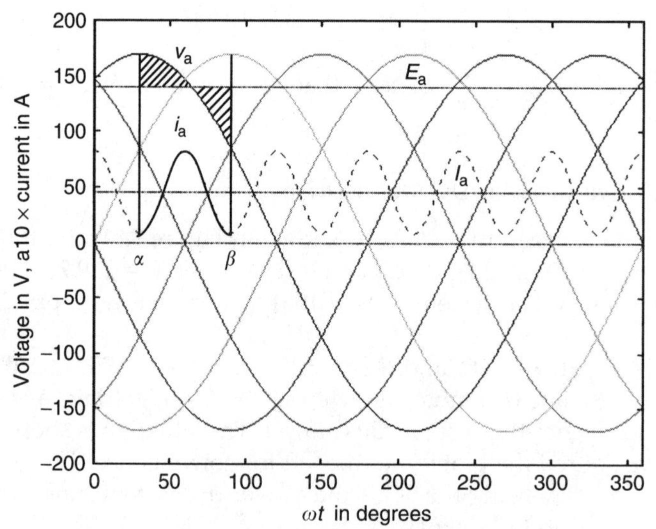

(a) Current and voltage waveforms: continuous conduction $\alpha = 30°$; $\beta = 90°$; $E_a = 140.3$ V; $I_a = 4.53$ A; $\omega_{rm} = 2020$ rpm

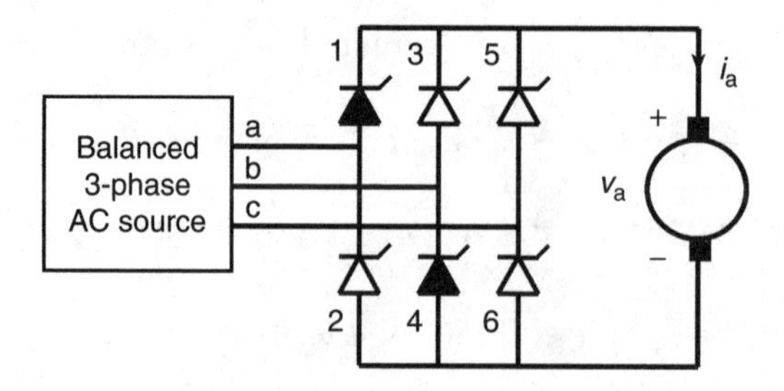

(b) Six-step thyristor converter

FIGURE 9.12. Six-step DC motor drive.

Consider that SCRs 5,4 are ON and we fire SCR #1 at $\alpha = 30°$, which momentarily shorts Phase a to Phase c. Since Phase a is positive relative to c, current surges from a to c backwards through SCR #5, which turns it OFF. Total current i_a is clamped because of the armature circuit inductance, and conduction transfers to SCR #1. The current level depends on the load. If enough current is needed, the current continuously flows over the entire 1/6 cycle and we are in the *continuous* current mode. The pattern repeats as we fire each SCR in the order indicated in the above table at equal 60° intervals. Since the average (DC) voltage across the armature inductance is zero, it follows that

$$E_a = \frac{1}{T/6}\int_{t_\alpha}^{t_\beta}(v_s - R_a i_a)\,dt = \frac{3}{\pi}\int_{\alpha}^{\alpha+\pi/3}(v_s - R_a i_a)\,d\theta$$

which for $R_a = 0$ produces $E_a = (3V_{max}\cos\alpha)/\pi$.

The speed is therefore

$$\omega_{rm} = \frac{E_a}{K_a\phi}$$

also

$$I_a = \frac{3}{\pi X_a}\int_{\alpha}^{\alpha+\pi/3}(v_s - R_a i_a - E_a)\,d\theta$$

$$T_{dev} = K_a\phi I_a = T_m + T_{RL}$$

For our demonstration machine and $\alpha = 30°$,

$$E_a = \frac{3V_{max}\cos\alpha}{\pi} = 140.3\text{ V}$$

$$\omega_{rm} = \frac{E_a}{K_a\phi} = 211.6\text{ rad/s} = 2020\text{ rpm}$$

also

$$I_a = \frac{3}{\pi X_a}\int_{\alpha}^{\alpha+\pi/3}(v_s - E_a)\,d\theta = 4.53\text{ A}$$

$$T_{dev} = K_a\phi I_a = 3\text{ N m}$$

Now consider $\alpha = 60°$. Specifically, consider the first cycle, starting in the interval ab:

$$E_a = \frac{3V_{max}\cos\alpha}{\pi} = 81.03\text{ V}$$

$$\omega_{rm} = \frac{E_a}{K_a\phi} = 122.2\text{ rad/s} = 1167\text{ rpm}$$

Now we reach the first current maximum at $169.7\cos(\omega t - 30°) = 81.03$; therefore $\omega t = 91.5°$ or $31.5°$ in the first interval.

The current reaches

$$i_a(\max) = \frac{1}{X_a} \int_{\pi/3}^{1.5966} (v_s - E_a)\, d\theta$$

$$= \frac{1}{7.54} \int_{\pi/3}^{1.5966} (169.7[\cos(\theta - \pi/6) - 81.03]\, d\theta = 2.62\ \text{A}$$

But the current must *average* to 4.52 A to supply the 3 N m load! Clearly, the system cannot run at 1167 rpm, it must slow down. As it slows down, the current will increase until we again reach torque balance, or stall if this proves impossible.

The problem is not particularly simple to solve because of its nonlinearities.

9.9 Four-Quadrant Performance

The circuit of Figure 9.12b can operate in either the MF or GB modes, since neither of these modes require a reversal of current. For this reason, the circuit is sometimes called a "full" converter. The GB mode is accessed if $\alpha > 90°$. To access the remaining two modes, we use two full converters arranged as shown in Figure 9.13, forming a so-called dual converter, operating according to

MF mode	Left converter ON:	$0° > \alpha > 90°$
GB mode	Left converter ON:	$90° > \alpha > 180°$
MB mode	Right converter ON:	$0° > \alpha > 90°$
GF mode	Right converter ON:	$90° > \alpha > 180°$

9.10 DC Motor Dynamic Performance

Our investigation to this point has been restricted to constant speed performance. Many applications require considering system dynamics as well. Recall Equation 3.1:

$$T_{dev} - T'_m = T_a = J\frac{d\omega_{rm}}{dt}$$

Experience shows that for many practical situations, excellent results can be obtained by using "quasi-DC" methods, neglecting armature inductance, provided that the voltage and

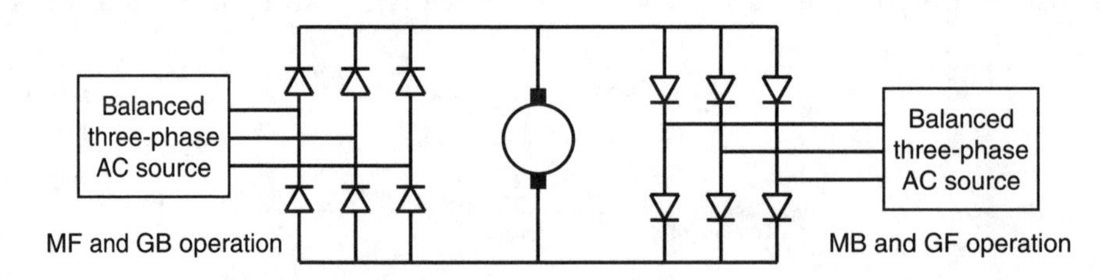

FIGURE 9.13. The dual converter.

speed do not change significantly within the time step selected to compute the dynamic response. Most loads and motors will have sufficient inertia to meet this restriction. The addition of the DC motor drive adds the additional requirement that the armature voltage can now vary in time. However, there are high-performance situations where this restriction is not met, and the reader is cautioned to be aware of this issue. Also, note that more data are needed, specifically the inertia constants for the motor and load.

Starting the system is a particularly important dynamic situation. Note that, unlike the AC induction motor, a full line voltage start in not an option for DC motors of any significant size. Consider the 50 hp motor of Example 9.3. If we were to apply rated voltage to the armature at standstill, the initial current would be

$$I_a = \frac{V_a}{R_a} = \frac{250}{0.0622} = 4019\text{ A}$$

or 2500% of rated current! Such a current would severely damage the motor as well as exceed the capabilities of the supply source. To effect a controlled and smooth start, the applied voltage must be controlled in such a way that the current and acceleration is kept within desirable limits. In effect, we need a source that has the V–I characteristics shown in Figure 9.14a, which approximates the characteristics of a DC motor drive. Note that

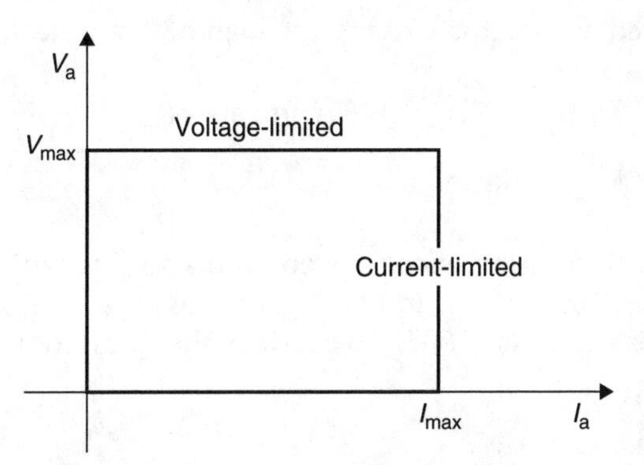

(a) V-I characteristics of an ideal current-limiting DC drive

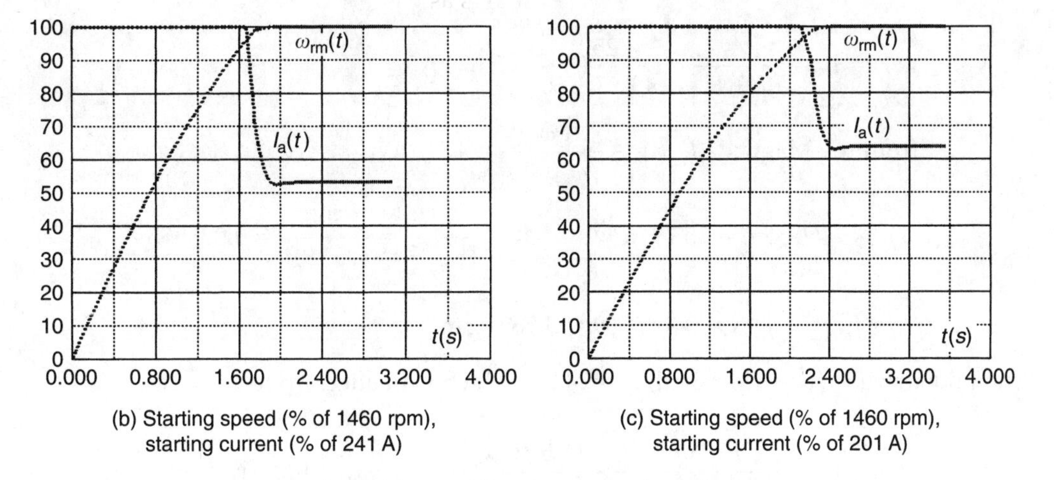

(b) Starting speed (% of 1460 rpm), starting current (% of 241 A)

(c) Starting speed (% of 1460 rpm), starting current (% of 201 A)

FIGURE 9.14. DC motor starting dynamics.

there are two distinct regions: current-limited and voltage-limited. We investigate system dynamic performance in Example 9.7.

Example 9.7
Data for the DC machine of Example 9.1 is repeated for convenience.

DC Machine Data

Ratings

Voltage (V_a) = 250.0 V;	horsepower = 50.0 hp
Current (I_a) = 160.7 A;	no. of poles = 4
Field current = 5.74 A;	rated speed = 2000 rpm = 209.4 rad/s
Base speed = 1150 rpm;	J motor = 0.569 kg m^2

Equivalent Circuit Values

$R_a = 0.0622\,\Omega$; $L_a = 0.3$ mH; $R_f = 43.57\,\Omega$
Rotational loss torque = $T_{RL} = K_{RL} * \omega_{rm} = 0.01871 * \omega_{rm}$ in N m
ω_{rm} = rotor speed in rad/s
MagC at 2000 rpm: $I_{FEQ} = E_a/K_{ag} + A_x * \exp(B_x * E_a)$.
$K_{ag} = 139.89\,\Omega$; $A_x = 0.0000580173$ A; $B_x = 0.025798$
Note: $A_x = 0$ indicates that the OCC has been linearized.

The machine is applied as a motor, driving the load of Example 9.4:

$$T_m = 200\text{ N m at 1460 rpm; also } T_m = A_2(\omega_{rm})^2$$

$$J_{load} = 3\text{ kg m}^2$$

Provide a plot of the starting transient current and speed when started at from an ideal drive, such that $V_{max} = 100\%\ V_{rated}$, and (a) $I_{max} = 100\%\ I_{rated}$; (b) $I_{max} = 150\%\ I_{rated}$.

The field rheostat is set at $R_X = 50\,\Omega$, to produce the same running speed (1460 rpm) as in Example 9.4.

SOLUTION

$$I_{FEQ} = I_F = \frac{V_a}{R_F + R_X} = \frac{250}{43.57 + 50} = 2.672\text{ A}$$

$$\text{from the MagC:} \quad K_a\phi = \frac{E_a \text{ at } \omega_{RAT}}{\omega_{RAT}} = 1.583\text{ Wb}$$

$$\omega_{rm} = 1460\text{ rpm} = 152.9\text{ rad/s} \quad A_2 = \frac{T_m}{(\omega_{rm})^2} = \frac{200}{(152.9)^2} = 0.008555$$

$$J_T = J_{motor} + J_{load} = 3 + 0.569 = 3.569\text{ kg m}^2$$

When the drive is current-limited, $I_a = 241$ A, which is in effect up to

$$241 = \frac{V_a - E_a(t)}{R_a} = \frac{250 - (1.583)\omega_0}{0.0622} \quad \text{or} \quad \omega_0 = 148.5\text{ rad/s}$$

$$\omega_{rm}\begin{cases} <\omega_0, & \text{current-limited} \\ >\omega_0, & \text{voltage-limited} \end{cases}$$

The equations to be solved are

$$3.569\frac{d\omega_{rm}}{dt} = (1.583)I_a(t) - [0.008555(\omega_{rm})^2 + 0.0187\omega_{rm}] \text{ and}$$

$$I_a(t) = \begin{cases} I_{max} & \text{current-limited} \\ \dfrac{V_a - E_a(t)}{R_a} = \dfrac{250 - (1.583)\omega_{rm}}{0.0622} & \text{Voltage-Limited} \end{cases}$$

The solution is best obtained with computer assistance:

Time (s)	Speed (rpm)	I_a (A)	V_a (V)	T_a (N m)	$R_a$sum (Ω)
0.000	0	241.1	15	382	0.0622
0.396	400	241.1	81	366	0.0622
0.824	800	241.1	148	320	0.0622
1.338	1200	241.1	214	244	0.0622
1.798	1452	148.7	250	35	0.0622
2.085	1460	127.8	250	−0	0.0622
2.285	1460	128.1	250	−0	0.0622
2.485	1460	128.1	250	−0	0.0622
2.685	1460	128.1	250	−0	0.0622
2.885	1460	128.1	250	−0	0.0622
3.085	1460	128.1	250	−0	0.0622

Graphical results are shown in Figure 9.14b. (b) The methodology is the same as for (a). Graphical results are shown in Figure 9.14c.

We can draw some general conclusions from Example 9.7.

- Analysis of the starting dynamics of a DC motor–load system requires full motor and load data. The inertia constants of all rotating parts are particularly significant.
- There is a trade-off between starting current and starting time. Faster starts require greater current. Currents in excess of rated can be tolerated for starting because the situation is transient.
- Full line voltage starts are not generally acceptable because of the extremely high corresponding currents.
- Quasi-DC analysis is appropriate as long as mechanical time constants are long compared to electrical time constants.
- The related differential equations are nonlinear, and generally should be solved using computer assistance.

Again, the reader is reminded that there are high-performance situations where "quasi-DC" analysis is inappropriate, and a full-scale transient analysis is required. For such work, a complete state model of the machine load and drive must be formulated, and solved. See Section 9.12 for more commentary on this point.

9.11 An Elevator Application

Recall that we examined a typical application, specifically the elevator that was investigated in Section 3.8. Review of this material is strongly recommended before proceeding. The mechanical design is as summarized in Figure 3.9.

Summarizing results from the load study:

- Required operating modes: MF, MR, GF, GR
- Gear ratio (motor: load speed) 40 : 1
- Maximum steady-state load power: 17.65 kW (23.66 hp)
- Maximum steady-state load torque: 3531 N m (88.25 N m)
- Maximum load steady-state speed: 5 rad/s (47.75 rpm)
- Maximum motor steady-state speed: 200 rad/s (1910 rpm)
- Maximum load inertia: 801 kg m^2 (0.5006 kg m^2)
- Maximum acceleration: 196.14 rad/s^2

Available power supply:

Three-phase four-wire 480 V 60 Hz 800 kVA: Utility panel (existing connected load 350 kVA)

Suppose we have decided to use a DC motor, and matching DC drive, supplied from the 480 V three-phase panel. Consider the following points:

1. Since four-quadrant operation is needed, we select a dual-converter drive, as shown in Figure 9.14a.
2. Based on 1, the maximum available DC voltage will be

$$V_{max} = 480\sqrt{2} = 679\text{ V}$$

$$\text{Unfiltered DC voltage} = \frac{3}{\pi}V_{max} = 648\text{ V}$$

3. Standard NEMA DC machine voltage ratings include
240, 250, 300, 500, 550 V
Base-rated speeds = 1150 to 2000 rpm; 1750 to 2300/2100 rpm
25, 30, 40, 50 , 60, 75, 100 hp

   We select a 550 V DC permanent magnetic field motor because

   - Since we plan to exclusively use armature control, no field control is necessary.
   - The drive can easily supply 550 V DC, even allowing for a 10% voltage reduction, in case of brownouts, and normal voltage drops in the drive circuitry.

4. Standard NEMA DC machine speed ratings include 1150, 1750, 2000, 2100, and 2300 rpm. We select a 2000 rpm rating because our maximum speed requirement is 1910 rpm.

5. The maximum load required was 23.66 hp, which might suggest a 25 hp motor. However, the elevator application requires constant starting and stopping. Hence, one can argue that the *normal* load condition is starting/stopping. We decide to select a motor of sufficient capacity so that rated current is only slightly exceeded, even under the worst case starting/stopping conditions. At this point, the motor inertia is unknown, since we have yet to select the motor. Example 9.7 suggests that 0.569 kg m^2 is a reasonable value for a 50 hp motor. Let us use 0.6 kg m^2 as an approximation, such that

$$T_a = J_T \frac{d\omega_{rm}}{dt} = (0.6 + 0.5)(196) = 215.6\,\text{N m}$$

Adding the maximum load torque,

$$T_{dev} = T_a + T_L = 215.6 + 88.3 = 303.9\,\text{N m}$$

Adding the maximum load torque,

$$K_a\phi = \frac{550}{200} = 2.75$$
$$I_a = \frac{T_{dev}}{K_a\phi} = \frac{304}{2.75} = 111\,\text{A}$$

A 550 V 75 hp DC motor would have a rated armature current of about

$$I_a = \frac{75(0.746)}{0.92(0.550)} = 111\,\text{A}$$

Thus, we tentatively select a 75 hp 550 V 2000 rpm DC motor.[11] Upon request, a vendor supplies the data in Table 9.4.

Note that the rated current is 108 A, which is close to our estimate of 111 A. To confirm that the machine can properly supply the steady-state load, a full load analysis is presented in Table 9.5.

It is confirmed that we operate well under the motor rating, supplying the maximum steady-state load.

TABLE 9.4
Machine Data

DC Machine Data	
Ratings	
Voltage (V_a) = 550.0 V;	horsepower = 75.0 hp
Current (I_a) = 108.0 A;	no. of poles = 4
$K_a\phi$ = 2.626 Wb;	rated, base speed = 2000 rpm = 209.4 rad/s
J motor = 0.828 kg-m^2	
Equivalent Circuit Values	
R_a = 0.2038 Ω; L_a = 11 mH	
Rotational loss torque = $T_{RL} = K_{RL}{*}\omega_{rm} = 0.02806{*}\omega_{rm}$ in N m	
ω_{rm} = rotor speed in rad/s	

[11] A more conservative choice would have been 100 hp, as we chose in Chapter 3.

TABLE 9.5
Full Load Analysis

Results of Steady-State Analysis	
Armature: V_a = 532.5 V; I_a = 35.75 A; E_a = 525.3 V	
Permanent magnet field: $K_a\phi$ = 2.626 Wb	
Load torque = 3531 N m; GR = load/motor speed = 0.0250;	
Motor inertia (JM) = 0.8282; Load inertia (JL) = 801.0	
JT = JM + sqr(GR) × JL = 0.8282 + 0.5006 = 1.3288 kg m^2	
Mode: motor forward;	speed = 1910 rpm = 200.0 rad/s
$K_a\phi$ = 2.626 Wb	
Torques (N m; RL = rotational loss):	
Developed = 93.89;	load = 88.27; T_{RL} = 5.613
Powers (magnitudes) in kW	
Input = 19.040 (exc. RXL);	developed = 18.779
Output = 17.656 (23.668 hp)	
Losses in W: RXL = field rheostat loss = 0.0	
AWL = armature winding loss;	FWL = field winding loss
RXL = field rheostat loss = 0.0	
AWL = 260.5; FWL = 0.0;	PRL = 1122.6
TOT = 1383.1 (exc. RXL);	efficiency = 92.74%

We now perform a worst-case starting analysis, which is presented in Figure 9.15. Under maximum load, we found that with drive settings of V_{max} = 96.8% of rated (533 V) and I_{max} = 125% of rated (136 A), the system smoothly accelerated to 98% of full speed in about 1 s ($\alpha = 0.98(200)/1 = 196\,\text{rad/s}^2$), meeting the specified acceleration requirement of 196 rad/s^2.

It is noted that we did exceed rated current by 25%. However, consider that:

- This was a worst-case situation, unlikely to occur continuously. Less than 15 people on the elevator will result in lower currents.
- This is a transient situation, of about 1 s duration. When averaged over a load step (start, run, stop, and hold) the current will be well under-rated, even in the maximum load case.

9.12 A More General DC Machine Model

If a true[12] transient DC shunt machine model is required, the equations (in motor coordinates) are

$$v_a = R_a i_a + L_a \frac{di_a}{dt} + e_a$$

$$v_F = (R_F + R_X)\, i_F + L_F \frac{di_F}{dt}$$

$$i_F = \frac{e_a}{K_{ag}} + A_x e^{B_x e_a}$$

[12] Once again, the reader is cautioned that there is no *exact* mathematical model for any physical device.

Gear ratio = GR = load/motor speed = 0.0250; J_s in kg m^2
Motor inertia (JM) = 0.8282; load inertia (JL) = 801.0000
JT = JM + sqr(GR) × JL = 0.8282 + 0.5006 = 1.3288

Load torque (N m) = TL = A0 + A1 * WL + ⋯ + AN * WL ** N
WL = load shaft speed in rad/s; order = 0

A0 = 3531.000; A1 = 0.000000; A2 = 0.000000000

Time (s)	Speed (rpm)	Current (A)	Voltage (V)	T_a (N m)	R_{sum} (Ω)
0.000	0	135.0	28	266	0.2038
0.209	400	135.0	138	265	0.2038
0.420	800	135.0	248	264	0.2038
0.631	1200	135.0	358	263	0.2038
0.843	1600	135.0	468	262	0.2038
0.978	1852	113.3	532	204	0.2038
1.005	1882	72.8	532	97	0.2038
1.139	1912	33.0	532	−7	0.2038
1.399	1909	35.8	532	−0	0.2038
1.719	1909	35.8	532	−0	0.2038
2.009	1909	35.8	532	−0	0.2038

(a) Starting transient

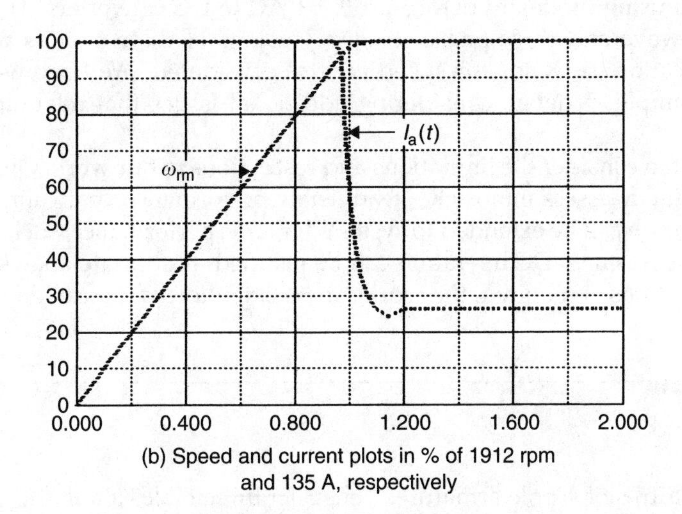

(b) Speed and current plots in % of 1912 rpm and 135 A, respectively

FIGURE 9.15. Worst-case starting analysis. Drive settings: V_{max} = 96.8%: I_{max} = 125% of rated.

$$
\begin{aligned}
e_a &= K_a \phi \omega_{rm} \\
T_{dev} &= K_a \phi i_a \\
T_{RL} &= K_{RL}\, \omega_{rm} \\
T_{dev} - T_m - T_{RL} &= J\,\frac{d\omega_{rm}}{dt} \\
T_m &= A_0 + A_1\omega_{rm} + A_2\omega_{rm}^2 + A_3\omega_{rm}^3 + \cdots
\end{aligned}
$$

The above model appears to consist of three coupled nonlinear differential equations, which though complex is manageable. However, the situation is actually more complicated. Even if the field were constant (v_F = constant), in general, under armature control, v_a and i_a would be outputs of a DC drive, which would itself be described in terms of a set of differential equations. Using computer assistance,[13] the task is certainly within most engineers' capabilities, but it is nontrivial and should not be casually undertaken.

9.13 Summary

Although DC machines are gradually being replaced with AC machines for most applications, the DC machine remains as an important device, and one of the three major EM machine types. The DC machine is still the EM device of choice in many low and medium power high-performance applications, because of its excellent speed and position control properties. Its primary disadvantages include its higher cost, and the maintenance and reliability problems associated with the commutator and brush structures.

We have considered DC machine structural details, some armature winding configurations, and various field options. Operation in all four steady-state operating modes (MF, MB, GF, and GB) was investigated, and we learned that equivalent circuit models could be used for accurate assessment of machine performance for a given application. We have considered speed control in some detail, investigating both field and armature control options. Armature control requires a variable voltage high-power DC source, which can be implemented using diode, thyristor, and IGBT AC to DC converters. The fundamental half-wave, full-wave and three-phase six-step versions of these circuits were presented and their application to DC motor speed control considered. We have worked through the elevator example in detail, considering additional issues that relate to DC machine performance.

It is important to consider the limitations and restrictions to our work. Our approach can be used to accurately assess motor DC steady-state performance, and running at constant speed. The approach can be extended to system dynamic performance, such as starting and stopping, as long as quasi-DC operation can be justified. If accurate analysis of high-performance applications is required, the work can be extended as indicated in Section 9.12.

Problems

9.1 Consider a 16-slot 4-pole armature. Construct three tables similar to Tables 9.1 to 9.3, providing details of coil locations and commutator connections appropriate to double-layer lap and wave DC armature windings.

9.2 Consider that the armature of Problem 9.1 is for a 125 V 3000 rpm DC machine. If the coils that make up the armature windings have five turns each, find K_a, and the required flux per pole, for

(a) The lap winding
(b) The wave winding.

9.3 The DC machine of Figure 9.5 operates as a generator at rated voltage, delivering 25 kW to an external load. The shunt field is supplied from a separate 250 V source. If it is driven at 1600 rpm, calculate all voltages, powers, and torques. Determine the field rheostat setting and efficiency.

[13] MATLAB©, MATHCAD©, MAPLE© and other tools would be ideal for this work.

9.4 The DC machine of Figure 9.5 operates as a shunt motor, at rated voltage, delivering 30 hp at 1600 rpm. Calculate all voltages, powers, and torques. Determine the field rheostat setting and efficiency.

9.5 Consider a permanent magnet field DC machine, described in motor coordinates. $R_a = 0.1\,\Omega$. Each row of the table defines a specific operating condition. Supply all missing entries.

V_a (v)	I_a (A)	T_{dev} (N m)	ω_{rm} (rad/s)	Mode
+200	+100	+150	+126.7	MF
+200		−150		
	−100		+140	
		−150	−140	

9.6 Consider the situation in Figure 9.16. The DC and AC machines are on the same shaft. The field rheostats are adjusted to $R_{ac} = R_{dc} = 50\,\Omega$, at which point $V_{aA} = V_{bB} = V_{cC} = 0$; $\omega_{rm} = 1800$ rpm, with switch S open. Then, switch S is closed. Each row of the table defines a specific operating condition. Supply all missing entries.

DC Machine			AC Machine			
R_{dc} (Ω)	I_a (A)	Mode (M, G, −)	R_{ac} (Ω)	I_a (A)	Mode (M, G, −)	Switch
50	+8.026	M	50	0	−	Open
50			50			Closed
50			70			Closed
50			30			Closed
52			50			Closed
48			50			Closed

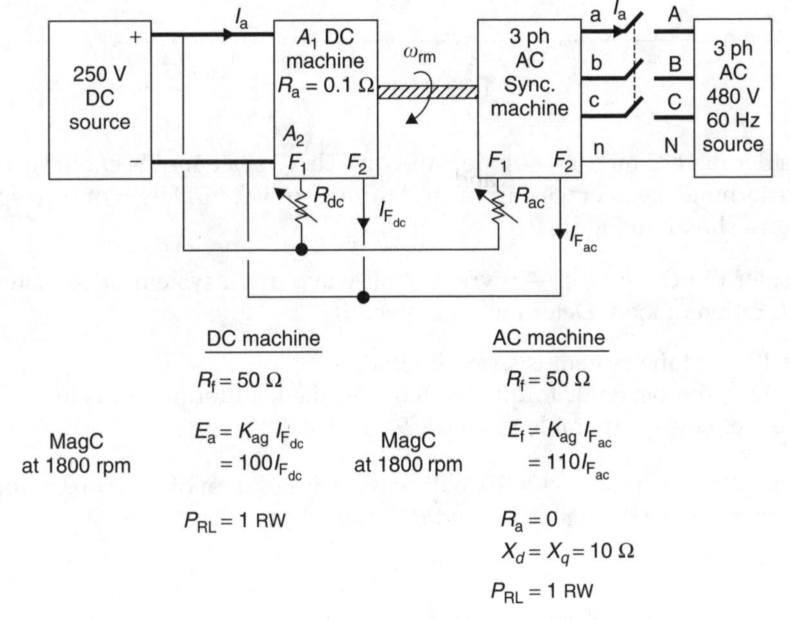

FIGURE 9.16. System for problem 9.6.

9.7 The following values were measured when performing the open–circuit test on a DC machine at 1800 rpm:

V_F (V)	I_F (A)	E_a (V)
43.6	1.00	140
79.5	1.82	250
112.3	2.58	325

(a) Suppose the test had been performed at 1500 rpm. What values would have been recorded for the same applied field voltages?
(b) Determine the constants for the MagC.
(c) Find R_F.

9.8 Suddenly applying 43.6 V to the field of the DC machine of Problem 9.7 produces the measured transient field current. Evaluate the field inductance.

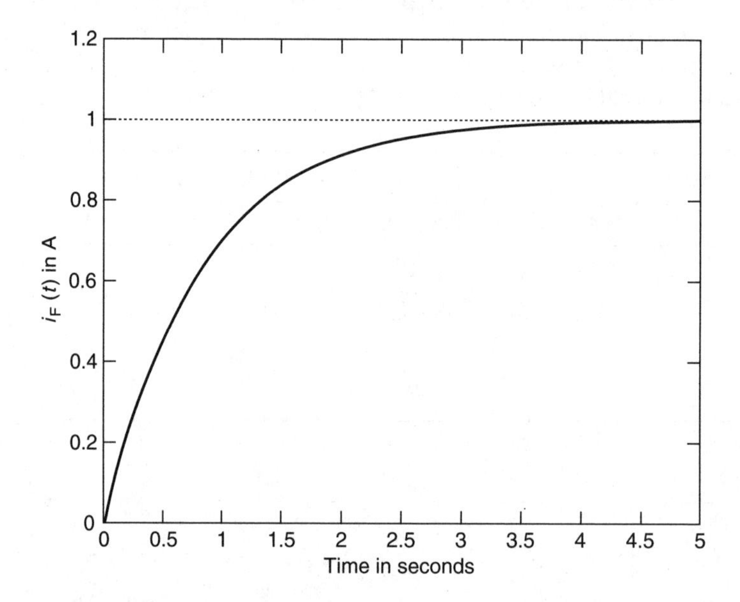

9.9 Consider the DC motor — diode half-wave drive system of Section 9.6. Suppose $\alpha_0 = 45°$. Determine the average current and torque, and accurately plot corresponding v_a e_a, and i_a, as shown in Figure 9.8.

9.10 Consider the DC motor — thyristor half-wave drive system of Section 9.6, but supplying a different load. Determine the speed if

(a) $\alpha = 45°$, and the system is speed limited,
(b) $\alpha = 120°$, the system is torque limited, and the load torque is 2 N m,
(c) produce plots comparable to Figure 9.9b.

9.11 Consider the DC motor — IGBT half-wave drive system of Section 9.6, supplying a constant 2 N m load. Given $\alpha = 60°$ and $\beta = 120°$, find

(a) the speed,
(b) β_{max},
(c) produce plots comparable to Figure 9.10b.

9.12 Consider the DC motor — diode full-wave drive system of Section 9.7, running at 1500 rpm. Determine α_0, the average current and torque, and produce plots comparable to Figure 9.12b.

9.13 Consider the DC motor — thyristor full-wave drive system of Section 9.7 running at 1500 rpm. For $\alpha = 90°$, find the average current and torque, and produce plots like in Figure 9.12d.

9.14 Consider a permanent magnetic field DC motor driving an inertial load, such that the motor and load inertia total to 2 kg m^2. The motor is to be started from an ideal current limited drive such that $I_{max} = 100$ A; $V_{max} = 250$ V. $R_a = 0.10\,\Omega$. The system has negligible losses and runs at a steady-state speed of 200 rad/s. Compute and plot the starting speed transients $I_a(t)$ and $\omega_{rm}(t)$.

9.15 A remote windy site uses a wind turbine-DC generator–battery system (respective efficiencies of 50, 93, and 95%) to provide a stand-alone 120 V DC electric power source, with respective efficiencies of 50, 93, and 95%. The monthly electric energy requirements at the site averages 1000 kW h. Assume that the wind blows at a constant velocity of 8 m/s. Determine (a) the requisite cross-sectional area of the turbine; and (b) the DC generator rating (kW).

10

Translational Electromechanical Machines

Our emphasis till this point has been on rotational EM machines, since the vast majority of existing EM conversion devices utilize this geometry. From one perspective, this is somewhat curious, since many, if not most, applications require translational motion as the end product. One of humankind's most elegant and useful inventions is the wheel, which fundamentally is a device that converts rotational to translational motion, or vice versa. The most obvious example is the automobile whose desired motion is translational, but is powered by an internal combustion engine, which is a rotational machine. Our premier application, the elevator, ultimately requires translational motion. It occurs to us that if a given application requires translational motion, force, and power, why not design a device that converts EM energy directly into this form? This would eliminate the need for gears, drums, and wheels, and the like.

The concept of translational (or linear)[1] EM machines is at least 150 years old, with patents dating back to 1841, and Girard's demonstration of his "floating train" at the 1869 World's Fair in Paris. The linear machine was revitalized by the development of Westinghouse's "electropult" in the 1940s, which could accelerate a 10 klb aircraft to 115 mph in 4.2 s. Advances in power electronics, superconducting magnets, and control theory have brought translational machines into prominence, particularly for transportation applications.

There are many applications for linear machines, including:

- *Railguns and catapults.* It is possible to achieve enormous accelerations (100 g), and extremely high velocity (in excess of 1000 m/s) in very short distances. The kinetic energies are sufficient to make the technology viable for weapon systems. Likewise, EM catapult technology is practical for launching aircraft from aircraft carriers.
- *Textile machines.* Looms and weaving machines applications require high-speed precision-controlled linear motion, making them an ideal match with linear motors.
- *Moving sidewalks and escalators.*
- *High-speed rail (HSR).* Above certain speeds, bearing and wheel design is increasingly costly and complicated. Linear machines, when combined with magnetic levitation, provide a viable technology.
- *Linear induction pumps.* It can be used to pump liquid metals, such as sodium, bismuth, and molten steel.
- *Automotive collision test facilities.* Linear motors provide excellent drivers for achieving precise acceleration and velocity-controlled experiments.

[1] The term "linear" is commonly, but rather inappropriately, used as a substitute for the more accurate term "translational," describing motion where every point in a rigid body moves parallel to every other point or "non-rotational" motion. We will conform to the accepted practice.

FIGURE 10.1. A LIM driven coaster ride. Photo by Theme Park Review, www.themeparkreview.com.

- *High performance amusement park rides.* A particular favorite of my granddaughters Jenny and Robyn (see Figure 10.1).
- *Elevators.* Can be powered with linear motors, particularly attractive for so-called ropeless designs, which are necessary in extremely tall buildings.

The Japanese *Handbook of Linear Motor Applications* lists over 50 applications for linear machines.

All of the basic rotational machines have their translational equivalents. We focus on two of these:

- the linear induction machine (LIM),
- the linear synchronous machine (LSM).

We start our study by examining the details of LIM construction.

10.1 Linear Induction Machine Construction

Our objective is not to provide detailed material on linear EM machine design. However, to understand the principles of operation and control, and to provide the fundamentals for validating mathematical models, one does need a basic understanding of how such machines are physically constructed.

In the conventional cage, or wound rotor, rotational induction machine (RIM), the stator contains the windings connected to the external electric source or "primary" windings. The voltages and currents in the rotor bars or windings are created by induction, and hence these paths are identified as "secondary" windings. Although the RIM virtually always is designed as discussed in Chapter 4 (i.e., primary mounted on the stator and the secondary on the rotor), this is not really necessary. The design can be reversed (primary on the rotor; secondary on the stator) with no impact on the machine model or the energy conversion process. That is, the primary windings can be located either on the stationary or on the moveable part of the machine.

Both designs are used for the LIM. Also, there are many different variations in design, depending on the application. For most and certainly for transportation applications, either the primary or secondary must be "short," no more than a few pole spans, and the opposite member "long," perhaps hundreds of kilometers. We shall focus on one of the several possible designs specific to HSR applications. A review of Section 4.1 might be useful before continuing.

10.1.1 Primary Design

Examine the structure of Figure 10.2a. It is composed of laminated ferromagnetic material, with equally spaced transverse slots positioned along the inner surface. Coils are placed in the slots, and are interconnected to form balanced three-phase windings. Consider a 24-slot section with six-windings (A1, B1, C1, A2, B2, C2) laid out as indicated in Table 10.1, forming two balanced three-phase windings.

When excited with balanced three-phase voltages, the structure produces a translational magnetic field, traveling a distance τ (the coil "pitch"), in one-half period. Hence, the primary field velocity is

$$v_s = \frac{\tau}{T/2} = 2\tau f = \frac{\tau\omega}{\pi} = \text{synchronous speed, in m/s} \tag{10.1}$$

If the primary is on the stator, it is laid out over the entire extent of the desired range of motion. This is the situation in the design we shall discuss.

Example 10.1

The primary of a LIM is to be appropriate for a 540 km/h HSR application.

(a) Determine coil pitch if the maximum frequency is to be 100 Hz.
(b) For the four-pole 24-slot design of Figure 10.1a, determine the distance between slots.

SOLUTION

(a) $v_s = 540\,\text{km/h} = \dfrac{540{,}000}{3600} = 150\,\text{m/s} = \dfrac{\tau\omega}{\pi} = \tau(2\times 100)$

$\tau = 0.75\,\text{m}$

(b) distance between slots = 12.5 cm.

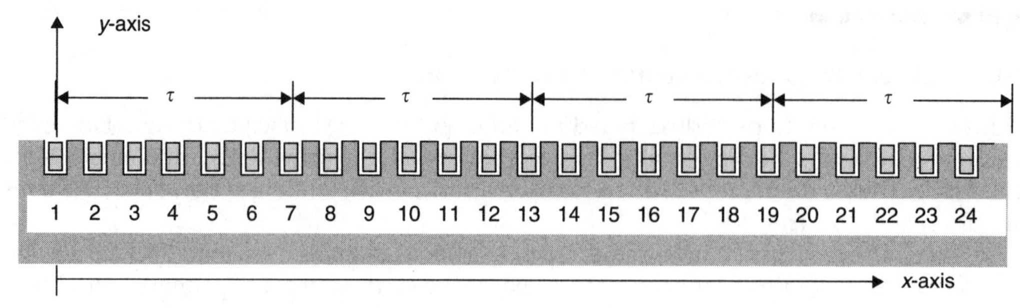

(a) 24-slot section of primary

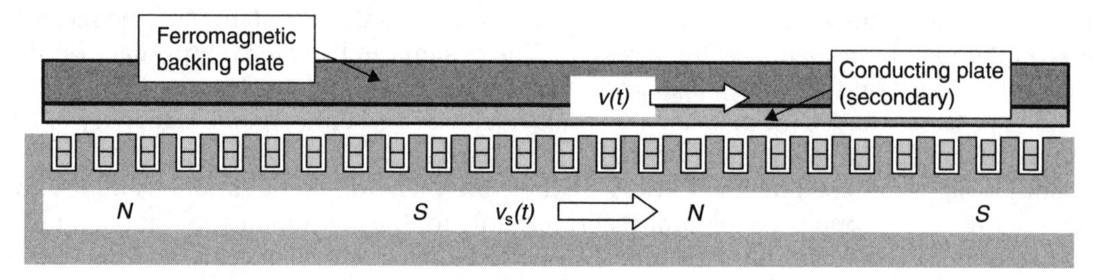

(b) Addition of secondary plate

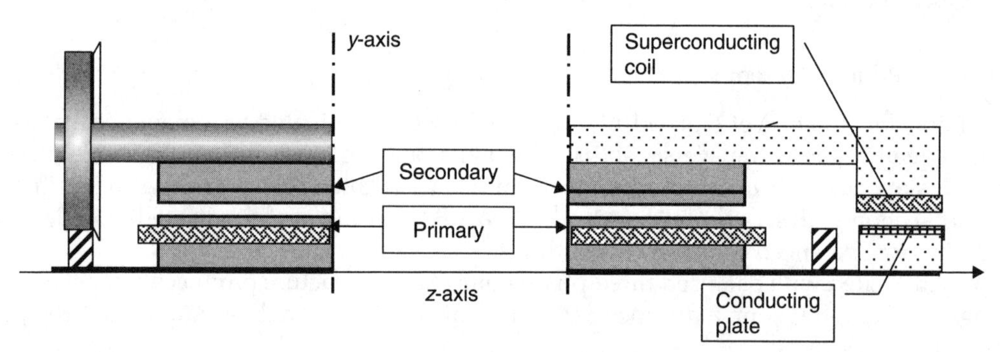

(c) Wheel suspension (left side only) (d) Electrodynamic levitation suspension (right side only)

FIGURE 10.2. A LIM design for HSR applications.

TABLE 10.1
Coil Side Slot Locations

Coil	Slot	Slot	Slot	Slot	Coil	Slot	Slot	Slot	Slot
A1	1	7	2	8	C1	3	9	4	10
B1	5	11	6	12	A2	13	19	14	20
C2	15	21	16	22	B2	17	23	18	24

10.1.2 Secondary Design

A secondary structure, consisting of a conducting aluminum plate, and backed by a ferromagnetic plate, with the same width as the primary structure, is shown in Figure 10.1b. The secondary structure is close and parallel to the primary, with one degree of mechanical freedom, so that it can move at velocity $v(t)$, as shown in Figure 10.1b. How will the two structures interact?

If $v(t) = v_s(t)$, the primary field pattern will appear motionless relative to the secondary, and no EM interaction occurs. However, if the secondary moves at $v(t) \neq v_s(t)$, voltages and currents will be induced in both the conducting plate and the ferromagnetic back plate. The secondary currents will interact with the primary field such that an EM force is produced in such a direction as to attempt to make the secondary structure synchronize with the primary field. As for the rotational device, it is expedient to define "slip" as

$$s = \frac{v_s - v}{v_s} = \text{slip}$$

If the secondary is on the moveable translator, as is the situation in the design we shall discuss, it must be supported such that the air gap is fixed, and that the translator can have unrestricted motion in one direction (x). There are three ways to support the translator:

- wheels
- air cushion
- magnetic levitation.

In the wheel design, for HSR, gravity provides the restraining force in the y-direction, keeping the vehicle in contact with the rails, as indicated in Figure 10.1c. Restraint in the z-direction is provided by flanges on the inner wheel rim, as in the case of conventional rail. Wheels have an economic advantage below 100 km/h and relatively short range (50 km and below), with reasonable reliability, noise, and loss characteristics.

Conventional wisdom suggests that while air cushion technology shows some promise and advantages for certain applications; it is generally too complex and lossy for high-speed applications, when compared with magnetic levitation (maglev) systems. In any case, the subject will not be discussed here.

For high-speed applications, maglev has the advantages of lower losses, higher reliability, easier maintenance of the railway, greater passenger comfort, lower pollution, smaller land usage profile, and smaller required curve radii at high speed. There are two common maglev designs: electromagnetic levitation (EML) and electrodynamic levitation (EDL), shown in Figure 10.3.

10.1.3 Electromagnetic Levitation (EML)

In the EML design, the proper EM levitating attractive force is maintained by supplying a DC current to a coil, driven by a regulated power supply, as shown in Figure 10.3a. Consider the steady state situation. If the bottom supported member moves a small distance Δ away from the top member, the air-gap **increases, weakening** the EM attractive force, so that the supported member falls away even further. If the bottom supported member moves a small distance closer to the top member, the air-gap **closes, strengthening** the EM attractive force, so that the supported member rises up and makes contact with the top. It is clear that unless we control the coil voltage (with feedback) the suspension system is stable.

Let us analyze the system. Neglecting all reluctance except the air-gap, we write:

$$\mathfrak{F} = Ni = \phi\mathfrak{R} = \phi\frac{2y}{\mu_0 A} \quad \text{from which it follows that:}$$

$$\lambda = N\phi = \frac{N^2\mu_0 A i}{2y} = Li$$

$$L = \frac{N^2\mu_0 A}{2y}$$

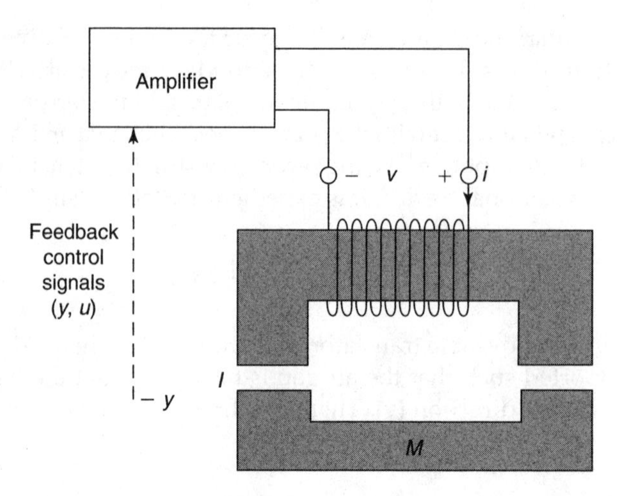

(a) Electromagnetic levitation (EML)

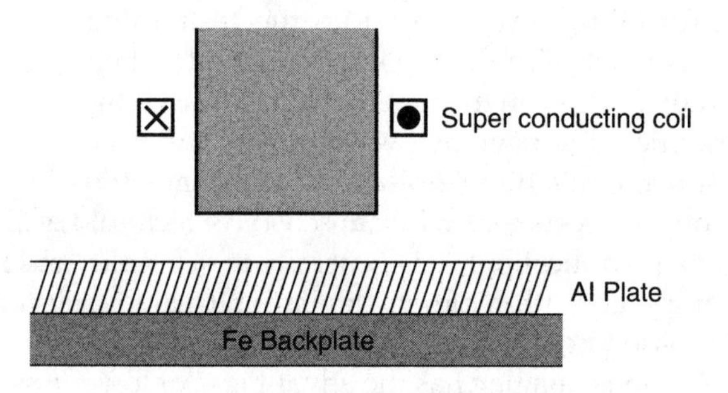

(b) Electrodynamic levitation (EDL)

FIGURE 10.3. Magnetic levitation.

Therefore $W_F = \frac{1}{2}\lambda i = \dfrac{N^2\mu_0 A i^2}{4y}$

and $F_{\text{dev}} = \dfrac{\partial W_F}{\partial y} = \dfrac{\partial}{\partial y}\left(\dfrac{N^2\mu_0 A i^2}{4y}\right) = -\dfrac{N^2\mu_0 A i^2}{4y^2}$

Define $u = py$

KVL at the coil terminals produces:

$$v = Ri + p\lambda = Ri + Lpi + (pL)i$$

$$pL = \frac{\partial L}{\partial y}\cdot\frac{\mathrm{d}y}{\mathrm{d}t} = -\frac{N^2\mu_0 A u}{2y^2}$$

$$v = Ri + \frac{N^2\mu_0 A}{2y}pi - \frac{N^2\mu_0 A u i}{2y^2} = Ri + Lpi - \frac{Lui}{y}$$

Summing "y" forces in the "y" direction ("down" is positive):

$$\begin{aligned} Mpu &= Mg - F_{\text{dev}} \\ &= Mg - \frac{N^2\mu_0 A i^2}{4y^2} = Mg - \frac{Li^2}{2y}, \text{ or} \end{aligned}$$

$$pu = g - \frac{Li^2}{2My}$$

Collecting the state equations:

$$\begin{aligned} py &= u \\ pu &= g - \frac{Li^2}{2My} \\ pi &= \frac{i}{L}u - \frac{R}{L}i + \frac{1}{L}v \end{aligned}$$

The equations are nonlinear. To linearize about a equilibrium point $y = y_0$

$$\begin{aligned} \Delta &= \frac{y}{y_0} \\ L_0 &= \frac{N^2\mu_0 A}{2y_0} \\ L &\cong (1-\Delta)L_0 \\ v &= V_0 + \Delta v \\ i &= I_0 + \Delta i \\ I_0 &= \frac{V_0}{R} = \sqrt{\frac{2Mg\,y_0}{L_0}} \end{aligned}$$

where Δ represents a small deviation from the equilibrium. Substituting

$$\begin{aligned} py &= u \\ pu &= g - \frac{L_0(1-\Delta)(I_0+\Delta i)^2}{2M(1+\Delta)y_0} \\ p(\Delta i) &= \frac{I_0+\Delta i}{L}u - \frac{R}{L}(I_0+\Delta i) + \frac{1}{L}(V_0+\Delta v) \\ &= \frac{I_0+\Delta i}{L}u - \frac{R\Delta i}{L} + \frac{\Delta v}{L} \end{aligned}$$

Expanding and rejecting second order effects:

$$\begin{aligned} p\Delta &= u \\ pu &= K_1\Delta - K_2\Delta i \\ p(\Delta i) &= \frac{K_1}{K_2}u - K_3\Delta i + \frac{1}{L_0}\Delta v \end{aligned}$$

where:

$$K_1 = \frac{L_0 I_0^2}{M y_0}$$

$$K_2 = \frac{L_0 I_0}{M y_0}$$

$$K_3 = -\frac{R}{L_0} + \frac{1}{y_0}$$

In matrix form:

$$\begin{bmatrix} p\Delta \\ pu \\ p\Delta i \end{bmatrix} = \begin{bmatrix} 0 & 1 & 0 \\ K_1 & 0 & -K_2 \\ 0 & I_0 & K_3 \end{bmatrix} \cdot \begin{bmatrix} \Delta \\ u \\ \Delta i \end{bmatrix} + \begin{bmatrix} 0 \\ 0 \\ (1/L_0)\Delta v \end{bmatrix}$$

The open loop characteristic equation derives to:

$$p^3 + K_3 p^2 - K_1 K_3 = 0$$

The missing "p" term, and the negative constant, reveal that the system has at least one pole in the right-half s-plane, indicating that the system is inherently unstable, as we had previously noted.

To stabilize the system, consider controlling the coil voltage using feedback. Specifically, let us control the coil voltage by feeding back a signal proportional to the air-gap length and rate of change, according to:

$$\Delta v = K_0 \Delta + K_u u$$

Observe that now there is at least a possibility of stability. As the air gap increases, the coil voltage increases (to provide more current). Also if the lower mass falls away, u increases, providing even more coil voltage to stop the mass. Since the EM force varies as the square of the current, we at least have a chance of maintaining stability. Substitute Δv into the open loop characteristic equation, the closed loop characteristic equation becomes:

$$p^3 + K_3 p^2 + \frac{K_u K_2}{L_0} p + \left(\frac{K_u K_2}{L_0} - K_1 K_3 \right) = 0$$

To clarify the issue, a numerical analysis would be useful.

Example 10.2

An EML levitation system must support 16 metric tons. The levitating coil has 100 turns; the total working area is 10 m^2. The normal design air-gap is 2 cm, and the coil resistance is 3 ohms.

(a) Determine the system values at equilibrium
(b) Determine K_1, K_2, and K_3.
(c) Determine the system poles for the open loop system.
(d) Determine the system poles for the closed loop system. Select K_0 and K_u for stability and critical damping.

SOLUTION

(a) $$L_0 = \frac{N^2\mu_0 A}{2y_0} = \frac{(100)^2(4\pi\times10^{-7})(10)}{2(0.02)} = 3.146\ H$$

$$F(i,y) = -\frac{N^2\mu_0 A I_0^2}{4y^2} = -\frac{4\pi\times10^{-2}I_0^2}{4(0.02)^2} = -78.54I_o^2$$

$$I_o = 44.7\ A \quad V_0 = I_0 R = 134.1\ V$$

(b) $$K_1 = \frac{L_0 I_0^2}{My_0} = \frac{3.142(44.7)^2}{16000(0.02)} = 19.62$$

$$K_2 = \frac{L_0 I_0}{My_0} = \frac{3.142(44.7)}{16000(0.02)} = 0.439$$

$$K_3 = -\frac{R}{L_0} + \frac{1}{y_0} = -\frac{3}{3.142} + 50 = 49$$

(c) The open loop characteristic equation is:
$p^3 + K_3p^2 - K_1K_3 = p^3 + 49p^2 - 962.3$
$= (p + 4.658)(p - 4.251)(p + 48.59)$

The system is clearly unstable (one pole in the right half plane).

(d) The closed loop characteristic equation is:

$$p^3 + K_3p^2 + \frac{K_uK_2}{L_0}p + \frac{K_0K_2}{L_0} - K_1K_3$$

Select $\frac{K_0K_2}{L_0} - K_1K_3 = 0$ to force a pole at the orgin

$$K_0 = \frac{K_1K_3L_0}{K_2} = 6890$$

Characteristic Function:

$p^3 + 49p^2 + 0.14K_up = p(p^2 + 49p + 0.14\,K_u)$

Solving the quardratic: $p = -24.5 \pm 0.5\sqrt{2401 - 0.56\,K_u}$

For critical damping select $K_u = 4288$, so that $p = -24.5$
Characteristic Equation: $p(p + 24.5)(p + 24.5) = 0$

and the system is stable and critically damped.

The main disadvantage of EML is now apparent. To maintain stability, a feedback controlled regulated power supply is required. If the feedback signals fail, we lose the suspension system, which could have catastrophic consequences in some applications. On the other hand, we can easily tune the suspension system by manipulating the feedback signals.

10.1.4 Electrodynamic Levitation

In the EDL design, the (shown in Figure 10.3b) EM levitating repulsive force is created by locating a superconducting magnet on the translator, which moves over a conducting plate on the stator. The superconducting field induces a voltage and current in the plate, which creates a opposing field. The interaction of these two fields creates a repulsive force, which acts upward on the translator. Above a certain speed (v_0), this lift is sufficient to lift the translator above the railway. As the translator rises, the air gap increases and the lift decreases, so that the system automatically adjusts to an equilibrium position slightly (within a few millimeters) above the railway. Below v_0, the lift is insufficient to lift the translator, so that wheels are required for low-speed operation. The speed v_0 depends on the design geometry, a typical value being around 22 m/s (80 km/h).

The main advantage to EDL is that it is inherently stable. However, there are significant problems with EDL systems:

- A drag force is also created, canceling some of the propulsion thrust.
- A strong normal force is created.
- EDL requires superconducting magnets.
- A secondary wheel suspension system is required at low speeds.
- A similar restraining system in the transverse (z) direction is required.
- The system has low damping in both the y and z directions, requiring damping plates; otherwise, passenger comfort would be seriously degraded by y and z oscillations.

These problems can be minimized with careful design, and the technology is quite practical. A full development of the subject is beyond the scope of our study.

10.2 Linear Induction Machine Modeling: The Equivalent Circuit

If the primary and secondary were of infinite length, the LIM equivalent circuit would be identical to that of the RIM. Unfortunately, either the primary, the secondary, or both must be "short," no more than a few pole pitches, so that end effects are quite significant and must be included. The main significant physical effects for the LIM are the same as those for the RIM. Primary and secondary winding resistance and leakage flux, and magnetizing current, must be accounted for. However, additional effects, such as A approximate circuit model for the LIM is provided in Figure 10.4.[2]

The series elements R_1 and X_1 correspond to those for the RIM model, and account for primary resistance and leakage flux, respectively. The shunt elements R_{FE} and X_{FE} also correspond to the RIM model, accounting for core losses and magnetizing current,

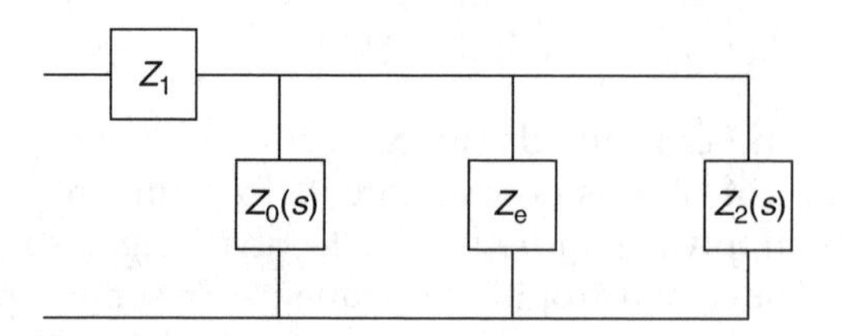

FIGURE 10.4. LIM per-phase equivalent circuit.

[2] See J. F. Gieras, *Linear Induction Drives*, Oxford Publications, 1995.

respectively. The shunt impedance Z_e is specific to the LIM, and accounts for longitudinal end effects.

The elements $R_2'(s)$ and $X_2'(s)$ account for secondary resistance and leakage flux, respectively.[3] Recall that the prime indicates that secondary quantities are reflected into the primary. They are functions of slip, and can be quite involved, in that they must account for the complex primary–secondary EM interactions. For example, for a flat, single-sided, double-layer (aluminum (AL) and iron (FE)) LIM secondary:

$$\bar{Z}'_{\mathrm{AL}}(s) = \frac{j\omega\mu_0}{k_{\mathrm{AL}}} \frac{k_{\mathrm{TR}}}{\tanh(k_{\mathrm{AL}} d_{\mathrm{AL}})} \left(\frac{L_i}{\tau}\right)$$

$$\bar{Z}'_{\mathrm{FE}}(s) = \frac{j\omega\mu_{\mathrm{FE}}}{k_{\mathrm{FE}}} \frac{k_{\mathrm{TR}} k_z}{\tanh(k_{\mathrm{FE}} d_{\mathrm{FE}})} \left(\frac{L_i}{\tau}\right)$$

$$k_{\mathrm{AL}} = \sqrt{js\omega\mu_0\sigma_{\mathrm{AL}} + \left(\frac{\pi}{\tau}\right)^2}$$

$$k_{\mathrm{FE}} = \sqrt{js\omega\mu_{\mathrm{FE}}\sigma_{\mathrm{FE}} + \left(\frac{\pi}{\tau}\right)^2}$$

$$\bar{Z}'_2(s) = \frac{\bar{Z}'_{\mathrm{AL}}(s) \cdot \bar{Z}'_{\mathrm{FE}}(s)}{\bar{Z}'_{\mathrm{AL}}(s) + \bar{Z}'_{\mathrm{FE}}(s)} = R'_2(s) + jX'_2(s)$$

where μ, σ, and d are the permeabilities, conductivities, and thicknesses of the AL and FE layers; τ is the coil pitch, L_i the effective core width, and ω the radian frequency. The power crossing the air gap and the EM developed force (thrust) are, respectively,

$$P_{\mathrm{ag}} = 3I_2'^2 R_2'(s); \qquad F_{\mathrm{dev}} = \frac{P_{\mathrm{ag}}}{v_{\mathrm{s}}}$$

Since the situation is asymmetrical, even if balanced three-phase voltages are applied, the currents will not be balanced. Further complicating the situation is the fact that the primary will be supplied with the output of a PWM inverter, which is not sinusoidal, and has time harmonics. Finally, the relevant magnetic fields are not sinusoidal in space, and the foregoing analysis should be independently performed for each space and time harmonic, using

$$s_n = \frac{nv_{\mathrm{s}} - v}{nv_{\mathrm{s}}} = \text{slip at the } n\text{th harmonic}$$

The forward and backward force components are computed by methods similar to that presented in Chapter 6. In spite of the complexities, the results are similar to what one might expect from experience with the RIM. Readers interested in pursuing the subject in analytical detail are referred to Gieras, *Linear Induction Drives*, Oxford Publications, 1995.

[3] Gieras writes the expression $R_2' = R_2'(s)/s$. Since R_2' is itself a function of s, use of conventional functional notation will lead to a logical symbolic contradiction. Here, $R_2'(s)$ is used in place of Gieras' R_2'.

10.3 The High Speed Rail (HSR) Application

A major application of linear machines is high speed rail (HSR), with research in progress in Japan, Germany, France and China. HSR offers significant advantages in speed and efficiency. Figure 10.5 shows the Japanese MLX01 MagLev Experimental Train which recently set the HSR world's speed record in excess of 500 km/hr.

The HSR dynamic performance equation of motion is:

$$F_{dev} - (F_{AD} + F_{LD} + F_G) = M\frac{dv}{dt}$$

where

$$\begin{aligned}
F_{dev} &= \text{EM developed force (thrust), in N} \\
F_{AD} &= k_{ad}v^2 = \text{aerodynamic drag, in N} \\
F_{LD} &= \frac{k_s}{|1+v|} = \text{suspension EM drag, in N} \\
F_G &= MgG_R = \text{gravitational gradient force, in N} \\
G_R &= \text{Grade, in \%} \\
v &= \text{velocity, in m/s} \\
k_{ad} &= \frac{c_d\rho_d A}{2} = \text{drag constant, in kg m/s}^2 \\
A &= \text{cross-sectional area, in m}^2 \\
c_d &= \text{drag coefficient, } 0.26 < c_d < 0.30 \text{ for HSR vehicles} \\
\rho_d &= \text{air density, 1.3 kg/m}^3 \text{ at 0°C; 1 atm,} \\
M &= M_V + M_L = \text{vehicle + payload mass, in kg}
\end{aligned}$$

We will consider HSR systems in the context of some examples.

FIGURE 10.5. The Japanese MLX01 MagLev experimental train.

Example 10.3
Consider the LIM of Example 10.1 to be applied to a HSR application, with a vehicle of the following design:

- Cross-section height × width = 4.0 × 3.8 m
- Length = 15 m
- Empty vehicle weight = 10 t = 10,000 kg
- Payload = 40 passengers + luggage = 40 × (75 + 75) = 6000 kg
- Drag coefficient = 0.26; air density = 1.2 kg/m^3 at 23°C.

Performance requirements:

- Top speed on level surface = 500 km/h
- Top speed climbing 6% grade = 450 km/h.

Neglect suspension drag at high speed. Determine

(a) The thrust and power required at top speed on level surface.
(b) The thrust and power required at top speed on grade.

SOLUTION

$$k_{ad} = \frac{c_d \rho_d A}{2} = \frac{0.26(1.2)(4.0)(3.8)}{2} = 2.371 \text{ N s}^2/\text{m}^2$$

(a) Aerodynamic drag at 500 km/h = $k_{ad}v^2 = 2.371(138.9)^2 = 45.74$ kN.
Power = $F_{AD}v = 45.74(138.9) = 6.353$ MW.
(b) Aerodynamic drag at 450 km/h = $k_{ad}v^2 = 2.371(125)^2 = 37.05$ kN.
Gravitational gradient force = $F_G = MgG_R = 16{,}000(9.802)(0.06) = 9.410$ kN

$$\text{Power} = (F_{AD} + F_G)v = (46.46)(125) = 5.807 \text{ MW}.$$

A LIM rating of 6.4 MW will meet the worst-case steady-state operating requirements.

Consider HSR starting and stopping.

Example 10.4
Continuing with the HSR system of Example 10.3:

$$k_s = 1 \text{ kN}$$

The thrust is controlled such that the system starts and accelerates at a uniform acceleration = 0.1 g to 500 km/h.

(a) What is the starting thrust?
(b) How long will it take to reach top speed?
(c) How far has the train traveled when we reach top speed?
(d) How long and far will a 0.1g stop take from 500 km/h?
(e) How long and far will an emergency stop at 0.5g take from 500 km/h?

SOLUTION

(a) At starting,

$$F_{dev} - (F_{AD} + F_{LD} + F_G) = F_{dev} - (0 + 1 + 0) = M\frac{dv}{dt} = 16(0.1\,g)$$

$$F_{dev} = 1 + 16(0.1\,g) = 16.69 \text{ kN}$$

(b) Time to start (T):

$$v = at = (0.1\,g)T_S = 138.9\text{ m/s}$$

$$T = \frac{v}{0.1\,g} = \frac{138.9}{0.9807} = 141.6\text{ s}$$

(c) Distance to start (x_1):

$$x_1 = \int_0^T (at)\,\mathrm{d}t = \frac{1}{2}aT^2 = \frac{1}{2}(0.1\,\mathrm{g})(141.6)^2 = 9.835\text{ km}$$

(d) Time and distance to stop at $0.1\,g$:

The results are the same as for (b) and (c); 141.6 s; 9.835 km

(e) Time and distance for emergency stop at $0.5\,g$:

$$T = \frac{v}{0.5\,\mathrm{g}} = \frac{138.9}{4.9035} = 28.33\text{ s}$$

$$x_1 = \frac{1}{2}aT^2 = \frac{1}{2}(0.5\,g)(28.33)^2 = 1967\text{ m}$$

Consider some issues in operating a HSR train system.

Example 10.5

Consider making a HSR train composed of cars of the design presented in Example 10.3, with the following operational data:

- Train consists of 8 cars.
- Line runs east to west; length = 600 km.
- There are 7 stations, equally spaced 100 km apart, numbered from S0 to S6, left to right. Track location points are designated with x values; e.g., $x = 325.4$, locates a point between S3 and S4, 25.4 km from S3, and 74.6 km from S4.
- All stations are at elevation 0, except S3, which is at elevation 500 m.
- Grade of 6% starting up at $x = 210.0$, and leveling off at $x = 218.333$.
- Grade of 6% starting down at $x = 381.667$ km, and leveling off at $x = 390$.
- 5 min stop at each station.
- 2 round trips per day; 365 days per year.

Neglect the small reduction and increase in speed going up and down a grade, since they will approximately offset each other.

(a) Construct an operational timetable.
(b) Determine the total energy required for a run from S0 to S6. Assume the gravitational grade energy is recovered. Assume that dynamic braking is used.
(c) Assuming that the EM conversion efficiency is 90%, and the cost of electrical energy is 0.10$ per kWh = 100$ per MWh, determine the energy cost per passenger.
(d) Compute the energy saved using regenerative braking, assuming that 80% of kinetic energy is recovered. Determine the annual cost savings.

SOLUTION

(a) At starting:
- Time to start at $0.1g = 141.6\,\text{s} = 2.36\,\text{min}$
- Time to run $100 - 2 \times 9.835\,\text{km}$ at $500\,\text{km} = 0.1607\,\text{h} = 9.64\,\text{min}$
- Time to stop at $0.1g = 2.36\,\text{min}$
- Subtotal $= 14.36\,\text{min}$

Station	X (km)	Eastbound		Westbound	
		Arrive	Depart	Arrive	Depart
S0	0	—	0.00	111.16	—
S1	100	14.36	19.36	91.80	96.80
S2	200	33.72	38.72	72.44	77.44
S3	300	53.08	59.08	53.08	59.08
S4	400	72.44	77.44	33.72	38.72
S5	500	91.80	96.80	14.36	19.36
S6	600	111.16	—	—	0.00

Note: All times in minutes.

(b) Energy computations (per car, with 100% payload):
Drag (running at 500 km/h) $= 6.353\,\text{MW} \times 0.1607\,\text{h} \times 6$ legs $= 6.126\,\text{MW h}$
Kinetic at 500 km/h $= \tfrac{1}{2}Mv^2 = 0.5(16{,}000)(138.9)^2 = 154.35\,\text{MJ} = 0.0429\,\text{MW h}$
6 starts $= 6 \times 0.0429 = 0.2573\,\text{MW h}$

Drag (accelerating/decelerating from 0 to 500 km/h):

$$dW = Pdt = Fv\,dt = k_{ad}v^3\,dt = k_{ad}(0.1\,gt)^3 dt$$

$$W = k_{ad}(0.1\,g)^3 \int_0^T (t)^3 dt = 2.371\,(0.9807)^3\,\frac{(141.6)^4}{4}$$

$$= 224.8\,\text{MJ} = 0.0624\,\text{MW h}$$

6 starts + 6 stops $= 12 \times 0.0624 = 0.7492\,\text{MW h}$
Total per car $= 6.126 + 0.2573 + 0.7492 = 7.133\,\text{MW h}$
Total per train $= 8 \times 7.133 = 57.06\,\text{MW h}$

(c) Electrical energy at 90% $= 63.40\,\text{MW h}$
Cost at 0.100\$ per kW h = 100\$ per MW h = \$6340
Cost per passenger $= \$6340/(8 \times 40) = \$19.81$

(d) With regenerative braking, energy saved is
$80\% \times 0.2573\,\text{MW h} \times 8$ cars $= 1.647\,\text{MW h}$

Annual savings at 4 trips per day × 365 days per year:

$1.647\,\text{MW h} \times \$100 \times 4 \times 365 = \$240{,}421$

10.4 Linear Synchronous Machine Construction

As in the conventional rotational synchronous machine (RSM), the translator of the LSM moves at synchronous speed, synchronized with the stator linear traveling magnetic wave. The primary–secondary designations are not so clear as with the LIM, since both windings can be energized. The terms "armature" and "field" are normally employed. The armature can be located on the stator or translator, with the field on the opposite structure. Both designs are used for the LSM. Also, there are many different variations in design, depending on the application. For transportation applications, either the armature or the field is normally "short," no more than a few pole spans, and the opposite member "long." Again

$$v_s = \frac{\tau}{T/2} = 2\tau f = \frac{\tau\omega}{\pi} = \text{synchronous speed, in m/s}$$

10.4.1 Linear Synchronous Machine Armature Design

The armature is basically the same as that used for the LIM, as shown in Figure 10.4a. It is composed of laminated ferromagnetic material, with equally spaced transverse slots positioned along the inner surface. Coils are placed in the slots, and are interconnected to form balanced three-phase windings. When excited with balanced three-phase voltages, the structure produces a translational magnetic field, traveling a distance τ (the coil "pitch") in one-half period. If the primary is on the stator, it is laid out over the entire extent of the desired range of motion. This is the situation in the design we shall discuss.

10.4.2 Linear Synchronous Machine Field Design

The LSM field structure has a design similar to that of the RSM. Its purpose is to provide a DC magnetic field, which can lock in (synchronize) with the armature-produced traveling magnetic field, and thereby produce thrust. The field may be produced either by windings carrying DC currents, or permanent magnets (PMs), the latter being the more common situation. Adjacent poles are of alternating north, south polarity, and are always salient.

10.5 Linear Synchronous Machine Nonsalient Pole Model

The LSM model for balanced three-phase constant speed operation is quite similar to that used for the RSM. The following terminology is specific to operation in the motor mode, and the coordinate system used is that appropriate to motor operation.

In the field circuit,

$$I_F = \frac{V_F}{R_F}$$

As for the RSM,

$$\hat{R} = \hat{F} + \hat{A}$$
$$\bar{E}_R = \bar{E}_F + \bar{E}_A$$

where the subscripts R, F, and A indicate the voltage caused by the respective field. We now return to consideration of the voltage (E_A), produced by the rotating stator field (A).

Since A is in direct proportion to I_a, and E_A to A, it follows that E_A is directly proportional to I_a. Therefore, let $E_A = XI_a$, where X is a constant of proportionality, and must be in ohms. The circuit appears in Figure 7.11d. Finally, applying KVL to the circuit:

$$\bar{V}_{an} = j\hat{X}_d\bar{I}_d + j\hat{X}_q\bar{I}_q + jX_\ell\bar{I}_a + R_a\bar{I}_a + \bar{E}_F$$
$$\bar{V}_{an} - R_a\bar{I}_a = j\hat{X}_d\bar{I}_d + j\hat{X}_q\bar{I}_q + jX_\ell\bar{I}_a + \bar{E}_F$$
$$\bar{V}_a = j\hat{X}_d\bar{I}_d + j\hat{X}_q\bar{I}_q + jX_\ell\bar{I}_a + \bar{E}_F$$
$$\bar{V}_a = jX_d\bar{I}_d + jX_q\bar{I}_q + \bar{E}_F$$
$$\bar{I}_a = \bar{I}_d + \bar{I}_q$$

where

$$X_d = \hat{X}_d + X_\ell$$
$$X_q = \hat{X}_q + X_\ell$$
$$\bar{V}_a = \bar{V}_{an} - R_a\bar{I}_a$$

As before, consider the projection of V_a onto the d and q axes as shown in Figure 7.12b:

$$V_d = V_a\sin(\delta), \qquad V_q = V_a\cos(\delta)$$

Likewise,

$$P = \frac{3E_fV_a}{X_d}\cdot\sin(\delta_{\max}) + \frac{3V_a^2(X_d - X_q)}{2X_dX_q}\cdot\sin(2\delta_{\max})$$
$$Q = \frac{3E_fV_a}{X_d}\cdot\cos(\delta_{\max}) + \frac{3V_a^2(X_d - X_q)}{2X_dX_q}\cdot\cos(2\delta_{\max}) - \frac{3V_a^2(X_d + X_q)}{2X_dX_q}$$

The key to employment of Equations 7 is to determine the location of the d and q axes, or alternatively the angle δ. Consider

$$\bar{V}_a - jX_q\bar{I}_a = \bar{E}_f + j(X_d - X_q)\bar{I}_d$$

But $\bar{E}_f$ and $j(X_d - X_q)I_d$ are both on the q-axis. Hence, $\bar{V}_a - jX_q\bar{I}_a$ must locate the q-axis. The remainder of the machine equations are essentially the same as for the RSM.

Also note that if $X_q = X_d$,

$$P_{\text{dev}} = 3\frac{E_fV_a\sin\delta}{X_d}$$
$$Q_{\text{dev}} = \frac{3E_fV_a\cos\delta}{X_d} - \frac{3V_a^2}{X_d}$$
$$\bar{V}_a = jX_d\bar{I}_a + \bar{E}_f$$

and the salient machine morphs into the nonsalient machine. The remaining machine equations are

$$P_{\text{stator}} = 3V_{an}I_a\cos\theta_a$$
$$\text{Stator winding loss} = \text{SWL} = 3I_a^2R_1$$
$$\text{Field winding loss} = \text{FL} = I_F^2R_F$$
$$\text{Translational loss} = P_{\text{TL}} = F_{\text{TL}}v_{\text{rm}} = K_{\text{TL}}(v_{\text{rm}})^2$$
$$\text{Sum of losses} = \Sigma L = \text{SWL} + \text{FL} + P_{\text{TL}}$$
$$P_{\text{dev}} = 3E_FI_a\cos(\delta - \theta_a)$$
$$P_m = P_{\text{out}} = P_{\text{dev}} - P_{\text{TL}}$$
$$P_{\text{in}} = P_{\text{out}} + \Sigma L$$
$$\text{Efficiency} = \eta = \frac{P_{\text{out}}}{P_{\text{in}}}$$
$$F_m = \frac{P_m}{v_{rm}} = F_{\text{out}}$$
$$\text{Translational loss force} = F_{\text{TL}} = K_{\text{TL}}v_{rm}$$
$$F_{\text{dev}} = F_m + F_{\text{TL}}$$

We will demonstrate the application of these concepts through an example.

Example 10.6

Consider the HSR vehicle of Example 10.3 to be driven by a LSM, defined in the following table:

Three-Phase Linear Synchronous Machine
Ratings
V_{Line} = 12 kV; I_{Line} = 336.8 A; $S_{3\text{ph}}$ = 7000 kVA
Primary frequency = 100 Hz; Synchronous speed = 540 km/h
Rotor type: Permanent magnet; E_F = 12 kV at 540 km/h
Synchronous speed = 540 km/h at 100 Hz;
Equivalent Circuit Values (R, X in Ω)
Ra = 0.0; K_{RL} = Ns/rad; P_{RL} = 325 kW at 540 km/h
X_l = 2.5; X_d = 25.0; X_q = 15.0

For operation as a motor, running at 500 km/h on a level surface, determine (a) the frequency and delta; (b) all LSM currents, voltages, powers, and forces; and the efficiency and pf.

SOLUTION

(a) 500 km/h is 92.59% of 540 km/h. For PWM operation,

$$\bar{V}_a = \bar{V}_{an} = 0.9259\left(\frac{12}{\sqrt{3}}\right) = 6.415\angle 0°\ \text{kV};\ f = 92.59\ \text{Hz}$$
$$X_d = 23.15\ \Omega;\ X_q = 13.89\ \Omega$$
$$E_f \text{ at } 500\ \text{km/h} = 11.1\ \text{kV}$$
$$\text{Translational Loss} = (0.9259)^2\,325 = 279\ \text{kW}$$
$$P_{\text{dev}} = 3\frac{E_fV_a\sin\delta}{X_d} + 3V_a^2\sin(2\delta)\left(\frac{X_d - X_q}{2X_dX_q}\right)$$

Solving for δ: $\delta = 32.9°$.

Hence, the q-axis is located at $-32.9°$ and the d-axis at $+57.1°$. The d, q currents can be calculated as

$$X_d I_d = 11.11 - 6.415\cos(32.9°) = 247.3\text{ A}; \qquad \bar{I}_d = 247.3\angle 57.1°\text{ A}$$

$$X_q I_q = 6.415\sin(32.9°) = 247.3\text{ A}; \qquad \bar{I}_d = 247.3\angle 57.1°\text{ A}$$

$$I_q = I_a\cos(30° + 24.5°) = 250.9\text{ A}; \qquad \bar{I}_q = 250.9\angle{-32.9°}\text{ A}$$

$$\bar{I}_a = \bar{I}_d + \bar{I}_q = 352.0\angle 11.8°\text{ A}$$

Solving for the powers:

$$\bar{S}_{3\text{ph}} = 3(6.415)(0.352)\angle{-11.8°} = 6.632\,\text{MW} - j1.385\,M\text{ var}$$

Primary winding loss = 0
Field winding loss = 0
Total losses = 0 + 0 + 279 = 279 kW
Output, input power = 6.353, 6.632 MW

Solving for the machine forces

Developed Force = $F_{\text{dev}} = P_{\text{dev}}/v = 6.632/138.9 = 47.75$ kN
Translational loss force = $F_{\text{TL}} = 279/138.9 = 2.009$ kN
Thrust = $T_{\text{m}} = T_{\text{dev}} - T_{\text{TL}} = 45.74$ kN
The relevant phasor diagram is shown in Figure 7.13a.
Efficiency = $\eta = P_{\text{out}}/P_{\text{in}} = 95.96\%$
pf = power factor = cos(11.8°) = 0.9791 leading.

10.6 Linear Electromechanical Machine Applications in Elevators

Elevators, as we have noted before, are a particularly attractive application to analyze because

- They may require machine four-quadrant operation.
- They are simple to understand. The load characteristic can be determined using elementary physics.
- They are practical and common.
- They are familiar. All of us have ridden in elevators.

Land in heavily populated cities is extremely expensive, pushing office and residential buildings to ever-increasing heights. To make such structures practical, it becomes necessary to provide vertical transportation (VT) systems, i.e., elevators, so that people and equipment can be efficiently, safely, and rapidly moved between levels. A well designed VT system must consider issues such as

- *Safety*. Systems must have highly reliable and redundant safety systems. Designs must feature simple and safe provisions for passenger rescue, in the event of electrical power failure, or mechanical lockup.

- *Traffic flow*. Capacity should be adequate to minimizing waiting time even during peak load conditions.
- *Passenger comfort*. This includes lighting, ventilation, motion discomfort, and ambiance.
- *Capital equipment costs*.
- *Operating costs*.
- *Footprint*. The system should occupy as little of building space as possible.

The most energy-efficient design normally uses a counterweight, connected to the elevator through a cable system, which mostly offsets the weight that must be moved by the EM machine. However, as we build taller buildings ("hyper-buildings" or "vertical cities"), very long cables are necessary. Since cables must support their own weight as well as the elevator, load, and counterweight, cable weight becomes a factor. Therefore, larger cables are required, which makes the cable weight even more significant. Also, long cables under tension are subject to vibration and harmonics (much like violin strings), which cause undesirable motion and noise.[4]

An attractive alternative employs the linear EM machine, applied to a so-called "ropeless" elevator design. Although not as energy efficient, and somewhat costly, this design has several advantages:

- Cables are eliminated completely, saving weight and cost, and eliminating all cable problems.
- Machine hoist rooms are eliminated.
- Multiple cars can share the same vertical shaft, reducing the number of shafts required. Since up to 30% of level space in very tall structure may be required for elevators, this is a significant and important savings.
- Down-going cars can operate in the generator mode, pumping energy back in the system, significantly improving the overall system efficiency.

Example 10.7

Consider an elevator system, designed to serve a 100-level building, as shown in Figure 10.6.

(a) Discuss the elevator system design.

(b) Consider a shaft in Bank A, which contains four cars. The system is controlled such that no two cars may get closer than two levels (8 m) of each other. Assume that a car is moving at top speed (10 m/s), and is overtaking a stopped car. If braking deceleration is to be no $>0.1g$, at what distance between cars should automatic braking be initiated?

(c) Determine if the wind drag is negligible at 10 m/s, or should be considered. The car cross-section is 10 m^2.

(d) Consider one shaft of Bank B composed of four cars, each with empty mass 1500 kg, including LSMs, and load capacity of 15 passengers at 75 each. Assuming a top speed of 8 m/s, determine the kW rating of each LSM.

(e) Estimate the external input current to the four car system of (d) from a 2.4 kV three-phase system, assuming all LSMs, plus drives, are 90% efficient, and draw power at unity pf. The load condition to be examined is

Car 1: going up at 2 m/s at the 66th level	3 passengers
Car 2: going down at 8 m/s at the 23rd level	9 passengers
Car 3: going up at 4 m/s at the 19th level	15 passengers
Car 4: going up at 4 m/s at the 57th level	10 passengers

[4] So can violins, if improperly played.

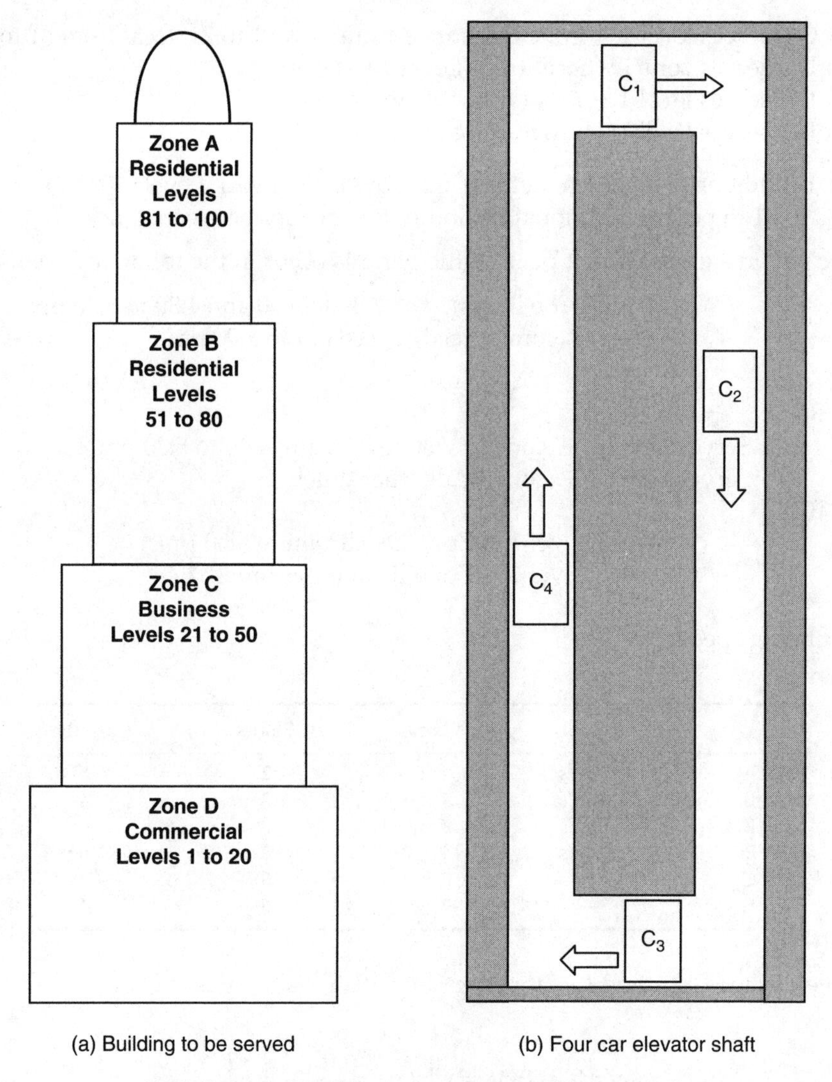

FIGURE 10.6. Ropeless elevator system employing LSM's.

SOLUTION

(a) In very tall structures, it is expedient to divide the structure into zones, as shown in Figure 10.2:

Zone A: residential	Levels 81 to 100
Zone B: residential	Levels 51 to 80
Zone C: business	Levels 21 to 50
Zone D: commercial	Levels 1 to 20

It is impractical to provide a single elevator car that stops at every level, because travel and waiting time required would be excessive. It is more logical to provide elevator banks that serve specific zones of the building. The number of shafts per bank is determined by the population to be served, and traffic patterns. For example,

Bank A: serves level 1 + Zone A (express from Level 1 to 81; local from 81 to 100)
Bank B: serves level 1 + Zone B (express from Level 1 to 51; local from 51 to 80)

Bank C: serves level 1 + Zone C (express from Level 1 to 21; local from 21 to 50)
Bank D: serves zone D (Local from Level 1 to Level 20)
Bank BC: serves level 1 + Programmable to Zone B or C
Bank BD: serves level 1 + Programmable to Zone B or D

Stairs (in addition to emergency stairs) are provided between Levels 20 to 21; 50 to 51; and 80 to 81 to permit additional flexibility for transfer between zones.

Traffic patterns are such that peak traffic periods occur in the following zones:

Zone C (business) 7:30 to 8:30 am; 4:30 to 5:30 pm
Zone D (commercial) 9:00 am to 8:00 pm

Therefore,

Bank BC:

Serves Level 1 + Zone C: 7:30 to 8:30 am; 4:30 to 5:30 pm
Serves Level 1 + Zone B: all other times

Bank BD:

Serves Level 1 + Zone D: 9:00 am to 8:00 pm
Serves Level 1 + Zone B: all other times

Top running speeds:

	Express	Local	Shafts	Cars/shaft
Bank A	10 m/s	4 m/s	2	6
Bank B	8 m/s	4 m/s	3	4
Bank BC	8 m/s	4 m/s	2	6
Bank BC	8 m/s	4 m/s	2	6
Bank C	6 m/s	3 m/s	6	4
Bank D	—	2 m/s	6	4

(b) The distance required for stopping from 10 m/s at 0.1 g is

$$x = \int_0^T v(t)\,\mathrm{d}t = \int_0^T at\,\mathrm{d}t = \left.\frac{at^2}{2}\right|_0^T = \frac{aT^2}{2}$$

But $aT = 10$ or $T = \dfrac{10}{a}$

$$x = \frac{a}{2}\left(\frac{10}{a}\right)^2 = \frac{50}{a} = \frac{50}{0.1\,g} = 51\,\text{m}$$

Therefore, stopping must be initiated no closer that 51 + 8 = 59 m between cars.
(c) As was the case for HSR:

$$k_{\text{ad}} = \frac{c_{\text{d}}\rho_{\text{d}}A}{2} = \frac{0.26(1.2)(10)}{2} = 1.56\,\text{N s}^2/\text{m}^2$$

Aerodynamic drag at 10 m/s = $k_{\text{ad}}v^2 = 1.56(10)^2 = 156\,\text{N}$
Power = $F_{\text{AD}}v = 156(10) = 1560\ \text{W} \approx 1.56\,\text{kW}$

As an approximation, raise the frictional loss to 1.4 + 1.6 = 3 kW at 10 m/s so that $P_{\text{TL}} = 0.03\,v^2\,\text{kW}$.

(d) Full load conditions:

$$P = Mgv = (1500 + 15 \cdot 75) \cdot (9.807) \cdot (8) = 205.9\,\text{kW}$$

Hence the LSM should be rated at about 206 kW.

(e) The load condition to be examined is

Car 1: stopped at the 66th level	3 passengers
Car 2: going down at 8 m/s at the 23rd level	9 passengers
Car 3: going up at 4 m/s at the 19th level	15 passengers
Car 4: going up at 4 m/s at the 57th level	10 passengers

Mechanical Powers:

Car 1: $P = Mgv = (1500 + 3 \cdot 75) \cdot (9.807) \cdot (2) = 33.8\,\text{kW}$
Car 2: $P = Mgv = (1500 + 9 \cdot 75) \cdot (9.807) \cdot (-8) = -170.6\,\text{kW}$
Car 3: $P = Mgv = (1500 + 15 \cdot 75) \cdot (9.807) \cdot (4) = 103.0\,\text{kW}$
Car 4: $P = Mgv = (1500 + 10 \cdot 75) \cdot (9.807) \cdot (-8) = 88.3\,\text{kW}$

$$\text{Electric power input} = \frac{\sum P_m}{0.9} = \frac{54.53}{0.9} = 60.59\,\text{kW}$$

$$I_L = \frac{P}{\sqrt{3}V_L} = \frac{60.59}{\sqrt{3}(2.4)} = 12.15\,\text{A}$$

10.7 Summary

If a given application requires translational motion as the end product, translational (or linear) EM machines should be considered. Although prohibitively expensive for some applications, advances in power electronics, superconducting magnets, and control theory make translational machines the preferred solution for other applications. Linear machines can eliminate the need for gears, drums, and wheels, and the like.

We have considered the armature and field design for LIMs and LSMs, observing that they are quite similar to "rolled-out" versions of RIM and RSM designs. Likewise, the equivalent circuit models are similar, particularly for the LSM. However, end effects do require significant modifications, and generally degrade performance somewhat.

The increasing traffic on the nation's roadways and airways justify the development of HSR, which proves to be a near-ideal application for linear EM machines. At this time, Europe and Japan are much further advanced in this technology than is the United States; however, the subject is of increasing interest in this country. Both LIMs and LSMs may be used with several variations in design, including one- and two-sided stators. For high-speed applications, maglev has advantages over wheels, including lower losses, higher reliability, easier maintenance of the railway, greater passenger comfort, lower pollution, smaller land usage profile, and smaller required curve radii at high speed. There are two common maglev designs: EML and EDL, and both are used in HSR. Superconducting magnets are commonly used in these designs.

The increasing population density in the world's cities have focused interest on very tall strucures ("hyper-buildings" or "vertical cities"). To make such structures habitable, it becomes necessary to provide VT systems, i.e., elevators, to transport people efficiently, safely, comfortably, and rapidly moved between levels. A well-designed VT system should occupy as little building space as possible. Above certain heights, cable systems are impractical due to excessive weight and vibration. This provides an excellent application for cableless LIMs and LSMs, which can accommodate multiple cars per shaft.

Problems

10.1 The primary of a LIM is to be appropriate for a 400 km/h HSR application.

(a) Determine coil pitch if the maximum frequency is to be 120 Hz.
(b) For a six-pole 36-slot armature design, determine the distance between slots.

10.2 Consider an HSR application, with a vehicle of the following design:

Cross-section height × width = 3.7 × 3.7 m
Length = 14 m
Empty vehicle weight = 10 t = 10,000 kg
Payload = 36 passengers + luggage = 36 × (75 + 75) = 5400 kg
Drag coefficient = 0.26; air density = 1.2 at 23°C

Neglect suspension drag at high speed. Determine

(a) The thrust and power required at 400 km/h on a level surface.
(b) The thrust and power required at 400 km/h climbing a 6% grade.
(c) The thrust and power required at 400 km/h going down a 6% grade.

10.3 An EML system must support 12 metric tons. The levitating coil has 100 turns; the total working area (0.5 × 13 m) is rectangular. The normal design air gap is 2 cm. The coil resistance is 2 Ω.

(a) Determine the coil current and voltage.
(b) Determine K_1, K_2, and K_3.
(c) Determine the system poles for the open-loop system.
(d) Determine the system poles for the closed-loop system, if $\Delta v = K_0\Delta + K_u u$. Select K_0 and K_u for stability and critical damping.

10.4 In the HSR system of Problem 10.2, the thrust is controlled such that the system starts and accelerates at a uniform acceleration = 0.1 g to 400 km/h.

(a) How long will it take to reach top speed?
(b) How far has the train traveled when we reach top speed?
(c) How long and far will a 0.1 g stop take from 400 km/h?
(d) How long and far will an emergency stop at 0.4 g take from 400 km/h.

10.5 Consider the system similar to that of Example 10.4, modified as follows:

- train consists of 6 cars.
- line runs east to west; length = 320 km.
- there are 5 stations, equally spaced 80 km apart, numbered from S0 to S4, left to right. All stations are at elevation 0.
- 5 min stop at each station.
- four round trips per day; 365 days per year.
- top speed 400 km/h; acceleration 0.1 g.

(a) Construct an operational timetable.
(b) Determine the total energy required for a run from S0 to S4. Assume that dynamic braking is used.
(c) Assuming that the EM conversion efficiency is 90%, and the cost of electrical energy is 0.10$ per kW h = 100$ per MW h, determine the energy cost per passenger.
(d) Compute the energy saved using regenerative braking, assuming that 80% of kinetic energy is recovered. Determine the annual cost savings.

10.6 Consider the elevator system of Example 10.7.

(a) Carefully read through the solution to Example 10.7a.
(b) Consider a car in Bank A. Suppose that no two cars may get closer than two levels (8 m) of each other, and that a car is moving at top speed (10 m/s), converging on a stopped car. If braking deceleration is to be no greater than 0.15 g, at what distance between cars should automatic braking be initiated?
(c) Determine the wind drag at 8 m/s. The car cross-section is 10 m^2.
(d) A redesign of the car produces an empty mass of 1200 kg, including LSMs, with load capacity of 12 passengers at 75 each. Assuming a top speed of 8 m/s, determine the kW rating of each LSM.
(e) Estimate the worst-case external input current to the four car system of (d) from a 2.4 kV three-phase system, assuming all LSMs, plus drives, are 90% efficient, and draw power at unity pf.
(f) Assume you live on the 100 level. What is the worst-case (i.e., longest) travel time from Level 1 to your level via Bank A, assuming a 12 passenger car, and that it takes an average of 12 s for doors to open, passenger(s) to disembark/board, and doors to close. What is the best (i.e., shortest) travel time?

10.7 Consider an elevator car with an empty mass of 1200 kg, and a load capacity of 12 passengers at 75 each. Consider supporting the elevator with five cables, each weighing 0.8 kg/m, as illustrated in Section 3.8. If the elevator is used in a 100 floor application (4 m between floors), determine the cable weight as a % of the loaded elevator weight. Comment the practically of this solution.

11

Special Purpose Machines and Sensors

We have investigated the three basic electromagnetic machine types in previous chapters. There are literally dozens, if not hundreds, of special types of EM machines specifically designed for certain restricted applications. Space does not permit an exhaustive investigation of all such machines. However, we shall consider several common and important special purpose machines.

11.1 The Universal Motor

A "universal" motor is one that runs on AC or DC. Consider the DC series motor, as shown in Figure 11.1a. Observe that the field polarity depends on the direction of the field current. But the field current is the armature current! Hence if the armature current reverses direction, as shown in Figure 11.1b, the field synchronously reverses polarity. If both the field and armature current reverse polarity, the developed torque (and hence rotation) is in the same direction. Thus such a motor is "universal" in that it runs on AC or DC.

However, it runs better on DC for the following reasons:

- Since the field throughout the stator and rotor iron is AC, both structures must be laminated to reduce core loss. Even so, core losses are greater for AC operation.
- On AC, the armature inductive reactance as well as the resistance impedes current flow. For a given voltage, this means less current, which means less torque and power.
- Arcing at the commutator is more severe on AC operation.

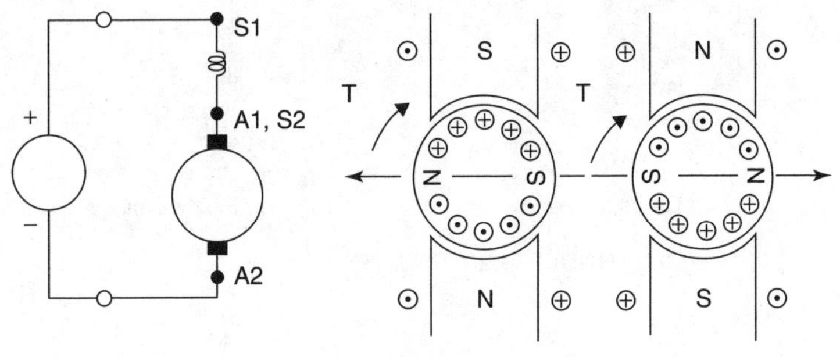

FIGURE 11.1. Universal motor.

Applications are limited to rather small sizes, and would include devices that require high torque at low speed, including electric drills, power screw drivers, portable saws, and kitchen appliances. Speed control can be achieved by adding a series SCR, and a parallel freewheeling diode.

11.2 The Shaded Pole Motor

A clever design for a small motor employs the so-called shaded pole concept. The shading winding is a single shorted turn, encircling part of the pole structure, as shown in Figure 11.2a. Assume the magnetic field in the pole (shown in Figure 11.2b) starts from zero and is increasing. A voltage will be induced in the shading winding, creating a current that opposes the change in the flux linkage, according to Lentz's law. Hence, the field will be stronger on the right and weaker on the left side, such that the "center of gravity" of the field is in position 1. As the field builds up and its rate of change decreases, the induced voltage in the shading winding becomes weaker, so that the position of the center of gravity moves to the center of the pole (position 2). As the field decreases, the induced voltage will oppose the change, inducing a current to support the field, shifting the field center of gravity moves to the left (position 3).

The overall effect is a moving field that sweeps across the pole face in one half cycle. But this is in effect a rotating magnetic field! The shading winding has created an approximate rotating field from single-phase excitation. Shaded pole motors are small (fractional horsepower), cheap motors used for applications with no critical speed control requirements. One example application is a box fan that usually features a three-speed switch (low, medium, high). The high-speed position creates the strongest magnetic field and the highest torque, driving the load at the highest speed.

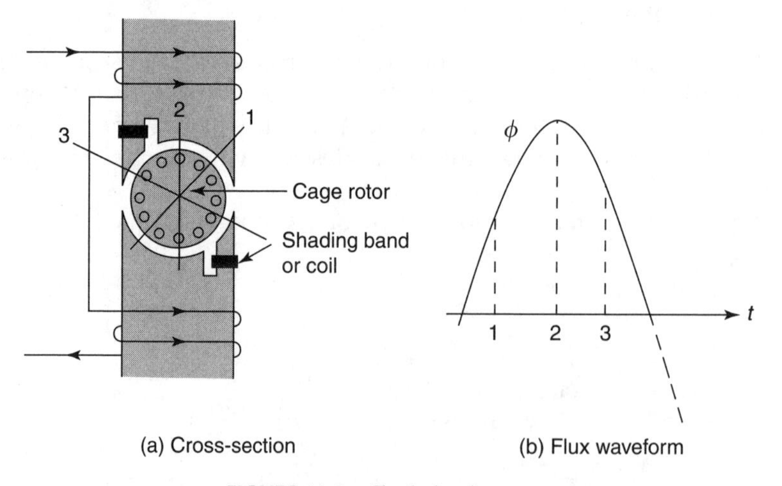

FIGURE 11.2. Shaded pole motor.

11.3 The Hysteresis Motor

Recall that all ferromagnetic structures penetrated by time-varying magnetic fields experience core (hysteresis and eddy current) losses, which are normally considered to be undesirable and hence to be minimized. But what if we intentionally made the rotor structure of an AC machine out of high hysteresis loss iron? We would create a structure with exaggerated hysteresis losses.

Consider an AC machine, with a three-phase P-pole stator, excited with balanced three-phase AC voltage, and with a solid stainless-steel rotor, running at less than synchronous speed. Such a machine would have very large hysteresis loss, eclipsing the other loss mechanisms. The core loss would be analogous to rotor winding loss in induction motors, since it is dissipated as internal rotor heat. Therefore

$$\text{RWL} \simeq P_{\text{h}} = K_{\text{h}} B_{\text{m}}^2 f_{\text{R}} = \text{rotor loss (in W)}^1$$

But recall from Chapter 4 that flux density is proportional to voltage, and:

$$f_{\text{R}} = sf$$

$$P_{\text{ag}} = \frac{\text{RWL}}{s}$$

Therefore

$$P_{\text{ag}} = \frac{KV^2(sf)}{s} = KV^2 f$$

The developed torque is the air-gap power divided by synchronous speed:

$$T_{\text{dev}} = \frac{KV^2 f}{\omega_{\text{sm}}} = \text{constant}$$

Based on the approximations stated, the developed torque will be constant up to synchronous speed! At synchronous speed the rotor frequency is 0, as must be the hysteresis loss. So how can the motor operate? Because of the large hysteretic effect, the rotor in effect becomes permanently magnetized, and synchronizes with the rotating magnetic field.

Usually hysteresis motors have a single-phase (not three-phase) shaded pole stator, excited from a single-phase AC voltage. Nonetheless, the machine operates essentially as discussed, and provides essentially constant torque over its entire speed range, up to synchronous speed, at which it synchronizes and runs as a synchronous device. Hysteresis motors feature smooth quiet operation, with ample starting torque required to accelerate large inertial loads. They are reliable and capable of being stalled without undesirable noise and excessive temperature rise. Hysteresis motors are available in ratings up to a few watts, and are normally equipped with built-in gearing and lifetime lubrication. Applications include small low-power devices, including chart recorders, valves, dampers, timers, and various industrial controls.

11.4 The Stepper Motor

We normally think of a motor running at some constant average speed, supplying power to a mechanical load. However, sometimes the application requires precise position control as well as speed control. An appropriate driver for such loads is the stepper motor. The stepper motor is basically a machine with a P-pole wound stator and a salient pole rotor, which may or may not include a permanent magnetic field.

[1] Using "2" for the Steinmetz exponent.

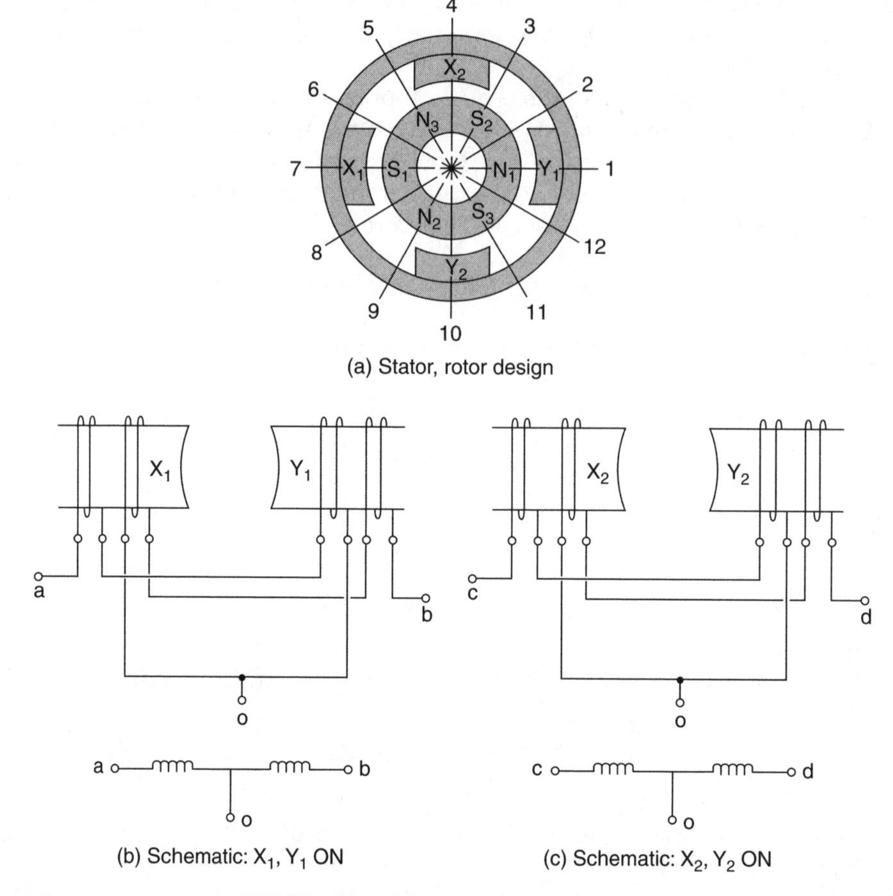

FIGURE 11.3. Stepper motor wiring details.

Consider the machine shown in Figure 11.3a, which has two pairs of stator poles (X, Y). Pair X_1, Y_1 is excited with windings a, b, and Pair X_2, Y_2 with windings c, d. Use the center of rotor pole N_1 as our reference to define rotor position, relative to the center of stator pole Y_1 (define as the angle θ, considering CCW to be positive). Now suppose current flows from o to a (binary signal oa = 1). Stator poles X_1, Y_1 will become N, S, respectively (verify this in Fig. 11.3b). Hence magnetic forces of attraction and repulsion will be developed such that the rotor will position itself as shown in Figure 11.3a (N_1 under Y_1 and S_1 under X_1)[2]: i.e. the rotor is in position 1 or $\theta = 0°$. Now if the oa current is turned off (oa = 0) and the oc current is turned on (oc = 1), stator poles X_2, Y_2 will be N, S, respectively, attracting rotor poles S_2, N_2. Hence, then rotor will rotate to position 2 or $\theta = 30°$. Next, if oc = 0 and ob = 1, stator poles X_1, Y_1 become S, N, pulling the rotor to position 3 or $\theta = 60°$. Thus the rotor moves through 360° in 30° steps. CCW and CW operational sequences are shown in Table 11.1a and Table 11.1b.

It is possible to increase the resolution by energizing two windings at a time, an operational condition called half-stepping. Suppose we start at position 1 (oa = 1) and turn on winding oc (oa = 1; oc = 1). Stator poles X_2, Y_2 become N, S, attracting S_2 and N_2. However, since oa = 1, stator poles X_1, Y_1 are still N, S, pulling back on S_1 and N_1. As θ increases, the

[2] Actually this is one of three possible equilibrium positions. $\theta = 120°$ or 240° are the other two. Thus it is possible for the rotor to "get lost," requiring that there must be some provision for resetting the system.

TABLE 11.1
Stepper Motor Operational Sequences

(a) Operational Sequence for CCW Rotation

θ	0	30	60	90	120	150	180	210	240	270	300	330
oa	1	0	0	0	1	0	0	0	1	0	0	0
ob	0	0	1	0	0	0	1	0	0	0	1	0
oc	0	1	0	0	0	1	0	0	0	1	0	0
od	0	0	0	1	0	0	0	1	0	0	0	1

(b) Operational Sequence for CW Rotation

θ	0	330	300	270	240	210	180	150	120	90	60	30
oa	1	0	0	0	1	0	0	0	1	0	0	0
ob	0	0	1	0	0	0	1	0	0	0	1	0
oc	0	0	0	1	0	0	0	1	0	0	0	1
od	0	1	0	0	0	1	0	0	0	1	0	0

(c) Half-Stepping for CCW Rotation

θ	0	15	30	45	60	75	90	105	120	135	150	165
oa	1	1	0	0	0	0	0	1	1	1	0	0
ob	0	0	0	1	1	1	0	0	0	0	0	1
oc	0	1	1	1	0	0	0	0	0	1	1	1
od	0	0	0	0	0	1	1	1	0	0	0	0

forward and the backward torque decreases, reaching a balance point halfway at $\theta = 15°$. A half-stepping sequence for CCW rotation is shown in Table 11.1c.

Suppose the stator voltage pulse rate is one pulse per millisecond. Since it takes 12 pulses to move the rotor through one complete revolution for full step operation (1 rev/12 ms), the speed is

$$\omega = 1\,\text{rev}/12\,\text{ms} = 83.33\,\text{rps} = 5000\,\text{rpm} = 523.6\,\text{rad/s}$$

For half-step operation,

$$\omega = 1\,\text{rev}/24\,\text{ms} = 41.67\,\text{rps} = 2500\,\text{rpm} = 261.8\,\text{rad/s}$$

Even with no field, salient rotor poles will tend to align with the stator field due to the reluctance torque produced. Permanent magnet field devices will tend to cog, even with no stator excitation. There are a variety of designs, with or without permanent magnet rotor fields. Various angular resolutions are commercially available, from 90 to 0.72° or even smaller. Stepper motors are normally small, with low inertia rotors and can operate at frequencies in the kilohertz range, allowing them to rotate at very high speeds, and may be started and stopped in milliseconds. They are ideally suited for applications that require precision position control.

11.5 Encoders

Encoders are sensors that provide an output analog or digital electrical signal that is proportional to shaft position and speed. They are used in conjunction with high-performance motor control applications, such as AC vector drives. Various technologies are employed,

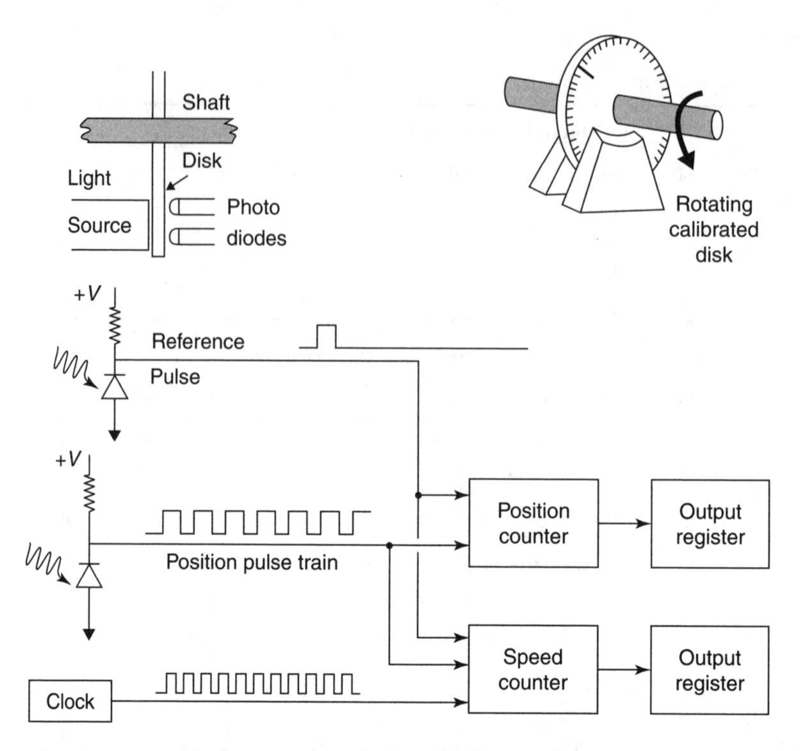

FIGURE 11.4. Digital rotary encoder.

including potentiometric, brush contact, magnetic, and synchromechanic. One of the more common designs is the optical encoder, illustrated in Figure 11.4. The light source provides a beam of light shining through a transparent disc onto a photo cell. If the disk is marked with opaque radial stripes, the circuit converts alternating light and dark bands into a pulse train. If the pulse train is then supplied to a counter, the count can be correlated to the disk position. If the counter output is correlated to a timer, the shaft speed can be determined. Consider an example.

Example 11.1

An optical encoder is to be used as a speed- and position-sensing device for the four-pole brushless DC motor application presented in Example 7.10.

(a) Using the design of Figure 11.4, specify the particulars of the position counter, the speed counter, and the clock, to provide capability of determining shaft position within one (electrical) degree and speed within 1 rpm for an operating range of 0 to 2400 rpm.
(b) The position counter is referenced to 0 when a north pole aligns with the Phase-a magnetic axis. Using typical data, explain how the encoder output is interpreted.
(c) Explain the relevance to the situation discussed in Example 7.10.

SOLUTION

(a) Suppose we design the disk to have 720 stripes, creating a square pulse train of 720 cycles per revolution. The counter receives the pulse train such that the leading edge constitutes a countable event. Hence, a count of 720 corresponds to a

2π rad or 360° physical (mechanical) rotation of the shaft. Since the application is four-pole,

$$\theta_{\text{Electrical}} = 2\theta_{\text{Mechanical}}$$

and the count will equal the angle in electrical degrees.

$$2^N > 720$$

so that the smallest acceptable $N = 10$, and a 10-bit counter is required.

The position counter starts when Phase-a current crosses through zero going positive. The counter stops when the reference point is detected on the disk. The disk reference point is aligned with a rotor north pole.

For speed, we need to count up to 2400, which corresponds to the time it takes to complete one revolution (at 2400 rpm). We will determine speed by counting timing pulses over one revolution (the speed counter starts and stops at the reference point on the disk). Let us select a 12-bit counter

$$2^{12} = 4096$$

At 2400 rpm, one revolution is executed in 25 ms, which is the time we shall allow to count to 2400. Let the clock frequency be 2400/25 ms = 96 kHz. Therefore, the speed counter contents will be the speed in rpm.

The counter outputs are transferred to buffer registers once per revolution, where they are fed back as control signals into the vector drive.

(b) Suppose the buffer register contents are

Position: 0010110100
Speed: 10000111000

Converting the position register contents into base 10

$$(0010110100)_2 = 1 \times 2^7 + 1 \times 2^5 + 1 \times 2^4 + 1 \times 2^2 = 180$$

Hence $\theta = 180$ electrical degrees
Converting the speed register contents into base 10

$$(10000111000)_2 = 2^{10} + 2^5 + 2^4 + 2^3 = 1024 + 32 + 16 + 8 = 1080$$

Hence, the shaft speed is 1080 rpm.

(c) Recall Example 7.11. The brushless DC motor was driving a pump which required 40 kW at 1080 rpm while it was also required that the system operate such that $\bar{E}_f$ and $\bar{I}_a$ were in phase, operating as a brushless DC motor. To meet these three conditions, the inverter must be controlled to:

- provide a frequency of 36 Hz (to force the speed to 1080 rpm)
- control the applied voltage rms magnitude (V_a) so that (a) $\bar{E}_f$ and $\bar{I}_a$ are in phase with E_f and (b) the output is 40 kW.

The inverter firing circuits can control two aspects of the machine applied voltage:

- the frequency
- the rms value

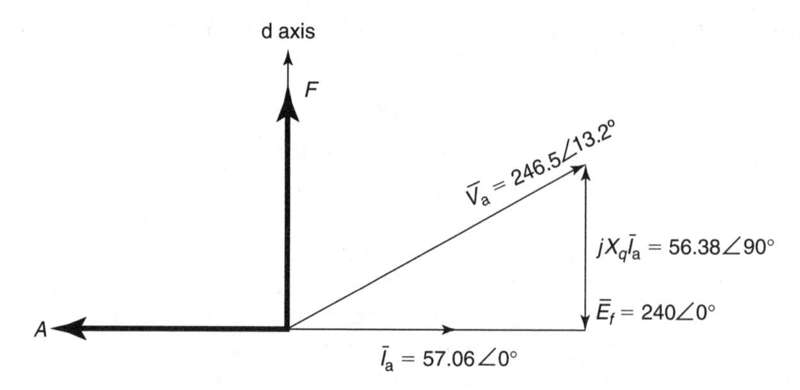

FIGURE 11.5. Phase-space relations for Example 11.1.

Frequency control is the simplest and most obvious. The user selects the desired speed, creating a reference signal (1080 rpm, 36 Hz, in our example). Since the machine is synchronous, the frequency uniquely determines the shaft speed and the speed feedback signal is unnecessary.

The load requires a specific torque and power at the selected speed (40 kW in our example, $P_{dev} = 41.08$ kW, adding on the rotational loss). Since we have a permanent magnet field, E_f is uniquely determined by speed (240 V in our example). If $\bar{E}_f$ and $\bar{I}_a$ are in phase, the current must be 57.06 A to meet the power requirement, and V_a must be 246.5 V to force the current into the proper phase position. Also recall the $\hat{F}$ leads E_f by 90° and $\hat{A}$ leads $\hat{F}$ by 90° for brushless DC operation. The necessary phase and position relations are summarized in Figure 11.5. Suppose we start the position counter by detecting a phase a current zero crossing going positive. We know that i_a reaches its maximum 90° later, at which point the position count would be "90". Let us terminate the count when we detect the rotor north pole, which would be the same location as the d-axis, or $\hat{A}$. If the count is position "180", we conclude that $\hat{F}$ lags $\hat{A}$, which is the desired condition.

If the position count is less than 180: $\hat{F}$ lags $\hat{A}$ by less than 90°; the current is in the lagging position, and the voltage is too large. If the position count is greater than 180: $\hat{F}$ lags $\hat{A}$ by more than 90°; the current is in the leading position, and the voltage is too small. Thus corrective actions are:

$$\begin{aligned} &\text{count} < 180: && \text{decrease } V_a \\ &\text{count} > 180: && \text{increase } V_a \end{aligned}$$

Hence both registers contain the proper signals to drive the 40 kW load @ 1080 rpm as a brushless dc motor.

11.6 Resolvers

Resolvers are sensors that provide output analog signals that can be used to determine shaft position and speed. Like encoders, they can be used in conjunction with high-performance motor control applications. Consider the device illustrated in Figure 11.6, which is structured like a tiny two-phase wound rotor induction motor: the device has two pairs of windings mounted 90 electrical degrees apart (one pair on the stator [1,2] and one pair

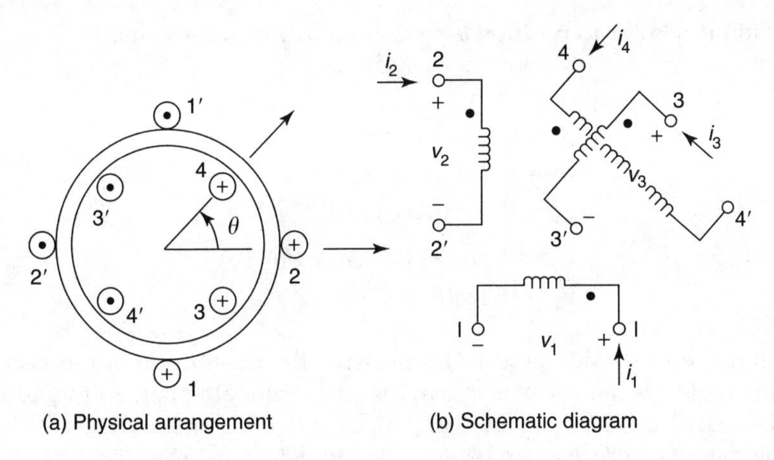

FIGURE 11.6. The resolver.

on the rotor [3,4]). Either the rotor or stator windings may be excited, with the other windings in effect terminated in an open circuit. The rotor winding terminals are accessible through slip rings. Consider the rotor to stator turns ratio to be 1:1, and the coupling to be perfect, although neither of these conditions is required for the device to function properly. In general, if the stator windings are excited with voltages of the radian frequency ω,

$$v_1(t) = V_1\sqrt{2}\cos(\omega t + \alpha), \qquad \bar{V}_1 = V_1\angle\alpha$$
$$v_2(t) = V_2\sqrt{2}\cos(\omega t + \beta), \qquad \bar{V}_2 = V_2\angle\beta$$

For a rotor speed of ω_R (electrical rad/s), the slip (s) is

$$s = \frac{\omega - \omega_R}{\omega}$$

The rotor frequency will be $s\omega$. The induced rotor voltages will be

$$v_3(t) = sV_1\sqrt{2}\cos\theta\cos(s\omega t + \alpha) - sV_2\sqrt{2}\sin\theta\cos(s\omega t + \beta)$$
$$v_4(t) = sV_1\sqrt{2}\sin\theta\cos(s\omega t + \alpha) + sV_2\sqrt{2}\cos\theta\cos(s\omega t + \beta)$$

θ defines rotor position relative to the stator.

Rotor speed and position may be derived from the measured rotor voltages. If we restrict the analysis to balanced two-phase stator excitation,

$$v_1(t) = V\sqrt{2}\cos(\omega t); \qquad \bar{V}_1 = V\angle 0$$
$$v_2(t) = V\sqrt{2}\sin(\omega t); \qquad \bar{V}_2 = V\angle{-90^\circ}$$
$$v_3(t) = sV\sqrt{2}\cos\theta\cos(s\omega t) - sV\sqrt{2}\sin\theta\sin(s\omega t) = sV\sqrt{2}\cos(s\omega t + \theta)$$
$$v_4(t) = sV\sqrt{2}\sin\theta\cos(s\omega t) + sV\sqrt{2}\cos\theta\sin(s\omega t) = sV\sqrt{2}\sin(s\omega t - \theta)$$

Move the stationary rotor to position $\theta = \alpha$, and apply a stator voltage:

$$v_1(t) = V\sqrt{2}\cos(\omega t); \qquad \bar{V}_1 = V\angle 0°$$
$$v_2(t) = 0 \quad \text{(winding shorted)}$$
$$v_3 = V\sqrt{2}\cos\alpha\cos(\omega t)$$
$$v_4 = V\sqrt{2}\sin\alpha\cos(\omega t)$$

$$\bar{V}_3 = \bar{V}_1\cos(\theta) = V\cos\alpha\angle 0°$$
$$\bar{V}_4 = \bar{V}_1\sin(\theta) = V\sin\alpha\angle 0°$$

Observe that if the rms value of v_1 (V) represents the magnitude of the vector, and the resolver rotor angle θ is the vector angle α, the rms terminal voltages of windings 3 and 4 represent the real and imaginary vector components of $\bar{V}_1$, respectively. This process is called "resolving." The process can be exploited to determine rotor position.

To demonstrate how resolvers can be used for speed and position sensing, consider Example 11.2.

Example 11.2

A two-pole resolver has stator ratings 60 Hz 120 V rms, and is to be used as a speed and position sensing device. The stator–rotor turns ratio is 1:1, and perfect coupling may be assumed.

(a) Suppose 120 V rms 60 Hz is applied to winding 1; winding 2 shorted; windings 1 and 3 are series-connected, as are windings 1 and 4: i.e., connect 1-1′–3-3′ and 1-1′–4-4′, and $V_X = V_{1\text{–}3'}$; $V_Y = V_{1\text{–}4'}$. With the rotor in an arbitrary unknown position and stationary, V_3, V_4, V_X, and V_Y were measured to be 103.9, 60.0, 223.9, and 60 V, respectively. What is the rotor position?

(b) Suppose balanced two-phase 120 V rms 60 Hz voltages are applied to stator windings 1 and 2, and winding 3 and 4 voltages are measured, each being 36 V rms. What is the rotor speed?

(c) Continuing (b), Suppose when the system speeds up, V_3 drops windings 1 and 3 voltages measured on an oscilloscope, such that

$$v_1(t) = 120\sqrt{2}\cos(377t)$$
$$v_3(t) = 36\sqrt{2}\cos(113.1t - 12.86°)$$

What is the rotor position?

SOLUTION

(a) We compute

$$|\cos\theta| = \frac{103.9}{120} = 0.866 \qquad \theta = +30°; -30°; +150°; -150°$$
$$|\sin\theta| = \frac{60}{120} = 0.5; \qquad \theta = +30°; -30°; +150°; -150°$$

We know that $\bar{V}_4$ and $\bar{V}_3$ are either in phase, or 180° out of phase, with $\bar{V}_1$.

$$\begin{array}{llll} \text{If } V_X > V_1: & \cos\theta > 0 & \text{If } V_X < V_1: & \cos\theta < 0 \\ \text{If } V_Y > V_1: & \sin\theta > 0 & \text{If } V_Y < V_1: & \sin\theta < 0 \end{array}$$

Since $V_X = 223.9$ V, $\cos\theta > 0$. Hence θ must be in the first or fourth quadrant.

Since $V_Y = 60\,\text{V}$, $\sin\theta < 0$. Hence θ must be in the third or fourth quadrant. Therefore θ must be in the fourth quadrant, and $\theta = -30°$.

(b) Observe that

$$|s| = \frac{V_3}{V_1} = \frac{36}{120} = 0.3, \qquad s = \pm 0.3$$
$$\omega_R = (1-s)\omega_S = 0.7(3600) = 2520\,\text{rpm}$$
$$\omega_R = (1-s)\omega_S = 1.3(3600) = 4680\,\text{rpm}$$

If, when the system speeds up, V_3 *drops*, 2520 rpm is the correct speed. If, when the system speeds up, V_3 *increases*, 4680 rpm is the correct speed. Since the latter is the case, the speed is 2520 rpm.

(c) As in (b)

$$|s| = \frac{V_3}{V_1} = \frac{36}{120} = 0.3, \qquad s = 0.3$$
$$\omega_R = (1-s)\omega_S = 263.9\,\text{rad/s} = 2520\,\text{rpm}$$

Now when V_1 reaches its positive maximum (which incidentally happens to be our time origin), the stator magnetic rotating magnetic field is positioned in the plane of winding 1-1′ (90° past the 1′1′ winding magnetic axis). From the measurement, V_3 will reach its positive maximum 12.86° (0.2244 rad or 1.984 ms) later in time. Since the rotor speed is 263.9 rad/s, at $t = 0$ winding 3 must have been located 0.5236 rad or 30° CW from winding 1.

Therefore, at $t = 0$,

$$\theta = -30°$$

11.7 Microelectromechanical Systems

Microelectromechanical systems (MEMS) are very small[3] electromechanical devices that are used as transducers. Beginning in about 1980, MEMS is a fascinating emerging technology that is rapidly developing, with a wealth of applications. Transducers are devices that convert energy from one form into another, and encompass sensors and actuators, the former making up the large majority of applications for MEMS. Sensors generally perform the energy conversion process for the sole purpose of providing information, whereas actuators generally involve motion, and may actually do mechanical work. As sensors, and because of their extremely small size, MEMS can be embedded in a wide variety of structures, from bridges and large buildings to various biological structures, including the human body.

Sensors can be classified according to the domain in which they operate:

- Thermal (heat energy, heat flow, or temperature, to electric, or vice versa)
- Mechanical (position, velocity, acceleration, force pressure)
- Chemical (composition, concentration, reaction rate)
- Magnetic (flux, flux density, polarization)
- Radiation (intensity, wave length, polarization, phase)

[3] MEMS are on a scale of conventional microelectronic circuits, and are manufactured by the same processes used to produce integrated circuits.

Conversion is from the relevant domain into electrical form (voltage, current, charge) or vice versa.

MEMS operate by exploiting certain properties of specific materials, including:

- *The piezoelectric effect*: Piezoelectricity, known for over 100 years, is literally "pressure–electricity." Certain materials, particularly those with crystalline structure, such as quartz crystals, and lead zirconate titanate ceramics, develop a charge across them that is proportional to an externally applied pressure. Such materials are sometimes called "electrets." Their response is linear for small deformations, and they can respond to frequencies up to 100 MHz. Applications include microphones, small earphone speakers, and strain gages.
- *The piezoresistive effect*: The resistance of certain materials change in proportion to an externally applied force, particularly semiconductors. Applications include strain gages.
- *The magnetostriction effect*: Magnetostriction is similar to piezoelectricity when a ferromagnetic material is penetrated by a magnetic field, a constricting force is produced that tends to squeeze and deform the material. When the field collapses, the force relaxes and the material returns to its former shape. If this is done at audible frequencies significant sound can be produced.
- *Thermo-electric effects*: When two dissimilar metal conductors are joined at two points to form a closed circuit, and the two junction points are maintained at two different temperatures, a current is observed to flow around the circuit in proportion to the difference in temperature. The current is produced by the so-called Seebeck voltage, which is composed of a combination of the junction voltages (the Peltier emf's), and an internal voltage induced within each conductor (the Thompson emf's). An example of metal compounds that exhibit this property most dramatically would be the NiCr–NiAl combination. A prime application is temperature measurement, and can be employed over very wide ranges (−250 to 1500°C). The effect is reversible, and an externally forced current can produce "Seebeck cooling" at the cold junction.
- *The thermo-resistive effect*: Certain materials (notably sintered mixtures of metallic oxides or "thermistors") exhibit large changes of ohmic resistance with temperature. Thermistors are available in the range from 100 Ω to 100 kΩ, in temperature ranges from about −50 to 300°C.
- *The photoelectric effect*: Electromagnetic radiation impinging on a P–N semiconductor junction creates hole–electron pairs.

Actuators generally use electric and magnetic signals to produce either lineal or rotary motion. Because of the extremely close spacing of parts, electrostatic designs may offer more efficacy than do electromechanical systems. A simple MEMS actuator is presented in Example 11.3.

Example 11.3

A to MEMS consists of the simple cantilever beam shown in Figure 11.7 and serves as a lineal actuator, the position of the end of the beam (x) controlled by the applied voltage. Assume that a linearized expression for the system capacitance is

$$C(x) = C_0\left(\frac{2d}{2d - x}\right), \quad C_0 = \frac{\varepsilon A}{d}, \quad 0 \le x \le d$$

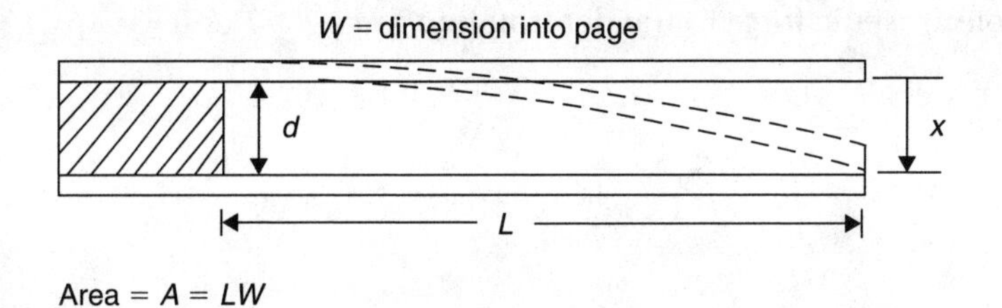

(a) Cantilevered beam as a lineal actuator

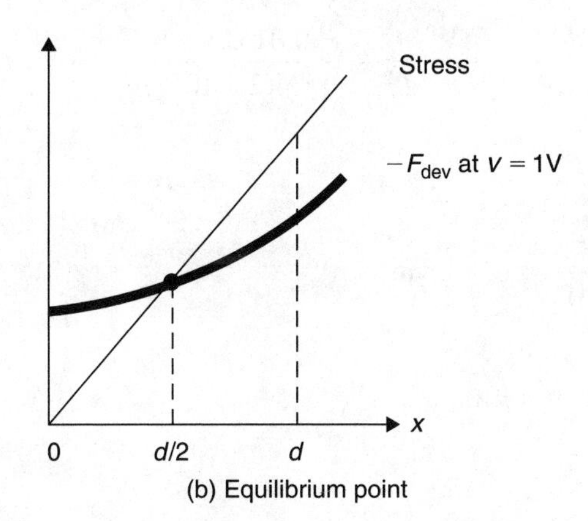

(b) Equilibrium point

$W = 30\ \mu\text{m} \quad L = 100\ \mu\text{m} \quad A = WL = 3 \times 10^{-9}\ \text{m}^2 \quad d = 2\ \mu\text{m}$

$$C_0 = \frac{\varepsilon A}{d} = \frac{\frac{1}{36\pi} \times 10^{-9}(3 \times 10^{-9})}{2 \times 10^{-6}} = 0.01326\ \text{pF}$$

$K_0 = 2.947\ \text{nN}/\mu\text{m}$

(c) System data

FIGURE 11.7. Simple MEMS actuator.

The beam end deflection is determined by the approximate linearized expression

$$F = K_0(d - x)$$

where K_0 is the beam modulus of elasticity relating the beam deformation to the distributed electrostatic force (F).

(a) Derive a relationship relating the beam equilibrium position to the applied voltage.
(b) Compute the beam end equilibrium position if the applied voltage is 1 V.

SOLUTION

(a) The electrostatic force is

$$F_{\text{dev}}(x, V) = +\frac{\partial W_f(x, V)}{\partial x} = \frac{\partial}{\partial x}\left[\frac{1}{2}CV^2\right] = \frac{C_0 V^2}{2}\frac{\partial}{\partial x}\left(\frac{1}{1 - x/2d}\right)$$

$$= \frac{-C_0 V^2}{4d(1 - x/2d)^2}$$

Matching the electrostatic force to the deformation force

$$K_0 x - \frac{C_0 V^2}{4d(1 - x/2d)^2} = 0$$

(b) $x = 0$

$$F_{\mathrm{dev}}(x, V) = \frac{-C_0 V^2}{4d(1 - x/2d)^2} = \frac{-0.01326 \times 10^{-12}}{4(2 \times 10^{-6})} = -1.658\,\mathrm{nN}$$

(c) $2.947x - \dfrac{1.658}{(1 - x/4)^2} = 0; \quad x \text{ in } \mu\mathrm{m}$

Solving for x, $x = 1\,\mu\mathrm{m}$

11.8 Summary

There are many and varied special types of EM machines available for specific purposes. Space does not permit an exhaustive investigation of all such machines; however, we have considered several common and important special purpose machines.

Universal motors can operate on AC and DC and are excellent drives for small fractional hp loads, particularly those which require high torque at low speed. Shaded pole motors are an ingenious and low-cost motor design, suitable for small loads that require minimal speed control.

Hysteresis motors feature smooth quiet operation, run at synchronous speed, and develop ample starting torque to accelerate large inertial loads. They are reliable and capable of being stalled without undesirable noise and excessive temperature rise. Hysteresis motors are available in ratings up to a few watts, and are normally equipped with built-in gearing and lifetime lubrication.

Loads that require precise speed and position control may use the stepper motor, which features a P-pole wound stator and a salient pole rotor, which may or may not include a permanent magnetic field.

Encoders are sensors that provide an output analog or digital electrical signal that is proportional to shaft position and speed. They are used in conjunction with high-performance motor control applications, such as AC vector drives. Various technologies are employed, including potentiometric, brush contact, magnetic, synchromechanic, and optical encoders. Resolvers are sensors that provide output analog signals that can be used to determine shaft position and speed. Like encoders, they can be used in conjunction with high-performance motor control applications. Microelectromechanical systems are very small electromechanical devices that are used as transducers. Transducers are devices that convert energy from one form into another, and encompass sensors and actuators, the former making up the large majority of applications for MEMS. Because of their extremely small size, MEMS can be embedded in a wide variety of structures, from bridges and large buildings to various biological structures, including the human body.

Problems

11.1 The universal motor develops unusually high torque at low speed. Give reasons.

11.2 The shaded pole induction motor is used for small low-power applications. Give reasons.

11.3 Explain the principle of operation of a hysteresis motor.

11.4 A 120 V 60 Hz six-pole hysteresis motor develops a stall torque of 20 N m. Suppose the motor drives a load described by

$$T'_{\mathrm{m}} = 0.08\,\omega_{\mathrm{rm}}; \qquad J_{\mathrm{Motor}} + J_{\mathrm{Load}} = 0.1\,\mathrm{kgm}^2$$

Determine and plot the starting speed characteristic (ω_{rm} versus t).

11.5 Extend your right hand in front of you, palm up (define the palm-up position as $\theta = 0°$). Without moving your arm, turn your palm to the left (as if you were going to shake hands: $\theta = 90°$). Return your palm to the original "up" position ($\theta = 0°$), then left again, and finally up. Repeat the whole sequence rapidly (up to left to up to left to up), which takes about 0.8 s.

(a) We wish to determine the sequence (for full step operation) of winding input currents (prepare a table in the manner of Table 11.1) to the stepper motor of Section 11.5 driving a robotic hand to replicate the maneuver. Also determine the switching time T (i.e., the pulse duration).
(b) Suppose the robotic hand requires twice rated motor torque. Specify a gear box between the motor and the hand, and specify the new time T.

11.6 Design an optical encoder of the type as shown in Figure 11.3 and discussed in Example 11.1, to be used as a speed and position sensing device for a six-pole brushless DC motor with an operating speed range of 0 to 1500 rpm. That is, specify the particulars of the position counter, the speed counter, and the clock, to provide capability of determining shaft position within one (electrical) degree and speed within 1 rpm.

11.7 The speed register contents of the optical encoder of Example 11.1 is

$$(001011001101)_2$$

Determine the encoder speed in rpm.

11.8 Consider Example 11.1. When the system supplies a load of 45 kW at 1500 rpm, determine the contents of the encoder position and speed registers.

11.9 A two-pole resolver has stator ratings 60 Hz 120 Vrms, and is to be used as a speed and position-sensing device. The stator–rotor turns ratio is 1:1, and perfect coupling may be assumed.

(a) Suppose 120 Vrms 60 Hz is applied to winding 1; winding 2 shorted; windings 1 and 3 are series connected, as are windings 1 and 4: i.e., connect 1-1′–3-3′ and 1-1′–4-4′, and $V_X = V_{1\text{–}3'}$; $V_Y = V_{1\text{–}4'}$. With the rotor in an arbitrary unknown position and stationary, V_3, V_4, V_X, and V_Y were measured to be 60, 103.9, 180, and 223.9 V, respectively. What is the rotor position?

(b) Suppose balanced two-phase 120 Vrms 60 Hz voltages are applied to stator windings 1 and 2, and windings 3 and 4 voltages are measured, each being 24 Vrms. What is the rotor speed?

(c) Suppose, as in (b), balanced two-phase 120 Vrms 60 Hz voltages are applied to stator windings 1 and 3 voltages are measured on an oscilloscope, such that

$$v_1(t) = 120\sqrt{2}\cos(377t)$$
$$v_3(t) = 24\sqrt{2}\cos(75.4t + 15°)$$

Also, when the system speeds up, V_3 drops. What is the rotor speed and position?

11.10 What is piezoelectricity?

11.11 What is magnetostriction?

11.12 For the MEMS of Example 11.3, determine the minimum applied voltage that drives x to its limiting value of $x = d$.

Epilogue

The study of electromagnetic machines is unique within the broader discipline of electrical engineering. The relevant engineering fundamentals have been studied for over 150 years, and still remain of interest today. The reasons are simple. The primary needs of human civilization, food, shelter, clothing, clean air, water, and transportation, remain unchanged. The conversion and control of bulk energy is basic to meeting these needs. The electromagnetic-mechanical (EM) energy conversion process and the devices or "machines," which implement that process, constitute an efficient, economical, reliable, green, safe, and in some cases, the only practical technological solution to this problem. The EM machine is to human muscle as the microprocessor is to the human brain.

From an engineering perspective, EM machines represent a particularly challenging field of study. One cannot solve even fundamental problems without considering the machine as a part of an integrated system consisting of an electrical source, a controller, the motor, and a mechanical termination. These studies require an integration of solid-state ("power") electronics with machine theory. The subject is intellectually four-dimensional, requiring the student to simultaneously consider electric, magnetic, mechanical, and thermal issues. We pursued a balanced approach, moving from the basic machine physics and principles of operation, through structural details of practical machine design into machine performance with realistic applications, considering relevant control issues. The digital computer permits integrated analyses that were impossible just a few years ago.

Contemporary engineering is at the beginning of a technological revolution that impacts directly on EM machines. Many ground and sea transportation systems are moving away from internal combustion engines toward EM machines. Vertical transportation systems, HSR, and hybrid vehicles are still in the early stages of development. Robotics will require more sophisticated EM machine design and control, and will create separate category of applications. MEMS creates an entirely new front upon which EM machine development will advance. Forecasting the future is perilous, but the author is confident in the following prediction: there is much more to know, to learn, to discover, to develop, and to use, with regard to EM machines. The mission of this book is to present a balanced, fundamental, organized basic study of EM machines. We wish readers Godspeed as they move into more advanced studies of the fascinating world of EM machines.

Appendix A: Units and Conversion Factors

This Appendix is provided as a convenient reference for units used in this book. For a comprehensive investigation of the subject, refer to the references cited at the end of the Appendix. Contemporary practice encourages usage of SI units (SI is the accepted abbreviation for the International Units, or more formally "Le Systeme International d'Unites").

Quantity	Typical Symbol(s)	SI Unit, abbreviation
mechanical...		
length	x, y, z, l, d	meter, m
mass	M	kilogram, kg
time	t	second, s
velocity	v, u	meter/second, m/s
force	F	newton, N
torque	T	newton-meter, N m
pressure	P	pascal, Pa
energy	W	joule, J
power	p, P	watt, W
angle	$\alpha, \beta, \gamma, \theta$	radian, rad
angular velocity	ω	radian/second, rad/s
electrical...		
charge	q, Q	coulomb, C
(electric) current	i, I	ampere, A
voltage	v, V, e, E	volt, V
frequency	f	hertz, Hz
radian frequency	ω	radian per second, rad/s
apparent power[1]	S	voltampere, VA
average power	P	watt, W
reactive power	Q	voltampere-reactive, var
resistance	R	ohm, Ω
inductance	L	henry, H
capacitance	C	farad, F
impedance, reactance	Z, X	ohm, Ω
conductance	G	siemens, S
admittance, susceptance	Y, B	siemens, S
thermal...		
temperature	θ, T	kelvin, K
thermal power	P, Q	watt, W
heat energy	W	joule, J
thermal capacitance	C	joule/kelvin, J/K
thermal conductance	G	watt/kelvin, W/K
specific heat	c_p	joule/(kelvin-kilogram), J/(K-kg)

(*Continued*)

[1] See Appendix B for further clarification of S, P, and Q power issues.

(*Continued*)

Quantity	Typical Symbol(s)	SI Unit, abbreviation
magnetic...		
flux	ϕ	weber, Wb
flux density	B	tesla, T
magnetic field intensity	H	ampere/meter, A/m
reluctance	$\mathfrak{R}$	1/henry, H^{-1}
permeance	$\mathfrak{P}$	henry, H
magnetomotive force	$\mathfrak{F}$	ampere-turns, A

SI Prefixes

yatto	Y	10^{24}	yocto	z	10^{-24}
zetta	Z	10^{21}	zepto	z	10^{-21}
exa	E	10^{18}	atto	a	10^{-18}
peta	P	10^{15}	femto	f	10^{-15}
tera	T	10^{12}	pico	p	10^{-12}
giga	G	10^{9}	nano	n	10^{-9}
mega	M	10^{6}	micro	μ	10^{-6}
kilo	k	10^{3}	milli	m	10^{-3}
hecto	h	10^{2}	centi	c	10^{-2}
deka	da	10^{1}	deci	d	10^{-1}

Example usage: $I = 115\,\text{A} = 0.115\,\text{kA} = 115000\,\text{mA}$

Selected Constants

permeability of free space = $0.4\,\pi\,\mu\text{H/m}$
permittivity of free space = 8.8542 pF/m
absolute zero (temperature) = 0 K = −273.2°C = −459.7°F = 0°R
freezing point of water = 273.2 K = 0°C = 32°F = 491.7°R
g = acceleration due to gravity = $9.807\,\text{m/s}^2 = 32.17\,\text{ft/s}^2$
G = gravitational constant = $66.72\,\text{pN}\,(\text{m}^2)\,(\text{kg}^{-2})$
c = speed of light in vacuum = 0.2998 m/ns = 299.8 km/s
Avogadro's constant = 1.602×10^{23} = molecules/mol
e = charge on electron = 0.1602 aC
m_0 = electron rest mass = 9.11×10^{-28} g
1 revolution = 2π radians = 360°
1 standard atmosphere = 760 mm Hg = 101 kPa
air density = $1.225\,\text{kg/m}^3$ @ 1 atmosphere, 15°C
1 toe (1 tonne oil equivalent) = 41.868 GJ = $1270\,\text{m}^3$ natural gas = 2.3 tonnes coal

Selected Conversion Factors

1 inch = 2.54 cm; 1 foot = 12 inches; 1 meter = 3.28084 feet
1 pound force (lbf) = 4.4482 newtons, 1 kilogram = 2.188 pound mass (lbm)
J in kg-m^2 = 0.0.04246 × J in lbm − ft^2
1 slug = 14.59 kilogram
1 foot-pound = 1.356 N-m, 1 horsepower = 746 watts
1 British Thermal Unit (BTU) = 1055 Joule
1 calorie = 4.19 Joule

Temperature in kelvin (K) = T_K
Temperature in degrees Celsius (°C) = $T_C = T_K - 273.2$
Temperature in degrees Fahrenheit (°F) = $T_F = 1.8T_C + 32$
Temperature in degrees Rankine (°R) = $T_R = T_F + 459.7$
1 milligauss = 1 mG = 0.1 microtesla = 0.1 μT
1 ton (refrigeration) = 3517 watts = 12 kBTU/h
1 tonne = 1 metric ton = 1000 kg
speed in rad/s = $(2\pi/60) \times$ speed in rpm (rpm = revolutions per minute)
1 day = 24 hours; 1 hour = 60 minutes; 1 minute = 60 seconds
1 nautical mile = 6000 ft = 1.1364 mile
1 knot = 1 nautical mile per hour = 1.1364 mph = 0.5080 m/s

Temperature Scales

Fahrenheit	Rankine	Centigrade	Kelvin
212°F	672°R	100°C	373 K
100°F	560°R	38°C	311 K
32°F	492°R	0°C	273 K
0°F	460°R	−18°C	255 K
−460°F	0°R	−273°C	0 K

Notes: 1. one unit C = one unit K; one unit F = one unit R; one unit K = 1.8 F units;
2. Absolute zero (0 K) = −459.67°F = −273.15°C;
3. Normal Human Body temperature is 98.2°F (36.8°C).

References

ANSI Metric Practice, ASTM E 380-76 IEEE Std 268-76 ANSI Z210.1-1976.
Reference Manual for SI Units, Inland Steel Company, 1976.
Robert A. Nelson, *Guide for metric practice, Physics Today*, 1996. http://www.physicstoday.org/guide/metric.pdf

Appendix B: A Review of Electrical Circuit Concepts

After the force of gravity is accounted for, when any atomic test particle is in the neighborhood of an electron, one of the following situations is observed:

Case 1. A force of attraction is observed
Case 2. A force of repulsion is observed
Case 3. No force is observed

These so-called electrostatic forces are accounted for by assigning the particle a property called "charge," such that the test particle is said to be

- "Positively charged" in Case 1
- "Negatively charged" in Case 2
- "Uncharged" in Case 3

If the test particle is itself an electron, Case 2 is observed and the charge (Q_e) can be quantified from physical experimentation. All atomic particles are observed to have integer multiples of Q_e.[1] The unit selected for charge is the coulomb (C).

The electron (and therefore charge) can freely move through certain materials, particularly metals. "Electric current" or simply "current" is defined as the flow rate of charge through a given path or

$$\text{Current} = i = \frac{dq}{dt} \quad (\text{C/s} \rightarrow \text{ampere, A})$$

The energy (per unit charge) required to move a unit charge from point a to point b in a material is defined as potential difference (from a to b) or voltage (between a and b):

$$\text{Voltage drop from a to b} = v_{ab} = \frac{dW}{dq} \quad (\text{J/C} \rightarrow \text{volt, V})$$

Current will flow through any path across which a source voltage (sometimes called an electromotive force, emf) is applied. The current will be limited by the combination of three basic passive properties:

Resistance R (in ohm, Ω)
A measure of the opposition in a material to the flow of current. The energy expended to overcome this effect is converted into thermal form (heat).

Inductance L (in henry, H)
A measure of the opposition to the *change* in current through a defined path. Currents create a corresponding magnetic field. When current is changed in a circuit, energy is taken from the forcing function and stored in this field. This energy may be recovered when the current returns to its original value.

[1] For our purposes. Quantum mechanics complicates the issue.

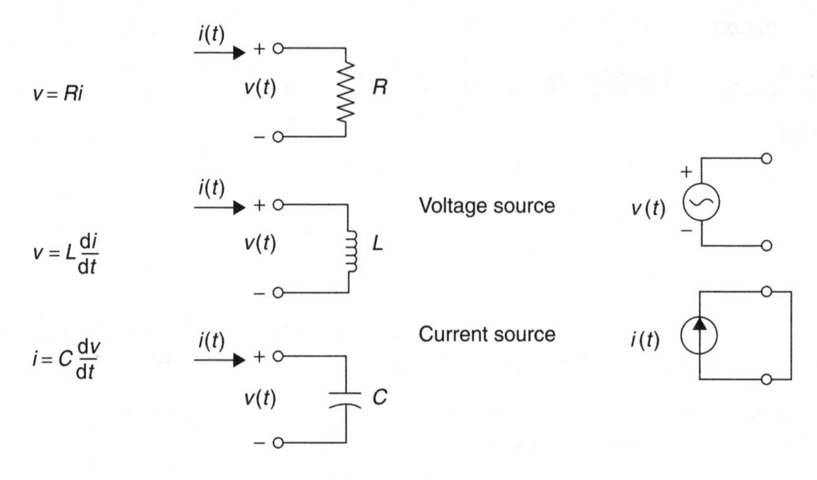

FIGURE B.1. Ideal circuit elements.

Capacitance C (in farad, F)
A measure of the circuit's capacity to store charge. Energy is required to store charge, but may be recovered when the charge is permitted to flow back to its source.

A *resistor* is simply an electrical element designed for a particular R value, with negligible L and C. An "ideal" resistor has zero L and C. Likewise an *inductor* has negligible R and C, and a *capacitor*, negligible R and L. An *ideal voltage source* maintains a known voltage $v(t)$ at its terminals, independent of its termination. An *ideal current source* maintains a known current $i(t)$ at its terminals, independent of its termination. The ideal elements are shown in Figure B.1. Circuit models are formed by interconnection of ideal elements. Basically, circuit analysis involves the systematic application of two fundamental laws:

Kirchhoff's current law (KCL). The sum of the currents out of every node in any circuit equals zero.
Kirchhoff's voltage law (KVL). The sum of the voltage drops around every closed path in any circuit equals zero.

For convenience, circuit analysis can be separated into three modes:

- DC (direct current), in which case all currents and voltages become constant in time: inductors become short circuits and capacitors open circuits.
- AC (alternating current), in which case all currents and voltages become sinusoidal at constant amplitude, frequency, and phase.
- Transient, in which case all currents and voltages are time varying, moving from some initial state to a final steady state.

All three modes of analysis are used to predict EM machine performance.

B.1 DC Circuit Concepts

- **Voltage source:** $v(t) = V$ (constant in time).
- **Current source:** $i(t) = I$ (constant in time).
- **Resistor:** $v = V = Ri = RI$.
- **Inductor:** $v(t) = 0, \quad i(t) = I$, short circuit.
- **Capacitor:** $v(t) = V, \quad i(t) = 0$, open circuit.

B.2 AC Circuit Concepts

All sinusoidal voltages and currents can be represented in the general forms:

$$v(t) = V_{max}\cos(\omega t + \alpha) \tag{B.1a}$$

$$i(t) = I_{max}\cos(\omega t + \beta) \tag{B.1b}$$

where

V_{max}, I_{max} = maximum (peak, amplitude) voltage, current (in V, A)
ω = radian frequency (in rad/s)
t = time (in s)
α, β = phase angles (formally in rad, but frequently given in deg)

Also

$f = \dfrac{\omega}{2\pi}$ = cyclic frequency (in Hz)

$T = \dfrac{1}{f}$ = period (in s)

Example: $v(t) = 169.7\cos(377t + 50°)$
where

$$V_{max} = 169.7\ V$$
$$\omega = 377\ \text{rad/s}$$
$$\alpha = 50$$
$$f = \frac{\omega}{2\pi} = \frac{377}{2\pi} = 60\ \text{Hz}$$
$$T = \frac{1}{f} = \frac{1}{60} = 16.67\ \text{ms}$$

The rms (root mean square or effective) value for any periodic function $v(t)$ with period T is

$$V_{rms} = V = \sqrt{\frac{1}{T}\int_0^T v(t)^2 dt}$$

which, for sinusoidal signals, becomes

$$V_{rms} = \frac{V_{max}}{\sqrt{2}}$$

Sinusoidal signals may be represented as complex numbers called "phasors," defined as

$$\bar{V} = V\angle\alpha = Ve^{j\alpha} = V\cos\alpha + jV\sin\alpha^{1} \tag{B.2}$$

[1] Equation B.2 is known as "Euler's identity" and is one of the most famous expressions in all of mathematics. The idea was first presented in Leonhard Euler's (1707–1783) book *Introduction in Analysis Infinitorum* in 1748. Note that "j" represents $\sqrt{-1}$ in electrical engineering practice; "i" is used more generally in formal mathematics, a symbol introduced by Euler in 1777.

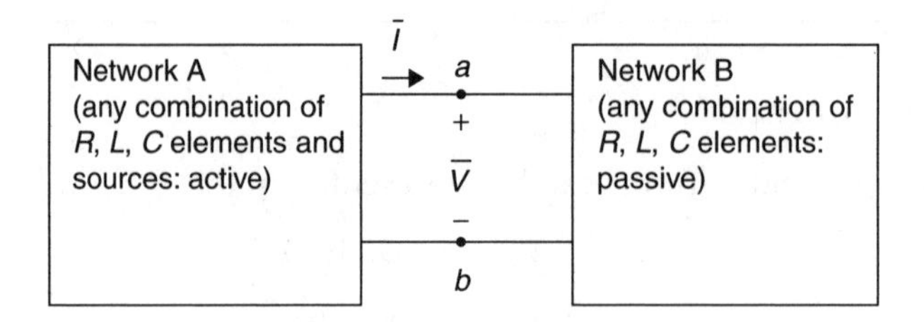

FIGURE B.2. Two interconnected one-port networks.

where

$\mathrm{Re}[\bar{V}]$ = real part of $[\bar{V}] = V\cos\alpha$

and

$\mathrm{Im}[\bar{V}]$ = imaginary part of $[\bar{V}] = V\sin\alpha$[2]

Example: $v(t) = 169.7\cos(377t + 50°)$ V

Rms value: $V_{\mathrm{rms}} = V = \dfrac{169.7}{\sqrt{2}} = 120\,\mathrm{V}$

Phasor: $\bar{V} = 120\,\angle 50° = 77.13 + j\,91.93\,\mathrm{V}$

Consider Figure B.2, where network B contains any combination of any number of R, L, and C elements. Such a network is said to be "passive." If network A applies a sinusoidal voltage, represented by the phasor $\bar{V} = V\angle\alpha$, the current may be represented by the phasor $\bar{I} = I\angle\beta$. We define the impedance ($\bar{Z}$) of network B at port ab:

$$\bar{Z} = \frac{\bar{V}}{\bar{I}} = R + jX \quad (\text{in } \Omega) \tag{B.3}$$

where

$R = \mathrm{Re}[\bar{Z}]$ = resistance of network B at port ab.
$X = \mathrm{Im}[\bar{Z}]$ = reactance of network B at port ab.

Likewise, the admittance ($\bar{Y}$) of network B at port ab:

$$\bar{Y} = \frac{\bar{I}}{\bar{V}} = \frac{1}{\bar{Z}} = G + jB \quad (\text{in S}) \tag{B.4}$$

where

$G = \mathrm{Re}[Y]$ = conductance of network B at port ab.
$B = \mathrm{Im}[Y]$ = susceptance of network B at port ab.

When network B is a single element, $\bar{Z}$ and $\bar{Y}$ values are as shown in Figures B.3a to B.3c. Elements are said to be in series when the same current flows through them, and in parallel when the same voltage appears across them, as shown in Figure B.3d and Figure B.3e, respectively.

[2] The descriptors "real" and "imaginary," introduced by René Descartes (1596–1650) in his work *La Geometrie* in 1637 are a most unfortunate choice of terminology. "Imaginary" numbers are no more or less "fictitious" than are "real" numbers: both are abstract mathematical concepts with clearly defined mathematical properties, and very useful for our purposes in machine theory and analysis. Even the greatest of us occasionally drop our sandwiches jelly-side down.

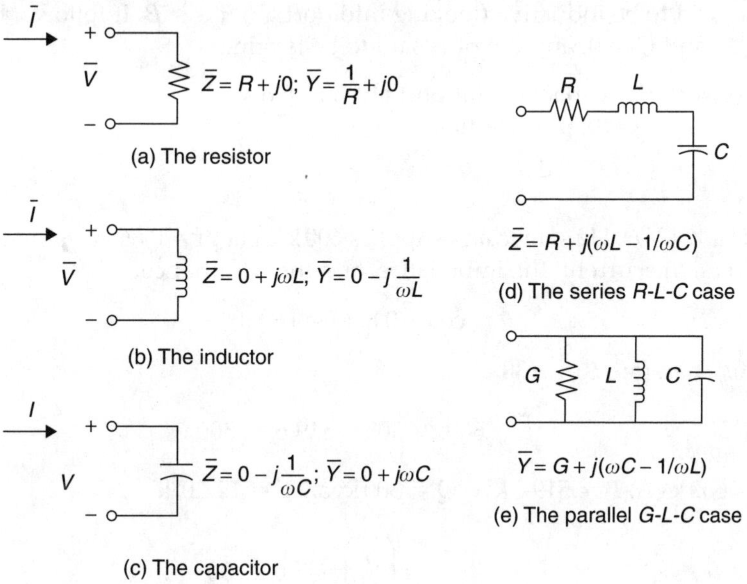

$\bar{Z} = R + j0;\ \bar{Y} = \frac{1}{R} + j0$

(a) The resistor

$\bar{Z} = 0 + j\omega L;\ \bar{Y} = 0 - j\frac{1}{\omega L}$

(b) The inductor

$\bar{Z} = 0 - j\frac{1}{\omega C};\ \bar{Y} = 0 + j\omega C$

(c) The capacitor

$\bar{Z} = R + j(\omega L - 1/\omega C)$

(d) The series *R-L-C* case

$\bar{Y} = G + j(\omega C - 1/\omega L)$

(e) The parallel *G-L-C* case

FIGURE B.3. Impedance, admittance for *R, L, C* elements.

Referring to Figure B.2, the instantaneous rate of energy transfer or instantaneous power (flow) from network A to network B is

$$p(t) = \frac{dw}{dt} = v(t) \cdot i(t) \quad (\text{W}) \tag{B.5}$$

Four other types of power are used in electric power engineering. "Complex power" is defined as

$$\bar{S} = \bar{V}\,\bar{I}^* = VI\angle(\alpha - \beta) = S\angle\theta \tag{B.6a}$$

Apparent power:

$$S = |\bar{S}| = VI \tag{B.6b}$$

Real (average) power:

$$P = \text{Re}(\bar{S}) = S\cos\theta = VI\cos\theta \tag{B.6c}$$

Reactive power:

$$Q = \text{Im}(\bar{S}) = S\sin\theta = VI\sin\theta \tag{B.6d}$$

It is traditional in power engineering to use "different" units for P, Q, and S, which are the watt (W), the voltampere-reactive (var), and the voltampere (VA), respectively. Actually, these are all physically equivalent to the watt; the distinction is one of accepted terminology and not of physics.

Power factor is defined as

$$\text{pf} = \text{power factor} = \frac{P}{S} \tag{B.7}$$

$$\text{pf} = \cos\theta \tag{B.8}$$

Network B is said to be inductive (looking into port ab) if $\alpha > \beta$. It follows that $\theta > 0$; $X > 0$, and $Q > 0$, and the pf is said to be lagging.

Likewise, Network B is capacitive (at port ab) if $\beta > \alpha$ with $\theta < 0$; $X < 0$, and $Q < 0$; pf is leading.

Example B.1

Network A is a 12 kV 60 Hz source and supplies 600 kVA at pf = 0.866 lagging to Network B. Find all powers, the current, the impedance, and the admittance.

$$\theta = \cos^{-1}(0.866) = (+)30°$$

Define $\alpha = 0°$; therefore $\beta = -30°$.

$$\bar{S} = 600\angle{+30°} = 519.6 + j300$$

so that $S = 600\,\text{kVA}$, $P = 519.6\,\text{kW}$, $Q = 300\,\text{kvar}$, $\bar{V} = 12\angle 0°\text{kV}$:

$$\bar{I} = \left(\frac{\bar{S}}{\bar{V}}\right)^* = 50\angle{-30°}$$

$$v(t) = 16.97\cos 377t\,\text{kV}$$
$$i(t) = 70.71\cos(377t - 30°)\,\text{A}$$

$$\bar{Z} = \frac{\bar{V}}{\bar{I}} = (12000\angle 0°)(50\angle{-30°}) = 240\angle{+30°} = 207.8 + j\,120\,\Omega$$

$$\bar{Y} = \frac{\bar{I}}{\bar{V}} = 4.167\angle{-30°} = 3.608 - j2.083\,\text{mS}$$

B.3 Balanced Three-Phase AC Circuit Concepts

Consider the four-wire system shown in Figure B.4a. Double-subscript notation applied to voltages are in "+, −" order, and to currents, in "from, to" order. Hence, $v_{an}(t)$ is positive when "a" is physically positive relative to "n," and $i_{an}(t)$ is positive when positive charge is physically flowing from "a" to "n." In a three-phase situation, "balanced" phase voltages are defined as "equal in magnitude, 120° phase-separated." Specifically, in a balanced situation, the phase voltages are

$$v_{an}(t) = V\sqrt{2}\cos(\omega t), \quad \bar{V}_{an} = \bar{V}_a = V\angle 0° \tag{B.9a}$$

$$v_{bn}(t) = V\sqrt{2}\cos(\omega t - 120°), \quad \bar{V}_{bn} = \bar{V}_b = V\angle{-120°} \tag{B.9b}$$

$$v_{cn}(t) = V\sqrt{2}\cos(\omega t + 120°), \quad \bar{V}_{cn} = \bar{V}_c = V\angle{+120°} \tag{B.9c}$$

We can determine the phase-to-phase (line) voltages by direct application of KVL:

$$\bar{V}_{ab} = \bar{V}_{an} - \bar{V}_{bn} = \sqrt{3}V\angle{+30°} \tag{B.10a}$$

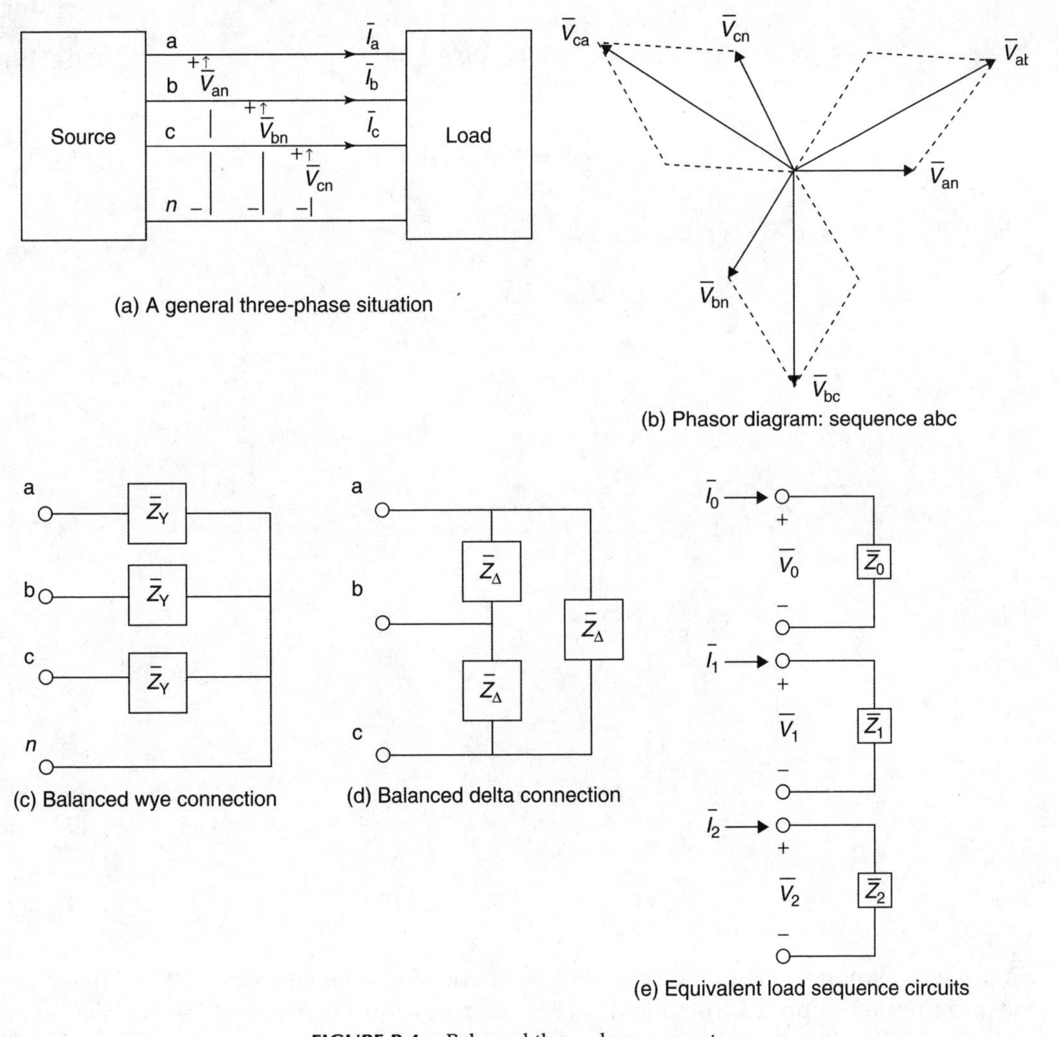

FIGURE B.4. Balanced three-phase concepts.

$$\bar{V}_{bc} = \bar{V}_{bn} - \bar{V}_{cn} = \sqrt{3}V\angle{-90°} \tag{B.10b}$$

$$\bar{V}_{ca} = \bar{V}_{cn} - \bar{V}_{an} = \sqrt{3}V\angle{+150°} \tag{B.10c}$$

The corresponding voltage phasor diagram is shown in Figure B.4b. If Phase b lags Phase "a" as shown, the phase sequence or rotation is said to be "abc." If Phases b and c are reversed (Phase c lags Phase a), the phase sequence or rotation is said to be "acb." In this book, the normal operating phase sequence for all balanced three-phase systems is assumed to be abc.

There are two, and only two, possible passive network terminations (in the section marked "load") that result in balanced three-phase currents. The wye connection ($\bar{Z}_Y = Z\angle\theta$), shown in Figure B.4c, produces

$$\bar{I}_a = \frac{\bar{V}_{an}}{\bar{Z}_Y} = I\angle{-\theta} \tag{B.11a}$$

$$\overline{I}_{\text{b}} = \frac{\overline{V}_{\text{bn}}}{\overline{Z}_{\text{Y}}} = I\angle(-\theta - 120^\circ) \tag{B.11b}$$

$$\overline{I}_{\text{c}} = \frac{\overline{V}_{\text{cn}}}{\overline{Z}_{\text{Y}}} = I\angle(-\theta + 120^\circ) \tag{B.11c}$$

The delta connection, shown in Figure B.3d, produces

$$\overline{I}_{\text{ab}} = \frac{\overline{V}_{\text{ab}}}{\overline{Z}_{\Delta}} = \frac{\sqrt{3}V}{\overline{Z}_{\Delta}}\angle{+30^\circ} \tag{B.12a}$$

$$\overline{I}_{\text{bc}} = \frac{\overline{V}_{\text{bc}}}{\overline{Z}_{\Delta}} = \frac{\sqrt{3}V}{\overline{Z}_{\Delta}}\angle{-90^\circ} \tag{B.12b}$$

$$\overline{I}_{\text{ca}} = \frac{\overline{V}_{\text{cb}}}{\overline{Z}_{\Delta}} = \frac{\sqrt{3}V}{\overline{Z}_{\Delta}}\angle{+150^\circ} \tag{B.12c}$$

and

$$\overline{I}_{\text{a}} = \overline{I}_{\text{ab}} - \overline{I}_{\text{ca}} = \frac{\sqrt{3}V}{Z_{\Delta}}\angle 0^\circ \tag{B.13a}$$

$$\overline{I}_{\text{b}} = \overline{I}_{\text{bc}} - \overline{I}_{\text{ab}} = \frac{\sqrt{3}V}{Z_{\Delta}}\angle{-120^\circ} \tag{B.13b}$$

$$\overline{I}_{\text{c}} = \overline{I}_{\text{ca}} - \overline{I}_{\text{bc}} = \frac{\sqrt{3}V}{Z_{\Delta}}\angle{+120^\circ} \tag{B.13c}$$

Consider the following situation: two three-phase loads, one wye-connected and the other delta-connected, supplied from identical balanced three-phase sources. The loads are said to be equivalent if the same currents flow to each. It is necessary to consider only the "a" phase current. For equivalence,

$$(\overline{I}_{\text{a}})_{\Delta} = (\overline{I}_{\text{a}})_{\text{Y}}$$
$$\frac{3V}{Z_{\Delta}} = \frac{V}{Z_{\text{Y}}}$$

or

$$\overline{Z}_{\Delta} = 3\overline{Z}_{\text{Y}} \tag{B.14}$$

Computation of the power transferred from source to load is straightforward:

$$\overline{S}_{\text{a}} = \overline{V}_{\text{a}}\overline{I}_{\text{a}}^{*} = P_{\text{a}} + jQ_{\text{a}}$$
$$\overline{S}_{\text{a}} = VI\angle\theta = S_{\text{a}}\angle\theta$$

$$S_{\text{a}} = VI$$

$$P_{\text{a}} = VI\cos\theta$$

$$Q_{\text{a}} = VI\sin\theta$$

The total or three-phase power is

$$\begin{aligned} \bar{S}_{3\text{ph}} &= \bar{S}_{\text{a}} + \bar{S}_{\text{b}} + \bar{S}_{\text{c}} = 3\bar{S}_{\text{a}} \\ &= 3VI\angle\theta = S_{\text{a}}\angle\theta \end{aligned} \quad \text{(B.15a)}$$

$$P_{3\text{ph}} = 3VI\cos\theta \quad \text{(B.15b)}$$

$$Q_{3\text{ph}} = 3VI\sin\theta \quad \text{(B.15c)}$$

An example would be instructive.

Example B.2

A balanced three-phase load of 600 kVA at pf = 0.866 lagging is supplied from a 12 kV 60 Hz three-phase source. Find all powers, voltages, currents, and wye and delta impedances.

We infer that 12 kV is the phase-to-phase (line) voltage, since that is the normal practice. Likewise, assume the phase sequence to be abc. Therefore

$$\begin{aligned} \bar{V}_{\text{an}} &= \bar{V}_{\text{a}} = \frac{12}{\sqrt{3}}\angle 0° = 6.928\angle 0°\text{ kV} \\ V_{\text{bn}} &= 6.928\angle{-120°}, \quad V_{\text{cn}} = 6.928\angle{+120°} \\ V_{\text{ab}} &= 12.00\angle{+30°} \\ V_{\text{bc}} &= 12.00\angle{-90°} \\ V_{\text{ca}} &= 1200\angle{+150°} \\ \theta &= \cos^{-1}(0.866) = (+)30° \\ \bar{S}_{3\text{ph}} &= 600\angle{+30°} = 519.6 + j300 \end{aligned}$$

so that $S_{3\text{ph}} = 600\text{ kVA}$, $P_{3\text{ph}} = 519.6\text{ kW}$, $Q_{3\text{ph}} = 300\text{ kvar}$.

$$\bar{S}_{\text{a}} = \bar{S}_{\text{b}} = \bar{S}_{\text{c}} = 200\angle{+30°} \quad \bar{I}_{\text{a}} = 200\angle 6.928 = 28.87\text{ A}$$

$$\bar{I}_{\text{a}} = 28.87\angle{-30°}\text{ A} \quad \bar{I}_{\text{b}} = 28.87\angle{-150°}\text{ A} \quad \bar{I}_{\text{c}} = 28.87\angle{+90°}\text{ A}$$

$$\begin{aligned} \bar{Z}_{\text{Y}} &= \frac{\bar{V}_{\text{an}}}{\bar{I}_{\text{a}}} = (6928\angle 0°/28.87\angle{-30°}) \\ &= 240\angle{+30°} = 207.8 + j120\ \Omega \end{aligned}$$

$$\bar{Z}_{\Delta} = 3Z_{\text{Y}} = 720\angle{+30°} = 623.5 + j360\ \Omega$$

In the delta load, the phase currents would be

$$\begin{aligned} \bar{I}_{\text{ab}} &= 16.67\angle 0°\text{ A} \\ \bar{I}_{\text{bc}} &= 16.67\angle{-120°}\text{ A} \\ \bar{I}_{\text{ca}} &= 16.67\angle{+120°}\text{ A} \end{aligned}$$

B.4 Symmetrical Components

In certain symmetrical situations in three-phase circuits, there is an advantage in transforming phase quantities into so-called sequence quantities, according to

$$\bar{V}_{\text{an}} = \bar{V}_0 + \bar{V}_1 + \bar{V}_2 \quad \text{(B.16a)}$$

$$\bar{V}_{bn} = \bar{V}_0 + a^2\bar{V}_1 + a\bar{V}_2 \tag{B.16b}$$

$$\bar{V}_{cn} = \bar{V}_0 + a\bar{V}_1 + a^2\bar{V}_2 \tag{B.16c}$$

where $a = 1\angle 120°$ or

$$\begin{bmatrix} \bar{V}_{an} \\ \bar{V}_{bn} \\ \bar{V}_{cn} \end{bmatrix} = \begin{bmatrix} 1 & 1 & 1 \\ 1 & a^2 & a \\ 1 & a & a^2 \end{bmatrix} \begin{bmatrix} \bar{V}_0 \\ \bar{V}_1 \\ \bar{V}_2 \end{bmatrix}$$

or

$$\hat{V}_{abc} = [T]\hat{V}_{012} \tag{B.17a}$$

where $\bar{V}_0$, $\bar{V}_1$, and $\bar{V}_2$ are called the zero, positive, and negative sequence voltages, respectively. Likewise

$$\hat{V}_{012} = [T]^{-1}\hat{V}_{abc} \tag{B.17b}$$

where

$$[T]^{-1} = \frac{1}{3}\begin{bmatrix} 1 & 1 & 1 \\ 1 & a & a^2 \\ 1 & a^2 & a \end{bmatrix} \tag{B.18}$$

The same transformation holds for currents:

$$\hat{I}_{abc} = [T]\hat{I}_{012} \tag{B.19a}$$

$$\hat{I}_{012} = [T]^{-1}\hat{I}_{abc} \tag{B.19b}$$

Impedance is different. Consider a generalized load, such that we write the circuit equations in phase coordinates:

$$\hat{V}_{abc} = [Z_{abc}]\hat{I}_{abc} \tag{B.20a}$$

Substituting sequence values:

$$[T]\hat{V}_{012} = [Z_{abc}]\ \{[T]\hat{I}_{012}\}$$

or

$$\hat{V}_{012} = \{[T]^{-1}[Z_{abc}][T]\}\hat{I}_{012}$$

Define the following:

$$[T]^{-1}[Z_{abc}][T] = [Z_{012}] \tag{B.20b}$$

such that

$$\hat{V}_{012} = [Z_{012}]\hat{I}_{012} \tag{B.20c}$$

For a symmetrical balanced three-phase load, $[Z_{\mathbf{abc}}]$ will be of the form

$$[Z_{abc}] = \begin{bmatrix} \bar{Z}_s & \bar{Z}_m & \bar{Z}_m \\ \bar{Z}_m & \bar{Z}_s & \bar{Z}_m \\ \bar{Z}_m & \bar{Z}_m & \bar{Z}_s \end{bmatrix} \tag{B.21a}$$

Therefore

$$[Z_{012}] = [T]^{-1}[Z_{abc}][T] = \begin{bmatrix} \bar{Z}_0 & 0 & 0 \\ 0 & \bar{Z}_1 & 0 \\ 0 & 0 & \bar{Z}_2 \end{bmatrix} \tag{B.21b}$$

where

$$\bar{Z}_0 = \bar{Z}_s + 2\bar{Z}_m \tag{B.22a}$$

$$\bar{Z}_1 = \bar{Z}_s - \bar{Z}_m \tag{B.22b}$$

$$\bar{Z}_2 = \bar{Z}_s - \bar{Z}_m \tag{B.22c}$$

For static systems (no moving parts), $\bar{Z}_2 = \bar{Z}_1$. However, when this approach is applied to machines, which always have moving parts, in general $\bar{Z}_2 \neq \bar{Z}_1$.

Since three-phase machines are always of symmetrical design, this last result is of special interest to us. It implies that a three-phase device may be represented as three decoupled single-phase circuits, as shown in Figure B.4e.

There is a special case of unbalance that deserves our attention. We consider a three-wire case (abc phases only, no neutral), where only magnitudes for $\bar{V}_{ab}$, $\bar{V}_{bc}$, and $\bar{V}_{ca}$ are known:

$$\begin{aligned} \bar{V}_{ab} &= \bar{V}_{an} - \bar{V}_{bn} = (\bar{V}_0 + \bar{V}_1 + \bar{V}_2) - (\bar{V}_0 + a^2\bar{V}_1 + a\bar{V}_2) \\ &= (1 - a^2)\bar{V}_1 + (1 - a)\bar{V}_2 \end{aligned} \tag{B.23a}$$

$$\begin{aligned} \bar{V}_{bc} &= \bar{V}_{bn} - \bar{V}_{cn} = (\bar{V}_0 + a^2\bar{V}_1 + a\bar{V}_2) - (\bar{V}_0 + a\bar{V}_1 + a^2\bar{V}_2) \\ &= (a^2 - a)\bar{V}_1 + (a - a^2)\bar{V}_2 \end{aligned} \tag{B.23b}$$

The phasor diagram can be drawn in closed form, as in Figure B.5. According to the law of cosines:

$$2V_{ab}V_{bc}\cos\theta = (V_{ab})^2 + (V_{bc})^2 - (V_{ca})^2 \tag{B.24}$$

Assigning the phase of V_{ab} to zero (temporarily), and assuming sequence abc, the phase of $\bar{V}_{bc}$ becomes $-(180 - \theta)$. We can solve (Equation B.23a and 23b) for $\bar{V}_1$ and $\bar{V}_2$. Since phase

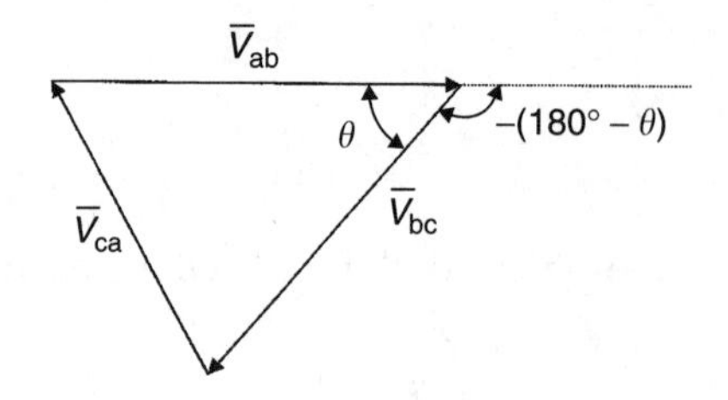

FIGURE B.5. Phasor diagram in closed form.

is relative, subtract the phase angle of $\bar{V}_1$ from all phasors, redefining $\bar{V}_1$ to be our phase reference. In a three-wire situation, the zero sequence current is forced to zero. Since the termination is passive, the zero sequence voltage must be also zero.

An example would be helpful.

Example B.3

Compute all unknown voltages in a three-wire system, given:

$$V_{ab} = 450\,\text{V} \qquad V_{bc} = 470\,\text{V} \qquad V_{ca} = 440\,\text{V}$$

SOLUTION

From Equation B.24

$$2(450)(470)\cos\theta = (450)^2 + (470)^2 - (440)^2$$

$\theta = 57.1°$ Assigning the phase of $\bar{V}_{ab}$ to 0°, the phase of $\bar{V}_{bc} = -(180° - 57.1°) = -122.9°$

$$\bar{V}_{ab} = 450\angle 0°$$
$$\bar{V}_{bc} = 470\angle -122.9°$$

Solving Equation B.23,

$$\bar{V}_1 = 261.6\angle -32.2°$$
$$\bar{V}_2 = 10.2\angle 129.1°.$$

Set $\bar{V}_0 = 0\angle 0°$

Solving Equation B.20a,

$$\bar{V}_{an} = 251.9\angle -31.5°$$
$$\bar{V}_{bn} = 269.4\angle -151.0°$$
$$\bar{V}_{cn} = 263.8\angle +85.6°$$

To redefine $\bar{V}_{an}$ as the phase reference, add 31.5° to the phase of all phasors:

$$\bar{V}_{ab} = 450.0\angle 31.5°, \qquad \bar{V}_{an} = 251.9\angle 0°$$
$$\bar{V}_{bc} = 470.0\angle -91.4°, \qquad \bar{V}_{bn} = 269.4\angle -119.3°$$
$$\bar{V}_{ca} = 440.0\angle 147.7°, \qquad \bar{V}_{cn} = 263.8\angle 117.1°$$
$$\bar{V}_0 = 0\angle 0°$$
$$\bar{V}_1 = 261.6\angle -0.7°$$
$$\bar{V}_2 = 10.2\angle 160.6°$$

Appendix C: Harmonic Concepts

C.1 Basic Concepts

Fourier[1] is credited with the insight that an arbitrary[2] periodic function $y(t)$ (see Figure C.1) may be approximated by a function $y^*(t)$, which is composed of a sum of sinusoidal signals of integer-multiple frequencies, according to

$$y^*(t) = \sum_{n=0}^{n=N} A_n \cos(n\omega_0 t) + B_n \sin(n\omega_0 t) \tag{C.1a}$$

or equivalently

$$y^*(t) = \sum_{n=0}^{n=N} C_n \cos(n\omega_0 t + \phi_n) \tag{C.1b}$$

such that

$$y(t) = \lim_{N\to\infty} y^*(t) \tag{C.1c}$$

$y^*(t)$ is sometimes called "the Fourier series." Equation C.1a is called the "sine, cosine form," and C.1b, the "single cosine or trigonometric form." The fundamental frequency is

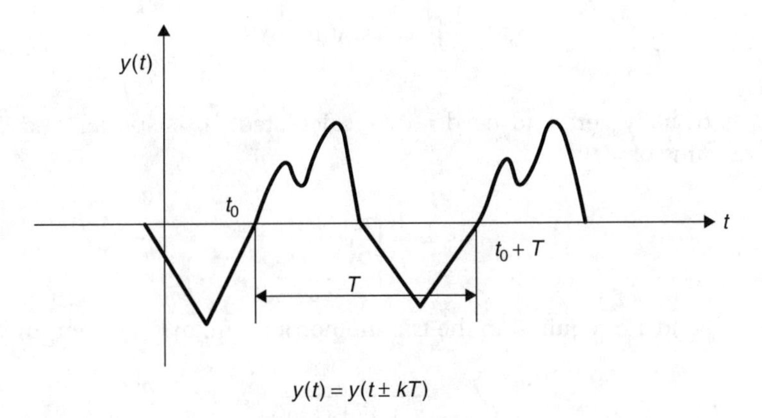

FIGURE C.1. A general periodic function.

[1] Jean Baptiste Joseph Fourier (1768–1830), a former Chair of Mathematics at the French École Normale Supérieure, and protegé of Napoleon, first presented the concept in an early 1807 paper.

[2] Actually the term "arbitrary" is a slight exaggeration. It suffices to say that all signals that result from physical processes are appropriate for Fourier analysis. For limitations and restrictions, the reader is referred to a rigorous mathematical general treatment of the subject.

related to the period of $y(t)$ according to

$$\text{period of } y(t) = T \tag{C.2a}$$

$$\text{fundamental frequency} = f_0 = \frac{1}{T} \tag{C.2b}$$

$$\text{fundamental radian frequency} = \omega_0 = 2\pi f_0 \tag{C.2c}$$

The nth harmonic frequency is

$$n\text{th harmonic frequency} = nf_0 \tag{C.3a}$$

$$n\text{th harmonic radian frequency} = n\omega_0 \tag{C.3b}$$

Because we are dealing with sines and cosines, many workers (your author, for example) prefer to use "θ" (an angle) for the independent variable, as opposed to t(time), where $\theta = \omega_0 t$.[3]

C.2 Coefficient Calculations

The constants A_n and B_n ($n \geq 1$) can be computed from

$$A_n = \frac{2}{T}\int_0^T y(t)\cos(n\omega_0 t)\,dt \tag{C.4a}$$

$$B_n = \frac{2}{T}\int_0^T y(t)\sin(n\omega_0 t)\,dt \tag{C.4b}$$

For $n = 0$, B_0 is trivially zero and need not be calculated. A_0 is special, and is called the average or DC value of $y(t)$:

$$A_0 = \frac{1}{T}\int_0^T y(t)\,dt \tag{C.5}$$

The constants C_n and ϕ_n, required in the trigonometric form, may be computed from

$$C_n = \sqrt{(A_n)^2 + (B_n)^2} \tag{C.6a}$$

$$\phi_n = \tan^{-1}\left(\frac{-B_n}{A_n}\right) \tag{C.6b}$$

[3] If this approach is taken, the nth harmonic is of the form $C_n \cos(n\theta + \phi_n)$. Equations C.4 to C.7 can be written in terms of angle; however, it is too cumbersome to provide "double" equations for all quantities, so only the time-domain expressions will be provided because of their greater generality.

A third form of the series is the exponential form:

$$y^*(t) = \sum_{n=-N}^{n=N} \bar{D}_n \exp(n\omega_0 t) \tag{C.7}$$

where the $\bar{D}_n$s are complex, and of the form

$$\bar{D}_n = D_n \varepsilon^{j\phi_n} = D_n \angle \phi_n \tag{C.8a}$$

$$\bar{D}_n = \frac{1}{T}\int_0^T y(t)\exp(-jn\omega_0 t)\,dt \tag{C.8b}$$

D_n may also be computed from A_n and B_n (and C_n):

$$\bar{D}_n = \frac{A_n - jB_n}{2} \tag{C.9a}$$

Also

$$D_0 = C_0 = A_0 \tag{C.9b}$$

$$D_n = \frac{C_n}{2}, \quad n \geq 1 \tag{C.9c}$$

C.3 Rms (Effective) Values

Several rms (effective) values are relevant to periodic functions:

$$\begin{aligned} Y_{RMS} &= \text{rms value of } y(t) \\ &= \sqrt{\int_0^T [y(t)]^2\,dt} \end{aligned} \tag{C.10a}$$

Also

$$\begin{aligned} Y_n &= \text{rms value of } n\text{th harmonic of } y(t) \\ &= \frac{C_n}{\sqrt{2}} = \sqrt{2}D_n, \quad n \geq 1 \end{aligned} \tag{C.10b}$$

$$\begin{aligned} Y_1 &= \text{rms value of the fundamental of } y(t) \\ &= \frac{C_1}{\sqrt{2}} = \sqrt{2}D_1 \end{aligned} \tag{C.10c}$$

$$\begin{aligned} Y_0 &= \text{rms value of DC term of } y(t) \\ &= C_0 = D_0 \end{aligned} \tag{C.10d}$$

$$Y^*_{\text{RMS}} = \text{rms value of } y^*(t) = \sqrt{\sum_{n=0}^{n=N} Y_n^2} \tag{C.10e}$$

$$Y^*_{\text{AC}} = \text{rms value of AC part of } y^*(t) = \sqrt{\sum_{n=1}^{n=N} Y_n^2} \tag{C.10f}$$

$$Y^*_{\text{HARM}} = \text{rms value of harmonic part of } y^*(t) = \sqrt{\sum_{n=2}^{n=N} Y_n^2} \tag{C.10g}$$

A figure of merit, called the total harmonic distortion (THD), measures the relative amount of harmonic content, relative to the fundamental, a given waveform possesses:

$$\text{THD} = \frac{Y^*_{\text{HARM}}}{Y_1} \times 100\% \tag{C.11}$$

C.4 Symmetries

Waveform symmetries can be related to the Fourier coefficients.

- If the periodic function $y(t) = y(-t)$, $y(t)$ is said to possess EVEN symmetry,

$$B_n = 0, n \geq 1 \quad \text{if } y(t) \text{ is EVEN}$$

- If the periodic function $y(t) = -y(-t)$, $y(t)$ is said to possess ODD symmetry,

$$A_n = 0, n \geq 0 \quad \text{if } y(t) \text{ is ODD}$$

- If the periodic function $y(t - T/2) = -y(t)$, $y(t)$ is said to possess HALF-WAVE symmetry,

$$C_n = 0 \text{ for all even } n \quad \text{if } y(t) \text{ is HALF-WAVE symmetric}$$

In the exponential form, D_n has EVEN symmetry (with respect to n) and ϕ_n has ODD symmetry (with respect to n).

C.5 Spectral Plots

It is useful to display the Fourier coefficients graphically. Several formats are used:

- The *single-sided amplitude spectrum* is a plot of C_n versus n, for $0 \geq n \geq N$. Sometimes the rms value is used for the ordinate (instead of C_n) and the harmonic frequency (radian or cyclic) is used for the abscissa (instead of n).
- The *single-sided phase spectrum* is a plot of ϕ_n versus n, for $0 \geq n \geq N$. Sometimes the harmonic frequency (radian or cyclic) is used for the abscissa (instead of n).
- The *double-sided amplitude spectrum* is a plot of D_n versus n, for $-N \geq n \geq N$. The plot has EVEN symmetry.

- The *double-sided phase spectrum* is a plot of ϕ_n versus n, for $-N \geq n \geq N$. The plot has ODD symmetry.

An example will demonstrate the application of these ideas.

Example C.1

Recall the single-phase inverter output $v_L(t)$ of Example 5.1, for $f = 50\,\text{Hz}$, $\Delta t = 2.50\,\text{ms}$. (shifted right by 3.75 ms). Determine the first 21 harmonics for $v_L(t)$, the rms value, and the THD. Provide single and double sided spectral plots.

SOLUTION

$v_L(t) = 0 \quad 0 < t < 7.5\text{ ms}$
$v_L(t) = 250 \quad 7.5 < t < 10\text{ ms}$
$v_L(t) = 0 \quad 10 < t < 17.5\text{ ms}$
$v_L(t) = -250 \quad 17.5 < t < 20\text{ ms}$

The analysis was performed by the computer program FSAP for the first 21 harmonics. Plots of $v_L(t)$ and $v_L(t)^*$ are provided in Figure C.2a. We plot $x = t/T$ on the horizontal axis.

Harm	A_n	B_n	C_n	D_n	Rms	ϕ_n
1	−112.552	46.621	121.826	60.913	86.143	−157.5
3	−37.552	90.658	98.128	49.064	69.386	−112.5
5	22.572	54.494	58.984	29.492	41.708	−67.5
7	16.167	6.697	17.499	8.750	12.374	−22.5
9	−12.621	5.228	13.661	6.830	9.659	−157.5
11	−10.374	25.044	27.108	13.554	19.168	−112.5
13	8.826	21.309	23.064	11.532	16.309	−67.5
15	7.699	3.189	8.333	4.167	5.892	−22.5
17	−6.844	2.835	7.408	3.704	5.238	−157.5
19	−6.175	14.907	16.135	8.068	11.409	−112.5
21	5.639	13.613	14.735	7.367	10.419	−67.5

RMC value = 125.0
DC, AC, fund, harmonic values: 0.0000, 125.0, 86.14, 90.58.
THD = 105.15%.

The single-sided amplitude spectrum is provided in Figure C.2b. Since the waveform has half-wave symmetry, the even harmonics are zero.

Checking results for the 3rd harmonic ($n = 3$):

$$A_3 = \frac{1}{\pi}\int_0^{2\pi} v(\theta)\cos(3\theta)\,d\theta$$

$$= \frac{250}{\pi}\left\{\int_{3\pi/4}^{\pi} v(\theta)\cos(3\theta)\,d\theta - \int_{7\pi/4}^{2\pi} v(\theta)\cos(3\theta)\,d\theta\right\}$$

$$= -37.552$$

$$B_3 = \frac{1}{\pi}\int_0^{2\pi} v(\theta)\sin(3\theta)\,d\theta$$

$$= \frac{250}{\pi}\left\{\int_{3\pi/4}^{\pi} v(\theta)\cdot\sin(3\theta)\cdot d\theta - \int_{7\pi/4}^{2\pi} v(\theta)\cdot\sin(3\theta)\cdot d\theta\right\}$$

$$= 90.66$$

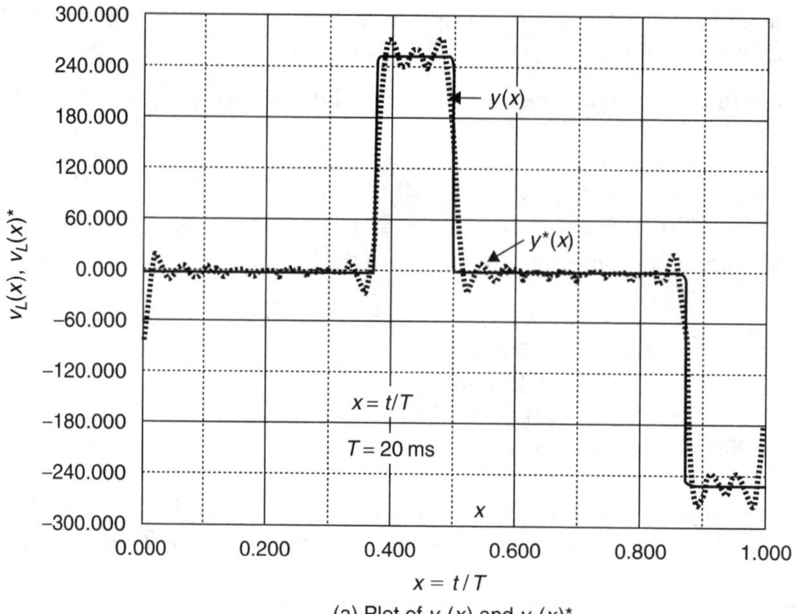

(a) Plot of $v_L(x)$ and $v_L(x)^*$

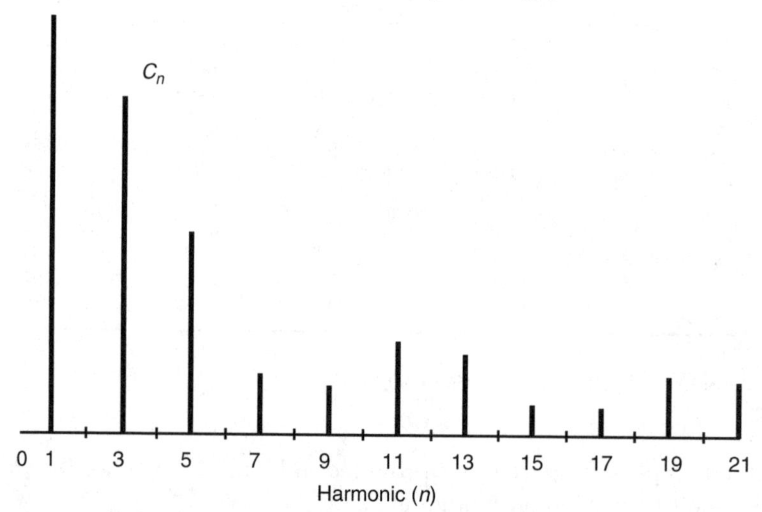

(b) Single-sided magnitude spectrum (normalized to the fundamental)

FIGURE C.2. Plots for Example C.1.

$$C_3 = \sqrt{A_3^2 + B_3^2} = \sqrt{(-37.55)^2 + (90.66)^2} = 98.13$$

$$\phi_3 = \tan^{-1}\left(\frac{-B_3}{A_3}\right) = \tan^{-1}\left(\frac{-90.66}{-37.55}\right) = -112.5^\circ$$

$$\bar{D}_3 = \frac{A_3 - jB_3}{2} = -18.78 - j45.33 = 49.7\angle{-112.5^\circ}$$

$$\text{Rms value of 3rd Harmonic} = \frac{C_3}{\sqrt{2}} = 69.39$$

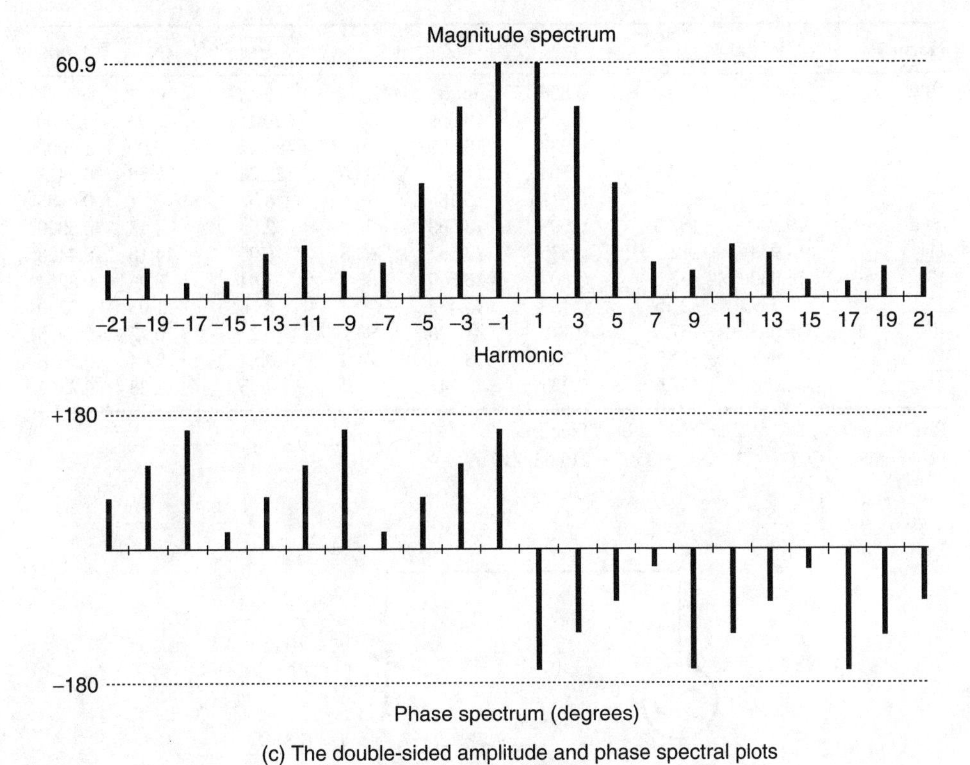

(c) The double-sided amplitude and phase spectral plots

FIGURE C.2. Continued

C.6 Response of Linear Circuits to Nonsinusoidal Excitation

The response of linear circuits to nonsinusoidal excitation can be determined by employing the principle of superposition. An example will demonstrate the application of these ideas.

Example C.2
Apply the voltage of Example C.1 to the *R-L-C* circuit of Figure C.3 and determine and plot the resulting current.

SOLUTION
The network impedance to the *n*th harmonic is

$$Z_n = R + jn\omega_0 L + 1/(jn\omega_0 C), \quad (\text{in } \Omega)$$

Series circuit analysis	$R = 10.000\ \Omega$;	$f_0 = 50.0\,\text{Hz}$
	$L = 6.366\,\text{mH}$;	X_L at $f_0 = 2.000\,\Omega$
	$C = 176.839\,\mu\text{F}$;	X_C at $f_0 = 18.000\,\Omega$

Harm	V_{rms} at V_{ang}		I_{rms}	Z_{mag} at Z_{ang}		P	Q	pf
0	0.000	0.0	0.000	open	—	0	0	—
1	86.136	−157.5	4.565	18.868	−58.0	208.41	−333.45	0.5300
3	69.324	−112.5	6.932	10.000	0.0	480.57	0.00	1.0000
5	41.603	−67.5	3.504	11.873	32.6	122.78	78.58	0.8423
7	12.313	−22.5	0.811	15.186	48.8	6.57	7.51	0.6585
9	9.580	−157.5	0.508	18.868	58.0	2.58	4.13	0.5300
11	18.933	−112.5	0.835	22.687	63.8	6.96	14.18	0.4408
13	16.030	−67.5	0.603	26.569	67.9	3.64	8.96	0.3764
15	5.759	−22.5	0.189	30.487	70.9	0.36	1.03	0.3280
17	5.085	−157.5	0.148	34.426	73.1	0.22	0.72	0.2905
19	10.995	−112.5	0.286	38.378	74.9	0.82	3.04	0.2606
21	9.958	−67.5	0.235	42.341	76.3	0.55	2.28	0.2362

Rms values V, I: 122.856, 9.129, True pf = 0.7431
Powers S, P, Q, D: 1121.61, 833.47, −213.02, 719.69

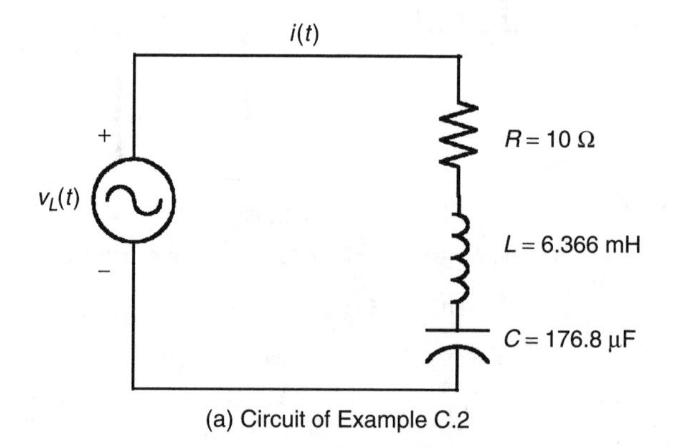

(a) Circuit of Example C.2

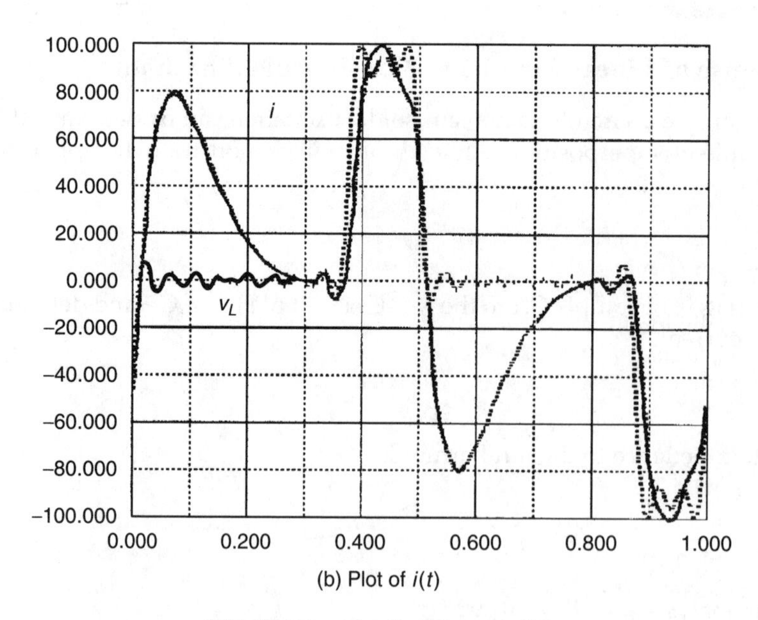

(b) Plot of $i(t)$

FIGURE C.3. Circuit of Example C.2.

References in Electric Machines, Transformers, and Power Electronics

Adkins, B., *The General Theory of Electrical Machines*. London: Pitman, 1962.

Adkins, B. and Gibbs, W.J., *Polyphase Commutator Machines*. Cambridge, MA: Cambridge, 1951.

Adkins, B. and Harley, R.G., *The General Theory of Alternating Current Machines*. London: Chapman and Hall, 1975.

AIEE Standards Committee Report, *Electrical Engineering*, 65(11), 512–516, 1946.

Alger, P., *Induction Machines*. 2nd ed. New York: Gordon and Breach, 1970.

Alger, P.D., *The Nature of Polyphase Induction Machines*. New York: Wiley, 1951.

ANSI/IEEE, *Standard No. C.56.20-1954: Test Code for Induction Motors*. New York: ANSI, 1954.

Ames, R.L., *A.C. Generators: Design and Application*. New York: Wiley, 1990.

Anderson, L.R., *Electric Machines and Transformers*. Reston, VA: Reston, 1981.

Anderson, P.M., *Analysis of Faulted Power Systems*. New York: IEEE Press, 1993.

Anderson, P.M. and Fouad, A.A., *Power System Stability and Control*. Ames, IA: Iowa State University, 1977.

Andreas, J.C., *Energy Efficient Electric Motors: Selection and Application*. New York: Marcel Dekker, 1982.

Baliga, B.J., *Modern Power Devices*. New York: Wiley, 1987.

Bartholomew, D., *Electrical Measurements and Instrumentation*. Boston: Allyn and Bacon, 1963.

Beale, E.M.L., *Mathematical Programming in Practice*. New York: Wiley, 1968.

Beaty, H.W. and Kirtley, J.L., *Electric Motor Handbook*. New York: McGraw-Hill, 1998.

Bedford, B.D. and Hoft, R.G., *Principles of Inverter Circuits*. New York: Wiley, 1964.

Bergen, A.R., *Power System Analysis*. Englewood Cliffs, NJ: Prentice-Hall, 1986.

Bergseth, F.R. and Venkata, S.S., *Introduction to Electric Energy Devices*. Englewood Cliffs, NJ: Prentice-Hall, 1987.

Bewley, L.V., *Alternating Current Machinery*. New York: Macmillan, 1949.

Binns, K.J. and Lawrenson, P.J., *Analysis and Computation of Electric and Magnetic Field Problems*. 2nd ed. New York: Pergamon, 1973.

Bird, B.M. and King, K.G., *An Introduction to Power Electronics*, New York: Wiley, 1983.

Blume, L.F. et al., *Transformer Engineering*. 2nd ed. New York: Wiley, 1951.

Boldea, I., *The Electric Generators Handbook*. Vols. 1 and 2. Boca Raton, FL: CRC Press, 2006.

Boldea, I. and Nasar, S.A., *Linear Motion Electromagnetic Systems*. New York: Wiley, 1985.

Bose, B.K., *Adjustable Speed A.C. Drive Systems*. New York: IEEE Press, 1981.

Bose, B.K., *Power Electronics and AC Drives*. Englewood Cliffs, NJ: Prentice-Hall, 1986.

Bose, B.K., *Microcomputer Control of Power Electronics and Drives*. New York: IEEE, 1987.

Brown, D. and Hamilton III, E.P., *Electromechanical Energy Conversion*. New York: Macmillan, 1984.

Bumby, J.R., *Superconducting Rotating Electrical Machines*. Oxford: Clarendon Press, 1983.

Byerly, R.T. and Kimbark, E.W., *Stability of Large Electric Power Systems*. Piscataway, NJ: IEEE Press, 1974.

Campbell, S.J., *Solid-State AC Motor Controls*. New York: Marcel Dekker, 1987.

Carr, L., *The Testing of Electrical Machines*. London: MacDonald, 1960.

Carry, C.C., *Electric Machinery: A Coordinated Presentation of AC and DC Machines*. New York: Wiley, 1958.

Chalmers, B. and Williamson, A., *AC Machines: Electromagnetics and Design*. New York: Wiley, 1991.

Chalmers, B.J., *Electromagnetic Problems of AC Machines*. London: Chapman and Hall, 1965.

Chapman, S.J., *Electric Machinery Fundamentals*. New York: McGraw-Hill, 1999.

Chaston, A.N., *Electric Machinery.* Reston, VA: Reston, 1986.
Clayton, A.E., *Performance and Design of DC Machines.* London: Pitman, 1938.
Cochran, P.L., *Polyphase Induction Motors.* New York: Marcel Dekker, 1989.
Concordia, C., *Synchronous Machines.* New York: Wiley, 1951.
Crary, S.B., *Power System Stability* (2 vols). New York: Wiley, 1947.
Crosno, C.D., *Fundamentals of Electromechanical Conversion.* New York: Harcourt, Brace 1968.
Cuk, S.M. and Middlebrook, R.D., *A New Optimum Topology Switching DC to DC Converter.* Power Electronics, 1977.
Dams, R.M., *Power Diode and Thyristor Circuits.* Stevenage, Herts, England: IEE, 1979.
Daniels, A.R., *The Performance of Electrical Machines.* London: McGraw-Hill, 1968.
Daniels, A.R., *Introduction to Electrical Machines.* London: Macmillan, 1976.
Datta, S.M., *Power Electronics & Control.* Reston, VA: Reston, 1985.
David C. and Woodson, H.H., *Electromechanical Energy Conversion.* New York: Wiley, 1959.
Davis, R.M., *Power Diode and Thyristor Circuits.* Stevenage, Herts, England: IEE, 1979.
DC Motors Speed Controls ServoSystem—An Engineering Handbook. 5th ed. Hopkins, MN: Electro-Craft Corporations, 1980.
DelToro, V., *Electric Machines and Power Systems.* Englewood Cliffs, NJ: Prentice-Hall, 1985.
Dewan, S.B., Slemon, G.R. and Straughen, A., *Power Semiconductor Drives.* New York: Wiley, 1984.
Dubey, G.K., *Power Semiconductor Controlled Drives.* Englewood Cliffs, NJ: Prentice-Hall, 1989.
Dudley, A.M. and Henderson, S.F., *Connecting Induction Motors: Operation and Practice.* 4th ed. New York: McGraw-Hill, 1960.
Edwards, J.D., *Electrical Machines.* 2nd ed. London: Macmillan, 1986.
Electrical Engineering Handbook. New York: Wiley, 1985.
Elgerd, O., *Basic Electric Power Engineering.* Reading, MA: Addison Wesley, 1977.
Elgerd, O.L., *Electric Energy Systems Theory.* 2nd ed. New York: McGraw-Hill, 1982.
El-Hawary, M.E., *Principles of Electric Machines with Power Electronic Applications.* Englewood Cliffs, NJ: Prentice-Hall, 1988.
Elliott, D.L., Holladay, C.G., Barchet, W.R., Foote, H.P., and Sandusky, W.F., *Wind Energy Resource Atlas of the United States* (Solar Energy Research Institute, 1986) http://rredc.nrel.gov/wind/pubs/atlas/titlepg.html (accessed August 16, 2006).
Ellison, A.J., *Electromechanical Energy Conversion.* London: George G. Harrap Co. Ltd., 1970.
Emanuel, P., *Motors, Generators, Transformers, and Energy.* Englewood Cliffs, NJ, Prentice-Hall, 1985.
Feinberg, R., *Modern Power Transformer Practice.* New York: Wiley, 1979.
Fink, D.G. and Beaty, H.W., Eds., *Standard Handbook for Electrical Engineers.* 11th ed. New York: McGraw-Hill, 1978.
Finney, D., *The Power Thyristor and Its Application.* UK: McGraw-Hill, 1980.
Finney, D., *Variable Frequency AC Motor Drive Systems.* London: Peregrinus, 1988.
Fitzgerald, A.E., Kingsley, Jr., C. and Umans, S.D., *Electric Machinery.* 6th ed. New York: McGraw-Hill, 2003.
Fitzgerald et al., *Electric Machinery.* 2nd ed. New York: McGraw-Hill, 1961.
Fitzgerald et al., *Electric Machinery.* 3rd ed. New York: McGraw-Hill, 1971.
Fitzgerald et al., *Electric Machinery.* 4th ed. New York: McGraw-Hill, 1983.
Fitzgerald et al., *Electric Machinery.* 5th ed. New York: McGraw-Hill, 1990.
Flanagan, W.M., *Handbook of Transformer Applications.* New York: McGraw-Hill, 1986.
Franklin, A.C. et al., *The J&P Transformer Book.* 11th ed. London: Butterworths, 1983.
Fransua, A. and Magureanu, R., *Electrical Machines and Drive Systems.* Oxford: Technical Press, 1984.
Garik, M.L. and Whipple, C.C., *Alternating-Current Machines.* 2nd ed. New York: Wiley, 1986.
Gehmlich, D.K. and Hammon, S.B., *Electromechanical Systems.* New York: McGraw-Hill, 1967.
Ghandhi, S.K., *Semiconductor Power Devices.* New York: Wiley, 1987,
Gibbs, W.J., *Tensors in Electrical Machine Theory.* New York: Chapman and Hall, 1952.
Gibbs, W.J., *Electric Machine Analysis Using Tensors.* London: Pitman, 1967.
Gieras, J.F., *Linear Induction Drives.* Oxford: Clarendon Press, 1994.
Gieras, J.F. and Piech, Z.J., *Linear Synchronous Motors: Transportation and Automation Systems.* Boca Raton, FL: CRC Press, 1999.
Gingrich, H.W., *Electrical Machinery, Transformers and Control.* Englewood Cliffs, NJ: Prentice-Hall, 1979.

Gourishankar, V. and Kelly, D.H., *Electromechanical Energy Conversion*. 2nd ed. New York: Intext Educational Publishers, 1973.
Griffith, D.C., *Uninterruptible Power Supplies*. New York: Marcel Dekker, 1989.
Gross, C.A., *Power System Analysis*. 2nd ed. New York: Wiley, 1986.
Gross, H., Ed., *Electrical Feed Drives for Machine Tools*. New York: Siemens and Wiley, 1983.
Grossner, N.R., *Transformers for Electronic Circuits*. New York: McGraw-Hill, 1967.
Grove, A.S., *Physics and Technology of Semiconductor Devices*. New York: Wiley, 1967.
Guru, B.S. and Hiziroglu, H.R., *Electric Machinery and Transformers*. Orlando, FL: Harcourt Brace Jovanocich, 1988.
Gyugyi, L. and Pelly, B.R., *Static Power Frequency Changers*. New York: Wiley, 1976.
Hancock, N.N., *Electric Power Utilization*. London: Sir Isaac Pitman, 1967.
Hancock, N.N., *Matrix Analysis of Electrical Machinery*. 2nd ed. Oxford: Pergamon, 1974.
Harris, M.R. et al., *Per Unit Systems: With Special Reference to Electrical Machines*. London: Cambridge University Press, 1970.
Harwood, P.B., *Control of Electric Motors*. New York: Wiley, 1952.
Hendershot, J.R., *Design of Brushless Permanent Magnet Motors*. Hillboro, OH: Magna Physics Corp., 1991.
Heumann, K., *Basic Principles of Power Electronics*. New York: Springer, 1986.
Hindmarsh, J., *Electrical Machines*. Oxford: Pergamon, 1965.
Hindmarsh, J., *Electrical Machines and Their Applications*. 4th ed. Oxford: Pergamon, 1984.
Hindmarsh, J., *Electrical Machines and Drives: Worked Examples*. 2nd ed. Oxford: Pergamon, 1985.
Hochart, B., Ed., *Power Transformer Handbook*. London: Butterworths, 1987.
Hoft, R.G., *Semiconductor Power Electronics*. New York: Van Nostrand Reinhold, 1986.
Hubert, C.L., *Preventive Maintenance of Electrical Equipment*. 2nd ed. New York: McGraw-Hill, 1969.
Ireland, J.R., *Ceramic Permanent-Magnet Motors*. New York: McGraw-Hill, 1968.
IEEE, *Standard No. 113-1962, Test Code for Direct-Current Machines*. NY: IEEE, 1962.
IEEE, *Standard No. 115-1995, Test Procedures for Synchronous Machines*. NY: IEEE, 1995.
IEEE, *Standard No. 114-2001: Test Procedures for Single-phase Induction Motors*. NY: IEEE, 2001.
IEEE, *Standard No. 112-2004: Test Procedures for Polyphase Induction Motors and Generators*. NY: IEEE, 2004.
IEEE Conference Publication No. 234, *Power Electronics and Variable-Speed Drives*. London: IEEE, 1984.
Jones, C.V., *The Unified Theory of Electrical Machines*. London: Butterworths, 1967.
Jordan, H.E., *Energy Efficient Electric Motors and Their Applications*. New York: Van Nostrand Reinhold, 1983.
Karsai, K., Kerenyi, D. and Kiss, L., *Large Power Transformers*. Amsterdam: Elsevier, 1987.
Kenjo, T., *Stepping Motors and Their Microprocessor Controls*. Oxford: Clarendon Press, 1985.
Kenjo, T. and Nagamori, S., *Permanent-Magnet and Brushless DC Motors*. Oxford: Clarendon Press, 1985.
Kilgenstein, O., *Switched-Mode Power Supplies in Practice*. New York: Wiley, 1989.
Kimbark, E.W., *Power System Stability: Synchronous Machines*. New York: Dover, 1968.
Kloeffler, R.G. et al., *Direct-Current Machinery*. New York: Macmillan, 1950.
Kloss, A., *A Basic Guide to Power Electronics*. New York: Wiley, 1984.
Knowlton, A.E., Ed., *Standard Handbook for Electrical Engineers*. 8th ed. New York: McGraw-Hill, 1949.
Kosow, I.L., *Electric Machinery and Transformers*. Englewood Cliffs, NJ: Prentice-Hall, 1972.
Kosow, I.L., *Control of Electric Machines*. Englewood Cliffs, NJ: Prentice-Hall, 1973.
Kostenko, M. and Piotrovsky L., *Electrical Machines*. Vols. 1 and 2. Moscow: Mir, 1974.
Krause, P.C. and Wasynczuk, O., *Electromechanical Motion Devices*. New York: McGraw-Hill 1989.
Krause, P.C., Wasynczuk, O. and Sudhoff, S.D., *Analysis of Electric Machinery*. New York: IEEE, 1995.
Kron, G., *Tensor Analysis of Network*. New York: Wiley, 1939.
Kron, G., *A Short Course in Tensor Analysis for Electrical Engineers*. New York: Wiley, 1942.
Kron, G., *Application of Tensors to the Analysis of Rotating Electrical Machinery*. G.E. Review, 1942.
Kron, G., *Equivalent Circuits of Electric Machinery*. New York: Wiley, 1951.
Kron, G., *Tensors for Circuits*. 2nd ed. New York: Dover, 1959.
Ku, Y.H., *Electric Energy Conversion*. New York: Ronald Press, 1959.
Kuhlmann, J.H., *Design of Electrical Apparatus*. 3rd ed. New York: Wiley, 1950.

Kuo, B.C., Ed., *Incremental Motion Control: Step Motors and Control Systems*. Champaign, IL: SRL, 1979.
Kusko, A., *Solid State DC Motor Drives*. Cambridge, MA: MIT, 1969.
Laithwaite, E.R., Ed., *Transport without Wheels*. Boulder, CO: Westview, 1977.
Laithwaite, E.R., *A History of Linear Electric Motors*. London: Macmillan, 1987.
Lander, C.W., *Power Electronics*. Maidenhead, Berkshire, England: McGraw-Hill 1981.
Langsdorf, A.S., *Principles of Direct-Current Machines*. New York: McGraw-Hill, 1959.
Larson, R.E., *State Increment Dynamic Programming*. New York: Elsevier, 1968.
Lawrence, R.R. and Richards, H.E., *Principles of Alternating Current Machinery*. 4th ed. New York: McGraw-Hill, 1953.
Leonhard, W., *Control of Electrical Drives*. Berlin: Springer, 1996.
Levi, E., *Polyphase Motors*. New York: Wiley, 1984.
Lewis, W.A., *The Principles of Synchronous Machines*. Chicago, IL: Institute of Technology, 1959.
Lienhard, J.H., *A Heat Transfer Textbook*. Englewood Cliffs, NJ: Prentice-Hall, 1981.
Lindsay, J.F. and Rashid, M.H., *Electromechanics and Electric Machinery*. Englewood Cliffs, NJ: Prentice-Hall, 1986.
Liwschitz-Garik, M. and Whipple, C., *Alternating-Current Machines*. 2nd ed. Princeton, NJ: Van Nostrand, 1961.
Loew, E.A. and Bergseth, F.R., *Direct and Alternating Currents: Theory and Machinery*. 4th ed. New York: McGraw-Hill, 1954.
Lye, R.W., *Power Converter Handbook*. Peterborough, ON.: Canadian General Electric Company Ltd., 1976.
Lyon, W.V., *Transient Analysis of Alternating Current Machinery: An Application of the Method of Symmetrical Components*. Cambridge, MA: MIT, 1954.
Mablekos, Van E., *Electric Machine Theory for Power Engineers*. New York: Harper & Row, 1980.
Magnetic Materials Producers Association, *Standard No. 0100-78: Standard Specifications for Permanent Magnet Materials*. Chicago, 1978.
Majmudar, H., *Electromechanical Energy Converters*. Boston: Allyn and Bacon, 1965.
Majmudar, H., *Introduction to Electrical Machines*. Worcester, MA: Worcester Polytechnic Institute, 1976.
Matsch, L.W., *Capacitors, Magnetic Circuits and Transformers*. Englewood Cliffs, NJ: Prentice-Hall, 1964.
Matsch, L.W. and Morgan, J.D., *Electromagnetic and Electromechanical Machines*. 3rd ed. NY: Harper & Row, 1986.
Mazda, F.F., *Thyristor Control*, London: Butterworths, 1973.
McCaig, M. and Clegg, A.G., *Permanent Magnets in Theory and Practice*. 2nd ed. New York: Wiley, 1987.
McIntyre, R.L., *Electric Motor Control Fundamentals*. 3rd ed. New York: McGraw-Hill, 1974.
McLaren, P.G., *Elementary Electric Power and Machines*. Chichester, West Sussex, UK: Ellis Horwood, 1984.
McLyman, Wm. T., *Transformer and Inductor Design Handbook*. New York: Marcel Dekker, 1978.
McMurray, W., *Theory and Design of Cycloconverters*. Cambridge, MA: MIT, 1972.
McPherson, G. and Laramore, R.D., *An Introduction to Electrical Machines and Transformers*. 2nd ed. New York: Wiley, 1990.
Meisel, J., *Principles of Electromechanical Energy Conversion*. New York: McGraw-Hill, 1966.
Middlebrook, R.D. and Cuk, S., *Advances in Switched-Mode Power Conversion*. Vols. I and II. Pasadena, CA: TESLAco, 1981.
Miller, T.J.E., *Brushless Permanent-Magnet and Reluctance Motor Drives*. New York: Oxford, 1989.
Millermaster, R., *Harwood's Control of Electric Motors*. 4th ed. New York: Wiley, 1970.
MIT Staff, *Magnetic Circuits and Transformers*. New York: Wiley, 1943.
Mohan, M., Undeland, M.T. and Robsins, W.P., *Power Electronics: Converters, Applications and Design*. New York: Wiley, 1989.
Molloy, E., Ed., *Small Motors and Transformers: Design and Construction*. London: George Newnes, 1953.
Morgan, A.T., *General Theory of Electrical Machines*. London: Heyden and Son, 1979.
Motto, J.W., Ed., *Introduction to Solid State Power Electronics*. Pittsburgh, PA: Westinghouse Electric Corp., 1977.

Murphy, J., *Power Semiconductor Control of Motors and Drives.* London: Pergamon, (to be published).
Murphy, J.M.D., *Thyristor Control of A.C. Motors.* London: Pergamon, 1973.
Murphy, J.M.D. and Turnbull, F.G., *Power Electronic Control of AC Motors.* London: Pergamon, 1988.
Nasar, S.A., *Electric Machines and Transformers.* New York: Macmillan, 1984.
Nasar, S.A., *Electric Energy Conversion and Transmission.* New York: Macmillan, 1985.
Nasar, S.A., Ed., *Handbook of Electric Machines.* New York: McGraw-Hill, 1987.
Nasar, S.A. and Boldea, I., *Linear Motion Electric Machines.* New York: Wiley, 1976.
Nassar, S.A. and Boldea, I., *Linear Electric Motors* Englewood Cliffs, NJ: Prentice-Hall, 1987.
Nasar, S.A. and Unnewehr, L.E., *Electromechanics and Electric Machines.* 2nd ed. New York: Wiley, 1983.
National Electrical Manufacturers Association, *Publication No. MG1-1972: Motors and Generators.* New York: NEMA, 1972.
O'Kelly, D. and Simmons, S., *Introduction to Generalized Electrical Machine Theory.* London: McGraw-Hill, 1968.
Ong, C.M., *Dynamic Simulation of Electric Machines.* Prentice-Hall, 1998.
Oxner, E.S., *Power FETs and Their Applications.* Prentice-Hall, 1982.
Parker, J.J. and Studders, R.J., *Permanent Magnets and Their Applications.* New York: Wiley, 1962.
Patrick, D.R. and Fardo, S.W., *Rotating Electrical Machines and Power Systems.* Englewood Cliffs, NJ: Prentice-Hall, 1985.
Pearman, R.A., *Power Electronic Solid State Motor Control.* Reston, VA: Reston, 1980.
Pelley, B.R., *Thyristor Phase-Controlled Converters and Cycloconverters.* New York: Wiley, 1971.
Perry, M.P., *Low-Frequency Electromagnetic Design.* New York: Marcel Dekker, 1987.
Peter, J.M., Ed., *The Power Transistor in its Environment.* France: Thomson Semiconductors, 1978.
Poloujadoff, M., *The Theory of Linear Induction Machinery.* Oxford: Clarendon Press, 1980.
Pressman, A.I., *Switching Power Supply Design.* New York: McGraw-Hill, 1991.
Puschstein, A.F., Lloyd, T.C. and Conrad, A.G., *Alternating Current Machines.* 3rd ed. New York: Wiley, 1954.
Rajagopalan, V., *Computer Aided Analysis of Power Electronic Systems.* New York: Marcel Dekker, 1987.
Ramamoorty, M., *An Introduction to Thyristors and Their Applications.* London: Macmillan, 1978.
Ramshaw, R.S., *Power Electronics: Thyristor Controlled Power for Electric Motors.* London: Chapman & Hall, 1982.
Ramshaw, R. and Van Heeswijk, R.G., *Energy Conversion: Electric Motors and Generators.* Orlando, FL: Saunders, 1990.
Rashid, M.H., *Power Electronics.* 2nd ed. Englewood Cliffs, NJ: Prentice-Hall, 1993.
Richardson, D.V., *Handbook of Rotating Electric Machinery.* Reston, VA: Reston, 1980.
Robertson, B.L. and Black, L.J., *Electric Circuits and Machines.* New York: Van Nostrand, 1949.
Rose, M.J., *Power Engineering Using Thyristors.* Vol. 1. London: Mullard, 1970.
Roters, H.C., *Electromagnetic Devices.* New York: Wiley, 1941.
Sanders, C.W., *Power Electronics.* New York: McGraw-Hill, 1981.
Say, M.G., *Introduction to the Unified Theory of Electromagnetic Machines.* London: Pitman, 1971.
Say, M.G., *Alternating Current Machines.* 5th ed. London: Pitman, 1983.
Say, M.G. and Taylor, E.O., *Direct Current Machines.* London: Pitman, 1980.
Schaefer, J., *Rectifier Circuits, Theory and Design.* New York: Wiley, 1965.
Scoll, A.W., *Cooling of Electric Equipment.* New York: Wiley, 1974.
Seely, S., *Electromechanical Energy Conversion.* New York: McGraw-Hill, 1962.
Sen, P.C., *Thyristor DC Drives.* New York: Wiley, 1981.
Sen, P.C., *Principles of Electric Machines and Power Electronics.* New York: Wiley, 1989.
Shepherd, W. and Hulley, L.N., *Power Electronics and Motor Control.* Cambridge, MA: Cambridge, 1987.
Siskind, C.S., *Electrical Machines, Direct and Alternating Current.* 2nd ed. New York: McGraw-Hill, 1959.
Skilling, H.H., *Electromechanics.* New York: Wiley, 1962.
Slemon, G.R., *Magnetoelectric Devices.* New York: Wiley, 1966.
Slemon, G.R., *Electric Machines and Drives.* Reading, MA: Addison-Wesley, 1992.
Slemon, G.R. and Straughen, A., *Electric Machines.* Reading, MA: Addison Wesley, 1981.
Smeaton, R.W., *Motor Applications and Maintenance Handbook.* 2nd ed. New York: McGraw-Hill, 1987.

Smith, J.L., *Superconductors in Large Synchronous Machines*. EPRI Research Report prepared at M.I.T, June 1975.

Smith, R.A., *Semiconductors*. Cambridge, MA: Cambridge, 1959.

Smith, R.T., *Analysis of Electrical Machines*. New York: Pergamon Press, 1982.

Smith, S., *Magnetic Components: Design and Applications*. New York: Van Nostrand, 1985.

Snelling, E.C., *Soft Ferrites-Properties and Applications*. London: Butterworths, 1988.

Sokira, T.J. and Jaffe, W., *Brushless DC Motors*. Blue Ridge Summit, PA: Tab Books, 1990.

Standard No. 393-1977: Test Procedures for Magnetic Cores. New York: 1977.

Stein, R. and Hunt, W.T., *Electric Power System Components: Transformers and Rotating Machines*. New York: Van Nostrand, 1979.

Steinmetz, C.P., *Theory and Calculation of Transient Electric Phenomena and Oscillations*. 3rd ed. New York: McGraw-Hill, 1920.

Steven, R.E., *Electrical Machines and Power Electronics*. Berkshire, UK: Van Nostrand, 1983.

Stigant, S.A. and Franklin, A.C., *J&P Transformer Book*. 10th ed. London: Butterworths, 1973.

Still, A. and Siskind, C.S., *Elements of Electrical Machine Design*. Tokyo: McGraw-Hill, 1954.

Streetman, B.G., *Solid State Electronic Devices*. 2nd ed. Englewood Cliffs, NJ: Prentice-Hall, 1980.

Sum, K.K., *Switch Mode Power Conversion-Basic Theory and Design*. New York: Marcel Dekker, 1984.

Sugandhi, K.K., *Thyristors-Theory and Applications*. New York: Halsted, 1984.

Takuechi, T.J., *Theory of SCR Circuits and Application to Motor Control*. Tokyo: Tokyo Electrical Engineering College Press, 1968.

Tarter, R.E., *Principles of Solid-State Power Conversion*. Indianapolis, IN: Howard W. Sams, 1985.

Teago, F.J., *The Commutator Motor*. London: Methuen, 1930.

Thaler, G.J. and Wilcox, M.L., *Electric Machines: Dynamics and Steady State*. New York: Wiley, 1966.

Thomas & Skinner, Inc., *Bulletin No. M303: Permanent Magnet Design*. Indianapolis, IN, 1967.

Thorborg, K., *Power Electronics*. UK: Prentice-Hall International, 1988.

Transformer and Inductor Design Handbook. 2nd ed. New York: Marcel Dekker, 1989.

Unnewehr, L.E. and Nasar, S.A., *Electric Vehicle Technology*. New York: Wiley, 1982.

Upson, A.R. and Batchelor, J.H., *Synchro Engineering Handbook*. London: Hutchison, 1965.

Van Valkenburgh, N. and Neville, Inc., *Basic Synchros and Servomechanisms*. New York: John F. Rider, 1955.

Veinott, C., *Theory and Design of Small Induction Motors*. New York: McGraw-Hill, 1959.

Veinott, C.A., *Spatial Harmonics (One-Phase and Polyphase)*. Sarasota, FL: C.A. Veinott, 1991.

Veinott, C.G. and Martin, J.E., *Fractional- and Subfractional-Horsepower Electric Motors*. 4th ed. New York: McGraw-Hill, 1986.

Vidyasagar, M., *Nonlinear Systems Analysis*. Englewood Cliffs, NJ: Prentice-Hall, 1978.

Wagner, C.F. and Evans, R.D., *Symmetrical Components*. New York: McGraw-Hill, 1935.

Walker, J.H., *Large Synchronous Machines*. Oxford, UK: Oxford, 1981.

Walsh, E.M., *Energy Conversion*. New York: Ronald Press, 1967.

Wells, R., *Static Power Converters*. New York: Wiley, 1962.

White, F.M., *Heat Transfer*. Reading, MA: Addison-Wesley, 1984.

White, D.C. and Woodson, H.H., *Electromechanical Energy Conversion*. New York: Wiley, 1959.

Wildi, T., *Electrical Machines, Drives and Power Systems*. Englewood Cliffs, NJ: Prentice-Hall.

Williams, B.W., *Power Electronics, Devices, Drivers and Applications*. New York: Halsted, 1987.

Wood, P., *Switching Power Converters*. New York: Van Nostrand, 1981.

Woodson, H.H. and Melcher, J.R., *Electromechanical Dynamics, Part I: Discrete Systems*. New York: Wiley, 1968.

Yamamura, S., *Theory of Linear Induction Motors*. 2nd ed. New York: Wiley, 1979.

Yamamura, S., *A.C. Motors for High-Performance Applications: Analysis and Control*. New York: Marcel Dekker, 1987.

Yamayee, Z.A., *Electromechanical Energy Devices and Power Systems*. New York: Wiley, 1994.

Yu, Y., *Electric Power Systems Dynamics*. New York: Academic Press, 1983.

Zorbas, D., *Electric Machines: Principles, Applications, and Control Schematics*. St. Paul, MN: West, 1989.

Index